SECOND EDITION

Introduction to COMPRESSIBLE FLUID FLOW

HEAT TRANSFER

A Series of Reference Books and Textbooks

SERIES EDITOR

Afshin J. Ghajar

Regents Professor
School of Mechanical and Aerospace Engineering
Oklahoma State University

Introduction to Compressible Fluid Flow, Second Edition, *Patrick H. Oosthuizen and William E. Carscallen*

Advances in Industrial Heat Transfer, *Alina Adriana Minea*

Introduction to Thermal and Fluid Engineering, *Allan D. Kraus, James R. Welty, and Abdul Aziz*

Thermal Measurements and Inverse Techniques, *Helcio R.B. Orlande, Olivier Fudym, Denis Maillet, and Renato M. Cotta*

Conjugate Problems in Convective Heat Transfer, *Abram S. Dorfman*

Engineering Heat Transfer: Third Edition, *William S. Janna*

To my wife, Jane, for all her loving encouragement and help.

P.H.O.

To my loving wife, Elizabeth, and our four children, Peter, Laleah, Emma, and Mather.

W.E.C.

Contents

Nomenclature

The following is a list of definitions of the main symbols used in this book:

A: Area
A^*: Area at section where $M = 1$
A_e: Exit plane area
A_{throat}: Throat area
a: Local speed of sound
a^*: Speed of sound at section where $M = 1$
a_0: Stagnation speed of sound
C: Constant
C_D: Drag coefficient
C_L: Lift coefficient
C_p: Pressure coefficient
c: Speed of light
c_m: Mean molecular speed
c_p: Specific heat at constant pressure
c_v: Specific heat at constant volume
D: Diameter
D_H: Hydraulic diameter of a duct
e: Internal energy
F: Force on system in direction indicated by subscript
F_D: Drag force
F_μ: Friction force
f: Friction factor
$\bar{f}$: Mean friction factor over length of duct considered
f_D: Darcy friction factor
h: Enthalpy or heat transfer coefficient
h_f: Enthalpy of formation
Kn: Knudsen number
K_p: Equilibrium constant
k: Thermal conductivity
L: Velocity component parallel to oblique shock wave or size of system
l^*: Length of duct required to produce choking
M: Mach number
M_{crit}: Critical Mach number
M_{des}: Design Mach number
M_{div}: Divergence Mach number
M_e: Exit plane Mach number
M_N: Component of Mach number normal to oblique shock wave
M_R: Reflected shock Mach number
M_S: Shock Mach number
m: Molar mass
$\dot{m}$: Mass flow rate
N: Velocity component normal to oblique shock wave

Nu: Nusselt number
n: Index of refraction
P: Perimeter
p: Pressure
p^*: Pressure at section where $M = 1$
p_0: Stagnation pressure
p_b: Back pressure
p_{bcrit}: Back pressure at which sonic flow first occurs
p_e: Exit plane pressure
Pr: Prandtl number
q: Heat transfer rate per unit mass flow rate or per unit area
Q: Heat transfer rate
R: Gas constant for gas being considered
$\Re$: Universal gas constant
Re: Reynolds number
r: Recovery factor
S: V/c
s: Entropy
s^*: Entropy at section where $M = 1$
s_0: Stagnation entropy
T: Temperature
T_0: Stagnation temperature
T_∞: Freestream temperature
T_w: Wall temperature
T_{wa}: Adiabatic wall temperature
T^*: Temperature at section where $M = 1$
t: Time
U_S: Shock velocity
U_{SR}: Reflected shock velocity
u: Velocity component in *x* direction
V: Velocity in one-dimensional flow
V^*: Velocity at section where $M = 1$
V_e: Exit plane velocity
$\hat{V}$: Maximum velocity
v: Velocity component in *y* direction
w: Rate at which work is done per unit mass flow rate
W: Rate at which work is done
x: Coordinate direction
y: Coordinate normal to *x*
Z: Compressibility factor, $p/R\rho T$
α: Mach angle
β: Oblique shock angle
γ: Specific heat ratio
δ: Turning angle
δ_{max}: Maximum oblique shock wave turning angle
θ: Prandtl–Meyer angle
θ_{vib}: Vibrational excitation temperature
λ: Mean free path
μ: Coefficient of viscosity

ν: Kinematic viscosity or Prandtl–Meyer angle
ρ: Density
ρ^*: Density at section where $M = 1$
ρ_0: Stagnation density
ρ_e: Exit plane density
τ_w: Wall shear stress
ψ: Dimensionless stream function
Φ: Potential function
Φ_p: Perturbation potential function
ω: Vorticity

Preface

Compressible flow occurs in many devices encountered in mechanical and aerospace engineering practice, and knowledge of the effects of compressibility on a flow is therefore required by many professional mechanical and aerospace engineers. Most conventional course sequences in fluid mechanics and thermodynamics deal with some aspects of compressible fluid flow, but the treatment is usually relatively superficial. For this reason, many mechanical and aerospace engineering schools offer a course dealing with compressible fluid flow at the senior undergraduate or at the graduate level. The purpose of such courses is to expand and extend the coverage given in previous fluid mechanics and thermodynamics courses. This book is intended to provide the background material for such courses. This book also lays the foundation for more advanced courses on specialized aspects of the subject such as hypersonic flow, thus complementing the more advanced books in this area.

The widespread use of computer software for the analysis of engineering problems has, in many ways, increased the need to understand the assumptions and the theory on which such analyses are based. Such an understanding is required to interpret the computed results and to judge whether a particular piece of software will give results that are of adequate accuracy for the application being considered. Therefore, while numerical solutions are discussed in Appendix B of this book, the major emphasis is on developing an understanding of the material and of the assumptions conventionally conventionally used in analyzing compressible fluid flows. Compared with available textbooks on the subject, then, this book is, it is hoped, distinguished by its attempt to develop a thorough understanding of the theory and of the assumptions on which this theory is based, by its attempt to develop in the student a fascination with the phenomena involved in compressible flow, and by the breadth of its coverage.

Our goal in writing this new text was to provide students with a clear explanation of the physical phenomena encountered in compressible flow, to develop in them an awareness of practical situations in which compressibility effects are likely to be important, to provide a thorough explanation of the assumptions conventionally used in the analysis of compressible flows, to provide a broad coverage of the subject, and to provide a firm foundation for the study of more advanced and specialized aspects of the subject. We have also tried to adopt an approach that will develop in the student a fascination with the phenomena involved in compressible flow.

The first seven chapters of this book deal with the fundamental aspects of the subject. They review some background material and discuss the analysis of isentropic flows, of normal and oblique shock waves, and of expansion waves. The next three chapters discuss the application of this material to the study of nozzle characteristics, of friction effects, and of heat exchange effects. Chapters dealing with the analysis of generalized one-dimensional flow, with simple numerical methods, and with two-dimensional flows are then given. The last three chapters in this book are interrelated and provide an introduction to hypersonic flow, to high-temperature gas effects, and to low-density flows. Some discussion of experimental methods is incorporated into this book, mainly to illustrate the theoretical material being discussed. However, because a number of shadowgraph and Schlieren photographs are given in the text, a separate discussion of these and other related methods of flow visualization is provided in Appendix D.

The first ten chapters together with selected material from the remaining chapters provide the basis of the typical undergraduate course in compressible fluid flow. A graduate-level course will typically cover more on numerical solutions, more material on two-dimensional flows and on high-temperature gas effects.

The subject of compressible fluid flow involves so many interesting physical phenomena, involves so many intriguing mathematical complexities, and utilizes so many interesting and novel numerical techniques that the difficulty faced in preparing a book on the subject is not that of deciding what material to include but rather in deciding what material to omit. In preparing this book, an attempt has been made to give a coverage that is broad enough to meet the needs of most instructors but at the same time to try to avoid losing the students' interest by going too deeply into the specialized areas of the subject.

Solution Manual

A manual that contains complete solutions to all of the problems in this textbook is available. In writing this manual, we have used the same problem-solving methodology as adopted in the worked examples in this book. The solution manual also provides summaries of the major equations developed in each chapter.

PowerPoint Presentations

A series of PowerPoint presentations covering most of the material in this book is also available.

Software

An interactive computer program, COMPROP2, for the calculation of the properties of various compressible flows was developed by A. J. Ghajar of the School of Mechanical and Aerospace Engineering at Oklahoma State University, and L. M. Tarn of the Mechanical Engineering Department of the University of Macau, to support this textbook. The program has modules for Isentropic Flow, Normal Shock Waves, Oblique Shock Waves, Fanno Flow, and Rayleigh Flow. The use of this software is described in Appendix A. The software is available free of charge to adopters of this book. It can be accessed at www.crcpress.com/product/isbn/9781439877913 under the Downloads Updates tab.

Patrick H. Oosthuizen
William E. Carscallen

Acknowledgments

This book is based on courses on Gas Dynamics and Compressible Fluid Flow taught by one of the authors (Patrick H. Oosthuizen) at the University of Cape Town and at Queen's University and by the other author (William E. Carscallen) at Carleton University. The students in these courses, by their questions, comments, advice, and encouragement, have had a major influence on the way in which the material is presented in this book, and their help is very gratefully acknowledged. The indirect influence of the authors' own professors on this book must also be gratefully acknowledged. In particular, R. Stegen at the University of Cape Town and I. I. Glass at the University of Toronto introduced one of the authors (Patrick H. Oosthuizen) to the exciting nature of the subject, as did W. Gilbert to the other author (William E. Carscallen). The authors would also like to gratefully acknowledge the advice and help they received from colleagues at Queen's University, at the National Research Council of Canada, and at Carleton University. Jane Paul undertook the tedious job of checking and editing most of the text, and her effort and her encouragement is also gratefully acknowledged. The encouragement, support, and tolerance of Jonathan Plant during the preparation of this book are also deeply appreciated. The authors would also like to express their gratitude to all of those, both instructors and students, who have used the book and who have informed us of errors in the first edition.

Authors

Patrick H. Oosthuizen is a professor of mechanical engineering at Queen's University in Kingston, Ontario, Canada. He received his BSc(Eng), MSc(Eng), and PhD degrees in mechanical engineering from the University of Cape Town, South Africa, and his MASc degree in aerospace engineering from the University of Toronto, Toronto, Ontario, Canada. He joined Queen's University after teaching for several years at the University of Cape Town. He does research in the areas of heat transfer, fluid mechanics, and energy systems. He authored more than 600 technical papers in journals and conference proceedings. He has received a number of teaching and research awards, has been involved with the organization of many national and international conferences, and has edited a number of conference proceedings.

William E. Carscallen was a principal research officer and manager of research and technology in the Gas Turbine Laboratory of the Institute for Aerospace Research, National Research Council of Canada (NRC) for many years. He has an honors diploma from the von Karman Institute for Fluid Dynamics and received his PhD degree from Queen's University in Kingston, Ontario, Canada. He is a recipient of an NRC President's Fund Award. Dr. Carscallen taught for a number of years as a sessional lecturer at Carleton University in Ottawa, Ontario, Canada, and is the author of numerous publications in journals and conference proceedings.

1

Introduction

Compressibility

The compressibility of a fluid is basically a measure of the change in density that will be produced in the fluid by a specified change in pressure. Gases are, in general, highly compressible whereas most liquids have a very low compressibility. Now, in a fluid flow there are usually changes in pressure associated, for example, with changes in the velocity in the flow. These pressure changes will, in general, induce density changes, which will have an influence on the flow, i.e., the compressibility of the fluid involved will have an influence on the flow. If these density changes are important, the temperature changes in the flow that arise due to the kinetic energy changes associated with the velocity changes also usually influence the flow, i.e., when compressibility is important, the temperature changes in the flow are usually important. Although the density changes in a flow field can be very important, there exist many situations of great practical importance in which the effects of these density and temperature changes are negligible. Classical incompressible fluid mechanics deals with such flows in which the pressure and kinetic energy changes are so small that the effects of the consequent density and temperature changes on the fluid flow are negligible, i.e., the flow can be assumed to be incompressible. There are, however, a number of flows that are of great practical importance in which this assumption is not adequate, the density and temperature changes being so large that they have a very significant influence on the flow. In such cases, it is necessary to study the thermodynamics of the flow simultaneously with its dynamics. The study of these flows in which the changes in density and temperature are important is basically what is known as *compressible fluid flow* or *gas dynamics*, it usually only being in gas flows that compressibility effects are important.

The fact that compressibility effects can have a large influence on a fluid flow can be seen by considering the three aircraft shown in Figure 1.1. The first of the aircraft shown in Figure 1.1 is designed for relatively low speed flight. It has straight wings, it is propeller driven, and the fuselage (the body of the aircraft) has a "rounded" nose. The second aircraft is designed for higher speeds. It has swept wings and tail surfaces and is powered by turbojet engines. However, the fuselage still has a "rounded" nose and the intakes to the engines also have rounded edges, which are approximately at right angles to the direction of flight. The third aircraft is designed for very-high-speed flight. It has highly swept wings and a sharp nose, and the air intakes to the engines have sharp edges and are of complex shape. These differences between the aircraft are mainly because compressibility effects become increasingly important as the flight speed increases.

Although the most obvious applications of compressible fluid flow theory are in the design of high speed aircraft, and this remains an important application of the subject, a

(a)

(b)

(c)

FIGURE 1.1
Aircraft designed to operate at different speeds: (a) De Havilland Canada Dash 8; (b) Canadair CL-600 Regional Jet; (c) Aerospatiale-BAC Concorde. (Courtesy of Eduard Marmet. Airliners.net.)

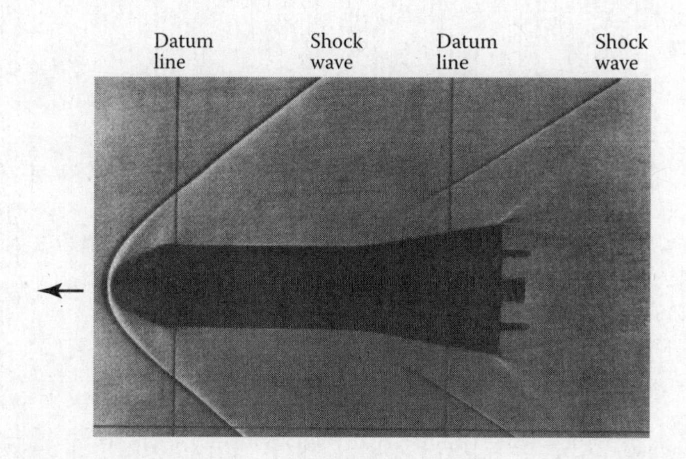

FIGURE 1.2
Photograph of supersonic flow over a body showing the presence of shock waves. Methods used to obtain such photographs are discussed in Appendix G. (Courtesy of NASA.)

knowledge of compressible fluid flow theory is required in the design and operation of many devices commonly encountered in engineering practice. Among these applications are

- Gas turbines: the flow in the blading and nozzles is compressible
- Steam turbines: here the flow in the nozzles and blades must be treated as compressible
- Reciprocating engines: the flow of the gases through the valves and in the intake and exhaust systems must be treated as compressible
- Natural gas transmission lines: compressibility effects are important in calculating the flow through such pipelines
- Combustion chambers: the study of combustion, in many cases, requires a knowledge of compressible fluid flow

Compressibility effects are normally associated with gas flows in which, as discussed in the next chapter, the flow velocity is relatively high compared with the speed of sound in the gas and if the flow velocity exceeds the local speed of sound, i.e., if the flow is "supersonic," effects may arise, which do not occur at all in "subsonic" flow, e.g., shock waves can exist in the flow as shown in Figure 1.2. The nature of such shock waves will be discussed in Chapters 5 and 6.

Example 1.1

Air flows down a variable area duct. Measurements indicate that the pressure is 80 kPa, the temperature is 5°C, and the velocity is 150 m/s at a certain section of the duct. Estimate, assuming incompressible flow, the velocity and pressure at a second section of the duct at which the duct area is half that of the section where the measurements were made. Comment on the validity of the incompressible flow assumption in this situation.

Solution

The situation considered is shown in Figure E1.1.

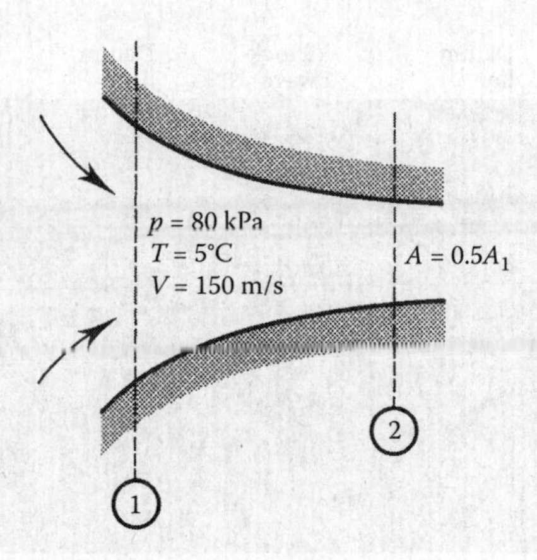

FIGURE E1.1
Flow situation considered.

Because the flow is assumed to be incompressible, i.e., because the density is assumed to remain constant, the continuity equation gives

$$V_1A_1 = V_2A_2$$

Hence, using the supplied information:

$$150 \times A_1 = V_2(A_1/2)$$

Therefore,

$$V_2 = 300 \text{ m/s}$$

Assuming that the effects of friction on the flow are negligible, the pressure change can be found using Bernoulli's equation, which gives

$$p_1 + \rho\frac{V_1^2}{2} = p_2 + \rho\frac{V_2^2}{2}$$

This can be rearranged to give

$$p_2 = \rho\left[\frac{V_1^2}{2} - \frac{V_2^2}{2}\right] + p_1 \quad \text{(a)}$$

The density, ρ, is evaluated using the initial conditions, i.e., using $p_1/\rho = RT_1$, which, since air flow is being considered, gives

$$\rho = \frac{80 \times 10^3}{(287 \times 278)} = 1.003 \text{ kg/m}^3$$

Substituting this back into Equation (a) then gives

$$p_2 = 1.003\left[\frac{150^2}{2} - \frac{300^2}{2}\right] + 80 \times 10^3$$

This gives

$$p_2 = 4.62 \times 10^4 \text{ Pa} = 46.2 \text{ kPa}$$

To check the validity of the assumption that the flow is incompressible, it is noted that if the flow can be assumed to be incompressible, the temperature changes in the flow will normally be negligible so the temperature at the exit will also be approximately 5°C. The equation of state therefore gives at the exit:

$$\rho_2 = \frac{p_2}{RT_2} = \frac{46.2 \times 10^3}{(287) \times (278)} = 0.579 \text{ kg/m}^3$$

Since this indicates that the density changes by more than 40%, the incompressible flow assumption is not justified.

Fundamental Assumptions

The concern in this book is essentially only with the flow of a gas. Now, to analyze an engineering problem, simplifying assumptions normally have to be made, i.e., a "model" of the situation being considered has to be introduced. The following assumptions will be adopted in the initial part of the present study of compressible fluid flow, some of these assumptions being relaxed in Chapters 5 and 6.

1. The gas is continuous, i.e., the motion of individual molecules does not have to be considered, the gas being treated as a continuous medium. This assumption applies in flows in which the mean free path of the gas molecules is very small compared with all the important dimensions of the solid body through or over which the gas is flowing. This assumption will, of course, become invalid if the gas pressure, and hence, density, becomes very low as is the case with spacecraft operating at very high altitudes and in low-density flows that can occur in high-vacuum systems.
2. No chemical changes occur in the flow field. Chemical changes influence the flow because they result in a change in composition with resultant energy changes. One common chemical change is that which results from combustion in the flow field. Other chemical changes that can occur when there are large pressure and temperature changes in the flow field are dissociation and ionization of the gas molecules. These can occur, for example, in the flow near a space vehicle during reentry to the earth's atmosphere.
3. The gas is perfect. This implies that
   (a) The gas obeys the perfect gas law, i.e.,

$$\frac{p}{\rho} = RT = \frac{\Re}{m}T \tag{1.1}$$

In Equation 1.1, $\Re$ is the universal gas constant, which has a value of 8314.3 J/kg molK, or 1545.3 ft-lbf/lbm mol°R, and m is the molar mass. R is the gas constant for a particular gas. For air, it is equal to 8314.3/28.966 = 287.04 J/kg K or 1545.3/28.966 = 53.3 ft-lbf/lbm°R. (A discussion of units will be given later in this chapter.)

(b) The specific heats at constant pressure and constant volume, c_p and c_v, are both constant, i.e., the gas is calorically perfect. The ratio of the two specific heats will be used extensively in the analyses presented in this book and is given by

$$\gamma = \frac{c_p}{c_v} \tag{1.2}$$

The symbol γ is used for the specific heat ratio instead of the frequently used symbol k because the symbol k is used for the thermal conductivity in this book. It should also be recalled that

$$R = c_p - c_v \tag{1.3}$$

The assumption that the gas is calorically perfect may not apply if there are very large temperature changes in the flow or if the gas temperature is high. A discussion of this will be given in Chapter 12. Although a calorically perfect gas has specific heats that are constant, a thermally perfect gas has specific heats that depend only on temperature and which are thus not necessarily constant.

4. Gravitational effects on the flow field are negligible. This assumption is almost always quite justified for gas flows.
5. Magnetic and electrical effects are negligible. These effects would normally only be important if the gas was electrically conducting, which is usually only true if the gas or a seeding material in the gas is ionized. In the so-called magneto-hydrodynamic (MHD) generator, a hot gas that has been seeded with a substance that easily ionizes and which therefore makes the gas a conductor is expanded through a nozzle to a high velocity. The high-speed gas stream is then passed through a magnetic field that generates an emf. This induces an electrical current flow in a conductor connected across the gas flow normal to the magnetic field. The arrangement of such a device is shown very schematically in Figure 1.3. Because there are no moving components in contact with the gas stream, this device can

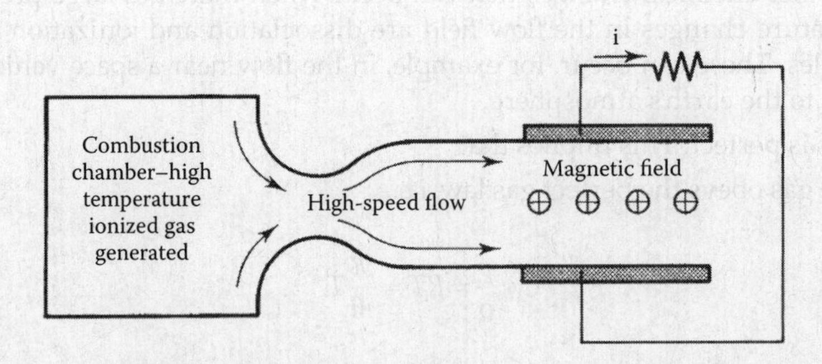

FIGURE 1.3
Magneto-hydrodynamic generator.

operate at very high gas temperatures. The magnetic and electrical effects that are important in this device are not considered in this book.

6. The effects of viscosity are negligible. This is never true close to a solid surface but in many cases the overall effects of viscosity remain small. The relaxation of this assumption and the effects of viscosity on compressible fluid flows will be discussed in Chapters 9 and 10.

When the above assumptions are adopted, the flow field is completely described by knowing the values of the following variables at all points in the flow field:

- Velocity vector, V
- Pressure, p
- Density, ρ
- Temperature, T

Therefore, to describe the flow field, four equations involving these four variables must be obtained. These equations are derived by applying the following principles:

- Conservation of mass (continuity equation)
- Conservation of momentum (Newton's law)
- Conservation of energy (first law of thermodynamics)
- Equation of state

The present book describes the application of these principles to the prediction of various aspects of compressible flow.

Example 1.2

The pressure and temperature in a gas in a large chamber are found to be 500 kPa and 60°C, respectively. Find the density if the gas is (a) air and (b) hydrogen.

Solution

The perfect gas law gives

$$\rho = \frac{pm}{\Re T}$$

Because m is 28.97 for air and 2 for hydrogen, this equation gives

$$\rho_{air} = \frac{500 \times 1000 \times 28.97}{8314 \times 333} = 5.23 \text{ kg/m}^3$$

and

$$\rho_{hydrogen} = \frac{500 \times 1000 \times 2}{8314 \times 333} = 0.36 \text{ kg/m}^3$$

Units

Although the SI system of units has virtually become the standard throughout the world, the Imperial or English system is still quite extensively used in industry, and the engineer still needs to be familiar with both systems. Examples and problems based on both of these systems of units are, therefore, incorporated into this book.

Now, in modern systems of units, the units of mass, length, time, and temperature are treated as fundamental units. The units of other quantities are then expressed in terms of these fundamental units. The relation between the derived units and the fundamental units is obtained by considering the "laws" governing the process and then equating the dimensions of the terms in the equation that follows from the "law," e.g., since

$$\text{force} = \text{mass} \times \text{acceleration}$$

it follows that the dimensions of force are equal to the dimensions of mass multiplied by the dimensions of acceleration, i.e., equal to the dimensions of mass multiplied by the dimensions of velocity divided by the dimensions of time. From this, it follows that in any system of units:

$$\text{units of force} = \text{units of mass} \times \text{units of length/units of time}^2$$

The fundamental units in the two systems used in this book are

SI Units

- Unit of mass = kilogram (kg)
- Unit of length = meter (m)
- Unit of time = second (s)
- Unit of temperature = kelvin or degree celsius (K or °C)

In this system, the unit of force is the newton, N, which is related to the fundamental units by

$$1\ \text{N} = 1\ \text{kg m/s}^2$$

English System

- Unit of mass = pound mass (lbm)
- Unit of length = foot (ft)
- Unit of time = second (sec)
- Unit of temperature = rankine or degree fahrenheit (R or F)

In this system of units, therefore, the unit of force should be 1 lbm ft/s^2. However, for historical reasons, it is usual in this system to take the unit of force as the pound force, lbf, which is related to the fundamental units by

$$1\ \text{lbf} = 32.2\ \text{lbm ft/s}^2$$

The force required to accelerate a 10-lbm mass at a rate of 10 ft/s² is, therefore, 100 lbm ft/s², i.e., 100/32.2 lbf. With the lbm as the unit of mass and the lbf as the unit of force, the mass of a body is equal in magnitude to the gravitational force acting on the body under standard sea-level gravitational conditions. This was a feature that was common to all earlier systems of units.

A list of derived units in both systems of units is given in Appendix J. This appendix also contains a list of conversion factors between the two systems of units. Whenever a pressure is given in this book, it should be assumed to be an absolute pressure unless gauge pressure is clearly indicated.

Example 1.3

A U-tube manometer containing water is used to measure the pressure difference between two points in a duct through which air is flowing. Under certain conditions, it is found that the difference in the heights of the water columns in the two legs of the manometer is 11 in. Find the pressure difference between the two points in the flow in psf, psi, and Pa.

Solution

The situation being considered is shown in Figure E1.3.

It is shown in most books on fluid mechanics that, for a manometer, the pressure difference Δp is related to the difference between the heights Δh of the liquid columns in the two legs of the manometer by

$$\Delta p = \rho_{man} g \Delta h$$

where ρ_{man} is the density of the liquid in the U-tube manometer. In the situation being considered, the manometer contains water so ρ_{man} = 62.4 lbm/ft³. Hence, since g = 32.2 ft/s², it follows that

$$\Delta p = 62.4 \times 32.2 \times (11/12)$$

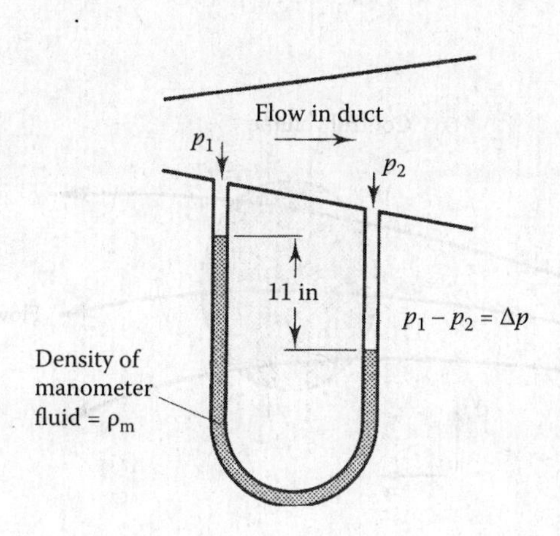

FIGURE E1.3
Situation considered.

the units being (lbm/ft³) × (ft/s²) × ft, i.e., (lbm ft/s²)/(ft²). Hence, since 1 lbf = 32.2 lbm ft/s²,

$$\Delta p = \frac{62.4 \times 32.2 \times (11/12)}{32.2} = 57.2 \text{ lbf/ft}^2 \quad \text{(i.e., psf)}$$

Since 1 lbf/in² (psi) = 144 psf it follows that

$$\Delta p = 57.2/144 = 0.397 \text{ lbf/in}^2 \text{ (i.e., psi)}$$

Lastly, since the conversion tables give 1 Pa = 0.020886 psf, it follows that

$$\Delta p = 57.2/0.020886 = 2738.7 \text{ Pa} = 2.739 \text{ kPa}$$

Conservation Laws

The analysis of compressible flows is, as mentioned previously, based on the application of the principles of conservation of mass, momentum, and energy to the flow. These principles, in relatively general form, will be discussed in this section. These conservation principles or laws will here be applied to the flow through a control volume which is defined by an imaginary boundary drawn in the flow as shown in Figure 1.4.

For such a control volume, conservation of mass requires that

Rate of increase of mass of fluid in control volume	=	Rate mass enters control volume	−	Rate mass leaves control volume

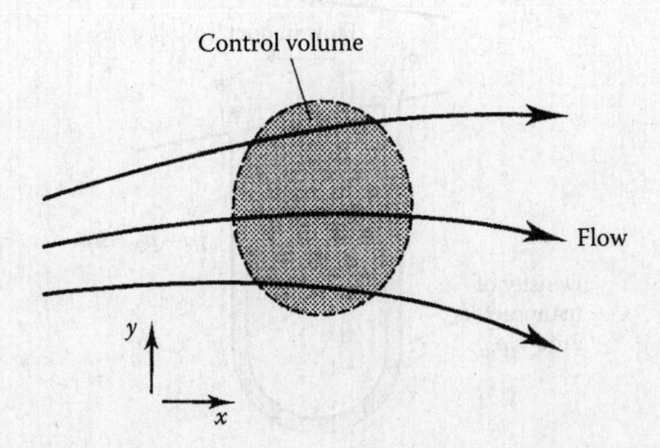

FIGURE 1.4
Control volume.

Next, consider conservation of momentum. Because momentum is a vector quantity, conservation of momentum must apply in any chosen direction. Hence, conservation of momentum requires that in any direction:

Net force on gas in control volume in direction considered	=	Rate of increase of momentum of fluid in control volume in direction considered

+	Rate momentum leaves control volume in direction considered	−	Rate momentum enters control volume in direction considered

Conservation of energy applied to the control volume requires that

Rate of increase in internal energy and kinetic energy of gas in control volume	+	Rate enthalpy and kinetic energy leave control volume	−	Rate enthalpy and kinetic energy enter control volume

=	Rate heat is transferred into control volume	−	Rate work is done by gas in control volume

Example 1.4

Air flows from a large chamber through a valve into an initially evacuated tank. The pressure and temperature in the large chamber are kept constant at 1000 kPa and 30°C, respectively, and the internal volume of the tank is 0.2 m^3. Find the time taken for the pressure in the initially evacuated tank to reach 160 kPa. Because of the values of the pressures existing in the two vessels, the mass flow rate through the valve can be assumed to remain constant (see discussion of choking in Chapter 8) and equal to 0.9×10^{-4} kg/s. Heat transfer from the tank can be neglected, and the kinetic energy of the gas in the chamber and in the tank can also be neglected.

Solution

The situation being considered is shown in Figure E1.4.

Conservation of energy requires that, since there is no flow out of the tank and no heat loss from the tank

Rate of increase of internal energy in tank = Rate of enthalpy flow into tank

i.e., since the enthalpy and internal energy per unit mass are equal to c_pT and c_vT, respectively:

$$\frac{d}{dt}[Mc_vT_T] = \dot{m}c_pT_I$$

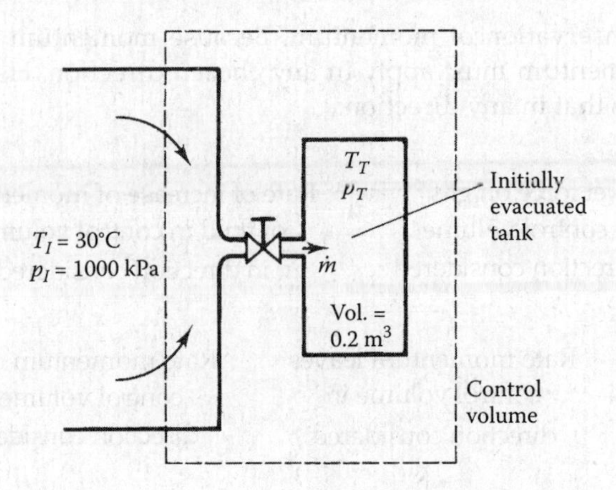

FIGURE E1.4
Flow situation considered and control volume used.

where M is the mass of air in the tank, T_T is the temperature of the air in the tank, m is the mass flow rate into the tank, and T_I is the constant temperature of the flow into the tank. Because $\dot{m}$ and T_I are constant, the above equation can be integrated to give

$$Mc_vT_T = \dot{m}c_pT_It$$

However, because $\dot{m}$ is constant and because the tank is initially evacuated,

$$M = \dot{m}t$$

Therefore,

$$\dot{m}tc_vT_T = \dot{m}c_pT_It$$

i.e.,

$$T_T = c_pT_I/c_v = \gamma T_I$$

Here, T_I is equal to 303 K so the above equation gives

$$T_T = 1.4 \times 303 = 424.2 \text{ K}$$

Next, it is noted that the perfect gas law gives for the tank

$$p_T = \rho_T RT_T = \frac{MRT_T}{V}$$

where V is the volume of the tank. Hence, again using

$$M = \dot{m}t$$

it follows that

$$p_T = \frac{mtRT_T}{V}$$

Therefore, the time taken to reach a pressure of 160 kPa is given by

$$t = \frac{160,000 \times 0.2}{0.00009 \times 287 \times 424.2} = 2921 \text{ s}$$

Hence, the time taken for the pressure in the initially evacuated tank to reach 160 kPa is 0.81 h.

Attention will mainly be given to flows in which the coordinate system can be so chosen that the flow is steady, i.e., to flows in which a coordinate system can be so chosen that none of the flow properties are changing with time. In real situations, even when the flow is steady on the average, the instantaneous flow is seldom truly steady. Instead, the flow variables fluctuate with time about the mean values either because the flow is turbulent or due to fluctuations in the system that produces the flow. It is assumed here that the mean values of the flow variables can be adequately described by equations that are based on the assumption of steady flow.

For steady flow, the above conservation equations give

Conservation of mass:

$$\begin{matrix}\text{Rate mass enters} \\ \text{control volume}\end{matrix} = \begin{matrix}\text{Rate mass leaves} \\ \text{control volume}\end{matrix}$$

Conservation of momentum:

$$\begin{matrix}\text{Net force on gas} \\ \text{in control volume} \\ \text{in direction considered}\end{matrix} = \begin{matrix}\text{Rate momentum leaves} \\ \text{control volume in} \\ \text{direction considered}\end{matrix} - \begin{matrix}\text{Rate momentum} \\ \text{enters control volume} \\ \text{in direction considered}\end{matrix}$$

Conservation of energy:

$$\begin{matrix}\text{Rate enthalpy and} \\ \text{kinetic energy leave} \\ \text{control volume}\end{matrix} - \begin{matrix}\text{Rate enthalpy and} \\ \text{kinetic energy enter} \\ \text{control volume}\end{matrix}$$

$$= \begin{matrix}\text{Rate heat is} \\ \text{transferred into} \\ \text{control volume}\end{matrix} - \begin{matrix}\text{Rate work is} \\ \text{done by gas in} \\ \text{control volume}\end{matrix}$$

In discussing the use of these conservation principles, attention will be directed to the flow through a system that, in general, has multiple inlets and outlets, an example of such a system being shown in Figure 1.5. Here there are two inlets to the system, 1 and 2, and two outlets, 3 and 4. A control volume, defined by an imaginary boundary drawn around the system, is introduced and the conservation principles are applied to the flow through this control volume.

As discussed above, the principle of conservation of mass requires that, since steady flow is being considered:

Rate mass leaves control volume = Rate mass enters control volume

Since the mass flow rate in a duct is given by

$$\dot{m} = \rho V A$$

where V is the mean velocity in the duct and A is its cross-sectional area, conservation of mass requires for the system shown in Figure 1.5:

$$\dot{m}_3 + \dot{m}_4 = \dot{m}_1 + \dot{m}_2 \tag{1.4}$$

i.e.,

$$\rho_3 V_3 A_3 + \rho_4 V_4 A_4 = \rho_1 V_1 A_1 + \rho_2 V_2 A_2$$

Attention will next be directed to the conservation of momentum principle. As discussed above, because momentum is a vector quantity, conservation of momentum in a particular direction will be considered. In any direction, it gives for steady flow

Net force on gas in control volume in direction considered	=	Rate momentum leaves control volume in direction considered	−	Rate momentum enters control volume in direction considered

Consider the situation shown in Figure 1.5. It will be assumed that the pressure around the surface of the control volume is the same everywhere except where the ducts carrying

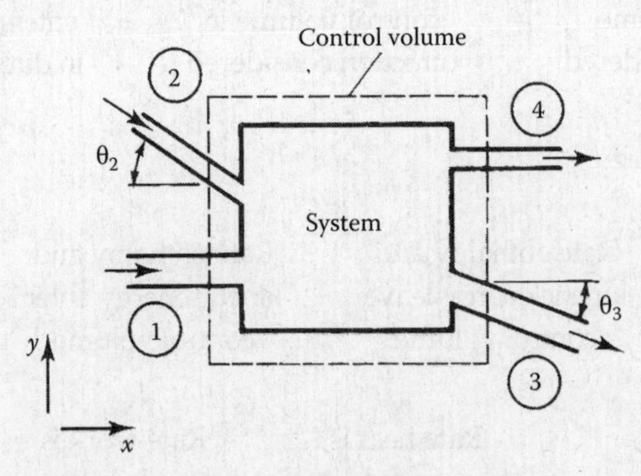

FIGURE 1.5
Type of system considered; the control volume is also shown.

the gas into and out of the control volume cross this surface. Since the momentum flux in any direction is equal to the product of the mass flow rate and the velocity component in the direction considered, it follows that for the control volume shown in Figure 1.5, conservation of momentum gives in the x-direction:

$$p_1A_1 + p_2A_2\cos\theta_2 - p_3A_3\cos\theta_3 - p_4A_4 + F_x = \dot{m}_3V_3\cos\theta_3 + \dot{m}_4V_4 - \dot{m}_1V_1 - \dot{m}_2V_2\cos\theta_2 \tag{1.5}$$

whereas in the y-direction, conservation of momentum gives

$$p_2A_2\sin\theta_2 - p_3\sin\theta_3 - F_y = \dot{m}_3V_3\sin\theta_3 - \dot{m}_2V_2\cos\theta_2 \tag{1.6}$$

In these equations, F_x and F_y are the x- and y-components of the force exerted by the system on the gas. This force is actually exerted on the gas at the surface of the system as shown in Figure 1.6. Since, as also illustrated in Figure 1.6, the gas will exert equal and opposite force components on the system, a force will have to be applied externally to the system, and therefore to the control volume, to keep it at rest.

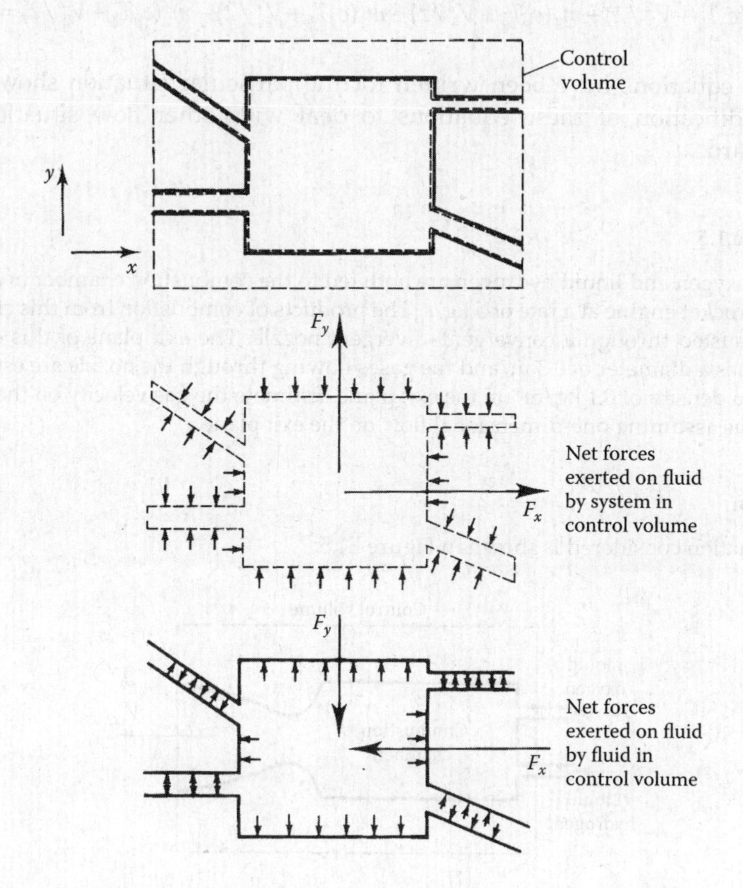

FIGURE 1.6
Force on fluid in control volume.

Lastly, the application of the conservation of energy principle to the flow through the control volume will be considered. Conservation of energy applied to the control volume requires that

Rate enthalpy and kinetic energy leave control volume	−	Rate enthalpy and kinetic energy enter control volume	=	Rate heat is transferred into control volume	−	Rate work is done by gas in control volume

Since gravitational effects are being neglected, this gives

$$\dot{m}_3(h_3 + V_3^2/2) + \dot{m}_4(h_4 + V_4^2/2) - \dot{m}_1(h_1 + V_1^2/2) - \dot{m}_2(h_2 + V_2^2/2) = Q - W \tag{1.7}$$

where h is the enthalpy of the gas per unit mass, Q is the rate of heat transfer to the system, and W is the rate at which the system is doing work. The effect of work transfer will not be considered in the present book, i.e., W will be taken as 0, and in most of the book, it will be assumed that h can be assumed to be equal to c_pT to an adequate degree of accuracy. Using this assumption, Equation 1.7 can be written as

$$\dot{m}_3(c_pT_3 + V_3^2/2) + \dot{m}_4(c_pT_4 + V_4^2/2) - \dot{m}_1(c_p\dot{T}_1 + V_1^2/2) - \dot{m}_2(c_pT_2 + V_2^2/2) = Q \tag{1.8}$$

The above equations have been written for the particular situation shown in Figure 1.5. The modification of these equations to deal with other flow situations is quite straightforward.

Example 1.5

Liquid oxygen and liquid hydrogen are both fed to the combustion chamber of a liquid fuelled rocket engine at a rate of 5 kg/s. The products of combustion from this chamber are exhausted through a convergent–divergent nozzle. The exit plane of this exhaust nozzle has a diameter of 0.3 m, and the gases flowing through the nozzle are estimated to have a density of 0.1 kg/m³ on the exit plane. Estimate the gas velocity on the nozzle exit plane assuming one-dimensional flow on the exit plane.

Solution

The situation considered is shown in Figure E1.5.

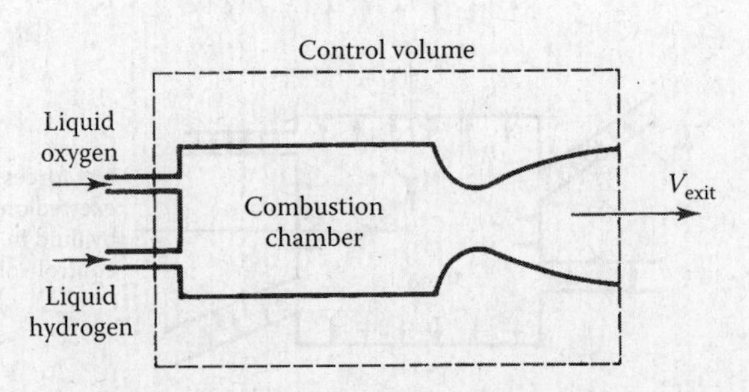

FIGURE E1.5
Control volume used.

Assuming that the flow is steady, the conservation of mass requires that the rate mass leaves the system must be equal to the rate at which mass enters the system. The rate at which mass leaves the nozzle is, therefore, 5 + 5 = 10 kg/s. But since the flow is assumed to be one-dimensional,

$$\dot{m} = \rho V A$$

Hence, the velocity on the nozzle exit plane is given by

$$V_{\text{exit}} = \frac{\dot{m}}{\rho A} = \frac{10}{0.1 \times \pi \times 0.3^2 / 4} = 1414.7 \text{ m/s}$$

Therefore, the gases leave the nozzle at a velocity of 1414.7 m/s.

Example 1.6

A solid fuelled rocket engine is fired on a test stand. The diameter of the exhaust nozzle on the discharge plane is 0.1 m. The exhaust plane velocity is estimated to be 650 m/s, and the pressure on this exhaust plane is estimated to be 150 kPa. An estimation of the rate at which combustion occurs indicates that the gas mass flow rate through the exhaust nozzle is 40 kg/s. If the ambient pressure is 100 kPa, estimate the force exerted on the test stand by the rocket.

Solution

The force T shown in Figure E1.6 is the force exerted on the test stand, i.e., the thrust. It is the force required to hold the engine in place. A force that is equal in magnitude to T but opposite in direction to T is exerted on the fluid in the engine.

Consider the application of the momentum equation to the control volume shown in Figure E1.6. Since no momentum enters this control volume, it follows that

$$\begin{array}{c}\text{Net force on control} \\ \text{volume in } x\text{-direction}\end{array} = \begin{array}{c}\text{Rate momentum} \\ \text{leaves control volume}\end{array}$$

However, the pressure is equal to ambient everywhere on the surface of the control volume except on the nozzle exit plane. Hence, if T' is the force exerted on the fluid by the system, it follows that

$$T' - (p_{\text{exit}} - p_{\text{ambient}})A_{\text{exit}} = \dot{m}V_{\text{exit}}$$

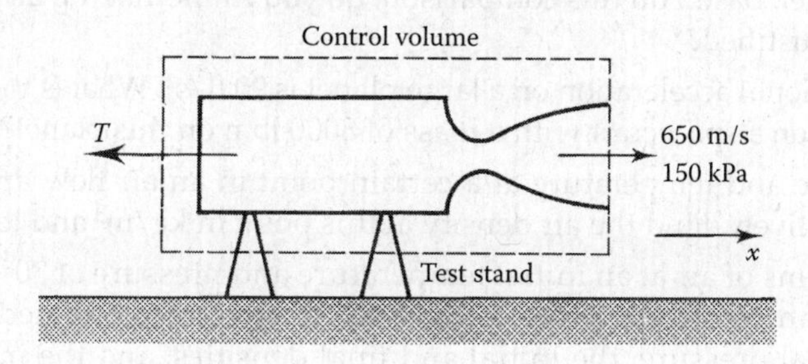

FIGURE E1.6
Control volume used.

This gives

$$T' - (150 \times 10^3 - 100 \times 10^3)\pi \frac{(0.1)^2}{4} = 40 \times 650$$

Hence,

$$T' = 26.4 \times 10^0 \text{ N} = 26.4 \text{ kN}$$

Therefore, the force exerted on the stand is

$$T = -T' = -26.4 \text{ kN}$$

The negative sign indicates that it acts in the negative x-direction, i.e., that it acts in the direction shown in Figure E1.6.

Concluding Remarks

Compressible flows are flows in which the density changes induced by the pressure changes through the flow field have a significant influence on the flow. Compressibility effects are usually associated with the high-speed flow of gases. Compressible gas flows are analyzed by applying the principles of conservation of mass, momentum, and energy together with the equation of state to deduce the variations of velocity, pressure, density, and temperature through the flow field.

PROBLEMS

1. An air stream enters a variable area channel at a velocity of 30 m/s with a pressure of 120 kPa and a temperature of 10°C. At a certain point in the channel, the velocity is found to be 250 m/s. Using Bernoulli's equation (i.e., $p + \rho V^2/2 = \text{constant}$), which assumes incompressible flow, find the pressure at this point. In this calculation, use the density evaluated at the inlet conditions. If the temperature of the air is assumed to remain constant, evaluate the air density at the point in the flow where the velocity is 250 m/s. Compare this density with the density at the inlet to the channel. Based on this comparison, do you think that the use of Bernoulli's equation is justified?
2. The gravitational acceleration on a large planet is 90 ft/s^2. What is the gravitational force acting on a spacecraft with a mass of 8000 lbm on this planet?
3. The pressure and temperature at a certain point in an air flow are 130 kPa and 30°C, respectively. Find the air density at this point in kg/m^3 and lbm/ft^3.
4. Two kilograms of air at an initial temperature and pressure of 30°C and 100 kPa undergoes an isentropic process, the final temperature attained being 850°C. Find the final pressure, the initial and final densities, and the initial and final volumes.

5. Two jets of air, each having the same mass flow rate, are thoroughly mixed and then discharged into a large chamber. One jet has a temperature of 120°C and a velocity of 100 m/s, whereas the other has a temperature of –50°C and a velocity of 300 m/s. Assuming that the process is steady and adiabatic, find the temperature of the air in the large chamber.
6. Two air streams are mixed in a chamber. One stream enters the chamber through a 5-cm diameter pipe at a velocity of 100 m/s with a pressure of 150 kPa and a temperature of 30°C. The other stream enters the chamber through a 1.5-cm-diameter pipe at a velocity of 150 m/s with a pressure of 75 kPa and a temperature of 30°C. The air leaves the chamber through a 9-cm-diameter pipe at a pressure of 90 kPa and a temperature of 30°C. Assuming that the flow is steady, find the velocity in the exit pipe.
7. The jet engine fitted to a small aircraft uses 35 kg/s of air when the aircraft is flying at a speed of 800 km/h. The jet efflux velocity is 590 m/s. If the pressure on the engine discharge plane is assumed to be equal to the ambient pressure and if effects of the mass of the fuel used are ignored, find the thrust developed by the engine.
8. The engine of a small jet aircraft develops a thrust of 18 kN when the aircraft is flying at a speed of 900 km/h at an altitude where the ambient pressure is 50 kPa. The air flow rate through the engine is 75 kg/s and the engine uses fuel at a rate of 3 kg/s. The pressure on the engine discharge plane is 55 kPa and the area of the engine exit is 0.2 m^2. Find the jet efflux velocity.
9. A small turbo-jet engine uses 50 kg/s of air, and the air/fuel ratio is 90:1. The jet efflux velocity is 600 m/s. When the afterburner is used, the overall air/fuel ratio decreases to 50:1 and the jet efflux velocity increases to 730 m/s. Find the static thrust with and without the afterburner. The pressure on the engine discharge plane can be assumed to be equal to the ambient pressure in both cases.
10. A rocket used to study the atmosphere has a fuel consumption rate of 120 kg/s and a nozzle discharge velocity of 2300 m/s. The pressure on the nozzle discharge plane is 90 kPa. Find the thrust developed when the rocket is launched at sea level. The nozzle exit plane diameter is 0.3 m.
11. A solid fuelled rocket is fitted with a convergent–divergent nozzle with an exit plane diameter of 30 cm. The pressure and velocity on this nozzle exit plane are 75 kPa and 750 m/s, respectively, and the mass flow rate through the nozzle is 350 kg/s. Find the thrust developed by this engine when the ambient pressure is (a) 100 kPa and (b) 20 kPa.
12. In a hydrogen-powered rocket, hydrogen enters a nozzle at a very low velocity with a temperature and pressure of 2000°C and 6.8 MPa, respectively. The pressure on the exit plane of the nozzle is equal to the ambient pressure, which is 10 kPa. If the required thrust is 10 MN, what hydrogen mass flow rate is required? The flow through the nozzle can be assumed to be isentropic and the specific heat ratio of the hydrogen can be assumed to be 1.4.
13. In a proposed jet propulsion system for an automobile, air is drawn in vertically through a large intake in the roof at a rate of 3 kg/s, the velocity through this intake being small. Ambient pressure and temperature are 100 kPa and 30°C, respectively. This air is compressed and heated and then discharged horizontally out of a nozzle at the rear of the automobile at a velocity of 500 m/s and a pressure

of 140 kPa. If the rate of heat addition to the air stream is 600 kW, find the nozzle discharge area and the thrust developed by the system.

14. Carbon dioxide flows through a constant area duct. At the inlet to the duct, the velocity is 120 m/s and the temperature and pressure are 200°C and 700 kPa, respectively. Heat is added to the flow in the duct and at the exit of the duct the velocity is 240 m/s and the temperature is 450°C. Find the amount of heat being added to the carbon dioxide per unit mass of gas and the mass flow rate through the duct per unit cross-sectional area of the duct. Assume that for carbon dioxide, $\gamma = 1.3$.
15. Air enters a heat exchanger with a velocity of 120 m/s and a temperature and pressure of 225°C and 2.5 MPa. Heat is removed from the air in the heat exchanger and the air leaves with a velocity 30 m/s at a temperature and pressure of 80°C and 2.45 MPa. Find the heat removed per kilogram of air flowing through the heat exchanger and the density of the air at the inlet and the exit to the heat exchanger.
16. The mass flow rate through the nozzle of a rocket engine is 200 kg/s. The areas of the nozzle inlet and exit planes are 0.7 and 2.4 m^2, respectively. On the nozzle inlet plane, the pressure and velocity are 1600 kPa and 150 m/s, respectively, whereas on the nozzle exit plane, the pressure and velocity are 80 kPa and 2300 m/s, respectively. Find the thrust force acting on the nozzle.

2

Equations for Steady One-Dimensional Compressible Fluid Flow

Introduction

Many of the compressible flows that occur in engineering practice can be adequately modeled as a flow through a duct or streamtube whose cross-sectional area is changing relatively slowly in the flow direction. A duct is here taken to mean a solid-walled channel, whereas a streamtube is defined by considering a closed curve drawn in a fluid flow. A series of streamlines will pass through this curve as shown in Figure 2.1. Further downstream, these streamlines can be joined by another curve as shown in Figure 2.1.

Since there is no flow normal to a streamline, in steady flow, the rate at which fluid crosses the area defined by the first curve is equal to the rate at which fluid crosses the area defined by the second curve. The streamlines passing through the curves effectively therefore define the "walls" of a duct and this "duct" is called a streamtube. Of course, in the case of a duct with solid walls, streamlines lie along the walls and the duct is effectively also a streamtube.

In the case of both flow through a streamtube and flow through a solid-walled duct, there can be no flow through the "walls" of the system, there being no flow through a solid wall and, by definition, no flow normal to a streamline. The two types of "duct" are shown in Figure 2.2.

Examples of the type of flow being considered in this chapter are those through the blade passages in a turbine and the flow through a nozzle fitted to a rocket engine, these being shown in Figure 2.2. In many such practical situations, it is adequate to assume that the flow is steady and one-dimensional. As discussed in the previous chapter, steady flow implies that none of the properties of the flow are varying with time. In most real flows that are steady on the average, the instantaneous values of the flow properties in fact fluctuate about the mean values. However, an analysis of such flows based on the assumption of steady flow usually gives a good description of the mean values of the flow variables. One-dimensional flow is strictly a flow in which the reference axes can be so chosen that the velocity vector has only one component over the portion of the flow field considered, i.e., if u, v, and w are the x, y, and z components of the velocity vector then strictly for the flow to be one-dimensional, it is necessary that it be possible for the x direction to be so chosen that the velocity components v and w are zero (see Figure 2.4).

In a one-dimensional flow, the velocity at a section of the duct will here be given the symbol V, as indicated in Figure 2.5.

Strictly speaking, the equations of one-dimensional flow are only applicable to flow in a straight pipe or streamtube of constant area. However, in many practical situations, the equations of one-dimensional flow can be applied with acceptable accuracy to flows with

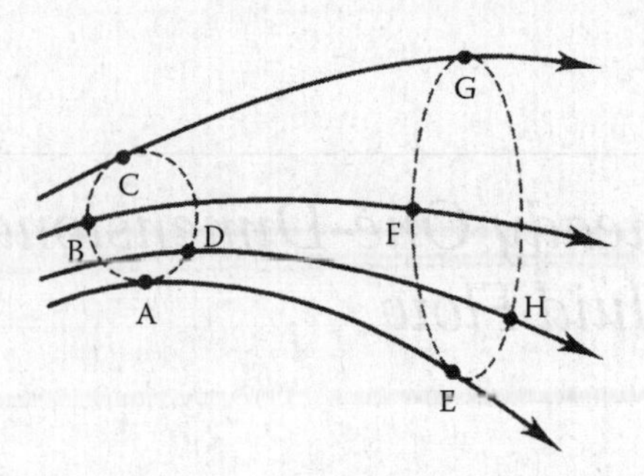

FIGURE 2.1
Definition of a streamtube.

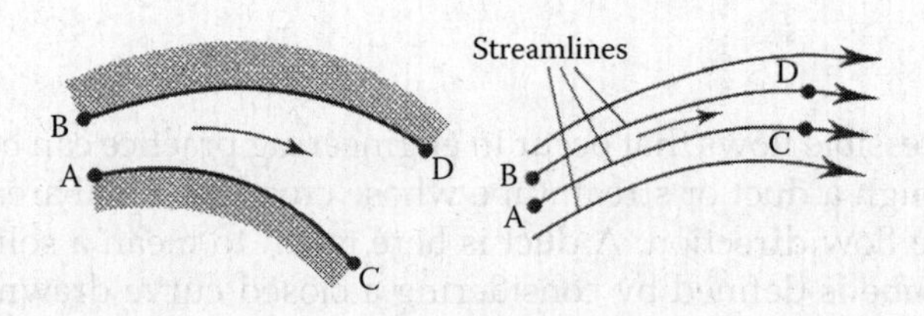

FIGURE 2.2
Solid-walled channel and streamtube.

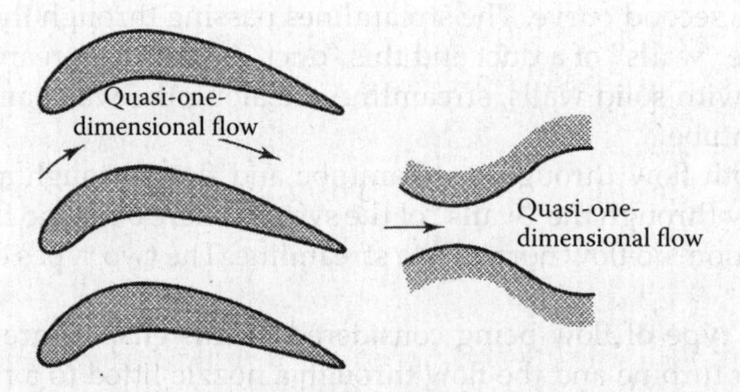

FIGURE 2.3
Typical duct flows.

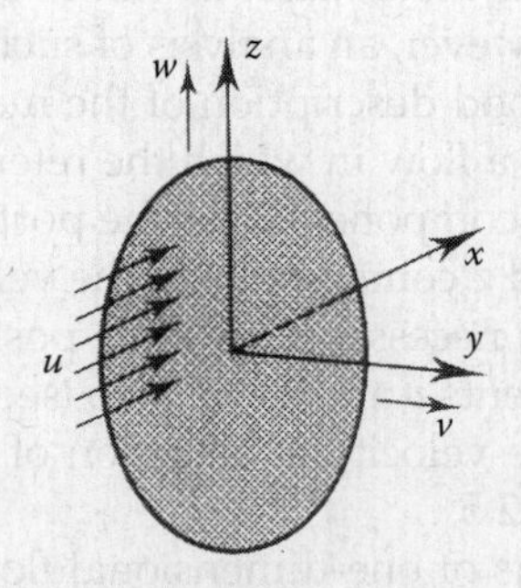

FIGURE 2.4
One-dimensional flow.

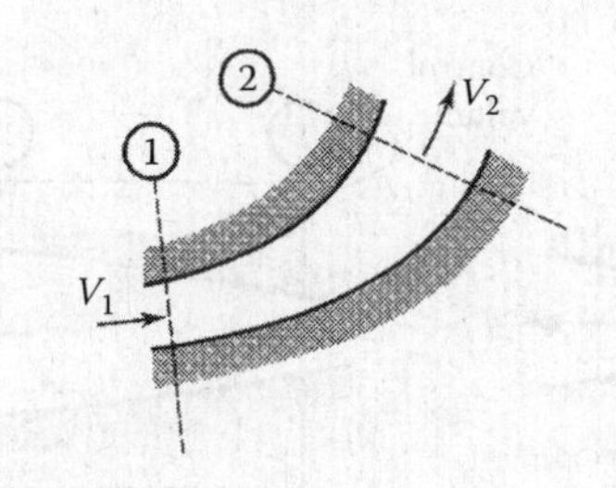

FIGURE 2.5
Definition of velocity V.

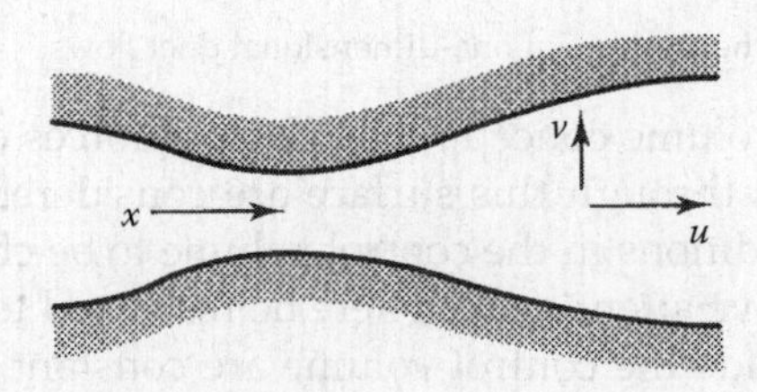

FIGURE 2.6
Flow situation that can be modeled as one-dimensional flow.

a variable area provided that the rate of change of area and the curvature of the system are small enough for one component of the velocity vector to remain dominant over the other two components. For example, although the flow through a nozzle of the type shown in Figure 2.6 is not strictly one-dimensional, because v remains very much less than u the flow can be calculated with sufficient accuracy for most purposes by ignoring v and assuming that the flow is one-dimensional, i.e., by only considering the variation of u with x. Such flows in which the flow area is changing but in which the flow at any section can be treated as one-dimensional, are commonly referred to as "quasi-one-dimensional" flows.

Control Volume

The concept of a control volume is used in the derivation and application of many equations of compressible fluid flow. As discussed in the previous chapter, a control volume is an arbitrary imaginary volume fixed relative to the coordinate system being used (the coordinate system can be moving) and bounded by a control surface through which fluid may pass as shown in Figure 2.7.

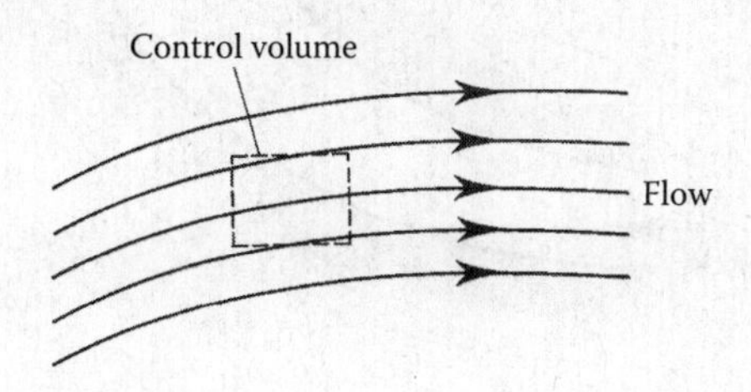

FIGURE 2.7
Control volume in a general two-dimensional flow.

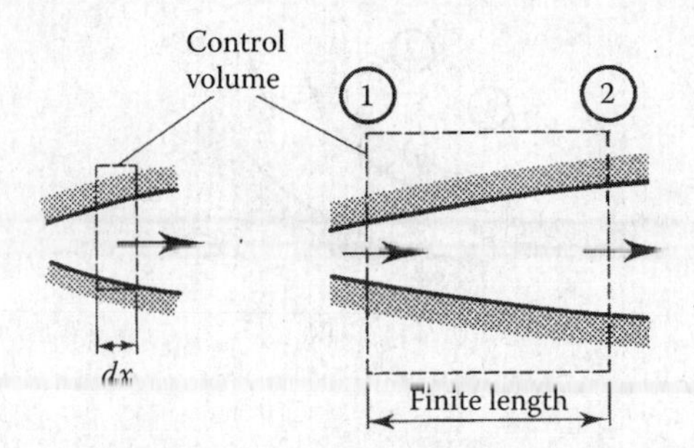

FIGURE 2.8
Types of control volume used in the analysis of one-dimensional duct flows.

In applying the control volume concept, the effects of forces on the control surface and mass and energy transfers through this surface are considered. In general, it should be noted, it is possible for conditions in the control volume to be changing with time, but for the reasons mentioned above attention will here be restricted to steady flow in which the conditions inside and outside the control volume are constant with time in terms of the coordinate system being used.

In the case of one-dimensional duct flow that is being considered here, control volumes of the type shown in Figure 2.8 are used. These control volumes cover either a differentially short length, dx, of the duct being considered or a finite length of this duct as shown in Figure 2.8.

In the case of the differentially short control volume, the changes in the flow variables through the a short control volume, such as those in velocity and pressure, i.e., dV and dp, will also be small, and in the analysis of the flow, the products of these differentially small changes such as $dV \times d\rho$ will be neglected, i.e., higher-order terms will be neglected.

Continuity Equation

The continuity equation is obtained by applying the principle of conservation of mass to the flow through a control volume. Consider the situation shown in Figure 2.9. The changes through this control volume are indicated in Figure 2.9, it being recalled that one-dimensional flow is being considered.

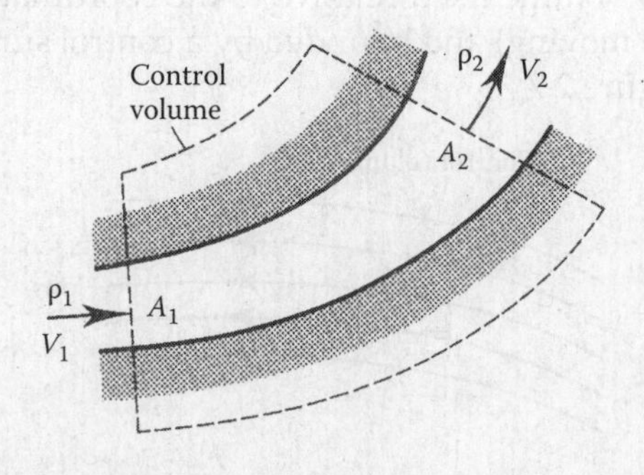

FIGURE 2.9
Control volume used in derivation of continuity equation.

Since there is no mass transfer across the "walls" of the duct or streamtube, the only mass transfer occurs through the ends of the control volume. If the possibility of a source of mass within the control volume is excluded, the principle of conservation of mass requires, for the steady flow case here being considered, that the rate at which mass enters through the left hand face of the control volume be equal to the rate at which mass leaves through the right hand face of the control volume, i.e., that

$$\dot{m}_1 = \dot{m}_2 \tag{2.1}$$

Since the rate at which mass crosses any section of the duct, i.e., $\dot{m}$, is equal to ρVA where A is the cross-sectional area of the duct at the section considered, Equation 2.1 gives

$$\rho_1 V_1 A_1 = \rho_2 V_2 A_2 \tag{2.2}$$

For the differentially short control volume indicated in Figure 2.10, this equation gives

$$\rho VA = (\rho + d\rho)\,(V + dV)\,(A + dA)$$

i.e., neglecting higher-order terms as discussed above

$$VA\,d\rho + \rho A\,dV + \rho V\,dA = 0$$

Dividing this equation by ρVA then gives

$$\frac{d\rho}{\rho} + \frac{dV}{V} + \frac{dA}{A} = 0 \tag{2.3}$$

This equation relates the fractional changes in density, velocity, and area over a short length of the control volume. If the density can be assumed constant, this equation indicates that the fractional changes in velocity and area are equal and have opposite signs, i.e., if the area increases the velocity will decrease, and vice versa. However, Equation 2.3 indicates that in compressible flow, where the fractional change in density is significant, no such simple relation between area and velocity changes exists.

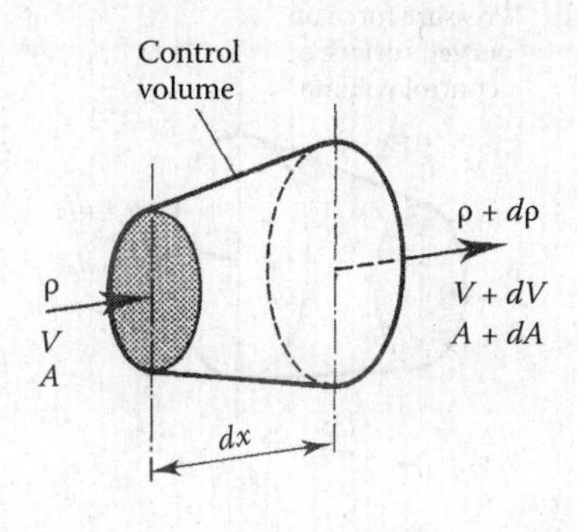

FIGURE 2.10
Differentially short control volume used in derivation of continuity equation.

Momentum Equation (Euler's Equation)

Euler's equation is obtained by applying the principle of conservation of momentum to a control volume that again consists of a short length, dx, of a streamtube. Steady flow is again assumed. The forces acting on the control volume are shown in Figure 2.11.

Because the flow is steady, the conservation of momentum requires that for this control volume the net force in direction x be equal to the rate at which momentum leaves the control volume in the x direction minus the rate at which momentum enters the control volume in the x direction. Since, by the fundamental assumptions previously listed, gravitational forces are being neglected, the only forces acting on the control volume are the pressure forces and the frictional force exerted on the surface of the control volume. Thus, the net force on the control volume in the x direction is

$$pA-(p+dp)(A+dA)+\frac{1}{2}[p+(p+dp)][(A+dA)-A]-dF_{\mu} \tag{2.4}$$

The third term in this equation represents the component of the force due to the pressure on the curved outer surface of the streamtube in the x direction. Since dx is small, it will be equal to the mean pressure on this curved surface multiplied by the projected area of this curved surface as illustrated in Figure 2.12. Since dx is small, the mean pressure on the curved surface can be taken as the average of the pressures acting on the two end surfaces, $0.5[p + (p + dp)]$, i.e., as $p + dp/2$ as indicated in Figure 2.12.

The term dF_{μ} in Equation 2.4 is the frictional force acting on the control surface in the x-direction.

Rearranging Equation 2.4 then gives the net force on the control volume in the x direction as

$$-Adp - dF_{\mu} \tag{2.5}$$

In writing this equation, the higher-order terms such as $dp\,dA$ have again been neglected since dx is taken to be small.

Since the rate at which momentum crosses any section of the duct is equal to $\dot{m}V$, the difference between the rate at which momentum leaves the control volume and the rate at which momentum enters the control volume is given by

$$\rho VA[(V + dV) - V] = \rho Va\, dV \tag{2.6}$$

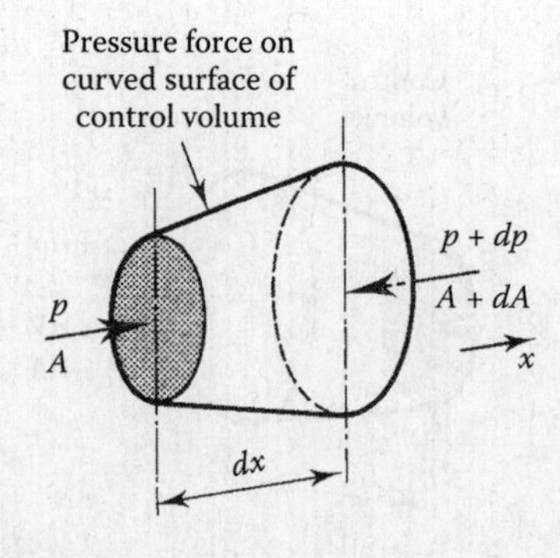

FIGURE 2.11
Differentially short control volume used in derivation of momentum equation.

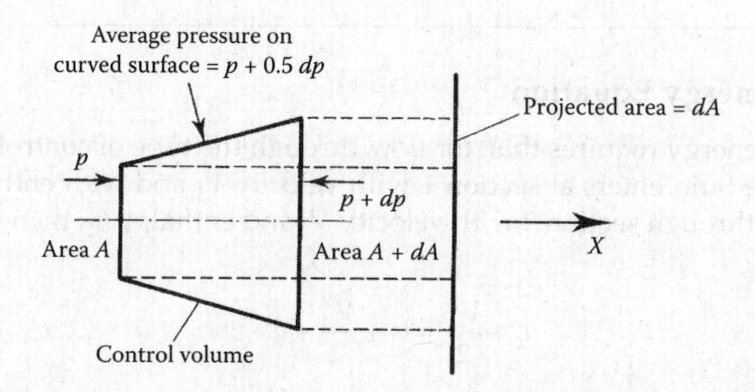

FIGURE 2.12
Pressure force on curved surface of control volume.

since no momentum enters through the curved walls of the control volume. Hence, since conservation of momentum requires that the net force on the control volume be equal to the rate at which momentum leaves the control volume minus the rate at which it enters the control volume, the above equations give

$$-A\,dp - dF_{\mu} = \rho VA\,dV \tag{2.7}$$

As discussed in the previous chapter, viscous (friction) effects will be neglected in the initial portion of this book, i.e., the term dF_{μ} in Equation 2.7 is assumed to be negligible. In this case, Equation 2.7 can be rearranged to give

$$-\frac{dp}{\rho} = V\,dV \tag{2.8}$$

This is Euler's equation for steady flow through a duct. Since V is, by the choice of the x direction, always positive, i.e., the x direction is taken in the direction of the flow, this equation indicates that an increase in velocity is always associated with a decrease in pressure and vice versa. This is an obvious result because the decrease in pressure is required to generate the force needed to accelerate the flow, i.e., to increase the velocity, and vice versa.

If Euler's equation is integrated in the x direction along the streamtube, it gives

$$\frac{V^2}{2} + \int \frac{dp}{\rho} = \text{constant} \tag{2.9}$$

To evaluate the integral, the variation of density with pressure has to be known. If the flow can be assumed to be incompressible, i.e., if the density can be assumed constant, this equation gives

$$\frac{V^2}{2} + \frac{p}{\rho} = \text{constant} \tag{2.10}$$

which is, of course, Bernoulli's equation. It should therefore be clearly understood that Bernoulli's equation only applies in incompressible flow.

Steady Flow Energy Equation

Conservation of energy requires that, for flow through the type of control volume considered above, if the fluid enters at section 1 with velocity V_1 and with enthalpy h_1 per unit mass and leaves through section 2 with velocity V_2 and enthalpy h_2, then

$$h_2 + \frac{V_2^2}{2} = h_1 + \frac{V_1^2}{2} + q - w \tag{2.11}$$

where q is the heat transferred into the control volume per unit mass of fluid flowing through it and w is the work done by the fluid per unit mass in flowing through the control volume. In this book, attention will be restricted to flows in which no work is done so that w is zero. Further, since only calorically perfect gases are being considered in this chapter

$$h = c_p T \tag{2.12}$$

Hence, the steady flow energy equation for the present purposes can be written as

$$c_p T_2 + \frac{V_2^2}{2} = c_p T_1 + \frac{V_1^2}{2} + q \tag{2.13}$$

Applying this equation to the flow through the differentially short control volume shown in Figure 2.13 gives

$$c_p T + \frac{V^2}{2} + dq = c_p(T + dT) + \frac{(V + dV)^2}{2} \tag{2.14}$$

If higher-order terms are again neglected, i.e., if dV^2 is neglected because the length of the control volume, dx, is very small, Equation 2.14 gives

$$c_p\, dT + V\, dV = dq \tag{2.15}$$

This equation indicates that in compressible flows, changes in velocity will, in general, induce changes in temperature and that heat addition can cause velocity changes as well as temperature changes.

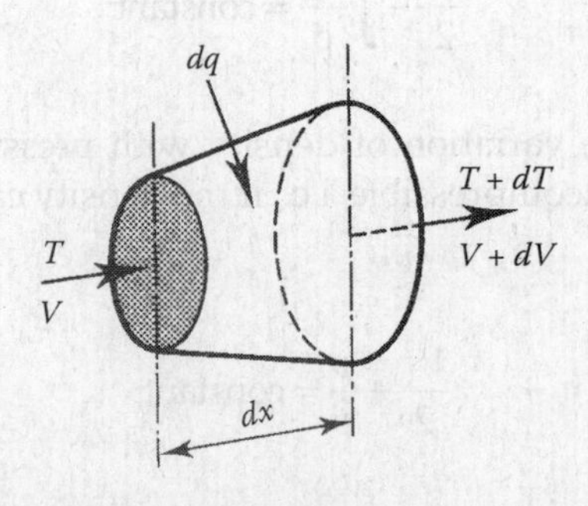

FIGURE 2.13
Differentially short control volume used in derivation of energy equation.

If the flow is adiabatic, i.e., if there is no heat transfer to or from the flow, Equation 2.11 gives

$$c_p T_2 + \frac{V_2^2}{2} = c_p T_1 + \frac{V_1^2}{2} \tag{2.16}$$

while Equation 2.15 gives for adiabatic flow

$$c_p dT + V\,dV = 0 \tag{2.17}$$

This equation shows that in adiabatic flow, an increase in velocity is always accompanied by a decrease in temperature.

Example 2.1

Air flows down a variable area duct. Measurements indicate that the temperature is 5°C and the velocity is 150 m/s at a certain section of the duct. Measurements at a second section indicate that the temperature has decreased to −20°C. Assuming that the flow is adiabatic and one-dimensional, find the velocity at this second section.

The energy equation gives

$$\frac{V_2^2}{2} = c_p T_1 + \frac{V_1^2}{2} - c_p T_2$$

Hence,

$$\frac{V_2^2}{2} = 1006 \times 278 + \frac{150^2}{2} - 1006 \times 253$$

c_p having been taken as 1006 J/kg°C. From the above equation it follows that

$$V_2 = 269.8 \text{ m/s}$$

Therefore, at the second section the velocity is 269.8 m/s.

Equation of State

When applied between any two points in the flow, the equation of state gives, since a perfect gas is being considered,

$$\frac{p_1}{\rho_1 T_1} = \frac{p_2}{\rho_2 T_2} \tag{2.18}$$

When applied between the inlet and the exit of a differentially short control volume, this equation becomes

$$\frac{p}{\rho T} = \frac{p + dp}{(\rho + d\rho)(T + dT)}$$

Since dp/p, $d\rho/\rho$, and dT/T are small, this equation gives when higher-order terms are neglected

$$\frac{p}{\rho T} = \frac{p}{\rho T}\left(1 + \frac{dp}{p}\right)\left(1 - \frac{d\rho}{\rho}\right)\left(1 - \frac{dT}{T}\right)$$

i.e.,

$$\frac{dp}{p} - \frac{d\rho}{\rho} - \frac{dT}{T} = 0 \tag{2.19}$$

This equation shows how the changes in pressure, density, and temperature are interrelated in compressible flows.

Example 2.2

Consider adiabatic air flow through a duct. At a certain section of the duct, the flow area is 0.2 m^2, the pressure is 80 kPa, the temperature is 5°C, and the velocity is 200 m/s. If, at this section, the duct area is changing at a rate of 0.3 m^2/m (i.e., $dA/dx = 0.3\ m^2/m$) find dp/dx, dV/dx, and $d\rho/dx$ (a) by assuming incompressible flow and (b) taking compressibility into account.

Solution

The continuity equation gives

$$\rho V \frac{dA}{dx} + \rho A \frac{dV}{dx} + VA \frac{d\rho}{dx} = 0$$

Hence, since using the information supplied

$$\rho = \frac{p}{RT} = \frac{80 \times 10^3}{287 \times 278} = 1.003\ \text{kg/m}^3$$

it follows that

$$1.003 \times 200 \times 0.3 + 1.003 \times 0.2 \times \frac{dV}{dx} + 200 \times 0.2 \times \frac{d\rho}{dx} = 0 \tag{a}$$

(a) Assuming incompressible flow, i.e., assuming that $d\rho/dx = 0$, the above equation gives

$$\frac{dV}{dx} = -\frac{200 \times 0.3}{0.2} = -300\ \text{(m/s)/m}$$

Also, since in incompressible flow, the conservation of momentum equation gives

$$-\frac{1}{\rho}\frac{dp}{dx} = V\frac{dV}{dx}$$

From this, it follows that

$$\frac{dp}{dx} = 200 \times 300 \times 1.003$$

i.e.,

$$\frac{dp}{dx} = 6.02 \times 10^4 \text{ Pa/m} = 60.2 \text{ kPa/m}$$

(b) With compressibility effects accounted for, i.e., when $d\rho/dx \neq 0$, the conservation of momentum equation still gives

$$-\frac{1}{\rho}\frac{dp}{dx} = V\frac{dV}{dx}, \quad \text{i.e.,} \quad \frac{dp}{dx} = -\rho V\frac{dV}{dx}$$

Conservation of energy gives

$$c_p\frac{dT}{dx} + V\frac{dV}{dx} = 0$$

From the perfect gas law, $p = \rho RT$, it follows that

$$\frac{dp}{dx} = R\left\{T\frac{d\rho}{dx} + \rho\frac{dT}{dx}\right\}$$

i.e., using the momentum equation result

$$-\rho V\frac{dV}{dx} = R\left\{T\frac{d\rho}{dx} + \rho\frac{dT}{dx}\right\}$$

However, the conservation of energy gives

$$\frac{dT}{dx} = -\frac{V}{c_p}\frac{dV}{dx}$$

so the above equation becomes

$$-\rho V\frac{dV}{dx} = R\left\{T\frac{d\rho}{dx} - \frac{\rho V}{c_p}\frac{dV}{dx}\right\}$$

which can be rearranged to give

$$\frac{d\rho}{dx} = \left\{\frac{V}{c_p} - \frac{V}{R}\right\}\frac{\rho}{T}\frac{dV}{dx}$$

Substituting the given values of the variables into this equation then gives

$$\frac{d\rho}{dx} = \left\{\frac{200}{1004} - \frac{200}{287}\right\}\frac{1.003}{278}\frac{dV}{dx} = -1.795 \times 10^{-3}\frac{dV}{dx} \tag{b}$$

Substituting this back into Equation (a) then gives

$$60.18 + 1.003 \times 0.2 \times \frac{dV}{dx} - 200 \times 0.2 \times (1.795 \times 10^{-3})\frac{dV}{dx} = 0$$

Therefore,

$$\frac{dV}{dx} = -467.3\,\frac{\text{m/s}}{\text{m}}$$

Substituting this result back into Equation (b) then gives

$$\frac{d\rho}{dx} = (-1.795 \times 10^{-3}) \times (-467.3) = 0.839\,\frac{\text{kg/m}^3}{\text{m}}$$

which then gives

$$\frac{dp}{dx} = -\rho V\frac{dV}{dx} = 9.37 \times 10^4\ \text{Pa/m} = 93.7\ \text{kPa/m}$$

The values obtained, taking compressibility into account, are therefore very different from those obtained when compressibility is ignored, i.e., compressibility effects cannot be ignored in the flow situation being considered here.

Entropy Considerations

In studying compressible flows, another variable, the entropy, s, needs, in general, to be introduced. The entropy basically places limitations on which flow processes are physically possible and which are physically excluded. The entropy change between any two points in the flow is given by

$$s_2 - s_1 = c_p \ln\left[\frac{T_2}{T_1}\right] - R\ln\left[\frac{p_2}{p_1}\right] \tag{2.20}$$

Since $R = c_p - c_v$, this equation can be written

$$\frac{s_2 - s_1}{c_p} = \ln\left[\left(\frac{T_2}{T_1}\right)\left(\frac{p_2}{p_1}\right)^{\frac{(\gamma-1)}{\gamma}}\right]$$

If there is no change in entropy, i.e., if the flow is isentropic, this equation requires that

$$\frac{T_2}{T_1} = \left(\frac{p_2}{p_1}\right)^{\frac{\gamma-1}{\gamma}} \tag{2.21}$$

Hence, since the perfect gas law gives

$$\frac{T_2}{T_1} = \frac{p_2}{p_1}\frac{\rho_1}{\rho_2}$$

it follows that in isentropic flow

$$\frac{p_2}{p_1} = \left(\frac{\rho_2}{\rho_1}\right)^{\gamma} \tag{2.22}$$

In isentropic flows, therefore p/ρ^{γ} is a constant.

If Equation 2.20 is applied between the inlet and the exit of a differentially short control volume, it gives

$$(s + ds) - s = c_p \ln\left[\frac{T + dT}{T}\right] - R\ln\left[\frac{p + dp}{p}\right]$$

Since, if ϵ is a small quantity, $\ln(1 + \epsilon)$ is to first order equal to ϵ, the above equation gives to first order of magnitude

$$ds = c_p\frac{dT}{T} - R\frac{dp}{p} \tag{2.23}$$

which can be written as

$$\frac{ds}{c_p} = \frac{dT}{T} - \left(\frac{\gamma - 1}{\gamma}\right)\frac{dp}{p} \tag{2.24}$$

Lastly, it is noted that in an isentropic flow, Equation 2.23 gives

$$c_p dT = \frac{RT}{p}dp$$

i.e., using the perfect gas law

$$c_p dT = \frac{dp}{\rho} \tag{2.25}$$

However, the energy equations for isentropic flow, i.e., for flow with no heat transfer, gives (see Equation 2.17)

$$c_p dT + V\,dV = 0$$

which, using Equation 2.25, gives

$$\frac{dp}{\rho} + V\,dV = 0$$

Comparing this with Equation 2.8 shows that this is identical to the result obtained using conservation of momentum considerations. In isentropic flow, it is then not necessary to consider both conservation of energy and conservation of momentum since, when the "isentropic equation of state," i.e., Equation 2.22, is used they give the same result.

Example 2.3

What is the form of the relation between the pressure and the velocity at two points in a flow if the flow is (a) isothermal and (b) isentropic?

The integrated Euler equation gives

$$\frac{V^2}{2} + \int \frac{dp}{\rho} = \text{constant}$$

In isothermal flow, the perfect gas equation gives $p/\rho = RT$ a constant. Hence, in this case, the Euler equation gives

$$\frac{V^2}{2} + RT \int \frac{dp}{p} = \text{constant}$$

i.e.,

$$\frac{V^2}{2} + RT \ln p = \text{constant}$$

Applying this equation between two points in a flow then gives

$$\frac{V_2^2}{2} - \frac{V_1^2}{2} = RT \ln\left(\frac{p_2}{p_1}\right)$$

Similarly, in isentropic flow, since $p \propto \rho^{\gamma}$, the integrated Euler equation gives

$$\frac{V^2}{2} + \int \frac{dp}{p^{1/\gamma}} = \text{constant}$$

i.e.,

$$\frac{V^2}{2} + \frac{\gamma}{\gamma - 1} p^{\frac{\gamma-1}{\gamma}} = \text{constant}$$

Applying this equation between two points in a flow then gives

$$\frac{V_2^2}{2} - \frac{V_1^2}{2} = \frac{\gamma}{\gamma - 1}\left(p_2^{\frac{\gamma-1}{\gamma}} - p_1^{\frac{\gamma-1}{\gamma}}\right)$$

Use of the One-Dimensional Flow Equations

The most obvious application of the quasi-one-dimensional equations is to flow through a solid-walled duct or a streamtube whose cross-sectional area is changing relatively slowly with distance as shown in Figure 2.14.

For the one-dimensional flow assumption to be valid, the rate of change of duct area with respect to the distance x along the duct must remain small. However, in applying the one-dimensional flow equations to the flow through a duct, it should be noted that the flow does not have to be one-dimensional at all sections of the duct to use the one-dimensional flow equations. For example, consider the situation shown in Figure 2.15. The flow at section 2 in Figure 2.15 cannot be assumed one-dimensional. However, the conditions at sections 1 and 3 can be related by the one-dimensional equation provided there is no reversed flow over section 3.

The one-dimensional equations also, as discussed above, apply to the flow through any streamtube. An example of a streamtube is shown in Figure 2.16. This figure shows streamlines in a two-dimensional low speed flow over a "streamlined" cylinder. The body is long in the direction normal to the page and the flow pattern therefore essentially depends

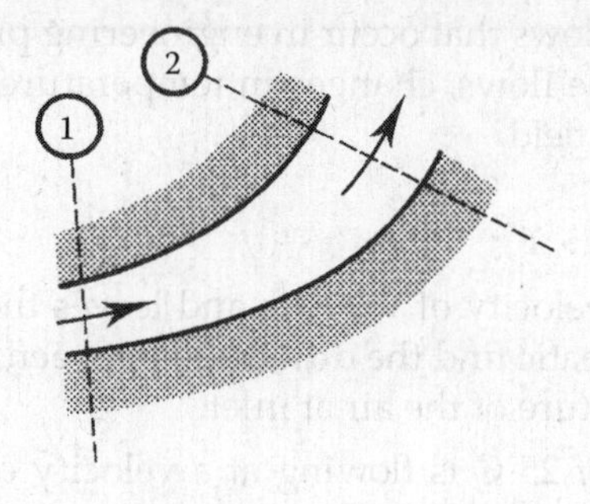

FIGURE 2.14
One-dimensional flow through a duct.

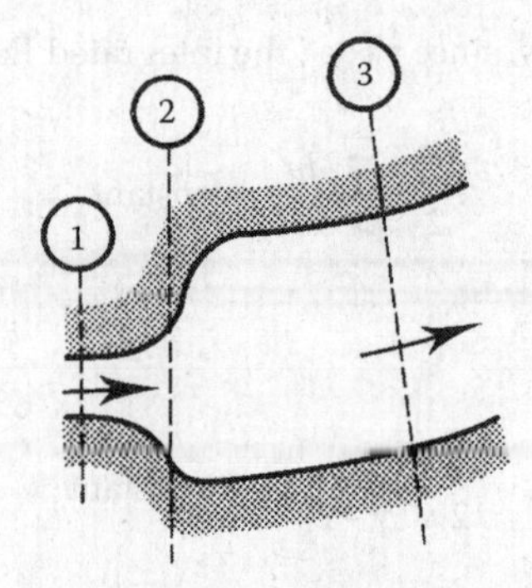

FIGURE 2.15
Duct in which one-dimensional flow assumptions are not valid throughout the flow.

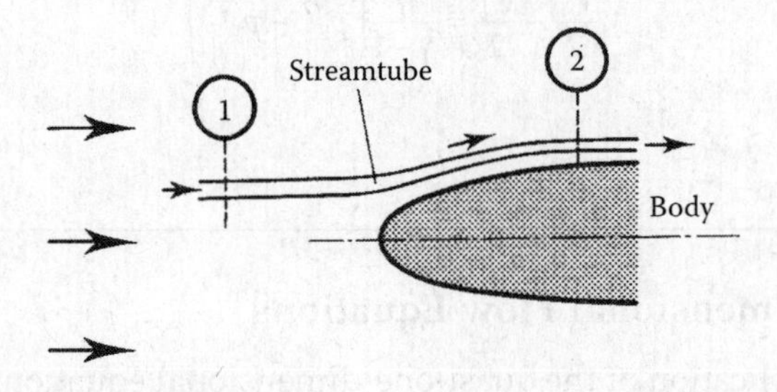

FIGURE 2.16
One-dimensional flow through a streamtube.

only on the x- and y-coordinates. The flow along the streamtube shown in Figure 2.16 will, however, be one-dimensional, the flow variables, such as velocity, depending only on the distance measured along the streamtube. As the fluid flows along this streamtube, its area changes, and there are associated changes in the pressure, temperature, and density.

Concluding Remarks

The equations discussed in this chapter, although only strictly applicable to flows that are one-dimensional, still form the basis of the analysis to an acceptable degree of accuracy of many compressible fluid flows that occur in engineering practice. The equations clearly indicate how, in compressible flows, changes in temperature and density are interlinked with changes in the velocity field.

PROBLEMS

1. Air enters a tank at a velocity of 100 m/s and leaves the tank at a velocity of 200 m/s. If the flow is adiabatic find the difference between the temperature of the air at exit and the temperature of the air at inlet.
2. Air at a temperature of 25°C is flowing at a velocity of 500 m/s. A shock wave (Chapters 5 and 6) occurs in the flow reducing the velocity to 300 m/s. Assuming

the flow through the shock wave to be adiabatic, find the temperature of the air behind the shock wave.

3. Air being released from a tire through the valve is found to have a temperature of 15°C. Assuming that the air in the tire is at the ambient temperature of 30°C, find the velocity of the air at the exit of the valve. The process can be assumed to be adiabatic.
4. A gas with a molecular weight of 4 and a specific heat ratio of 1.67 flows through a variable area duct. At some point in the flow the velocity is 180 m/s and the temperature is 10°C. At some other point in the flow, the temperature is –10°C. Find the velocity at this point in the flow assuming that the flow is adiabatic.
5. At a section of a circular duct through which air is flowing the pressure is 150 kPa, the temperature is 35°C, the velocity is 250 m/s, and the diameter is 0.2 m. If, at this section, the duct diameter is increasing at a rate of 0.1 m/m, find dp/dx, dV/dx, and $d\rho/dx$.
6. Consider an isothermal air flow through a duct. At a certain section of the duct the velocity, temperature, and pressure are 200 m/s, 25°C, and 120 kPa, respectively. If the velocity is decreasing at this section at a rate of 30% per m, find dp/dx, ds/dx, and $d\rho/dx$.
7. Consider adiabatic air flow through a variable area duct. At a certain section of the duct the flow area is 0.1 m^2, the pressure is 120 kPa, the temperature is 15°C, and the duct area is changing at a rate of 0.1 m^2/m. Plot the variations of dp/dx, dV/dx, and $d\rho/dx$ with the velocity at the section for velocities between 50 and 300 m/s.
8. Methane flows through a circular pipe that has a diameter of 4 cm. The temperature, pressure, and velocity at the inlet to the pipe are 200 K, 250 kPa, and 30 m/s, respectively. Assuming that the flow is steady and isothermal calculate the pressure on the exit plane and the heat added to the methane in the pipe if the velocity on the pipe exit plane is 35 m/s. Assume that the methane can be treated as a perfect gas with a specific heat ratio of 1.32 and a molar mass of 16.

3

Some Fundamental Aspects of Compressible Flow

Introduction

It was indicated in the previous chapters that compressibility effects become important in a gas flow when the velocity in the flow is high. An attempt will be made in this chapter to show that it is not the value of the gas velocity itself but rather the ratio of the gas velocity to the speed of sound in the gas that determines when compressibility is important. This ratio is termed the Mach number, M, i.e.,

$$M = \frac{\text{gas velocity}}{\text{speed of sound}} = \frac{V}{a} \tag{3.1}$$

where a is the speed of sound.

If $M < 1$, the flow is said to be subsonic, whereas if $M > 1$, the flow is said to be supersonic. If the Mach number is near 1 and there are regions of both subsonic and supersonic flow, the flow is said to be transonic. If the Mach number is very much greater than 1, the flow is said to be hypersonic. Hypersonic flow is normally associated with flows in which $M > 5$.

As will be shown later in this chapter, the speed of sound in a perfect gas is given by

$$a = \sqrt{\frac{\gamma p}{\rho}} = \sqrt{\gamma R T} \tag{3.2}$$

The speed of sound in a gas depends therefore only on the absolute temperature of the gas.

Isentropic Flow in a Streamtube

To illustrate the importance of the Mach number in determining the conditions under which compressibility must be taken into account, isentropic flow, i.e., frictionless adiabatic flow, through a streamtube will be first considered. The changes in the flow variables over a short length, dx, of the streamtube shown in Figure 3.1 are considered.

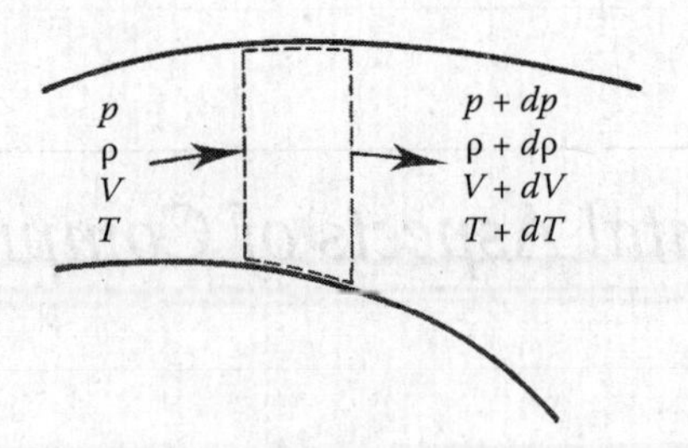

FIGURE 3.1
Portion of streamtube considered.

The Euler equation, Equation 2.8, derived in the previous chapter by applying the conservation of momentum principle and ignoring the effects of friction gives

$$\frac{dp}{p} = -\frac{\rho V^2}{p}\frac{dV}{V} \tag{3.3}$$

Using the expression for the speed of sound, a, given in Equation 3.2 allows this equation to be written as

$$\frac{dp}{p} = -\gamma\frac{V^2}{a^2}\frac{dV}{V} \tag{3.4}$$

but $M = V/a$, so this equation can be written as

$$\frac{dp}{p} = -\gamma M^2\frac{dV}{V} \tag{3.5}$$

This equation shows that the magnitude of the fractional pressure change, dp/p, induced by a given fractional velocity change, dV/V, depends on the square of the Mach number.

Next consider the energy equation. Since adiabatic flow is being considered, Equation 2.17 gives

$$\frac{dT}{T} = -\frac{V^2}{c_p T}\frac{dV}{V} \tag{3.6}$$

which can be rearranged to give

$$\frac{dT}{T} = -\frac{\gamma R}{c_p}M^2\frac{dV}{V} \tag{3.7}$$

However, since $R = c_p - c_v$, i.e., since $R/c_p = 1 - 1/\gamma$, it follows that

$$\frac{\gamma R}{c_p} = \gamma - 1 \tag{3.8}$$

Equation 3.7 can therefore be written as

$$\frac{dT}{T} = -(\gamma - 1)M^2 \frac{dV}{V} \tag{3.9}$$

This equation shows that the magnitude of the fractional temperature change, dT/T, induced by a given fractional velocity change, dV/V, also depends on the square of the Mach number.

Lastly, consider the equation of state. As shown in the previous chapter, this gives

$$\frac{dp}{p} = \frac{d\rho}{\rho} + \frac{dT}{T} \tag{3.10}$$

Combining this equation with Equations 3.5 and 3.9 then gives

$$\frac{d\rho}{\rho} = -\gamma M^2 \frac{dV}{V} + (\gamma - 1)M^2 \frac{dV}{V} = -M^2 \frac{dV}{V} \tag{3.11}$$

This equation indicates that

$$\frac{d\rho/\rho}{dV/V} = -M^2$$

From this equation, it will be seen that for a given fractional change in velocity, i.e., for a given dV/V, the corresponding induced fractional change in density will also depend on the square of the Mach number. For example, at Mach 0.1, the fractional change in density will be 1% of the fractional change in velocity; at Mach 0.33, it will be about 10% of the fractional change in velocity; whereas at Mach 0.4, it will be 16% of this fractional change in velocity. Therefore, at low Mach numbers, the density changes will be insignificant but as the Mach number increases, the density changes, i.e., compressibility effects, will become increasingly important. Hence, compressibility effects become important in high Mach number flows. The Mach number at which compressibility must start to be accounted for depends very much on the flow situation and the accuracy required in the solution. As a rough guide, it is sometimes assumed that if $M > 0.5$, then there is a possibility that compressibility effects should be considered.

It should also be noted that Equation 3.9 gives

$$\frac{dT/T}{dV/V} = -(\gamma - 1)M^2$$

This indicates that if the Mach number is high enough for density changes in the flow to be significant, the temperature changes in the flow will also be important.

It should be clear from the above results that the Mach number is the parameter that determines the importance of compressibility effects in a flow.

Example 3.1

Consider the isentropic flow of air through a duct whose area is decreasing. Find the percentage changes in velocity, density, and pressure induced by a 1% reduction in area for Mach number between 0.1 and 0.95.

Solution

The continuity equation gives

$$\frac{d\rho}{\rho}+\frac{dV}{V}+\frac{dA}{A}=0$$

i.e., using the relation for $d\rho/\rho$ given above

$$(1-M^2)\frac{dV}{V}+\frac{dA}{A}=0$$

Therefore,

$$\frac{dV}{V}=-\frac{1}{(1-M^2)}\frac{dA}{A}$$

Again, using the expression for the density change given above gives

$$\frac{d\rho}{\rho}=+\frac{M^2}{(1-M^2)}\frac{dA}{A}$$

Then, using the expression for the pressure change gives

$$\frac{dp}{p}=+\frac{\gamma M^2}{(1-M^2)}\frac{dA}{A}$$

In the present case, $dA/A = -0.01$, so the above equations give

$$\frac{dV}{V}=+\frac{0.01}{(1-M^2)}$$

$$\frac{d\rho}{\rho}=-\frac{0.01M^2}{(1-M^2)}$$

$$\frac{dp}{p}=-\frac{0.01\gamma M^2}{(1-M^2)}$$

Using these relations then gives the results shown in Table E3.1.

The percentage change in density will thus be seen to increase from 0.01% at Mach 0.1 to 9% at Mach 0.95. It will also be noted that there is a singularity at $M = 1$. The consequences of this will be discussed later.

TABLE E3.1

Percentage Changes in Velocity, Density and Pressure Produced by 1% Reduction in Area

M	dV/V	$d\rho/\rho$	dp/p
0.1	0.01010	−0.00010	−0.00014
0.2	0.01042	−0.00042	−0.00058
0.3	0.01099	−0.00099	−0.00139
0.4	0.01191	−0.00191	−0.00267
0.5	0.01333	−0.00333	−0.00467
0.6	0.01563	−0.00563	−0.00788
0.7	0.01961	−0.00961	−0.01345
0.8	0.02778	−0.01778	−0.02489
0.9	0.05263	−0.04263	−0.05968
0.95	0.10256	−0.09256	−0.12959

Speed of Sound

The importance of the Mach number in determining the characteristics of high-speed gas flows was discussed in the preceding section. To calculate M, the local speed of sound in the gas has to be known. An expression for the speed of sound, which was used in the previous section, will therefore be derived in the present section.

The speed of sound is, of course, the speed at which very weak pressure waves are transmitted through a gas, this speed being dependent on the type of gas. Consider a plane infinitesimally weak pressure wave propagating through a gas. It could, for example, be thought of as a wave that is propagating down a duct after being generated by giving a small velocity to a piston at one end of a duct containing the gas as shown in Figure 3.2.

More generally, a plane wave will be a small, effectively plane, portion of a spherical wave moving outward through the gas from a point source of disturbance as indicated in Figure 3.3.

Now from experience (this will actually be proved later in this book), it is known that the wave remains steep, i.e., the longitudinal distance over which the changes produced by the wave occur remains small as indicated in Figures 3.2 and 3.3.

Let the pressure change across the wave be dp and let the corresponding density and temperature changes be $d\rho$ and dT, respectively. The gas into which the wave is propagating is assumed to be at rest. The wave will then induce a gas velocity, dV, behind it as it moves through the gas. The changes across the wave are therefore as shown in Figure 3.4.

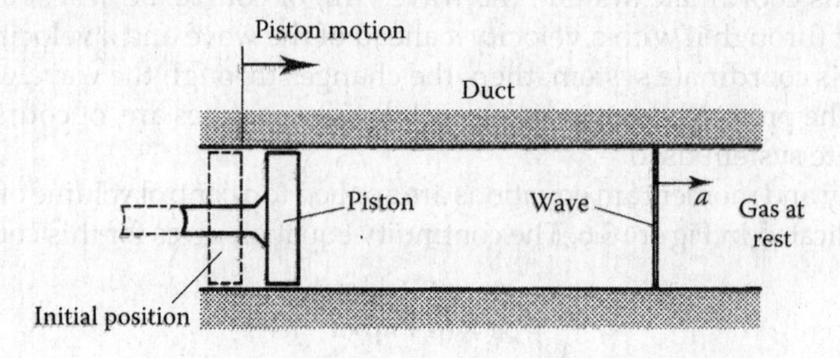

FIGURE 3.2
Generation of a weak pressure wave by the motion of a piston in a duct.

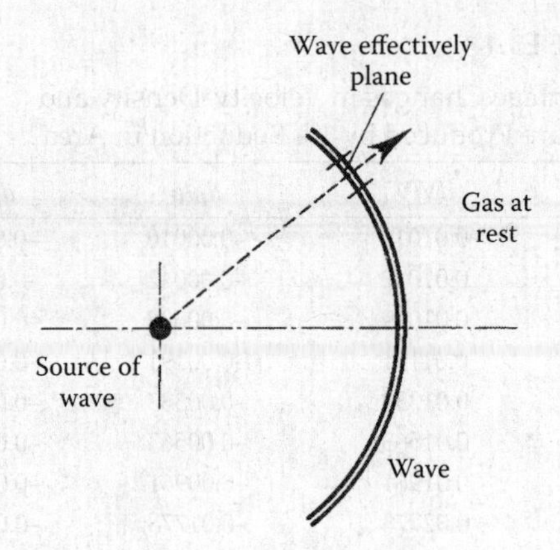

FIGURE 3.3
Portion of spherical pressure wave considered.

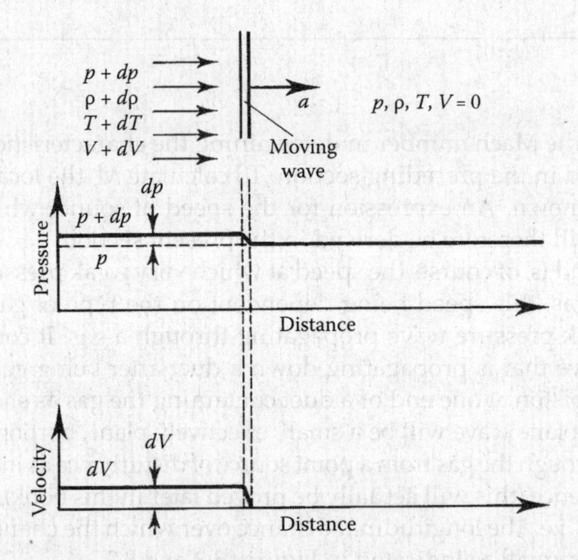

FIGURE 3.4
Changes through propagating weak pressure wave.

To analyze the flow through the wave and thus to determine a, it is convenient to use a coordinate system that is attached to the wave, i.e., a coordinate system that is moving with the wave. In this coordinate system, the wave will, of course, be at rest and the gas will effectively flow through it with a velocity a ahead of the wave and a velocity $a - dV$ behind the wave. In this coordinate system, then, the changes through the wave will be as shown in Figure 3.5. The pressure, temperature, and density changes are, of course, independent of the coordinate system used.

The continuity and momentum equations are applied to a control volume of unit area across the wave as indicated in Figure 3.6. The continuity equation gives for this control volume

$$\frac{\dot{m}}{A} = \rho a = (\rho + d\rho)(a - dV) \tag{3.12}$$

where $\dot{m}/A$ is the mass flow rate per unit area through the wave.

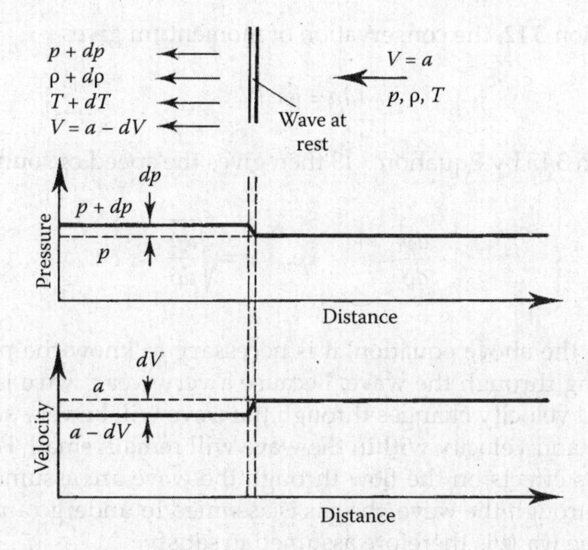

FIGURE 3.5
Changes relative to the weak pressure wave.

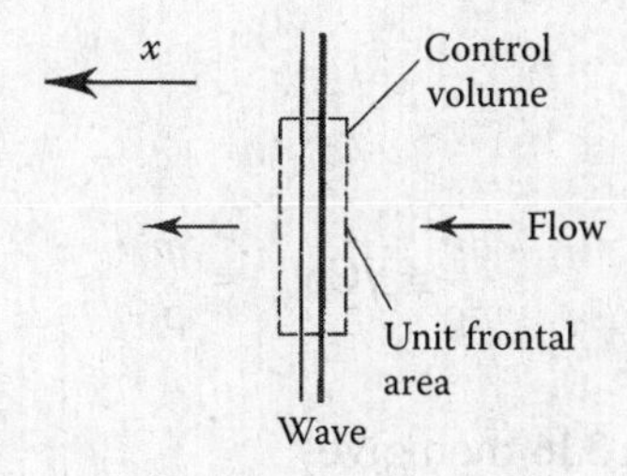

FIGURE 3.6
Control volume used in analysis of flow through a weak pressure wave.

Since the case of a very weak wave is being considered, the second-order term, i.e., $d\rho\, dV$, that arises in Equation 3.12 can be neglected, and this equation then gives

$$d\rho = \frac{\rho}{a} dV \tag{3.13}$$

The conservation of momentum is next considered. The only forces acting on the control volume are the pressure forces. The momentum equation therefore gives, since a control volume with unit area normal to the flow direction is being considered

$$p - (p + dp) = \left(\frac{\dot{m}}{A}\right)[(a - dV) - a]$$

which leads to

$$dp = \left(\frac{\dot{m}}{A}\right) dV \tag{3.14}$$

Hence, using Equation 3.12, the conservation of momentum gives

$$dp = \rho a\, dV \tag{3.15}$$

Dividing Equation 3.15 by Equation 3.13 then gives the speed of sound as

$$\frac{dp}{d\rho} = a^2, \quad \text{i.e.,} \quad a = \sqrt{\frac{dp}{d\rho}} \tag{3.16}$$

To evaluate a using the above equation, it is necessary to know the process that the gas undergoes in passing through the wave. Because a very weak wave is being considered, the temperature and velocity changes through the wave will be very small and the gradients of temperature and velocity within the wave will remain small. For this reason, heat transfer and viscous effects on the flow through the wave are assumed to be negligible. Hence, in passing through the wave, the gas is assumed to undergo an isentropic process. The flow through the wave is therefore assumed to satisfy

$$\frac{p}{\rho^\gamma} = C \quad \text{(a constant)} \tag{3.17}$$

Differentiating this then gives

$$\frac{dp}{d\rho} = \gamma C \rho^{\gamma-1} = \frac{\gamma p}{\rho} \tag{3.18}$$

Substituting this into Equation 3.16 then gives

$$a = \sqrt{\frac{\gamma p}{\rho}} = \sqrt{\gamma R T} \tag{3.19}$$

The experimental measurements of the speed of sound in gases are in good agreement with the values given by this equation, thus confirming that the assumptions made in its derivation are justified.

Therefore, for a given gas, the speed of sound depends only on the square root of the absolute temperature. Further, since the equation for the speed of sound can be written as

$$a = \sqrt{\gamma \frac{\Re}{m} T} \tag{3.20}$$

where m is the molar mass of the gas, it follows that, since γ does not vary greatly between gases, the speed of sound of a gas at a given temperature is approximately inversely proportional to its molar mass.

Some typical values for the speed of sound at 0°C are shown in Table 3.1.

TABLE 3.1

Speed of Sound Values for Various Gases

Gas	Molar Mass	γ	Speed of Sound at 0°C (m/s)
Air	28.960	1.404	331
Argon (Ar)	39.940	1.667	308
Carbon dioxide (CO_2)	44.010	1.300	258
Freon 12 ($CC1_2F_2$)	120.900	1.139	146
Helium (He)	4.003	1.667	970
Hydrogen (H_2)	2.016	1.407	1270
Xenon (Xe)	131.300	1.667	170

Example 3.2

An aircraft is capable of flying at a maximum Mach number of 0.91 at sea level. Find the maximum velocity at which this aircraft can fly at sea level if the air temperature is (a) 5°C and (b) 45°C.

Solution

Since

$$M_{max} = \frac{V_{max}}{a}$$

it follows that

$$V_{max}\Big|_{\text{sea level}} = M_{max} \times a_{\text{sea level}} = 0.91\sqrt{\gamma R T_{\text{sea level}}}$$

When $T_{\text{sea level}} = 5°\text{C} = 278$ K, the above equation gives

$$V_{max}\Big|_{\text{sea level}} = 0.91\sqrt{1.4 \times 287 \times 278} = 304 \text{ m/s}$$

Similarly, when $T_{\text{sea level}} = 45°\text{C} = 318$ K, the above equation gives

$$V_{max}\Big|_{\text{sea level}} = 0.91\sqrt{1.4 \times 287 \times 318} = 325 \text{ m/s}$$

In the days when world speed records for aircraft were established when flying at ground level, if there was a limiting Mach number on the aircraft, it therefore would pay to attempt to establish the record on a hot day or in a high-temperature geographic region such as in a desert.

Example 3.3

An aircraft is driven by propellers with a diameter of 4 m. At what engine speed will the tips of the propellers reach sonic velocity if the air temperature is 15°C?

Solution

Since the tip moves through a distance of πD in one revolution, the tip speed is given by

$$V = n\pi D$$

where n is the rotational speed in revolutions per second. Hence, if the Mach number at the tip is 1,

$$M = 1, \qquad V = a = \sqrt{\gamma RT}$$

Combining the above results then gives

$$n = \frac{\sqrt{\gamma RT}}{\pi D} = \frac{\sqrt{1.4 \times (8314/29) \times 288}}{\pi \times 4} = 27.06 \text{ revs/sec} = 1623 \text{ revs/min}$$

Example 3.4

In evaluating the performance of an aircraft, a "standard atmosphere" is usually introduced. The conditions in the "standard atmosphere" are meant to represent average conditions in the atmosphere. In the U.S. Standard Atmosphere, the temperature in the inner portion of the atmosphere is defined by the following two equations:

For altitudes, H, of from 0 m (sea level) to 11,019 m

$$T = 288.16 - 0.0065H$$

Above an altitude, H, of 11,019 m

$$T = 216.66$$

The altitude, H, is measured in meters, and the temperature, T, is in kelvin.

Plot a graph showing how the speed of sound varies with altitude in this atmosphere for altitudes from sea level to 12,000 m.

Solution

Using $a = \sqrt{\gamma RT}$, the following is obtained

$$a = \sqrt{1.4 \times 287 \times (288.16 - 0.0065H)} \quad \text{for } 0 \le H \le 11{,}019 \text{ m}$$

$$a = \sqrt{1.4 \times 287 \times 216.66} \quad \text{for } H > 11{,}019 \text{ m}$$

This has been used to derive the variation of the speed of sound, a, with altitude, H. The result is shown in Figure E3.4. It will be noted that

$$a\big|_{H=0} = 340 \text{ m/s}, \qquad a\big|_{H=11{,}019} = 294.9 \text{ m/s}$$

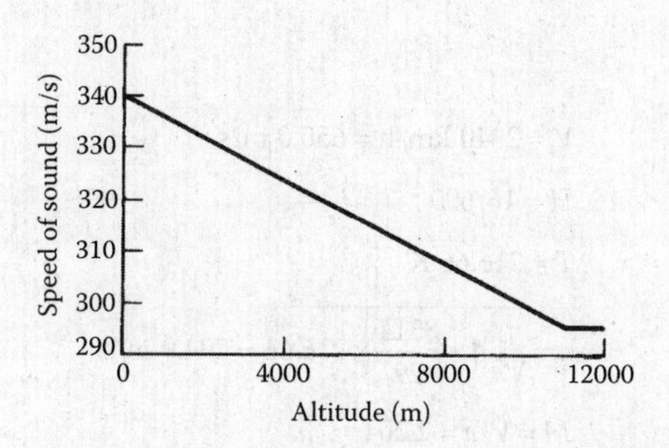

FIGURE E3.4
Variation of speed of sound with altitude.

Example 3.5

Typical cruising speeds and altitudes for three commercial aircraft are

- Dash 8: Cruising speed: 500 km/h at an altitude of 4570 m.
- Boeing 747: Cruising speed: 978 km/h at an altitude of 9150 m.
- Concorde: Cruising speed: 2340 km/h at an altitude of 16,600 m.

Find the Mach number of these three aircraft when flying at these cruise conditions. Use the properties of the standard atmosphere discussed in the previous problem.

Solution

Dash 8

$$V = 500\ \text{km/h} = 138.9\ \text{m/s}$$

$$H = 4570\ \text{m}$$

$$T = 288.16 - 0.0065 \times 4570 = 258.4\ \text{K}$$

$$a = \sqrt{1.4 \times \frac{8314}{29} \times 258.4} = 322.0\ \text{m/s}$$

$$M = V/a = 138.9/322.0 = 0.431$$

Boeing 747

$$V = 978\ \text{km/h} = 271.7\ \text{m/s}$$

$$H = 9150\ \text{m}$$

$$T = 288.16 - 0.0065 \times 9150 = 228.7\ \text{K}$$

$$a = \sqrt{1.4 \times \frac{8314}{29} \times 228.7} = 303.0\ \text{m/s}$$

$$M = V/a = 0.897$$

Concorde

$$V = 2340 \text{ km/h} = 650.0 \text{ m/s}$$

$$H = 16,600$$

$$T = 216.66 \text{ K}$$

$$a = \sqrt{1.4 \times \frac{8314}{29} \times 216.66} = 294.9 \text{ m/s}$$

$$M = V/a = 2.204$$

Example 3.6

A weak pressure wave (a sound wave) across which the pressure rise is 0.05 kPa is traveling down a pipe into air at a temperature of 30°C and a pressure of 105 kPa. Estimate the velocity of the air behind the wave.

Solution

Consider the analysis of a sound wave presented above. As discussed in this analysis, the conservation of mass gives for the flow through the control volume shown in Figure 3.6, which has unit frontal area

$$\dot{m} = \rho a = [(\rho + d\rho)(a - dV)]$$

whereas the conservation of momentum gives

$$p - (p + dp) = \dot{m}[(a - dV) - a]$$

Therefore, $dp = \dot{m}\, dV = \rho a\, dV$. The velocity behind the sound wave, dV, is therefore given by

$$dV = \frac{1}{\rho a} dp$$

However, for the values given

$$\rho = \frac{p}{RT} = \frac{105 \times 10^3}{287 \times 303} = 1.209 \text{ kg/m}^3$$

and

$$a = \sqrt{\gamma RT} = \sqrt{1.4 \times 287 \times 303} = 348.7 \text{ m/s}$$

Therefore, using the equation for dV given above

$$dV = \frac{1}{1.209 \times 348.7} \times 0.05 \times 10^3 = 0.119 \text{ m/s}$$

The velocity behind the wave is therefore 0.119 m/s.

Mach Waves

Consider a small solid body moving relative to a gas. For the gas to pass smoothly over the body, disturbances tend to be propagated ahead of the body to "warn" the gas of the approach of the body, i.e., because the pressure at the surface of the body is greater than that in the surrounding gas, pressure waves spread out from the body. Since these pressure waves are very weak except in the immediate vicinity of the body, they effectively move outward at the speed of sound.

To illustrate the effect of the velocity of the body relative to the speed of sound on the flow field, consider the small body, i.e., essentially a point source of disturbance, to be moving at a uniform linear velocity, u, through the gas and let the speed of sound in the gas be a. Although the body is essentially emitting waves continuously, a series of waves emitted at time intervals t will be considered. Since the body is moving through the gas, the origin of these waves will be continually changing. Waves generated at times 0, t, $2t$, and $3t$ will be considered. First, consider the case where the speed of the body is less than the speed of sound, i.e., the Mach number is less than 1. The body is at positions a, b, c, and d at the four times considered. Since the waves spread radially outward from their point of origin at the speed of sound, the wave pattern at time $3t$ will be as shown in Figure 3.7.

Next consider the case where the small body is moving faster than the speed of sound, i.e., where the Mach number is greater than 1. In this case, the wave pattern at time $3t$ will be as shown in Figure 3.8. It will be seen that in this supersonic case all the waves lie within the cone indicated which has its vertex at the body at the time considered. Only the gas that lies within this cone is "aware" of the presence of the body. This cone has a vertex angle, α, which will be seen to be given by

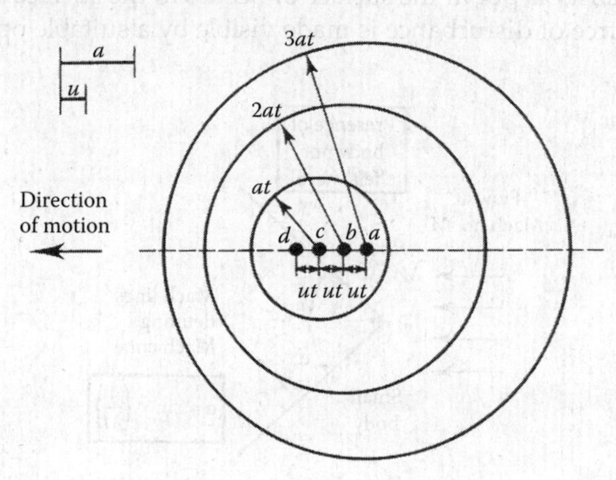

FIGURE 3.7
Spread of waves from source of disturbance moving at a subsonic speed.

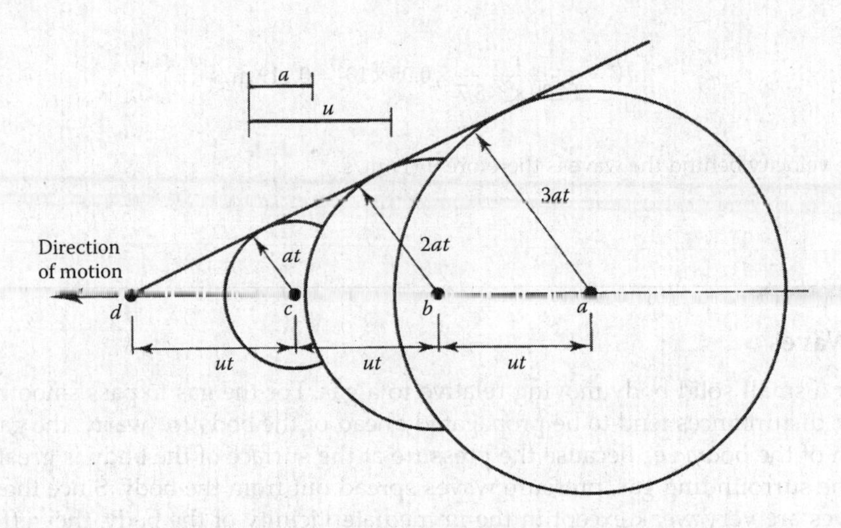

FIGURE 3.8
Spread of waves from source of disturbance moving at a supersonic speed.

$$\sin\alpha = \frac{a}{V} = \frac{1}{M} \tag{3.21}$$

The angle α is termed the Mach angle.

A comparison of the results shown in Figures 3.7 and 3.8 shows the importance of the Mach number in determining the nature of the flow field.

If the body is at rest and the gas is moving over it at a supersonic velocity, all the disturbances generated by the body are swept downstream and lie within the Mach cone shown in Figure 3.9. There will be essentially jumps in the values of the flow variables when the flow reaches the cone. The cone is therefore termed a conical Mach wave.

Similarly, in two-dimensional flow, all waves originating at a weak line source of disturbance will all lie behind a plane wave inclined at the Mach angle to the flow as shown in Figure 3.10. This result is sometimes used in the measurement of the Mach number of a gas flow. A small irregularity is put in the surface or occurs in the surface and the Mach wave generated at this source of disturbance is made visible by a suitable optical technique. By

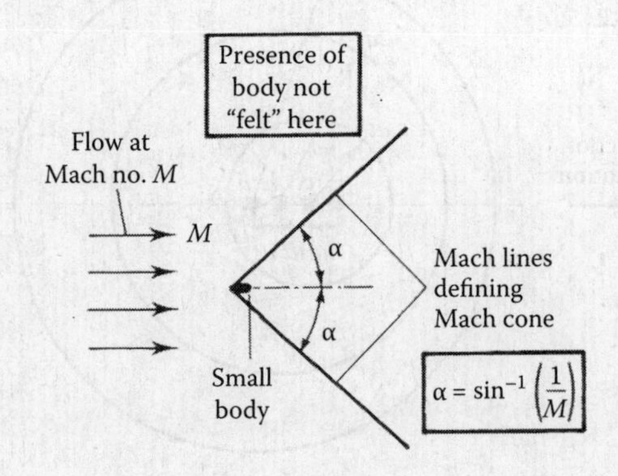

FIGURE 3.9
Conical Mach wave.

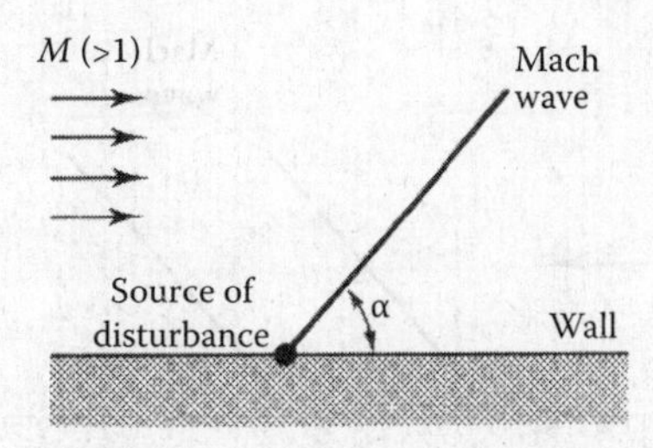

FIGURE 3.10
Plane Mach wave.

measuring the angle made by the wave generated at the disturbance to the oncoming flow, the Mach number can be found using

$$M = \frac{1}{\sin \alpha} \tag{3.22}$$

Mach waves originating from small changes in wall shape are shown in Figure 3.11. In addition to the Mach waves, it will be seen from Figure 3.11 that strong waves are generated near the leading edge of the body. The analysis of such waves will be discussed later.

Example 3.7

Air at a temperature of –10°C flows through a supersonic wind tunnel. A shadowgraph (see Appendix K) photograph of the flow reveals weak waves originating at imperfections on the walls. These weak waves are at an angle of 40° to the flow (see Figure E3.7). Find the Mach number and velocity in the wind tunnel.

Solution

$$M = \frac{1}{\sin \alpha} = \frac{1}{\sin 40°} = 1.556$$

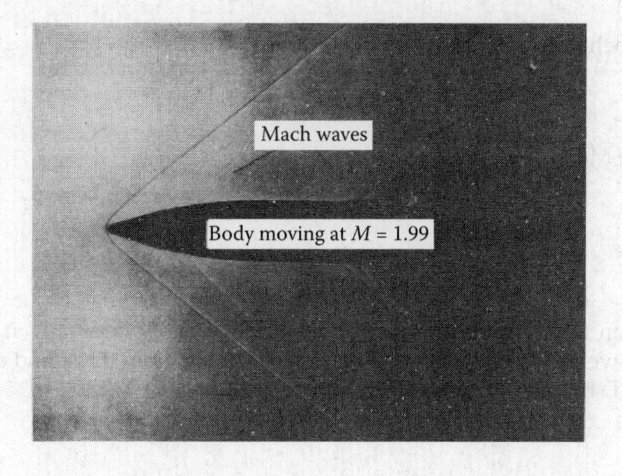

FIGURE 3.11
Mach waves originating at the surface of a body moving at supersonic velocity. (From J. Black, *An Introduction to Aerodynamic Compressibility*, Bunhill Publications, London, 1947. Plate IV.)

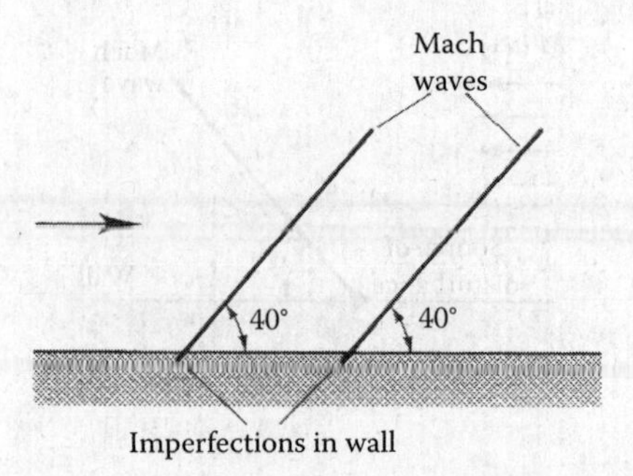

FIGURE E3.7
Weak waves considered.

Using this then gives

$$V = Ma = M\sqrt{\gamma RT} = 1.556\times\sqrt{1.4\times\frac{8314}{29}\times 263} = 505.5 \text{ m/s}$$

Therefore, the Mach number and velocity are 1.556 and 505.5 m/s, respectively.

Example 3.8

A gas has a molar mass of 44 and a specific heat ratio of 1.3. Find the speed of sound in this gas if the gas temperature is −30°C. If this gas is flowing at a velocity of 450 m/s, find the Mach number and the Mach angle.

Solution

The speed of sound is given by

$$a = \sqrt{\gamma RT} = \sqrt{1.3\times\frac{8314}{44}\times 243} = 244.3 \text{ m/s}$$

Hence, the Mach number is given by

$$M = V/a = 1.842$$

Therefore, the Mach angle is given by

$$\alpha = \sin^{-1}(1/M) = 32.9°$$

Example 3.9

An observer on the ground finds that an airplane flying horizontally at an altitude of 5000 m has traveled 12 km from the overhead position before the sound of the airplane is first heard. Estimate the speed at which the airplane is flying.

Solution

It is assumed that the net disturbance produced by the aircraft is weak, i.e., that, as indicated by the wording of the question, basically what is being investigated is how far

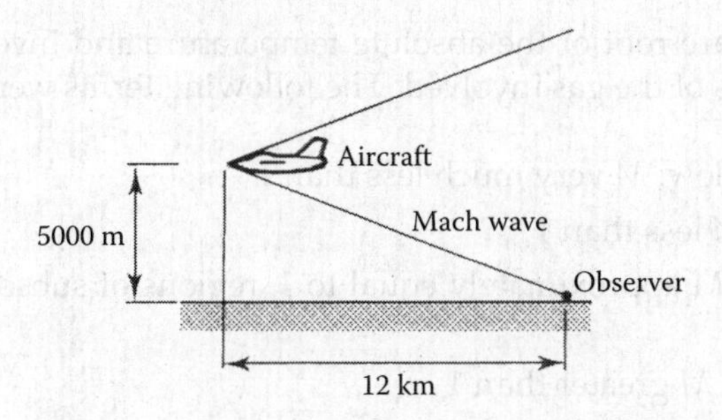

FIGURE E3.9
Situation considered.

the aircraft will have traveled from the overhead position when the sound waves emitted by the aircraft are first heard by the observer. If the discussion of Mach waves given above is considered, it will be seen that, as indicated in Figure E3.9, the aircraft will first be heard by the observer when the Mach wave emanating from the nose of the aircraft reaches the observer.

Now, since the temperature varies through the atmosphere, the speed of sound varies as the sound waves pass down through the atmosphere, which means that the Mach waves from the aircraft are actually curved. This effect is, however, small and will be neglected here, the speed of sound at the average temperature between the ground and the aircraft being used to describe the Mach wave.

Now as discussed in Example 3.3, for altitudes, H, of from 0 m (sea level) to 11,019 m the temperature in the atmosphere is given by $T = 288.16 - 0.0065H$ so, at the mean altitude of 2500 m, the temperature is $288.16 - 0.0065 \times 2500 = 271.9$ K. Hence, the mean speed of sound is given by

$$a = \sqrt{\gamma RT} = \sqrt{1.4 \times 287.04 \times 271.9} = 330.6 \text{ m/s}$$

From Figure E3.9, it will be seen that if α is the Mach angle based on the mean speed of sound, then

$$\tan \alpha = 5000/12{,}000 = 0.417$$

However, since $\sin \alpha = 1/M$, it follows that $\tan \alpha = 1/\sqrt{M^2 - 1}$; thus,

$$M = \sqrt{(1/0.417)^2 + 1} = 2.6$$

Hence, it follows that

$$\text{Velocity of aircraft} = 2.6 \times 330.6 = 859.6 \text{ m/s}$$

Concluding Remarks

The discussion presented in this chapter, which was essentially only concerned with the flow of a gas, indicates that the Mach number M is the parameter that determines the importance of compressibility effects in a flow. The speed of sound was shown to vary directly with the square root of the absolute temperature and inversely with the square root of the molar mass of the gas involved. The following terms were introduced

- Incompressible flow: M very much less than 1
- Subsonic flow: M less than 1
- Transonic flow: M approximately equal to 1, regions of subsonic and supersonic flow existing
- Supersonic flow: M greater than 1
- Hypersonic flow: M very much greater than 1

It was shown that in supersonic flow, weak disturbances are propagated along lines, termed Mach waves, and an expression for the angle such waves make to the flow was derived.

PROBLEMS

1. The velocity of an air flow changes by 1%. Assuming that the flow is isentropic, plot the percentage changes in pressure, temperature, and density induced by this change in velocity with flow Mach number for Mach numbers between 0.2 and 2.
2. Calculate the speed of sound at 288 K in hydrogen, helium, and nitrogen. Under what conditions will the speed of sound in hydrogen be equal to that in helium?
3. Find the speed of sound in carbon dioxide at temperatures of 20°C and 600°C.
4. A very weak pressure wave, i.e., a sound wave, across which the pressure rise is 30 Pa moves through air which has a temperature of 30°C and a pressure of 101 kPa. Find the density change, the temperature change, and the velocity change across this wave.
5. An airplane is traveling at 1500 km/h at an altitude where the temperature is –60°C. What is the Mach at which the airplane is flying?
6. An airplane is flying at 2000 km/h at an altitude where the temperature is –50°C. Find the Mach number at which the airplane is flying.
7. An airplane can fly at a speed of 800 km/h at sea level where the temperature is 15°C. If the airplane flies at the same Mach number at an altitude where the temperature is –44°C, find the speed at which the airplane is flying at this altitude.
8. The test section of a supersonic wind tunnel is square in cross section with a side length of 1.22 m. The Mach number in the test section is 3.5, the temperature is –100°C, and the pressure is 20 kPa. Find the mass flow rate of air through the test section.
9. A certain aircraft flies at the same Mach number at all altitudes. If it flies at a speed that is 120 km/h slower at an altitude of 12,000 m than it does at sea level, find the Mach number at which it flies. Assume standard atmospheric conditions.
10. Air at a temperature of 45°C flows in a supersonic wind tunnel over a very narrow wedge. A shadowgraph photograph of the flow reveals weak waves emanating from the front of the wedge at an angle of 35° to the undisturbed flow. Find the Mach number and velocity in the flow approaching the wedge.
11. Air at a temperature of –10°C flows through a supersonic wind tunnel. A Schlieren photograph of the flow reveals weak waves originating at imperfections on the

walls. These weak waves are at an angle of 35° to the flow. Find the air velocity in the wind tunnel.

12. A gas with a molar mass of 44 and a specific heat ratio 1.67 flows through a channel at supersonic speed. The temperature of the gas in the channel is 10°C. A photograph of the flow reveals weak waves originating at imperfections in the wall running across the flow at an angle to 45° to the flow direction. Find the Mach number and the velocity in the flow.
13. Air at 80°F is flowing at Mach 1.9. Find the air velocity and the Mach angle.
14. Air at a temperature of 25°C is flowing with a velocity of 180 m/s. A projectile is fired into the air stream with a velocity of 800 m/s in the opposite direction to that of the air flow. Calculate the angle that the Mach waves from the projectile make to the direction of motion.
15. An observer at sea level does not hear an aircraft that is flying at an altitude of 7000 m until it is a distance of 13 km from the observer. Estimate the Mach number at which the aircraft is flying. In arriving at the answer, assume that the average temperature of the air between sea level and 7000 m is –10°C.
16. An aircraft is flying at an altitude of 6 km at Mach 3. Find the distance behind the aircraft at which the disturbances created by the aircraft reach sea level.
17. An observer on the ground finds that an airplane flying horizontally at an altitude of 2500 m has traveled 6 km from the overhead position before the sound of the airplane is first heard. Assuming that, overall, the aircraft creates a small disturbance, estimate the speed at which the airplane is flying. The average air temperature between the ground and the altitude at which the airplane is flying is 10°C. Explain the assumptions you have made in arriving at the answer.

4

One-Dimensional Isentropic Flow

Introduction

Many flows that occur in engineering practice can be adequately modeled by assuming them to be steady, one-dimensional, and isentropic. The equations that describe such flows will be discussed in this chapter. The applicability of the one-dimensional flow assumption was discussed in Chapter 2 and will not be discussed further here. An isentropic flow is, of course, an adiabatic flow (a flow in which there is no heat exchange) in which viscous losses are negligible, i.e., it is an adiabatic frictionless flow. Although no real flow is entirely isentropic, there are many flows of great practical importance in which the major portion of the flow can be assumed to be isentropic. For example, in internal duct flows there are many important cases where the effects of viscosity and heat transfer are restricted to thin layers adjacent to the walls, i.e., are only important in the wall boundary layers, and the rest of the flow can be assumed to be isentropic as indicated in Figure 4.1.

Similarly in external flows, the effects of viscosity and heat transfer can be assumed to be restricted to the boundary layers, wakes, and shock waves, and the rest of the flow can be treated with adequate accuracy by assuming it to be isentropic as indicated in Figure 4.2.

Even when nonisentropic effects become important, it is often possible to calculate the flow by assuming it to be isentropic and to then apply an empirical correction factor to the solution so obtained to account for the nonisentropic effects. This approach has been frequently adopted in the past, for example, in the design of nozzles.

Governing Equations

By definition, the entropy remains constant in an isentropic flow. Using this fact, it was shown in Chapter 2 that in such a flow:

$$\frac{p}{\rho^{\gamma}} = c \quad \text{(a constant)} \tag{4.1}$$

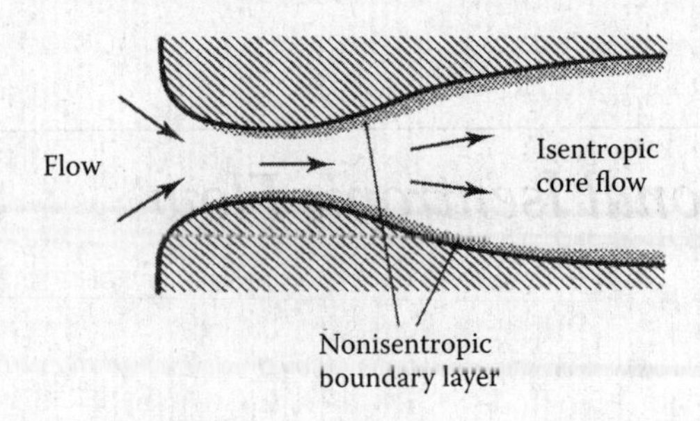

FIGURE 4.1
Region of duct flow that can be assumed to be isentropic.

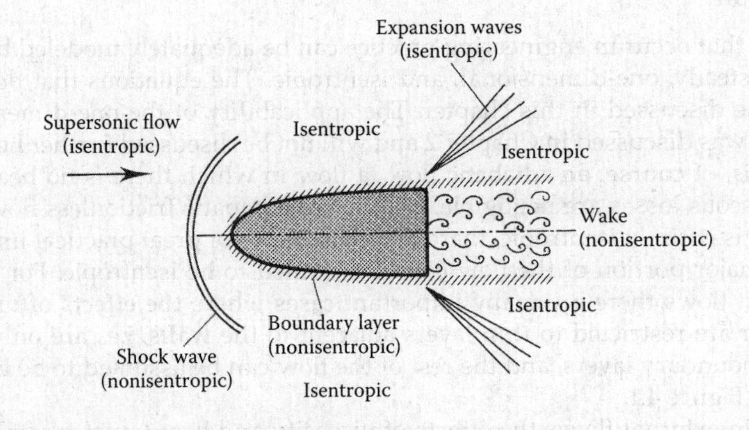

FIGURE 4.2
Region of external flow that can be assumed to be isentropic.

If any two points, such as 1 and 2 shown in Figure 4.3, in an isentropic flow are considered, it follows from Equation 4.1 that

$$\frac{p_2}{p_1} = \left(\frac{\rho_2}{\rho_1}\right)^{\gamma} \tag{4.2}$$

Hence, since the general equation of state gives

$$\frac{p_1}{\rho_1 T_1} = \frac{p_2}{\rho_2 T_2}, \quad \text{i.e.,} \quad \frac{T_2}{T_1} = \frac{p_2}{p_1}\frac{\rho_1}{\rho_2} \tag{4.3}$$

it follows that in isentropic flow

$$\frac{T_2}{T_1} = \left(\frac{\rho_2}{\rho_1}\right)^{\gamma-1} = \left(\frac{p_2}{p_1}\right)^{\frac{\gamma-1}{\gamma}} \tag{4.4}$$

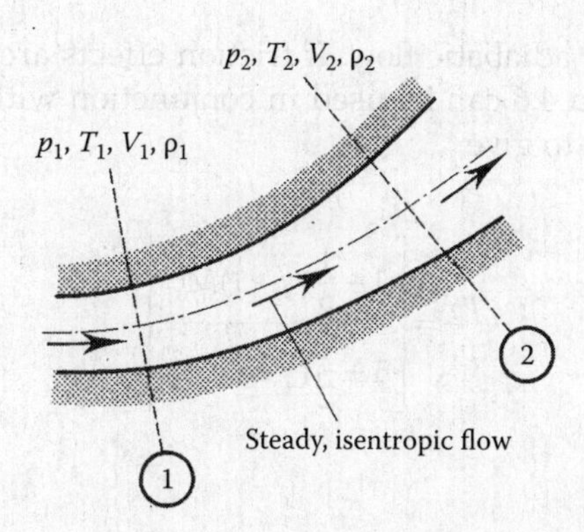

FIGURE 4.3
Flow situation considered.

From this, it then follows, recalling that $a = \sqrt{\gamma RT}$, that

$$\frac{a_2}{a_1} = \left(\frac{T_2}{T_1}\right)^{\frac{1}{2}} = \left(\frac{\rho_2}{\rho_1}\right)^{\frac{\gamma-1}{2}} = \left(\frac{p_2}{p_1}\right)^{\frac{\gamma-1}{2\gamma}} \tag{4.5}$$

The steady flow adiabatic energy equation is next applied between points 1 and 2. This gives

$$c_p T_1 + \frac{V_1^2}{2} = c_p T_2 + \frac{V_2^2}{2}$$

i.e.,

$$\frac{T_2}{T_1} = \frac{1 + (V_1^2 / 2c_p T_1)}{1 + (V_2^2 / 2c_p T_2)}$$

However,

$$\frac{V^2}{2c_p T} = \left[\frac{V^2}{2\gamma RT}\right]\left[\frac{\gamma R}{c_p}\right] = \frac{\gamma - 1}{2} M^2$$

Thus, it follows that

$$\frac{T_2}{T_1} = \frac{1 + \left(\dfrac{\gamma - 1}{2}\right) M_1^2}{1 + \left(\dfrac{\gamma - 1}{2}\right) M_2^2} \tag{4.6}$$

This equation applies in adiabatic flow. If friction effects are also negligible, i.e., if the flow is isentropic, Equation 4.6 can be used in conjunction with the isentropic state relations given in Equation 4.5 to give

$$\frac{p_2}{p_1} = \left[\frac{1+\frac{1}{2}(\gamma-1)M_1^2}{1+\frac{1}{2}(\gamma-1)M_2^2}\right]^{\frac{\gamma}{\gamma-1}} \tag{4.7}$$

and

$$\frac{\rho_2}{\rho_1} = \left[\frac{1+\frac{1}{2}(\gamma-1)M_1^2}{1+\frac{1}{2}(\gamma-1)M_2^2}\right]^{\frac{1}{\gamma-1}} \tag{4.8}$$

Lastly, it is recalled that the continuity equation gives

$$\rho_1 V_1 A_1 = \rho_2 V_2 A_2$$

which can be rearranged to give

$$\left(\frac{\rho_2}{\rho_1}\right)\left(\frac{V_2}{V_1}\right) = \frac{A_1}{A_2} \tag{4.9}$$

The above equations are, together, sufficient to determine all the characteristics of one-dimensional isentropic flow. It will be noted that the momentum equation was not used in the above analysis of isentropic flow. As discussed in the previous chapter, in isentropic flow, the momentum equation will always give the same result as the energy equation. This can be illustrated using the integrated Euler equation (2.9), i.e.,

$$\frac{V_2^2}{2} - \frac{V_1^2}{2} + \int_1^2 \frac{dp}{\rho} = 0 \tag{4.10}$$

Since the relation between ρ and p is known in isentropic flow, the integral can be evaluated as follows

$$\int_1^2 \frac{dp}{\rho} = \int_1^2 \frac{dp}{(p/p_1)^{\frac{1}{\gamma}}\rho_1} = \left(\frac{p_1}{\rho_1^{\gamma}}\right)^{\frac{1}{\gamma}}\left(\frac{\gamma}{\gamma-1}\right)\left(p_2^{\left(1-\frac{1}{\gamma}\right)} - p_1^{\left(1-\frac{1}{\gamma}\right)}\right)$$

$$= \left(\frac{\gamma}{\gamma-1}\right)\left(\frac{p_1}{\rho_1}\right)\left[\left(\frac{p_2}{p_1}\right)^{\frac{\gamma-1}{1}} - 1\right] \tag{4.11}$$

Substituting this result into Equation 4.10 then gives

$$V_2^2 - V_1^2 + \left(\frac{2}{\gamma - 1}\right) a_1^2 \left[\left(\frac{p_2}{p_1}\right)^{\frac{\gamma-1}{\gamma}} - 1\right] = 0 \tag{4.12}$$

However, it was shown before that the isentropic equation of state gives

$$\left(\frac{a_2}{a_1}\right)^2 = \left(\frac{p_2}{p_1}\right)^{\frac{\gamma-1}{\gamma}}$$

Substituting this into Equation 4.12 then gives

$$\left(\frac{p_2}{p_1}\right)^{\frac{\gamma-1}{\gamma}} \left[1 + \frac{\gamma - 1}{2} M_2^2\right] = 1 + \left(\frac{\gamma - 1}{2}\right) M_1^2 \tag{4.13}$$

which can be rearranged as

$$\frac{p_2}{p_1} = \left[\frac{1 + \frac{1}{2}(\gamma - 1) M_1^2}{1 + \frac{1}{2}(\gamma - 1) M_2^2}\right]^{\frac{\gamma}{\gamma-1}} \tag{4.14}$$

This is the same as the result that was derived previously using the energy equation. Thus, in the analysis of isentropic flow, either the momentum or the energy equation may be used. The energy equation is usually simpler to apply than the momentum equation.

Example 4.1

A gas that has a molar mass of 39.9 and a specific heat ratio of 1.67 is discharged from a large chamber in which the pressure is 500 kPa and the temperature is 30°C through a nozzle. Assuming one-dimensional isentropic flow, find

a. If the pressure at some section of the nozzle is 80 kPa, the Mach number, temperature, and velocity at this section.

b. If the nozzle has a circular cross section and if its diameter is 12 mm at the section discussed in (a) above, the mass flow rate through the nozzle.

Solution

The flow situation is shown in Figure E4.1.

The following are given: $\gamma = 1.67$, $m = 39.9$, $p_0 = 500$ kPa, $T_0 = 303$ K. Here p_0 and T_0 are the pressure and temperature in the large chamber, it being assumed that the velocity in this chamber is effectively zero. Steady, one-dimensional, isentropic flow is assumed.

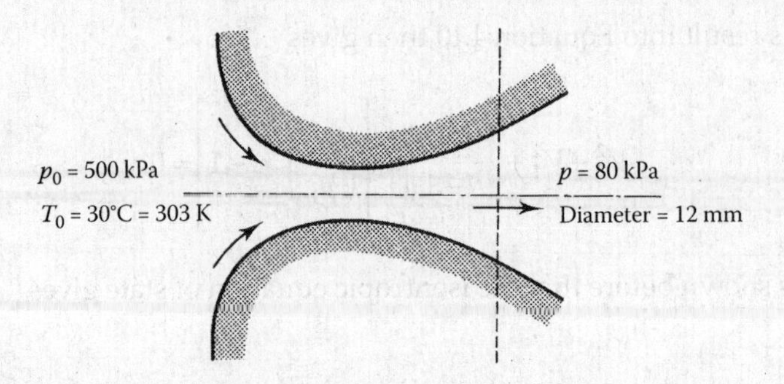

FIGURE E4.1
Flow situation considered.

a. Isentropic relations, Equation 4.7, give the following, since the velocity in the chamber and therefore the Mach number in the chamber are assumed to be zero:

$$M = \sqrt{\left[\frac{2}{\gamma - 1}\right]\left\{\frac{p_0}{p}^{\frac{\gamma-1}{\gamma}} - 1\right\}}$$

Hence,

$$M = \sqrt{\frac{2}{0.67}\left\{\left[\frac{500}{80}\right]^{0.4} - 1\right\}} = 1.8$$

Also, since the Mach number in the chamber is assumed to be zero, Equation 4.6 gives

$$T = T_0\left\{1 + \frac{\gamma - 1}{2}M^2\right\}^{-1}$$

Hence,

$$T = 303\text{ K}\left\{1 + \frac{0.67}{2}1 \times 8^2\right\}^{-1} = 145.3\text{ K}$$

Using this value of the temperature then gives

$$a = \sqrt{\frac{\gamma \Re T}{m}} = \sqrt{1.67\frac{8314}{39.9}145.3} = 224.9\text{ m/s}$$

Hence,

$$V = aM = (1.8)(224.9\text{ m/s}) = 404.7\text{ m/s}$$

Therefore, the Mach number is 1.8, the temperature is 145.3 K (–127.7°C), and the velocity is 404.7 m/s.

b. The mass flow rate is given by $\dot{m} = \rho V A$. Using the values of temperature and pressure established in (a), the density can be found as follows:

$$\rho = \frac{p}{RT} = \frac{80 \times 10^3}{(8314/39.9) \times 145.3} = 2.64 \text{ kg/m}^3$$

Hence,

$$\dot{m} = 2.64 \times 404.7 \frac{\pi \times (0.012)^2}{4} = 0.121 \text{ kg/s}$$

Therefore, the mass flow rate through the nozzle is 0.121 kg/s.

Example 4.2

A gas with a molar mass of 4 and a specific heat ratio of 1.3 flows through a variable area duct. At some point in the flow, the velocity is 150 m/s, the pressure is 100 kPa, and the temperature is 15°C. Find the Mach number at this point in the flow. At some other point in the flow, the temperature is found to be –10°C. Find the Mach number, pressure, and velocity at this second point in the flow assuming the flow to be isentropic and one-dimensional.

Solution

The flow situation being considered is shown in Figure E4.2.

For the gas being considered

$$R = \mathfrak{R}/m = 8314/4 = 2078.5$$

Hence, the Mach number at section 1 is given by

$$M_1 = \frac{V_1}{a_1} = \frac{150}{\sqrt{1.3 \times 2078.5 \times 288}} = \frac{150}{882.2} = 0.17$$

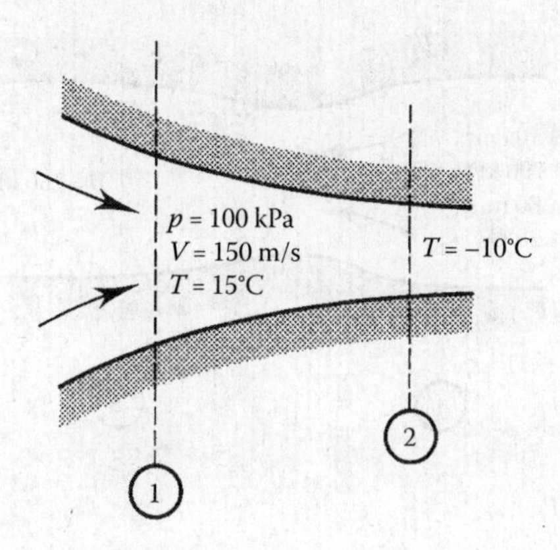

FIGURE E4.2
Flow situation considered.

The speed of sound at section 2 will be given by

$$a_2 = a_1(T_2/T_1)^{1/2} = 882.2(263/288)^{1/2} = 843 \text{ m/s}$$

Now isentropic relations give

$$\frac{T_2}{T_1} = \frac{1+(\gamma-1/2)M_1^2}{1+(\gamma-1/2)M_2^2}$$

which can be arranged to give

$$M_2 = \sqrt{\frac{(1+(\gamma-1/2)M_1^2)T_1/T_2 - 1}{(\gamma-1)/2}} = 0.8157$$

However, as derived above, the speed of sound at section 2 is equal to 843 m/s. Hence,

$$V_2 = M_2 a_2 = 687.6 \text{ m/s}$$

Isentropic relations also give

$$p_2 = \left(\frac{a_2}{a_1}\right)^{2\gamma/\gamma-1} \times p_1 = \left(\frac{843}{882.15}\right)^{8.67} \times 100 = 0.6747 \times 100 = 67.5 \text{ kPa}$$

Example 4.3

Air flows through a nozzle that has an inlet area of 10 cm². If the air has a velocity of 80 m/s, a temperature of 28°C, and a pressure of 700 kPa at the inlet section, and a pressure of 250 kPa at the exit, find the mass flow rate through the nozzle and, assuming one-dimensional isentropic flow, the velocity at the exit section of the nozzle.

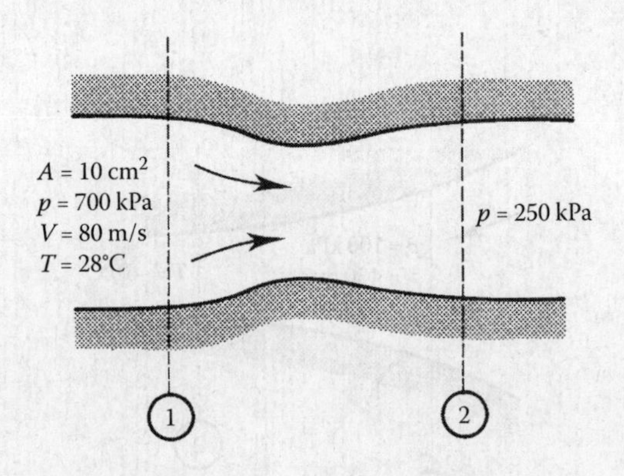

FIGURE E4.3
Flow situation considered.

Solution

The flow situation being considered is shown in Figure E4.3.

The mass flow rate is given by

$$\dot{m} = \rho_1 V_1 A_1 = \frac{p_1}{RT_1} \times V_1 \times A_1 = \frac{700 \times 10^3}{287 \times 301} \times 80 \times 10^{-3} = 0.648 \text{ kg/s}$$

Also,

$$M_1 = \frac{V_1}{a_1} = \frac{80}{\sqrt{\gamma RT}} = \frac{80}{\sqrt{1.4 \times 287 \times 301}} = \frac{80}{347.77} = 0.23$$

Therefore, since isentropic relations give

$$\frac{p_2}{p_1} = \left[\frac{1 + \frac{\gamma - 1}{2} M_1^2}{1 + \frac{\gamma - 1}{2} M_2^2}\right]^{\frac{\gamma}{\gamma - 1}}$$

It follows that

$$\frac{250}{700} = \left[\frac{1 + \frac{1.4 - 1}{2} 0.23^2}{1 + \frac{1.4 - 1}{2} M_2^2}\right]^{\frac{1.4}{1.4-1}}$$

Solving this equation for M_2 then gives $M_2 = 1.335$. However, since the flow is assumed isentropic

$$\frac{T_2}{T_1} = \left(\frac{p_2}{p_1}\right)^{\frac{\gamma - 1}{\gamma}}$$

so $T_2 = 301 \times (250/700)^{1/3.5} = 224.3$ K. Hence,

$$V_2 = M_2 \times a_2 = 1.335 \times \sqrt{1.4 \times 287 \times 224.3 \text{ K}} = 400.8 \text{ m/s}$$

Therefore, the exit velocity is 400.8 m/s.

Stagnation Conditions

Stagnation conditions are those that would exist if the flow at any point in a fluid stream was isentropically brought to rest. (To define the stagnation temperature, it is actually only necessary to require that the flow be adiabatically brought to rest. To define the stagnation pressure and density, it is necessary, however, to require that the flow be brought to rest isentropically.)

If the entire flow is essentially isentropic and if the velocity is essentially zero at some point in the flow, then the stagnation conditions existing at all points in the flow will be those existing at the zero velocity point as indicated in Figure 4.4.

However, even when the flow is nonisentropic, the concept of the stagnation conditions is still useful, the stagnation conditions at a point then being the conditions that would exist if the local flow were brought to rest isentropically as indicated in Figure 4.5.

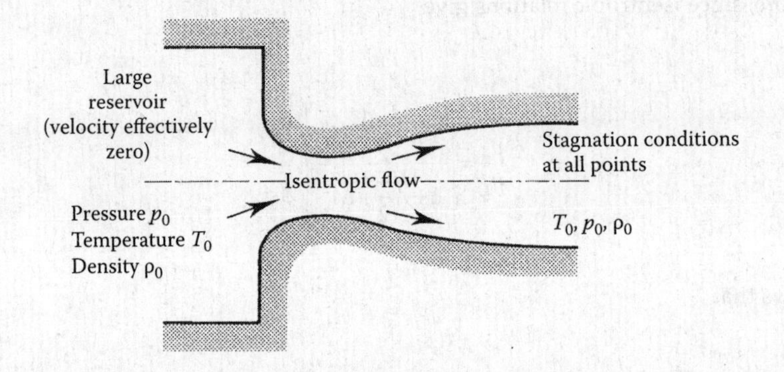

FIGURE 4.4
Stagnation conditions in an isentropic flow.

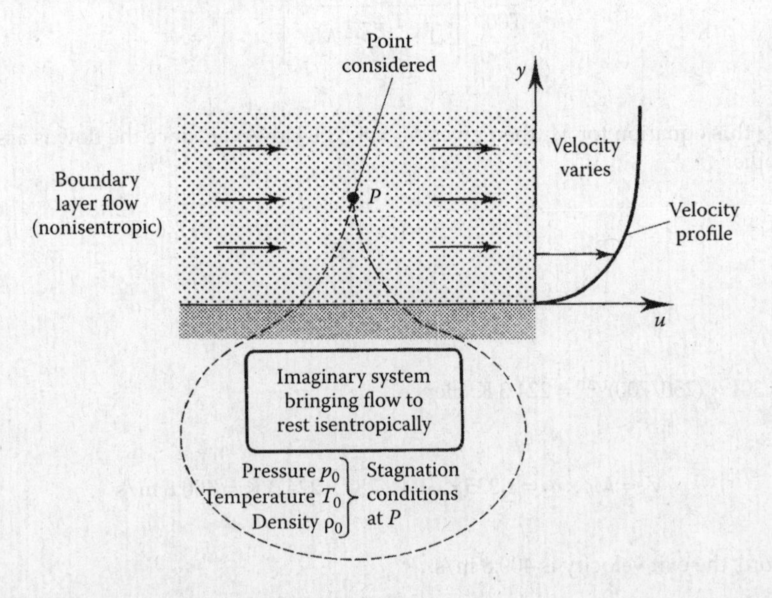

FIGURE 4.5
Stagnation conditions at a point in a nonisentropic flow.

If the equations derived in the previous section are applied between a point in the flow where the pressure, density, temperature, and Mach number are p, ρ, T, and M, respectively, and a point in the flow where the velocity is zero then if the stagnation conditions are denoted by the subscript 0, the stagnation pressure, density, and temperature will, since the Mach number is zero at the point where the stagnation conditions exist, be given by

$$\frac{p_0}{p} = \left[1 + \frac{\gamma - 1}{2} M^2\right]^{\frac{\gamma}{\gamma - 1}} \quad (4.15)$$

$$\frac{\rho_0}{\rho} = \left[1 + \frac{\gamma - 1}{2} M^2\right]^{\frac{1}{\gamma - 1}} \quad (4.16)$$

$$\frac{T_0}{T} = \left[1 + \frac{\gamma - 1}{2} M^2\right] \quad (4.17)$$

The variations of p_0/p and T_0/T with M for the particular case of $\gamma = 1.4$ as given by Equations 4.15 and 4.17 are shown in Figure 4.6.

The stagnation conditions will effectively exist at the leading edge of a bluff body in subsonic flow, the deceleration of the flow ahead of the body being essentially isentropic. The situation is shown in Figure 4.7.

This is not true in supersonic flow because, in this case, shock waves form ahead of the body and produce part of the deceleration of the flow. This is shown in Figure 4.8. As discussed in the next chapter, the flow through the shock wave is not isentropic, which means that the stagnation conditions for the flow ahead of the shock wave are different from those downstream of the shock wave. This supersonic flow situation will be discussed in the next chapter.

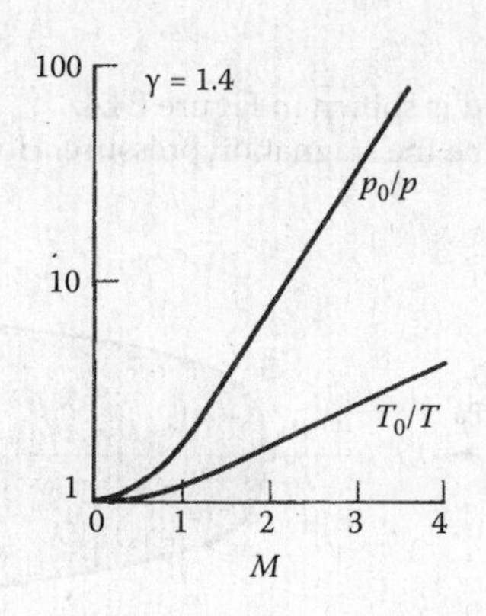

FIGURE 4.6
Variation of stagnation pressure and temperature ratios with Mach number.

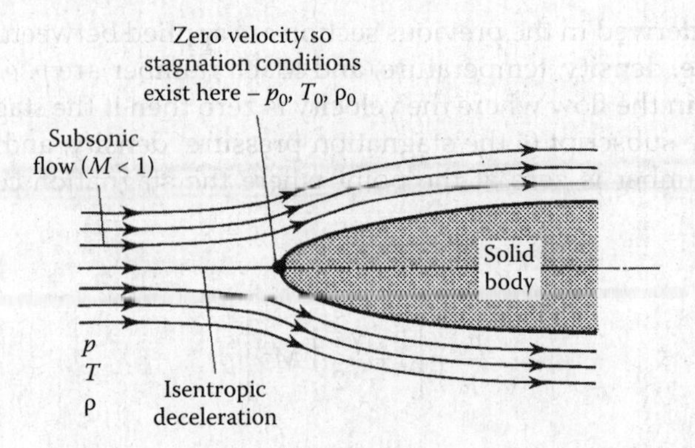

FIGURE 4.7
Stagnation conditions at leading edge of a "submerged" body.

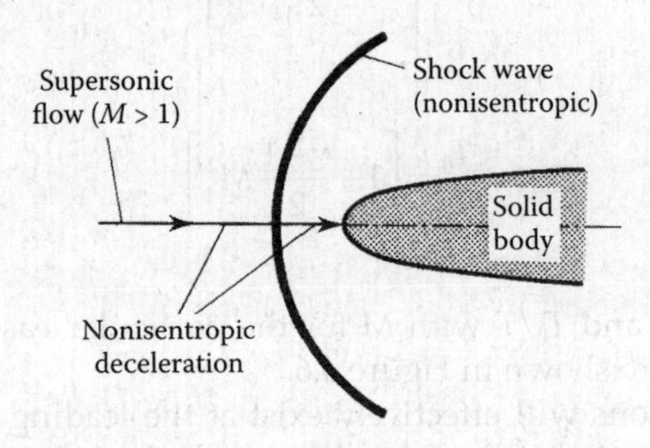

FIGURE 4.8
Supersonic flow near the leading edge of a "submerged" body.

Example 4.4

Air flows over a body. The air flow in the freestream ahead of the body has Mach 0.85 and a static pressure of 80 kPa. Find the highest pressure acting on the surface of the body.

Solution

The flow situation considered is shown in Figure E4.4.

The highest pressure will be the stagnation pressure. However, for $M = 0.85$, isentropic flow relations give

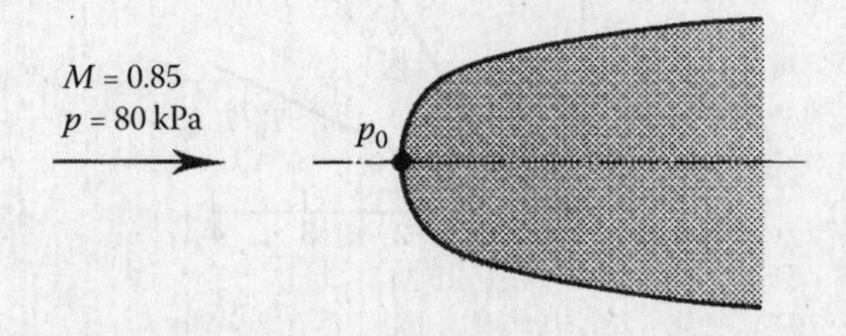

FIGURE E4.4
Flow situation considered.

$$\frac{p_0}{p} = \left[1 + \frac{\gamma - 1}{2} M^2\right]^{\frac{\gamma}{\gamma - 1}}$$

i.e.,

$$\frac{p_0}{p} = \left[1 + \frac{1.4 - 1}{2} 0.85^2\right]^{\frac{1.4}{1.4-1}} = 1.604$$

Therefore,

$$p_0 = 80 \text{ kPa} \times 1.604 = 128.3 \text{ kPa}$$

Therefore, the highest pressure acting on the surface of the body is 128.3 kPa.

Returning to a consideration of subsonic stagnation point flow, it follows from the above discussion that a pitot tube placed in a subsonic compressible flow will register the stagnation pressure. This is illustrated in Figure 4.9.

Since the disturbance produced by the tube is restricted to the stagnation point region, a pitot–static tube can be used to measure the Mach number in subsonic flow, the static hole measuring essentially the static pressure existing in the free stream ahead of the pitot-static tube. This is shown in Figure 4.10.

Since

$$\frac{p_0}{p} = \left[1 + \frac{\gamma - 1}{2} M^2\right]^{\frac{\gamma}{\gamma - 1}}$$

it follows that the Mach number is given by

$$M = \sqrt{\left(\frac{2}{\gamma - 1}\right)\left[\left(\frac{p_0}{p}\right)^{\frac{\gamma-1}{\gamma}} - 1\right]} \tag{4.18}$$

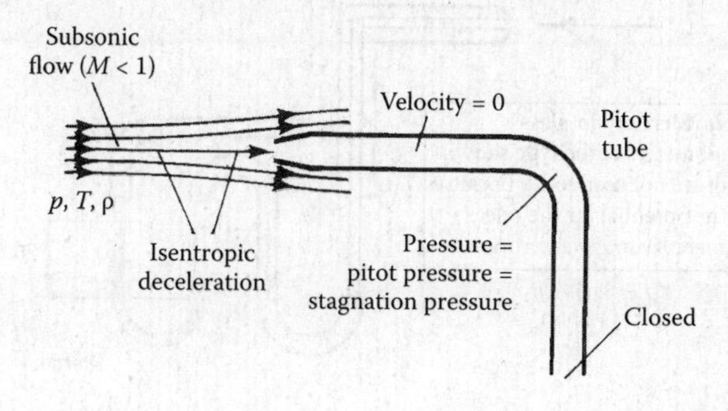

FIGURE 4.9
Subsonic flow over a pitot tube.

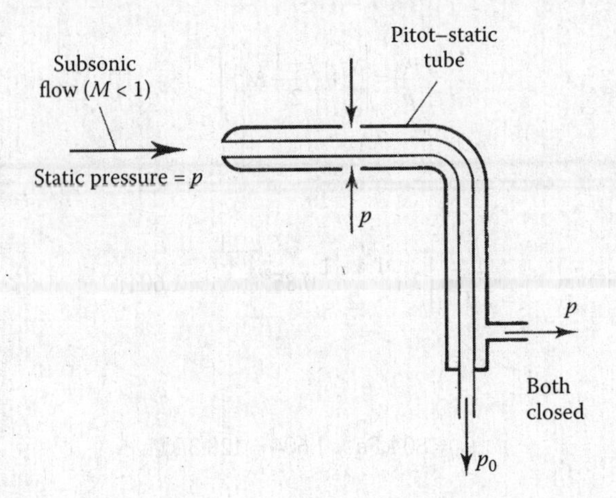

FIGURE 4.10
Flow over a pitot–static tube.

Example 4.5

A pitot–static tube is placed in a subsonic air flow. The static pressure and temperature in the flow are 80 kPa and 12°C, respectively. The difference between the pitot and static pressures is measured using a manometer and found to be 200 mm Hg. Find the air velocity and the Mach number.

Solution

The flow situation considered here is shown in Figure E4.5.

The pressure difference is found from the manometer reading. This gives

$$p_0 - p = \rho_m \times g \times \Delta H$$

where ρ_m is the density of the liquid in the manometer, which for mercury is 13,580 kg/m^3. Therefore,

$$p_0 - p = 13{,}580 \times 9.81 \times 0.2 = 26.64 \text{ kPa}$$

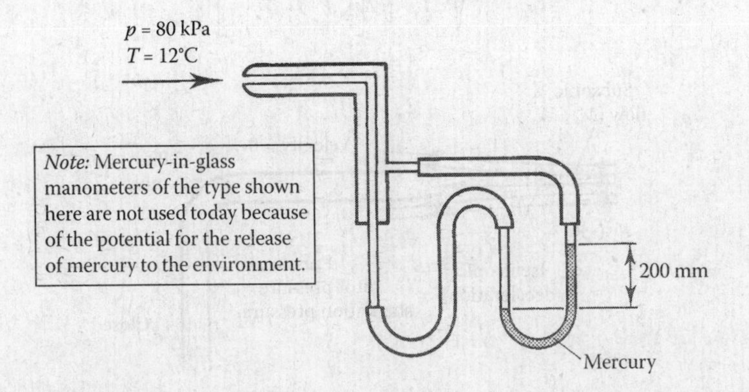

FIGURE E4.5
Pitot-static tube arrangement considered.

However,

$$\frac{p_0}{p} - 1 = \frac{p_0 - p}{p} = \frac{26.64}{80}$$

Hence,

$$\frac{p_0}{p} = 1.3331$$

Now, isentropic flow relations give, as shown above,

$$M = \sqrt{\left(\frac{2}{\gamma - 1}\right)\left[\left(\frac{p_0}{p}\right)^{\frac{\gamma-1}{\gamma}} - 1\right]}$$

Therefore, in the flow here being considered,

$$M = \sqrt{\left(\frac{2}{0.4}\right)[(1.3331)^{0.4/1.4} - 1]} = 0.654$$

Hence, the Mach number is 0.654. Since the temperature is given as 12°C = 285 K, the velocity can be found using

$$V = Ma = M\sqrt{\gamma RT} = 0.654 \times \sqrt{1.4 \times 287 \times 285} = 221.3 \text{ m/s}$$

Therefore, the velocity is 221.3 m/s.

NOTE: Mercury-in-glass manometers of the type considered are not used today because of the potential for the release of mercury to the environment.

It will be noted that to find the Mach number using Equation 4.18, p_0 and p have to be separately measured, and if the velocity is required, the temperature will normally also have to be measured to find the speed of sound. In incompressible flow, of course, the Bernoulli equation gives

$$V = \sqrt{2\frac{(p_0 - p)}{\rho}} \tag{4.19}$$

which indicates that to determine the velocity in incompressible flow, only the difference between p_0 and p has to be measured and not their individual values.

It is convenient for many purposes to know the error that would be incurred using the incompressible pitot tube equation in a compressible flow. The magnitude of this error will also illustrate the magnitude of compressibility effects in a subsonic flow. Now, using the full compressible flow pitot tube equation, it follows that

$$p_0 - p = p\left[\frac{p_0}{p} - 1\right]$$

$$= \left(\frac{1}{2}\rho V^2\right)\left(\frac{2p}{\rho V^2}\right)\left\{\left[1 + \left(\frac{\gamma - 1}{2}\right)M^2\right]^{\frac{\gamma}{\gamma-1}} - 1\right\}$$

$$= \frac{1}{2}\rho V^2\left\{\left(\frac{2}{\gamma M^2}\right)\left[\left(1 + \frac{\gamma - 1}{2}M^2\right)^{\frac{\gamma}{\gamma-1}} - 1\right]\right\} \tag{4.20}$$

which gives the actual velocity as

$$V = \sqrt{\frac{2(p_0 - p)}{\rho}}\left\{\left(\frac{2}{\gamma M^2}\right)\left[\left(1 + \frac{\gamma - 1}{2}M^2\right)^{\frac{\gamma}{\gamma-1}} - 1\right]\right\}^{-\frac{1}{2}} \tag{4.21}$$

However, if the incompressible flow pitot tube equation is used, the velocity would be given by

$$V = \sqrt{\frac{2(p_0 - p)}{\rho}} \tag{4.22}$$

Therefore, the error incurred in using the incompressible flow equation to find the velocity from the measured pressure difference, i.e.,

$$\epsilon = \left|\frac{(V_{\text{actual}} - V_{\text{incom}})}{V_{\text{actual}}}\right| = \left|1 - \frac{V_{\text{incom}}}{V_{\text{actual}}}\right| \tag{4.23}$$

is given by

$$\epsilon = \left|1 - \left\{\left(\frac{2}{\gamma M^2}\right)\left[\left(1 + \frac{\gamma - 1}{2}M^2\right)^{\frac{\gamma}{\gamma-1}} - 1\right]\right\}^{\frac{1}{2}}\right| \tag{4.24}$$

Thus, the error depends only on the Mach number, and its variation is indicated in Figure 4.11.

From Figure 4.11, it follows that the incompressible flow equation can be used to determine velocity with errors of less than 1% if the Mach number is less than ~0.3. However, the error rises to almost 5% when the Mach number is 0.6.

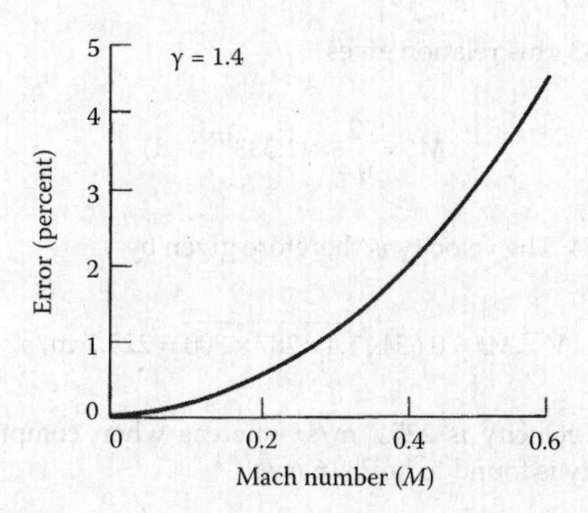

FIGURE 4.11
Variation of error incurred by the use of the incompressible pitot–static tube equation with Mach number.

Example 4.6

A pitot–static tube is placed in a subsonic airflow. The static pressure and temperature in the flow are 96 kPa and 27°C, respectively. The difference between the pitot and static pressures is measured and found to be 32 kPa. Find the air velocity (a) assuming an incompressible flow and (b) assuming compressible flow.

Solution

The density in the flow is given by

$$\rho = \frac{p}{RT} = \frac{96}{287 \times 300\ \mathrm{K}} = 1.115\ \mathrm{kg/m^3}$$

a. If incompressible flow is assumed, the velocity is given by

$$V = 2\sqrt{\frac{p_0 - p}{\rho}} = 2\sqrt{\frac{32 \times 10^3}{1.115}} = 239.6\ \mathrm{m/s}$$

b. When compressibility is accounted for, the velocity is found by noting that

$$\frac{p_0 - p}{p} = \frac{p_0}{p} - 1 = \frac{32}{96}$$

Hence, $p_0/p = 1.3333$. However,

$$\frac{p_0}{p} = \left[1 + \frac{\gamma - 1}{2} M^2\right]^{\frac{\gamma}{\gamma-1}}$$

Thus, for $p_0/p = 1.3333$, this relation gives

$$M^2 = \frac{2}{0.4} \times (1.333^{1/3.5} - 1)$$

which gives $M = 0.654$. The velocity is therefore given by

$$V = Ma = 0.654\sqrt{1.4 \times 287 \times 300} = 225.7 \text{ m/s}$$

Hence, the actual velocity is 225.7 m/s, whereas when compressibility effects are neglected, the velocity is found to be 239.6 m/s.

Critical Conditions

The critical conditions are those that would exist if the flow was isentropically accelerated or decelerated until the Mach number was unity, i.e., they are the conditions that would exist if the Mach number was isentropically changed from M to 1. These critical conditions are usually denoted by an asterisk, i.e., they are denoted by the symbols V^*, p^*, ρ^*, T^*, and A^*. Using Equations 4.6 through 4.8 and setting M_2 equal to 1 then gives the following relations for the critical conditions:

$$\frac{T^*}{T} = \left[\frac{2}{\gamma+1} + \frac{\gamma-1}{\gamma+1}M^2\right] \tag{4.25}$$

$$\frac{a^*}{a} = \left[\frac{2}{\gamma+1} + \frac{\gamma-1}{\gamma+1}M^2\right]^{\frac{1}{2}} \tag{4.26}$$

$$\frac{p^*}{p} = \left[\frac{2}{\gamma+1} + \frac{\gamma-1}{\gamma+1}M^2\right]^{\frac{\gamma}{\gamma-1}} \tag{4.27}$$

$$\frac{\rho^*}{\rho} = \left[\frac{2}{\gamma+1} + \frac{\gamma-1}{\gamma+1}M^2\right]^{\frac{1}{\gamma-1}} \tag{4.28}$$

The relation between the critical conditions and the stagnation conditions can be found by setting M equal to zero in the above equations. This gives

$$\frac{T^*}{T_0} = \frac{2}{\gamma+1} \tag{4.29}$$

$$\frac{a^*}{a_0} = \sqrt{\frac{2}{\gamma+1}} \tag{4.30}$$

$$\frac{p^*}{p_0} = \left(\frac{2}{\gamma+1}\right)^{\frac{\gamma}{\gamma-1}} \tag{4.31}$$

$$\frac{\rho^*}{\rho_0} = \left(\frac{2}{\gamma+1}\right)^{\frac{1}{\gamma-1}} \tag{4.32}$$

For the case of air flow, these equations give

$$\frac{T^*}{T_0} = 0.833, \quad \frac{p^*}{p_0} = 0.528, \quad \frac{\rho^*}{\rho_0} = 0.634$$

Example 4.7

A gas is contained in a large vessel at a pressure of 300 kPa and a temperature of 50°C. The gas is expanded from this vessel through a nozzle until the Mach number reaches a value of 1. Find the pressure, temperature, and velocity at this point in the flow if the gas is (a) air and (b) helium.

Solution

It was shown above that

$$\frac{p^*}{p_0} = \left(\frac{2}{\gamma+1}\right)^{\frac{\gamma}{\gamma-1}}$$

Hence, since $\gamma = 1.4$ for air and 1.667 for helium and since $p_0 = 300$ kPa, it follows that the pressure p^* at the point where $M = 1$ is $300 \times (2/2.4)^{(1.4/0.4)} = 158.5$ kPa for air and $300 \times (2/2.667)^{(1.667/0.667)} = 146.1$ kPa for helium.

It was also shown above that

$$\frac{T^*}{T_0} = \frac{2}{\gamma+1}$$

Hence, since $T_0 = 323$ K, it follows that the temperature T^* at the point where $M = 1$ is $323 \times (2/2.4) = 269.2$ K $= -3.8$°C for air and $323 \times (2/2.667) = 242.2$ K $= -30.8$°C for helium.

Lastly, since at the point considered $M = 1$,

$$V = Ma = a$$

Hence, using the temperature values already found,

$$V = \sqrt{1.4 \times (8314.3/28.97) \times 269.2} = 328.9 \text{ m/s}$$

for air and

$$V = \sqrt{1.667 \times (8314.3/4) \times 242.2} = 916.1 \text{ m/s}$$

for helium.

Maximum Discharge Velocity

The "maximum discharge velocity" or "maximum escape velocity" is the velocity that would be generated if a gas was adiabatically expanded until its temperature had dropped to absolute zero. Using the adiabatic energy equation gives the maximum discharge velocity as

$$\frac{\hat{V}^2}{2} = \frac{V^2}{2} + c_p T = c_p T_0 \tag{4.33}$$

This can be rearranged to give

$$\hat{V} = \sqrt{(V^2 + 2c_p T)} = \sqrt{2c_p T_0} = \sqrt{\left(V^2 + \frac{2a^2}{\gamma - 1}\right)} = \sqrt{\frac{2a_0^2}{\gamma - 1}} \tag{4.34}$$

There is therefore a definite maximum velocity that can be generated in a gas having a given stagnation temperature. However, since the temperature is zero when this maximum velocity is reached, the Mach number will be infinite since, under these conditions, the speed of sound is 0. It should be noted that the maximum discharge velocity given by the above equation could not be obtained in reality because at very low temperatures the assumptions used in deriving the above equations cease to apply, i.e., the properties of the gas change and once the temperature gets low enough the gas will liquefy. It should also be noted that the maximum discharge velocity has nothing to do with the existence of a maximum velocity at which a body can move relative to a gas, no such limit existing according to the laws of conventional mechanics.

Example 4.8

Consider the flow situation described in Example 4.7. What is the maximum velocity that could be generated by expanding the gas through a nozzle system?

Solution

It was shown above that

$$\hat{V} = \sqrt{\frac{2a_0^2}{\gamma - 1}}$$

However,

$$T_0 = 273 + 50 = 323 \text{ K}$$

so

$$a_0 = \sqrt{1.4 \times (8314.3/28.97) \times 323} = 360.3 \text{ m/s}$$

for air and

$$a_0 = \sqrt{1.667 \times (8314.3/4) \times 323} = 1057.9 \text{ m/s}$$

for helium. Therefore,

$$\hat{V} = \sqrt{2 \times 360.3^2/0.4} = 805.7 \text{ m/s}$$

for air and

$$\hat{V} = \sqrt{2 \times 1057.9^2/0.667} = 1831.9 \text{ m/s}$$

for helium.

Isentropic Relations in Tabular and Graphical Form and from Software

The equations derived above for one-dimensional, isentropic flow are relatively easily solved using a calculator or computer. Traditionally, however, isentropic flow calculations have been undertaken using sets of tables or graphs that give the variations of such quantities as p_0/p, and T_0/T with M in isentropic flow for a fixed value of the specific heat ratio γ. Such tables and graphs are available for various values of the specific heat ratio but care must be taken to ensure that a table for the correct value of γ is used.

A typical set of tables would have the following headings

$$M \quad \frac{p_0}{p} \quad \frac{T_0}{T} \quad \frac{\rho_0}{\rho} \quad \frac{a_0}{a} \quad \frac{A}{A^*}$$

The meaning of the entry A/A^* will be discussed in Chapter 8. Such a table can be conveniently used in the calculation of the properties of a one-dimensional isentropic flow. For example, if the flow through a variable area channel is being considered and if the Mach number and pressure at one section are known, say M_1 and p_1, and if the pressure at some other section is known, say p_2, then to find the Mach number at the second section the value of M_1 is used with the table to find p_1/p_0. Then, since the stagnation pressure is constant in isentropic flow,

$$\frac{p_2}{p_0} = \frac{p_2}{p_1} \times \frac{p_1}{p_0} \tag{4.35}$$

which allows p_0/p_2 to be found. Then using the tables, the value of M_2 corresponding to this value of p_0/p can be found.

A set of isentropic tables for air, i.e., $\gamma = 1.4$, is given in Appendix B.

As mentioned above, today with the widespread availability of programmable calculators and computers there has been a considerable reduction in the use of isentropic flow tables and charts for calculating the properties of isentropic flows. For example, the software COMPROP provided to support this book allows the conditions in isentropic flow to be determined for any value of γ. Where applicable, the worked examples in this book will be presented in such a way that they could have been solved using the equations directly or using isentropic tables or the software.

Example 4.9

Air flows from a large vessel in which the pressure is 300 kPa and the temperature is 40°C through a nozzle. If the pressure at some section of the discharge nozzle is measured as 200 kPa, find the temperature and velocity at this section. If the Mach number at some other section of the nozzle is 1.5, find the pressure, temperature, and velocity at this section. Assume that the flow is steady, isentropic, and one-dimensional.

Solution

In this case, because the vessel is large and the velocity in it effectively zero,

$$p_0 = 300 \text{ kPa}, \quad T_0 = 313 \text{ K}$$

At the first section, here termed section 2, $p_0/p = 300/200 = 1.5$; thus, isentropic relations or tables for $\gamma = 1.4$ or software gives $M = 0.78$ and $T_0/T = 1.12$, and so at this section

$$T = 313/1.12 = 279 \text{ K} = 6°\text{C}$$

Hence, using $V = Ma$, it follows that

$$V = 0.78 \times \sqrt{1.4 \times (8314.3/28.97) \times 279.2} = 261.1 \text{ m/s}$$

At the second section $M = 1.5$ so isentropic relations or tables for $\gamma = 1.4$ or software give $T_0/T = 1.45$ and $p_0/p = 3.67$, and so at this section

$$T = 313/1.45 = 216 \text{ K} = -57°\text{C}$$

$$p = 300/3.67 = 81.7 \text{ kPa}$$

$$V = 1.5 \times \sqrt{1.4 \times (8314.3/28.97) \times 216} = 441.9 \text{ m/s}$$

Example 4.10

The velocity, pressure, and temperature at a certain point in a steady air flow are 600 m/s, 70 kPa, and 5°C, respectively. If the pressure at some other point in the flow is 30 kPa, find the Mach number, temperature, and velocity that exist at this second point. Assume that the flow is isentropic and one-dimensional.

Solution

In this case,

$$M_1 = 600/\sqrt{1.4 \times (8314.3/28.97) \times (273 + 5)} = 1.80$$

At section 1, using this value of M, isentropic relations, or tables for $\gamma = 1.4$ or the software give $T_0/T = 1.65$ and $p_0/p = 5.75$. However, because p_0 does not change in isentropic flow, it follows that

$$\frac{p_2}{p_1} = \frac{p_0/p_1}{p_0/p_2}$$

Hence, since $p_2/p_1 = 30/70 = 0.4286$, it follows that

$$\frac{p_0}{p_2} = \frac{5.75}{0.4286} = 13.42$$

For this value of p_0/p, isentropic relations or tables for $\gamma = 1.4$ or the software give $M = 2.345$ and $T_0/T = 2.10$. Hence, because T_0 also does not change in isentropic flow, it follows that

$$\frac{T_2}{T_1} = \frac{T_0/T_1}{T_0/T_2}$$

Hence, $T_2 = 278 \times 1.65/2.10 = 218.4$ K. From this and the fact that M_2 is 2.345, it follows that

$$V_2 = 2.345 \times \sqrt{1.4 \times (8314.3/28.97) \times 218.4} = 694.7 \text{ m/s}$$

Concluding Remarks

Although no flow is, of course, truly isentropic, the main characteristics of many practically significant flows can be predicted using the equations presented in this chapter. The concepts of stagnation point conditions and critical conditions have also been introduced in this chapter. The use of isentropic tables for the calculation of one-dimensional, steady, isentropic flows was also discussed. Such tables are sometimes convenient to use, but it must always be remembered that a given table applies only to a specific value of γ. The use of software for the calculation of one-dimensional, steady, isentropic flows was also discussed.

PROBLEMS

1. A gas with a molar mass of 4 and a specific heat ratio of 1.67 flows through a variable area duct. At some point in the flow, the velocity is 200 m/s and the temperature is 10°C. Find the Mach number at this point in the flow. At some other point in the flow the temperature is –10°C. Find the velocity and Mach number at this point in the flow assuming that the flow is isentropic.
2. Air flows through a convergent–divergent duct with an inlet area of 5 cm^2 and an exit area of 3.8 cm^2. At the inlet section, the air velocity is 100 m/s, the pressure is 680 kPa, and the temperature is 60°C. Find the mass flow rate through the nozzle and, assuming isentropic flow, the pressure, and velocity at the exit section.
3. The exhaust gases from a rocket engine can be assumed to behave as a perfect gas with a specific heat ratio of 1.3 and a molecular weight of 32. The gas is expanded from the combustion chamber through the nozzle. At a point in the nozzle where the cross-sectional area is 0.2 m^2, the pressure, temperature, and Mach number are 1500 kPa, 800°C, and 0.2, respectively. At some other point in the nozzle, the pressure is found to be 80 kPa. Find the Mach number, temperature, and cross-sectional area at this point. Assume a one-dimensional, isentropic flow.
4. The exhaust gases from a rocket engine have a molar mass of 14. They can be assumed to behave as a perfect gas with a specific heat ratio of 1.25. These gases are accelerated through a nozzle. At some point in the nozzle where the cross-sectional area of the nozzle is 0.7 m^2, the pressure is 1000 kPa, the temperature is 500°C, and the velocity is 100 m/s. Find the mass flow rate through the nozzle and the stagnation pressure and temperature. Also, find the highest velocity that could be generated by expanding this flow. If the pressure at some other point in the nozzle is 100 kPa, find the temperature and velocity at this point in the flow assuming the flow to be one-dimensional and isentropic.
5. A gas has a molar mass of 44 and a specific heat ratio of 1.3. At a certain point in the flow, the static pressure and temperature are 80 kPa and 15°C, respectively and the velocity is 100 m/s. The gas is then isentropically expanded until its velocity is 300 m/s. Find the pressure, temperature, and Mach number that exist in the resulting flow.
6. Carbon dioxide flows through a variable area duct. At a certain point in the duct the velocity is 200 m/s and the temperature is 60°C. At some other point in the duct, the temperature is 15°C. Find the Mach numbers and stagnation temperatures at the two points. Assume that the flow is adiabatic.
7. At a certain point in a gas flow, the velocity is 900 m/s, the pressure is 150 kPa, and the temperature is 60°C. Find the stagnation pressure and temperature if the gas is air and if it is carbon dioxide.
8. Helium, at a pressure of 120 kPa and a temperature of 20°C, flows at a velocity of 800 m/s. Find the Mach number and the stagnation temperature and the stagnation pressure.

9. In an argon flow the temperature is 40°C and the pressure is half the stagnation pressure. Find the Mach number and the velocity in the flow.
10. An aircraft is flying at Mach 2.2 at an altitude of 10,000 m in the standard atmosphere, find the stagnation pressure and temperature for the flow over the aircraft.
11. If a gas is flowing at 300 m/s and has a pressure and temperature of 90 kPa and 20°C, find the maximum possible velocity that could be generated by expansion of this gas if the gas is air and if it is helium.
12. A pitot–static tube is placed in a subsonic air flow. The static temperature and pressure in the air flow are 30°C and 101 kPa, respectively. The difference between the pitot and static pressures is measured using a manometer and is found to be 250 mm Hg. Find the air velocity, assuming the flow to be incompressible and taking compressibility effects into account.
13. A pitot–static tube is placed in a subsonic air flow. The static pressure and temperature are 101 kPa and 30°C, respectively. The difference between the pitot and static pressures is measured and was found to be 37 kPa. Find the air velocity.
14. A pitot tube placed in an air stream indicates a pressure of 186 kPa. If the local Mach number is 0.8, determine the static pressure.
15. A pitot tube indicates a pressure of 155 kPa when placed in an air stream in which the temperature is 15°C and the Mach number is 0.7. Find the static pressure in the flow. Also, find the stagnation temperature in the flow.
16. A pitot tube is placed in a stream of carbon dioxide in which the pressure is 60 kPa and the Mach number is 0.9. What will the pitot pressure be?
17. Consider a one-dimensional isentropic flow through a duct. At a certain section of this duct, the velocity is 360 m/s, the temperature is 45°C, and the pressure is 120 kPa. Find the Mach number and the stagnation temperature and pressure at this point in the flow. If the temperature at some other point in the flow is 90°C, find the Mach number and pressure at this point in the flow.
18. A liquid fuelled rocket is fired on a test stand. The rocket nozzle has an exit diameter of 30 cm, and the combustion gases leave the nozzle at a velocity of 3800 m/s and a pressure of 100 kPa, which is the same as the ambient pressure. The temperature of the gases in the combustion area is 2400°C. Find the temperature of the gases on the nozzle exit plane, the pressure in the combustion area, and the thrust developed. Assume that the gases have a specific heat ratio of 1.3 and a molar mass of 9. Assume that the flow in the nozzle is isentropic.
19. The pressure, temperature, and Mach number at the entrance to a duct through which air is flowing are 250 kPa, 26°C, and 1.4, respectively. At some other point in the duct, the Mach number is found to be 2.5. Assuming isentropic flow, find the temperature, velocity, and pressure at the second section. Also, find the mass flow rate through the duct per square meter at the second section.
20. An aircraft is flying at Mach 0.95 at an altitude where the pressure is 30 kPa and the temperature is –50°C. The diffuser at the intake to the engine decreases the Mach number to 0.3 at the inlet to the engine. Find the pressure and temperature at the inlet to the engine.
21. A conical diffuser has an inlet diameter of 15 cm. The pressure, temperature, and velocity at the inlet to the diffuser are 70 kPa, 60°C, and 180 m/s, respectively. If the pressure at the diffuser exit is 78 kPa, find the exit diameter of the diffuser.

22. The control system for some smaller space vehicles uses nitrogen from a high-pressure bottle. When the vehicle has to be maneuvered, a valve is opened allowing nitrogen to flow out through a nozzle thus generating a thrust in the direction required to maneuver the vehicle. In a typical system, the pressure and temperature in the system ahead of the nozzle are about 1.6 MPa and 30°C, respectively, whereas the pressure in the jet on the nozzle exit plane is about 6 kPa. Assuming that the flow through the nozzle is isentropic and the gas velocity ahead of the nozzle is negligible, find the temperature and the velocity of the nitrogen on the nozzle exit plane. If the thrust required to maneuver the vehicle is 1 kN, find the area of the nozzle exit plane and the required mass flow rate of nitrogen. It can be assumed that the vehicle is effectively operating in a vacuum.
23. Hydrogen enters a nozzle with a very low velocity and at a temperature and pressure of 3800°R and 1000 psia, respectively. The pressure on the exit plane of the nozzle is 2 psia. Calculate the hydrogen flow rate per unit nozzle exit area. The flow through the nozzle can be assumed to be isentropic.
24. An aircraft flies at sea level at a speed of 220 m/s. What is the highest pressure that can be acting on the surface of the aircraft?
25. Consider an air flow with a speed of 650 m/s, a pressure of 100 kPa, and a temperature of 20°C. What is the stagnation pressure and the stagnation temperature in the flow?
26. When an aircraft is flying at subsonic velocity, the pressure at its nose, i.e., at the stagnation point, is found to be 160 kPa. If the ambient pressure and temperature are 100 kPa and 25°C, respectively, find the speed and the Mach number at which the aircraft is flying.
27. A body moves through air at a velocity of 200 m/s. The pressure and temperature in the air upstream of the body are 100 kPa and 30°C, respectively. Find the pressure at a point on the body where the velocity of the air relative to the body is zero (1) accounting for compressibility and (2) assuming incompressible flow. Assume that the flow is isentropic.
28. Air enters a duct at a pressure of 30 psia, a temperature of 100°F, and a velocity of 580 ft/sec. At some other point in the duct, the pressure is found to be 12 psia. Assuming that the flow is isentropic, find the temperature and Mach number at this point in the flow.
29. Consider a rocket engine that burns hydrogen and oxygen. The combustion chamber temperature and pressure are 3800 K and 1.5 MPa, respectively, the velocity in the combustion chamber being very low. The pressure on the nozzle exit plane is 1.5 kPa. Assuming that the flow is isentropic, find the Mach number and the velocity on the exit plane. Assume that the products of combustion behave as a perfect gas with $\gamma = 1.22$ and $R = 519.6$ J/kg · K.
30. At a point in a supersonic air flow, the pressure and temperature are 5 kPa and −80°C. If the stagnation pressure at this point is 100 kPa, find the Mach number and the stagnation temperature.
31. Air flows through a circular pipe that has a diameter of 45 cm at Mach 0.3. The stagnation temperature and stagnation pressure are 500 K and 250 kPa, respectively. Calculate the air mass flow rate through the pipe.

32. An aircraft is flying at an altitude of 12,000 m, the atmospheric air pressure at this altitude being 19.39 kPa. The internal volume of the aircraft is 860 m^3. If the 12-cm-diameter window in an exit door failed and blew out how long would it take for the pressure inside the aircraft take to drop from its initial value of 101 kPa to a value that is equal to 40% of this initial value? Assume that the hole in the door acts as a converging nozzle and that the temperature inside the aircraft remains constant at 20°C.

5

Normal Shock Waves

Shock Waves

It has been found experimentally that, under some circumstances, it is possible for an almost spontaneous change to occur in a flow, with the velocity decreasing and the pressure increasing through this region of sharp change. The possibility that such a change can actually occur follows from the analysis given below. It has been found experimentally, and it also follows from the analysis given below, that such regions of sharp change can only occur if the initial flow is supersonic. The extremely thin region in which the transition from the initial supersonic velocity, relatively low-pressure state to the state that involves a relatively low velocity and high pressure is termed a shock wave. The changes that occur through a normal shock wave, i.e., a shock wave that is straight and at right angles to the flow direction, are shown in Figure 5.1. A photograph of a normal shock wave is shown in Figure 5.2.

A shock wave is extremely thin, usually only a few mean free paths thick. A shock wave is analogous in many ways to a "hydraulic jump," which occurs in free-surface liquid flows (shown schematically in Figure 5.3). A hydraulic jump occurs, for example, in the flow downstream of a weir.

A shock wave is, in general, curved. However, many shock waves that occur in practical situations are straight, being either at right angles (i.e., normal) to or at an angle to the upstream flow (see Figure 5.4). A straight shock wave that is at right angles to the upstream flow is, as noted above, termed a normal shock wave, whereas a straight shock wave that is at an angle to the upstream flow is termed an oblique shock wave.

In the case of a normal shock wave, the velocities both ahead (i.e., upstream) of the shock and after (i.e., downstream) of the shock wave are at right angles to the shock wave. In the case of an oblique shock wave, there is a change in flow direction across the shock. This is illustrated in Figure 5.5.

A complete shock wave may be effectively normal in part of the flow, curved in other parts of the flow, and effectively oblique in other parts of the flow, as shown in Figure 5.6.

Because of its own importance and because, as will be shown later, the oblique shock relations can be deduced from those for a normal shock wave, the normal shock wave will be considered first, such waves being the subject of this chapter. Oblique shock waves will then be discussed in the next chapter. Curved shock waves are relatively difficult to analyze and they will not be discussed in detail in this book.

Normal shock waves occur in a number of practical situations such as, for example, in the intakes to the engines in some supersonic aircraft, in the exhaust system of reciprocating engines, in long distance gas pipelines, and in mine shafts as a result of the use of explosives.

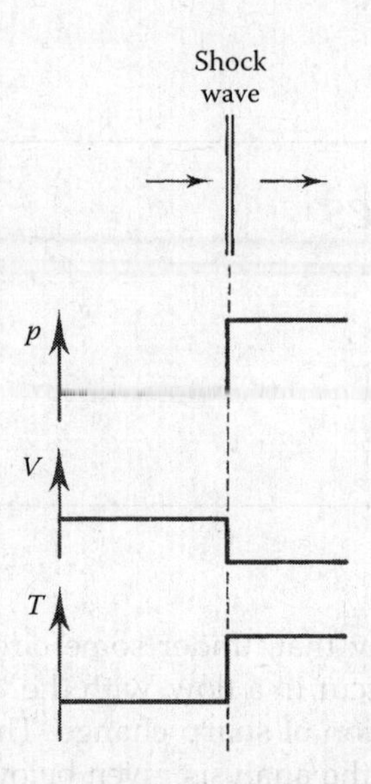

FIGURE 5.1
Changes through a normal shock wave.

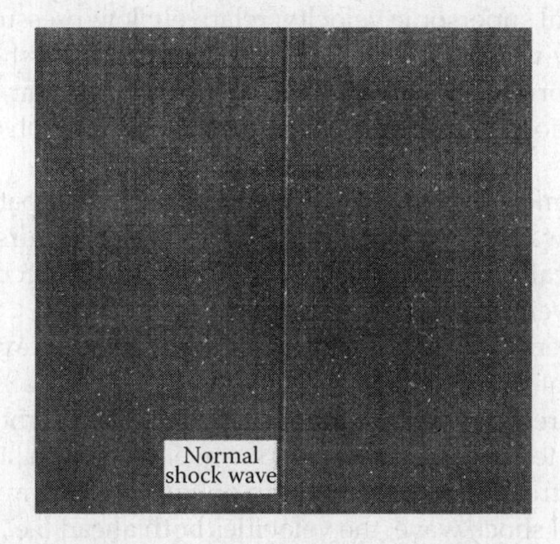

FIGURE 5.2
Photograph of a normal shock wave. (Reprinted with permission from W. Bleakney, D. K. Weimer, and C. H. Fletcher, The Shock Tube: A Facility for Investigations in Fluid Dynamics, *Rev. Sci. Instr.*, 20(11), pp. 807–815. Copyright 1949, American Institute of Physics.)

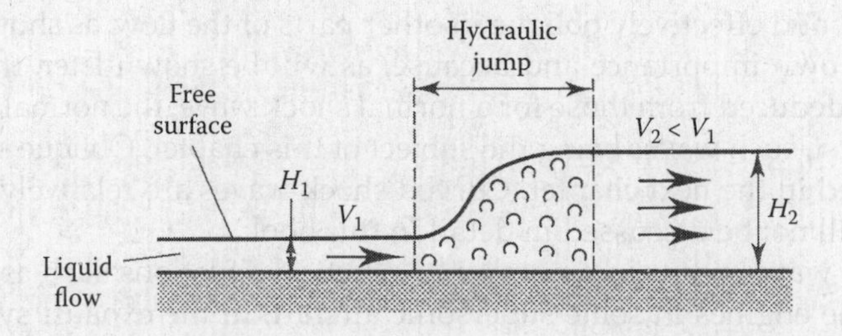

FIGURE 5.3
Hydraulic jump.

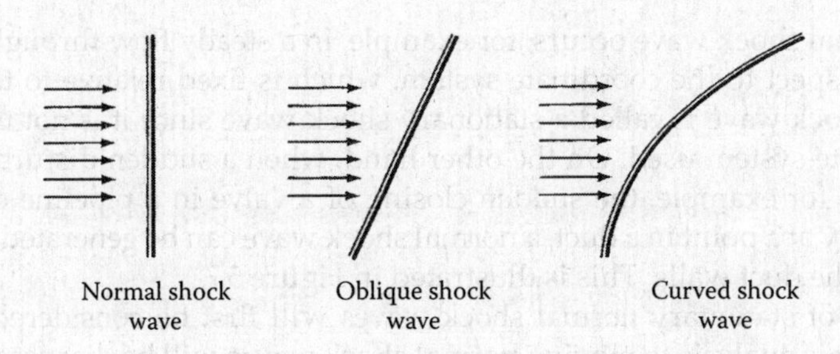

FIGURE 5.4
Curved, normal, and oblique shock waves.

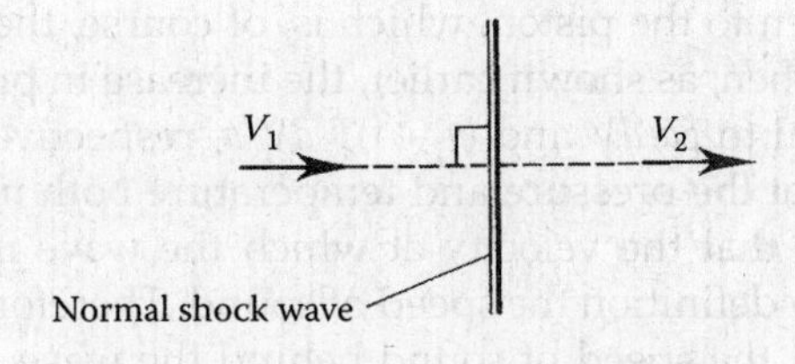

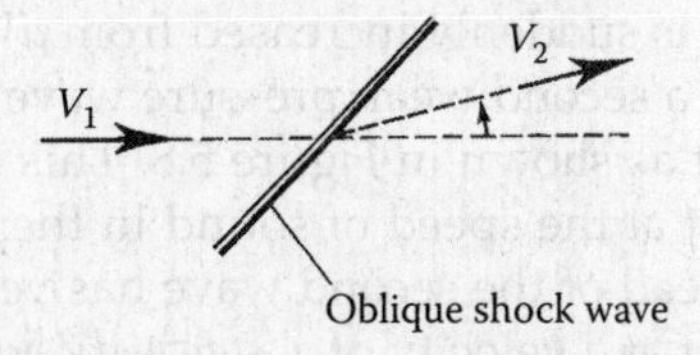

FIGURE 5.5
Velocity changes across normal and oblique shock waves.

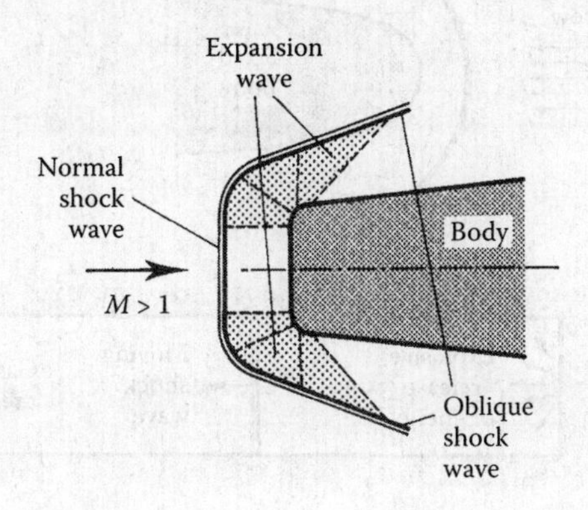

FIGURE 5.6
Shock wave with changing shape along wave.

When a normal shock wave occurs, for example, in a steady flow through duct, it is stationary with respect to the coordinate system, which is fixed relative to the walls of the duct. Such a shock wave is called a stationary shock wave since it is not moving relative to the coordinate system used. On the other hand, when a sudden disturbance occurs in a flow, such as, for example, the sudden closing of a valve in a pipeline or an explosive release of energy at a point in a duct, a normal shock wave can be generated, which is moving relative to the duct walls. This is illustrated in Figure 5.7.

The analysis of stationary normal shock waves will first be considered and then the application of this analysis to moving normal shock waves will be discussed.

To illustrate how a shock wave can form, consider the generation of a sound wave as discussed in Chapter 3. There it was assumed that there was a long duct containing a gas at rest and that there was a piston at one end of this duct that was initially at rest. Then, at time 0, the piston was given a small velocity into the duct, giving rise to a weak pressure pulse, i.e., a sound wave, that propagated down the duct into the gas (see Figure 5.8).

If dV is the velocity given to the piston, which is, of course, the same as the velocity of the gas behind the wave, then, as shown earlier, the increase in pressure and temperature behind the wave are equal to $\rho a\, dV$ and $(\gamma - 1)T\, dV/a$, respectively. Since ρ, a, and T are all positive, this shows that the pressure and temperature both increase across the wave. It was also shown earlier that the velocity at which the wave moves down the duct is equal to $\sqrt{\gamma RT}$, which is by definition the speed of sound. Therefore, since the temperature increases across the wave, the speed of sound behind the wave will be $a + da$, where da is positive. Now consider what happens if some time after the piston is given velocity dV into the duct, its velocity is suddenly increased from dV to $2\, dV$. As a result of the second increase in piston speed, a second weak pressure wave will be generated that follows the first wave down the duct as shown in Figure 5.8. This second wave will be moving relative to the gas ahead of it at the speed of sound in the gas through which it is propagating. However, the gas ahead of the second wave has velocity dV. Hence, the second wave moves relative to the duct at a velocity of $a + da + dV$, whereas the first wave is moving at

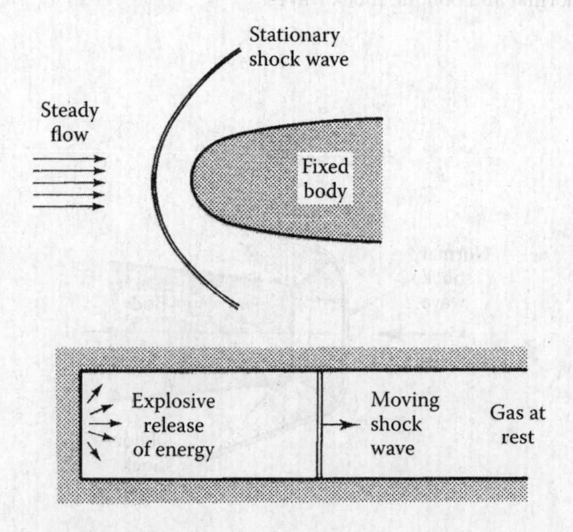

FIGURE 5.7
Stationary (top) and moving (bottom) normal shock waves.

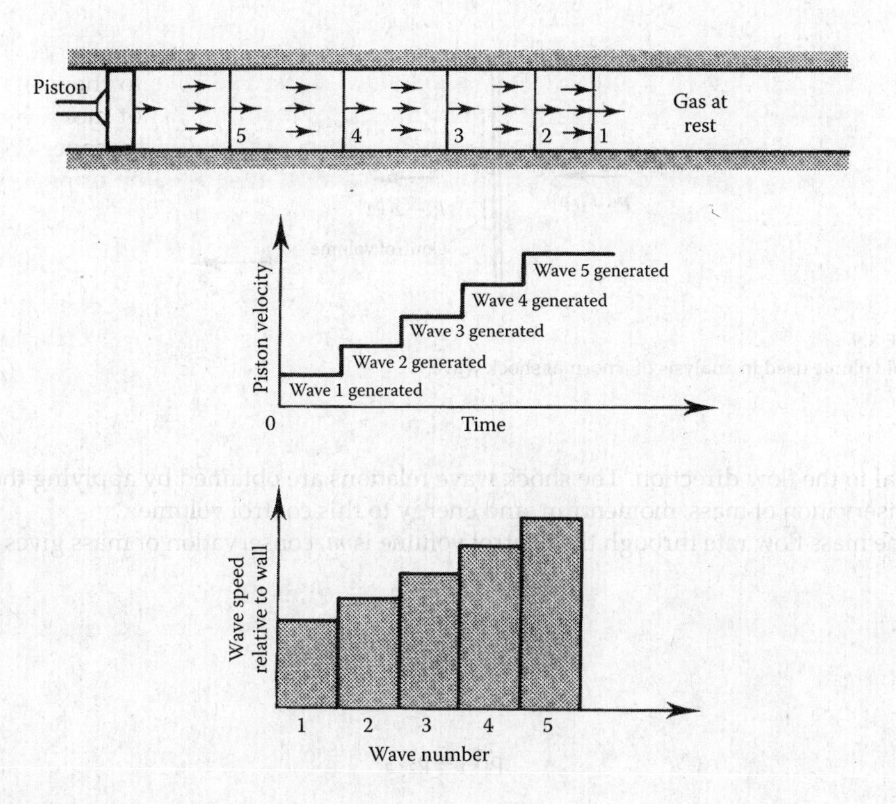

FIGURE 5.8
Generation of weak waves by piston movement.

a velocity of a relative to the duct. Therefore, since both da and dV are positive, the second wave is moving faster than the first wave, and if the duct is long enough, the second wave will overtake the first wave. The second wave cannot pass through the first. Instead, the two waves merge into a single stronger wave. If therefore the piston is given a whole series of step increases in velocity, a series of weak pressure waves will be generated, which will all eventually overtake each other and merge into a single strong wave if the duct is long enough. Since the "back" of this wave is always trying to move faster than the "front" of this wave, the wave will remain thin. Because the change in pressure across the merged wave, i.e., $dp + dp + dp + \ldots$ will in general be large, the temperature gradients in the wave will not be small, and the flow process, unlike that across a single weak wave, cannot be assumed to be isentropic. This thin merged single wave across which large changes in pressure, temperature, etc. occur and across which the flow is not isentropic is a shock wave.

Stationary Normal Shock Waves

Attention will first be given to the changes that occur through a stationary normal shock wave. To analyze the flow through a stationary normal shock wave, consider a control volume of the form indicated in Figure 5.9. This control volume has a cross-sectional area of A

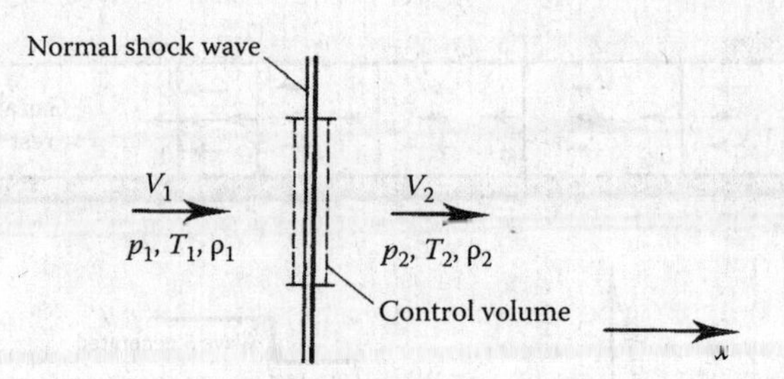

FIGURE 5.9
Control volume used in analysis of a normal shock wave.

normal to the flow direction. The shock wave relations are obtained by applying the laws of conservation of mass, momentum, and energy to this control volume.

If the mass flow rate through the control volume is $\dot{m}$, conservation of mass gives

$$\dot{m} = \rho_1 V_1 A = \rho_2 V_2 A$$

i.e.,

$$\rho_1 V_1 = \rho_2 V_2 \tag{5.1}$$

Since the only forces acting on the control volume in the flow direction are the pressure forces, conservation of momentum applied to the control volume gives

$$p_1 A - p_2 A = \dot{m}(V_2 - V_1) \tag{5.2}$$

Combining this with Equation 5.1 then gives

$$p_1 - p_2 = \rho_1 V_1 (V_2 - V_1) \tag{5.3}$$

or

$$p_1 - p_2 = \rho_2 V_2 (V_2 - V_1) \tag{5.4}$$

These two equations can be rearranged to give

$$V_1 V_2 - V_1^2 = \frac{p_1 - p_2}{\rho_1} \tag{5.5}$$

or

$$V_2^2 - V_2 V_1 = \frac{p_1 - p_2}{\rho_2} \tag{5.6}$$

Adding these two equations together then gives

$$V_2^2 - V_1^2 = (p_1 - p_2)\left(\frac{1}{\rho_1} + \frac{1}{\rho_2}\right) \tag{5.7}$$

Lastly, consider the application of conservation of energy to the flow across the shock wave. Because a one-dimensional flow is being considered, there are no changes in the flow properties in any direction that is normal to that of the flow, and because the upstream and downstream surfaces of the control volume lie upstream and downstream of the shock wave, there are no temperature gradients normal to any surface of the control volume. The flow through the control volume is therefore adiabatic and the energy equation therefore gives

$$\frac{V_1^2}{2} + c_p T_1 = \frac{V_2^2}{2} + c_p T_2 = c_p T_0 = \text{constant} \tag{5.8}$$

The stagnation temperature therefore does not change across the shock. Using, as before, $p/\rho = RT$ and $R = c_p - c_v$, this equation can be written as

$$V_1^2 + \left(\frac{2\gamma}{\gamma - 1}\right)\frac{p_1}{\rho_1} = V_2^2 + \left(\frac{2\gamma}{\gamma - 1}\right)\frac{p_2}{\rho_2} \tag{5.9}$$

which can be rearranged to give

$$V_2^2 - V_1^2 = \left(\frac{2\gamma}{\gamma - 1}\right)\left(\frac{p_1}{\rho_1} - \frac{p_2}{\rho_2}\right) \tag{5.10}$$

Combining Equations 5.10 and 5.7 then gives

$$\left(\frac{2\gamma}{\gamma - 1}\right)\left(\frac{p_2}{\rho_2} - \frac{p_1}{\rho_1}\right) = (p_2 - p_1)\left(\frac{1}{\rho_1} + \frac{1}{\rho_2}\right) \tag{5.11}$$

Now, a relationship between the density ratio, ρ_2/ρ_1, and the pressure ratio, p_2/p_1, is being sought. This can be obtained by multiplying Equation 5.11 by ρ_2/p_1 to give

$$\left(\frac{2\gamma}{\gamma - 1}\right)\left(\frac{p_2}{p_1} - \frac{\rho_2}{\rho_1}\right) = \left(\frac{p_2}{p_1} - 1\right)\left(\frac{\rho_2}{\rho_1} + 1\right) \tag{5.12}$$

which can be rearranged to give

$$\frac{p_2}{p_1} = \frac{\left[\left(\frac{\gamma + 1}{\gamma - 1}\right)\frac{\rho_2}{\rho_1} - 1\right]}{\left[\left(\frac{\gamma + 1}{\gamma - 1}\right) - \frac{\rho_2}{\rho_1}\right]} \tag{5.13}$$

Alternatively, Equation 5.12 could have been arranged to give

$$\frac{\rho_2}{\rho_1} = \frac{\left[\left(\frac{\gamma+1}{\gamma-1}\right)\frac{p_2}{p_1}+1\right]}{\left[\left(\frac{\gamma+1}{\gamma-1}\right)+\frac{p_2}{p_1}\right]} \tag{5.14}$$

It is next noted that the continuity equation, i.e., Equation 5.1, gives

$$\frac{V_1}{V_2} = \frac{\rho_2}{\rho_1} \tag{5.15}$$

Hence, using Equation 5.14 gives

$$\frac{V_1}{V_2} = \frac{\left[\left(\frac{\gamma+1}{\gamma-1}\right)\frac{p_2}{p_1}+1\right]}{\left[\left(\frac{\gamma+1}{\gamma-1}\right)+\frac{p_2}{p_1}\right]} \tag{5.16}$$

The temperature ratio across the shock wave is obtained by noting that the equation of state gives

$$p_1 = \rho_1 R T_1, \qquad p_2 = \rho_2 R T_2 \tag{5.17}$$

which together give

$$\frac{T_2}{T_1} = \frac{p_2}{p_1}\frac{\rho_1}{\rho_2} \tag{5.18}$$

Using Equation 5.14, this equation gives

$$\frac{T_2}{T_1} = \frac{\left[\left(\frac{\gamma+1}{\gamma-1}\right)+\frac{p_2}{p_1}\right]}{\left[\left(\frac{\gamma+1}{\gamma-1}\right)+\frac{p_1}{p_2}\right]} \tag{5.19}$$

Equations 5.14, 5.16, and 5.19 relate the density, velocity, and temperature ratios across a normal shock wave to the pressure ratio across the shock wave. The pressure ratio, p_2/p_1, is often termed the strength of the shock wave. These equations therefore give ρ_2/ρ_1, V_2/V_1, and T_2/T_1 in terms of the shock strength. This set of equations is often termed the Rankine–Hugoniot normal shock wave relations.

Now it will be noted that Equation 5.7 can be rearranged to give

$$V_1^2\left[\left(\frac{V_2}{V_1}\right)^2+1\right]=\left(\frac{p_1}{\rho_1}\right)\left(1-\frac{p_2}{p_1}\right)\left(1+\frac{\rho_1}{\rho_2}\right)$$

Because ρ_2/ρ_1 and V_2/V_1 have been shown to be functions of p_2/p_1, it follows from this equation that for a particular value of p_2/p_1 there is an associated particular value of

$$\frac{V_1^2}{(p_1/\rho_1)}=\frac{V_1^2}{a_1^2/\gamma}=\gamma M_1^2$$

i.e., a particular shock strength is associated with a particular upstream Mach number. This will be discussed further in the next section. Before doing this, the entropy changes across the shock will be discussed.

Although the application of conservation of mass, momentum, and energy principles shows that a shock wave can exist, it does not indicate whether the shock can be either compressive (i.e., $p_2/p_1 > 1$) or expansive (i.e., $p_2/p_1 < 1$). To examine this, the second law of thermodynamics must be used. Now the entropy change across the shock wave is given by

$$\begin{aligned} s_2-s_1 &= c_p\ln\left(\frac{T_2}{T_1}\right)-R\ln\left(\frac{p_2}{p_1}\right) \\ &= (R+c_v)\ln\left(\frac{p_2}{p_1}\frac{\rho_1}{\rho_2}\right)-R\ln\left(\frac{p_2}{p_1}\right) \end{aligned} \tag{5.20}$$

This equation can be rearranged to give

$$\begin{aligned} \frac{s_2-s_1}{R} &= \left(1+\frac{1}{\gamma-1}\right)\ln\left(\frac{p_2}{p_1}\frac{\rho_1}{\rho_2}\right)-\ln\left(\frac{p_2}{p_1}\right) \\ &= \ln\left[\left(\frac{p_2}{p_1}\right)^{\frac{1}{\gamma-1}}\left(\frac{\rho_2}{\rho_1}\right)^{\frac{-\gamma}{\gamma-1}}\right] \end{aligned} \tag{5.21}$$

Equation 5.14 can then be substituted into this equation to give the entropy increase as a function of the shock strength, p_2/p_1. This gives

$$\frac{s_2-s_1}{R}=\ln\left\{\left(\frac{p_2}{p_1}\right)^{\frac{1}{\gamma-1}}\left[\frac{\left(\frac{\gamma+1}{\gamma-1}\right)\frac{p_2}{p_1}+1}{\left(\frac{\gamma+1}{\gamma-1}\right)+\frac{p_2}{p_1}}\right]^{\frac{-\gamma}{\gamma-1}}\right\}$$

i.e.,

$$\frac{s_2 - s_1}{R} = \ln\left\{\left(\frac{p_2}{p_1}\right)^{\frac{1}{\gamma-1}}\left[\frac{(\gamma+1)\frac{p_2}{p_1}+(\gamma+1)}{(\gamma+1)+(\gamma-1)\frac{p_2}{p_1}}\right]^{\frac{-\gamma}{\gamma-1}}\right\} \tag{5.22}$$

Now, the second law of thermodynamics requires that the entropy must remain unchanged or must increase, i.e., it requires that

$$\frac{s_2 - s_1}{R} \geq 0 \tag{5.23}$$

The variation of $(s_2 - s_1)/R$ with p_2/p_1 for two values of γ (γ is always greater than 1) as given by Equation 5.22 is shown in Figure 5.10.

It will be seen from the results given in Figure 5.10 that for Equation 5.23 to be satisfied, it is necessary that

$$p_2/p_1 \geq 1 \tag{5.24}$$

It therefore follows that the shock wave must always be compressive, i.e., that p_2/p_1 must be greater than 1, i.e., the pressure must always increase across the shock wave. Using Equations 5.14, 5.16, and 5.19 then indicates that the density always increases, the velocity always decreases, and the temperature always increases across a shock wave.

The entropy increase across the shock is basically the result of the fact that, because the shock wave is very thin, the gradients of velocity and temperature in the shock are very

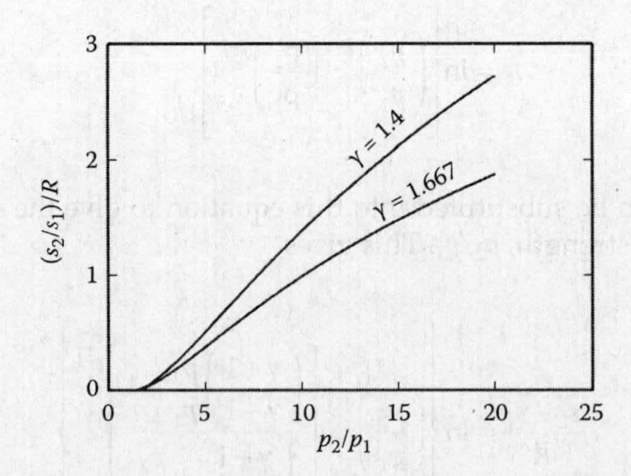

FIGURE 5.10
Effect of pressure ratio on entropy change across a normal shock wave.

high. As a result, the effects of viscosity and heat conduction are important within the shock, leading to the entropy increase across the shock wave.

Because the flow across a shock is adiabatic, the stagnation temperature does not change across a shock wave (see Equation 5.8). However, because of the entropy increase across a shock, the stagnation pressure always decreases across a shock wave. This is perhaps most easily seen by considering the flow situation shown in Figure 5.11.

In the situation being considered, a gas flows from a large reservoir in which the velocity is effectively zero and is isentropically expanded until the Mach number is M_1. A normal shock wave then occurs. After the shock, the flow is isentropically decelerated until the velocity is again effectively zero in a second large reservoir. Since the flow is isentropic everywhere except across the shock wave, the pressure in the first reservoir, p_{01}, is the stagnation pressure everywhere in the flow ahead of the shock wave, whereas the pressure in the second reservoir, p_{02}, is the stagnation pressure everywhere in the flow downstream of the shock wave. Now, Equation 5.20 applies between any two points in the flow. It can therefore be applied between a point in the first reservoir and a point in the second reservoir to give

$$s_{02} - s_{01} = c_p \ln\left(\frac{T_{02}}{T_{01}}\right) - R\ln\left(\frac{p_{02}}{p_{01}}\right) \tag{5.25}$$

However, the stagnation temperature does not change across the shock, so the first term on the right-hand side of Equation 5.25 is zero, i.e., Equation 5.25, gives

$$s_{02} - s_{01} = -R\ln\left(\frac{p_{02}}{p_{01}}\right) \tag{5.26}$$

However, since the flow is isentropic before and after the shock wave

$$s_1 = s_{01} \quad \text{and} \quad s_2 = s_{02} \tag{5.27}$$

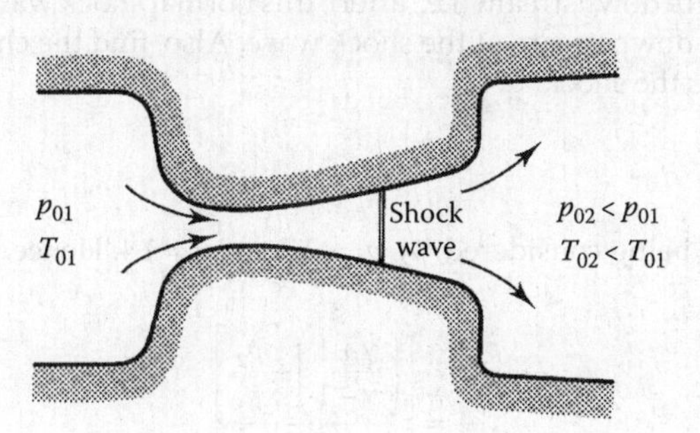

FIGURE 5.11
Flow situation used in analysis of stagnation pressure change across a normal shock wave.

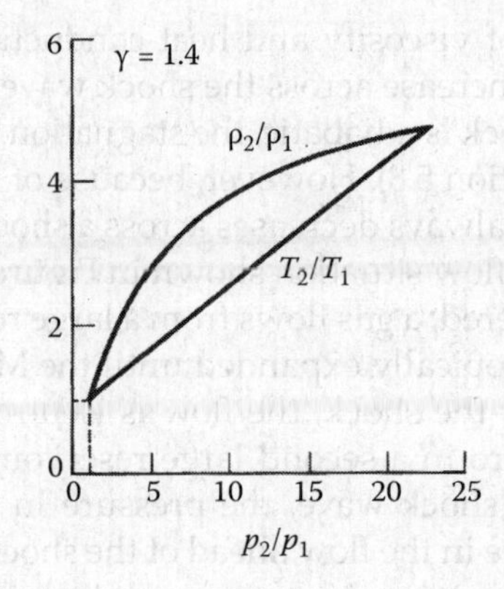

FIGURE 5.12
Variation of changes across a normal shock wave with the pressure ratio.

Therefore, Equation 5.26 gives

$$\frac{s_2 - s_1}{R} = -\ln\left(\frac{p_{02}}{p_{01}}\right) \tag{5.28}$$

or

$$\frac{p_{02}}{p_{01}} = \exp[-(s_2 - s_1)/R] \tag{5.29}$$

Because the entropy must increase across the shock wave, this equation shows that the stagnation pressure must decrease across a shock.

The variations of the changes that occur across a normal shock wave with shock strength are illustrated in Figure 5.12.

Example 5.1

A normal shock occurs at a point in an air flow where the pressure is 30 kPa and the temperature is –30°C. If the pressure ratio across this shock wave is 2.7, find the pressure and temperature downstream (i.e., after), this normal shock wave and the velocities both upstream and downstream of the shock wave. Also, find the change in the stagnation pressure across the shock.

Solution

For the shock wave being considered, $p_2/p_1 = 2.7$ and $\gamma = 1.4$. Hence, since

$$\frac{T_2}{T_1} = \frac{\left[\left(\dfrac{\gamma+1}{\gamma-1}\right) + \dfrac{p_2}{p_1}\right]}{\left[\left(\dfrac{\gamma+1}{\gamma-1}\right) + \dfrac{p_1}{p_2}\right]}$$

it follows that since for air $\gamma = 1.4$

$$T_2/T_1 = (6 + 2.7)/(6 + 1/2.7) = 1.366$$

Also, since

$$\frac{\rho_2}{\rho_1} = \frac{\left[\left(\frac{\gamma+1}{\gamma-1}\right)\frac{p_2}{p_1}+1\right]}{\left[\left(\frac{\gamma+1}{\gamma-1}\right)+\frac{p_2}{p_1}\right]}$$

it follows that

$$\rho_2/\rho_1 = (6 \times 2.7 + 1)/(6 + 2.7) = 1.977$$

Further, since $V_1/V_2 = \rho_2/\rho_1$, it follows that

$$V_1/V_2 = 1.977$$

From these results, it follows that $p_2 = 2.7 \times 30 = 81$ kPa and $T_2 = (273 - 30) \times 1.366 = 331.9$ K $= 58.9$°C.

One way to find the velocities is to recall that the energy equation gives

$$V_2^2 - V_1^2 = \left(\frac{2\gamma}{\gamma-1}\right)\left(\frac{p_1}{\rho_1} - \frac{p_2}{\rho_2}\right)$$

However, the perfect gas law gives $\rho_1 = p_1/RT_1 = 30{,}000/(287.04 \times 243) = 0.43$ kg/m³. Therefore, since $V_1/V_2 = 1.977$, it follows that

$$\left(\frac{1}{1.977^2} - 1\right)V_1^2 = 7\left(\frac{30{,}000}{0.43} - \frac{81{,}000}{1.977 \times 0.43}\right)$$

This equation gives $V_1 = 489.9$ m/s and so $V_2 = 489.9/1.977 = 247.8$ m/s. Hence,

$$M_1 = 489.9/\sqrt{1.4 \times 287.04 \times 243} = 1.568$$

and

$$M_2 = 247.8/\sqrt{1.4 \times 287.04 \times 331.9} = 0.679$$

Therefore, since for Mach 1.568, $p_0/p = 4.057$, whereas for Mach 0.679, $p_0/p = 1.361$, these values being obtained either using the relationship given in the previous chapter or using isentropic tables or the software, the change in stagnation pressure across the shock wave is given by

$$\Delta p_0 = \left(\frac{p_{02}}{p_2}\frac{p_2}{p_1} - \frac{p_{01}}{p_1}\right)p_1 = (1.361 \times 2.7 - 4.057) \times 30 = -11.47 \text{ kPa}$$

i.e., the stagnation pressure decreases by 11.47 kPa across the shock wave. Because the flow through the wave is adiabatic, there is, of course, no change in stagnation temperature through the wave.

Normal Shock Wave Relations in Terms of Mach Number

Although the relations derived in the previous section for the changes across a normal shock in terms of the pressure ratio across the shock, i.e., in terms of the shock strength, are the most useful form of the normal shock wave relations for some purposes, it is often more convenient to have these relations in terms of the upstream Mach number, M_1. To obtain these forms of the normal shock wave relations, it is convenient to start again with a control volume across the shock wave such as that shown in Figure 5.13, and to again apply the conservation of mass, momentum, and energy to this control volume, but in this case, to rearrange the relations in terms of Mach number.

In writing the conservation laws, no generality is lost by taking the area of the control volume parallel to the wave as unity. Conservation of mass then gives

$$\rho_1 V_1 = \rho_2 V_2 \tag{5.30}$$

Dividing this equation by a_1 then gives

$$\rho_1 \frac{V_1}{a_1} = \rho_2 \frac{V_2}{a_2} \frac{a_2}{a_1}$$

which can be rewritten in terms of Mach numbers as

$$\frac{\rho_2}{\rho_1} = \frac{M_1}{M_2} \frac{a_1}{a_2} \tag{5.31}$$

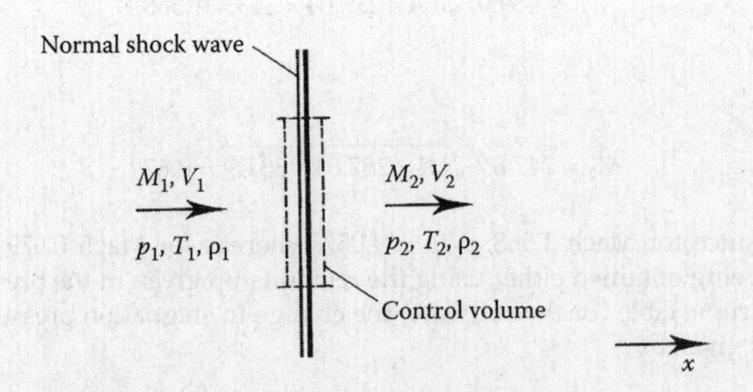

FIGURE 5.13
Control volume used in deriving expressions for the changes across a normal shock wave in terms of the upstream Mach number.

Next, consider the conservation of momentum. This gives for the control volume shown in Figure 5.13

$$p_1 - p_2 = \rho_2 V_2^2 - \rho_1 V_1^2 \tag{5.32}$$

Hence, since

$$a^2 = \frac{\gamma p}{\rho}, \quad \text{i.e.,} \quad p = \frac{a^2 \rho}{\gamma}$$

Equation 5.32 becomes

$$\frac{a_1^2 \rho_1}{\gamma} + \rho_1 V_1^2 = \frac{a_2^2 \rho_2}{\gamma} + \rho_2 V_2^2$$

Dividing this through by a_1^2/γ then gives

$$\rho_1 + \gamma \rho_1 M_1^2 = \rho_2 \left(\frac{a_2}{a_1}\right)^2 + \gamma \rho_2 M_2^2 \left(\frac{a_2}{a_1}\right)^2$$

which can be rearranged to give

$$\frac{\rho_2}{\rho_1} = \left(\frac{1 + \gamma M_1^2}{1 + \gamma M_2^2}\right)\left(\frac{a_1}{a_2}\right)^2 \tag{5.33}$$

Lastly, consider the application of the conservation of energy principle to the control volume. This gives

$$V_1^2 + \left(\frac{2}{\gamma - 1}\right) a_1^2 = V_2^2 + \left(\frac{2}{\gamma - 1}\right) a_2^2 \tag{5.34}$$

Dividing this equation by $2a_1^2/(\gamma - 1)$ then gives

$$\left(\frac{\gamma - 1}{2}\right) M_1^2 + 1 = \left(\frac{\gamma - 1}{2}\right) M_2^2 \left(\frac{a_2}{a_1}\right)^2 + \left(\frac{a_2}{a_1}\right)^2$$

which can be rearranged to give

$$\left(\frac{a_2}{a_1}\right)^2 = \left[\frac{2 + (\gamma - 1) M_1^2}{2 + (\gamma - 1) M_2^2}\right] \tag{5.35}$$

The density ratio, ρ_2/ρ_1, is now eliminated between Equations 5.31 and 5.33, giving

$$\left(\frac{a_2}{a_1}\right)=\left(\frac{1+\gamma M_1^2}{1+\gamma M_2^2}\right)\left(\frac{M_2}{M_1}\right) \tag{5.36}$$

The speed of sound ratio, a_2/a_1, is next eliminated between Equations 5.35 and 5.36 to give

$$\frac{2+(\gamma-1)M_1^2}{2+(\gamma-1)M_2^2}=\left(\frac{1+\gamma M_1^2}{1+\gamma M_2^2}\right)^2\left(\frac{M_2}{M_1}\right)^2$$

This equation can be rearranged to give

$$(\gamma-1)(M_2^4-M_1^4)-2\gamma M_2^2M_1^2(M_2^2-M_1^2)+2(M_2^2-M_1^2)=0 \tag{5.37}$$

However, $(M_2^2-M_1^2)$ cannot be zero, as this would imply that there was no change in the Mach number across the shock wave. This term can therefore be cancelled out of Equation 5.37, giving

$$M_2^2=\frac{\left[M_1^2+\left(\dfrac{2}{\gamma-1}\right)\right]}{\left[\left(\dfrac{2\gamma}{\gamma-1}\right)M_1^2-1\right]}=\left[\frac{(\gamma-1)M_1^2+2}{2\gamma M_1^2-(\gamma-1)}\right] \tag{5.38}$$

This equation relates the downstream Mach number to the upstream Mach number, and the result can be used to derive expressions for the pressure ratio, the temperature ratio, and the density ratio in terms of the upstream Mach number. It is first noted that substituting Equation 5.38 into Equation 5.35 gives

$$\left(\frac{a_2}{a_1}\right)^2=\frac{T_2}{T_1}=\frac{[2+(\gamma-1)M_1^2]}{\left\{2+(\gamma-1)\left[\dfrac{(\gamma-1)M_1^2+2}{2\gamma M_1^2-(\gamma-1)}\right]\right\}}$$

$$=\left\{\frac{[2\gamma M_1^2-(\gamma-1)][2+(\gamma-1)M_1^2]}{(\gamma+1)^2M_1^2}\right\} \tag{5.39}$$

To obtain an expression for the pressure ratio, it is noted that the equation of state gives

$$\frac{p_2}{p_1}=\left(\frac{\rho_2}{\rho_1}\right)\left(\frac{T_2}{T_1}\right)=\left(\frac{\rho_2}{\rho_1}\right)\left(\frac{a_2}{a_1}\right)^2 \tag{5.40}$$

Hence, using Equation 5.31, the following is obtained:

$$\frac{p_2}{p_1} = \frac{1+\gamma M_1^2}{1+\gamma M_2^2} = \frac{(1+\gamma M_1^2)}{\left\{1+\gamma\left[\frac{(\gamma-1)M_1^2+2}{2\gamma M_1^2-(\gamma-1)}\right]\right\}} \tag{5.41}$$

The right-hand side of this equation can be written as

$$\frac{(1+\gamma M_1^2)[2\gamma M_1^2-(\gamma-1)]}{[2\gamma M_1^2-(\gamma-1)]+\gamma[(\gamma-1)M_1^2+2]}$$

i.e., as

$$\frac{(1+\gamma M_1^2)[2\gamma M_1^2-(\gamma-1)]}{2\gamma M_1^2-\gamma+1+\gamma^2 M_1^2-\gamma M_1^2+2\gamma}$$

i.e., as

$$\frac{(1+\gamma M_1^2)[2\gamma M_1^2-(\gamma-1)]}{\gamma M_1^2+\gamma^2 M_1^2+\gamma+1}$$

i.e., as

$$\frac{(1+\gamma M_1^2)[2\gamma M_1^2-(\gamma-1)]}{(\gamma+1)(1+\gamma M_1^2)}$$

Hence, Equation 5.41 gives the pressure ratio as

$$\frac{p_2}{p_1} = \frac{2\gamma M_1^2-(\gamma-1)}{(\gamma+1)} \tag{5.42}$$

The density ratio can now be directly obtained by again noting that the equation of state gives

$$\frac{\rho_2}{\rho_1} = \left(\frac{p_2}{p_1}\right)\left(\frac{T_1}{T_2}\right)$$

Hence, using Equations 5.42 and 5.39, the following is obtained

$$\frac{\rho_2}{\rho_1} = \frac{(\gamma+1)M_1^2}{2+(\gamma-1)M_1^2} \tag{5.43}$$

The stagnation pressure ratio across a normal shock wave is obtained by noting that

$$\frac{p_0}{p} = \left(1 + \frac{(\gamma - 1)}{2} M^2\right)^{\gamma/(\gamma-1)} \tag{5.44}$$

Hence, since

$$\frac{p_{02}}{p_{01}} = \frac{p_{02}/p_2}{p_{01}/p_1}\frac{p_2}{p_1}$$

using Equations 5.42 and 5.44, the stagnation pressure change across a normal shock is given by

$$\frac{p_{02}}{p_{01}} = \left\{\frac{\left(1 + \frac{(\gamma - 1)}{2} M_2^2\right)}{\left(1 + \frac{(\gamma - 1)}{2} M_1^2\right)}\right\}^{\gamma/(\gamma-1)} \left\{\frac{2\gamma M_1^2 - (\gamma - 1)}{(\gamma + 1)}\right\}$$

i.e., using Equation 5.38 to give M_2, and rearranging gives

$$\frac{p_{02}}{p_{01}} = \left\{\frac{(\gamma - 1)}{2} \frac{M_1^2}{\left(1 + \frac{(\gamma - 1)}{2} M_1^2\right)}\right\}^{\gamma/(\gamma-1)} \left\{\left(\frac{2\gamma}{\gamma + 1}\right) M_1^2 - \left(\frac{\gamma - 1}{\gamma + 1}\right)\right\}^{-1/(\gamma-1)} \tag{5.45}$$

The stagnation temperature does not, of course, as mentioned before, change across the shock wave.

The above equations, some of which are summarized below, give the pressure ratio, the density ratio, the temperature ratio, the downstream Mach number, etc., in terms of the upstream Mach number for any gas, i.e., for any value of γ.

$$\frac{p_2}{p_1} = \frac{2\gamma M_1^2 - (\gamma - 1)}{(\gamma + 1)}, \quad \frac{\rho_2}{\rho_1} = \frac{(\gamma + 1) M_1^2}{2 + (\gamma - 1) M_1^2}$$

$$\frac{T_2}{T_1} = \left\{\frac{[2\gamma M_1^2 - (\gamma - 1)][2 + (\gamma - 1) M_1^2]}{(\gamma + 1)^2 M_1^2}\right\}$$

$$M_2^2 = \frac{(\gamma - 1) M_1^2 + 2}{2\gamma M_1^2 - (\gamma - 1)}$$

The variations of pressure ratio, density ratio, temperature ratio, and downstream Mach number with upstream Mach number given by these equations are shown in Figure 5.14 for the case of $\gamma = 1.4$.

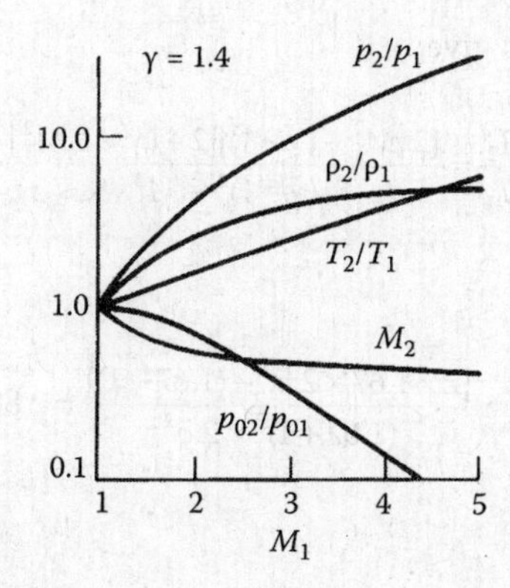

FIGURE 5.14
Variation of changes across normal shock with upstream Mach number.

Example 5.2

A gas that has a molar mass of 39.9 and a specific heat ratio of 1.67 is discharged through a nozzle. A normal shock wave occurs at a section of the flow at which the Mach number is 2.5, the pressure is 40 kPa, and the temperature is –20°C. Find the Mach number, pressure, and temperature downstream (i.e., after) this normal shock wave.

Solution

The flow situation being considered is shown in Figure E5.2.
Normal shock relations give

$$M_2^2 = \frac{M_1^2 + \dfrac{2}{(\gamma-1)}}{\dfrac{2\gamma}{(\gamma-1)}M_1^2 - 1} = \frac{2.5^2 + \dfrac{2}{(1.67-1)}}{\dfrac{2\times 1.67}{(1.67-1)}\times 2.5^2 - 1} = 0.306$$

Hence,

$$M_2 = 0.553$$

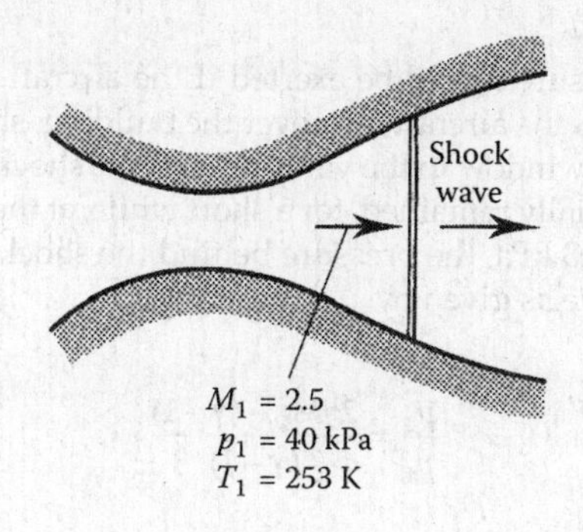

FIGURE E5.2
Flow situation considered.

Normal shock relations also give

$$\frac{T_2}{T_1} = \frac{[2\gamma M_1^2 - (\gamma - 1)][2 + (\gamma - 1)M_1^2]}{(\gamma - 1)^2 \times M_1^2}$$

Hence,

$$\frac{T_2}{T_1} = \frac{[2 \times 1.67 \times 2.5^2 \quad (1.67 - 1)]}{(1.67 + 1)^2 \times 2.5^2} = 2.806$$

Thus,

$$T_2 = 710 \text{ K} = 437°\text{C}$$

Also, using normal shock relations gives

$$\frac{p_2}{p_1} = \frac{2\gamma M_1^2 - (\gamma - 1)}{(\gamma + 1)}$$

$$= \frac{2 \times 1.67 \times 2.5^2 - (1.67 - 1)}{1.67 + 1}$$

$$= 7.5674$$

Thus,

$$p_2 = 303 \text{ kPa}$$

Therefore, the Mach number, pressure, and temperature behind the shock wave are 0.553, 303 kPa, and 437°C, respectively.

Example 5.3

An aircraft flying at Mach 1.2 at sea level passes over a building. Estimate the highest force that could be exerted on a 1-m-wide × 2-m-high window in this building as a result of the aircraft flying over it.

Solution

The highest possible pressure would be exerted if the aircraft essentially had a normal shock ahead of it which, as the aircraft flew over the building, sharply increased the pressure on the outside of the window to the value behind the shock wave and if the pressure inside the building essentially remained, for a short while, at the initial ambient pressure.

Assuming that $p_1 = 101.3$ kPa, the pressure behind the shock wave, which is assumed to be a normal shock wave, is given by

$$\frac{p_2}{p_1} = \frac{2\gamma M_1^2 - (\gamma - 1)}{(\gamma + 1)}$$

i.e., $p_2/p_1 = 1.513$. Hence, the maximum force on the window is given by $(p_1 - p_2)$ × window area, i.e., by (1.513 × 101.3 – 101.3) × (1 × 2) = 103.9 kN.

If it is recalled that the average weight of a person is approximately 0.67 kN, it will be realized that the force exerted on the window by the passage of the aircraft has the potential to break the window.

Lastly, consideration should be given to the change in entropy across a normal shock wave in terms of the upstream Mach number. Now, as shown above (see Equation 5.21), the change in entropy is given by

$$\frac{s_2 - s_1}{R} = \left(1 + \frac{1}{\gamma - 1}\right) \ln\left(\frac{p_2}{p_1}\frac{\rho_1}{\rho_2}\right) - \ln\left(\frac{p_2}{p_1}\right) = \ln\left[\left(\frac{p_2}{p_1}\right)^{\frac{1}{\gamma-1}}\left(\frac{\rho_2}{\rho_1}\right)^{\frac{-\gamma}{\gamma-1}}\right] \tag{5.46}$$

The right-hand side of this equation can be expressed in terms of the upstream Mach number using the relationships derived above for the pressure and density ratios. Using these gives

$$\frac{s_2 - s_1}{R} = \ln\left\{\left[\frac{2\gamma}{\gamma+1}(M_1^2 - 1) + 1\right]^{\frac{1}{\gamma-1}}\left[\frac{(\gamma+1)M_1^2}{2 + (\gamma-1)M_1^2}\right]^{\frac{-\gamma}{\gamma-1}}\right\} \tag{5.47}$$

The variation of $(s_2 - s_1)/R$ with M_1 as given by this equation for various values of γ is shown in Figure 5.15.

Now, the entropy must remain unchanged or must increase across the shock wave. It will be seen that this can only be the case if

$$M_1 \geq 1 \tag{5.48}$$

It therefore follows that the Mach number ahead of a shock wave must always be greater than 1 and that the shock wave must therefore, as discussed above, be always compressive,

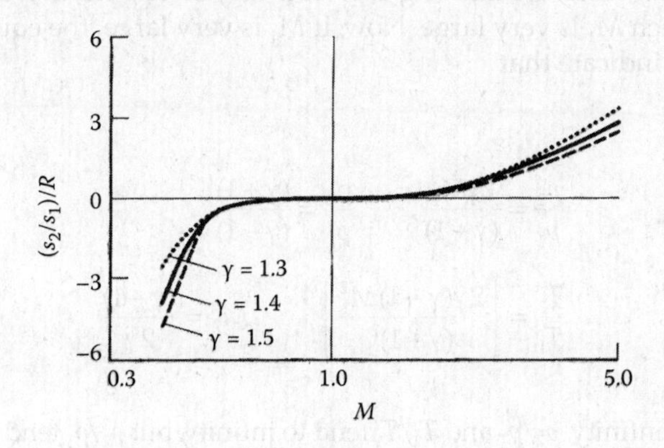

FIGURE 5.15
Variation of entropy change across normal shock with upstream Mach number.

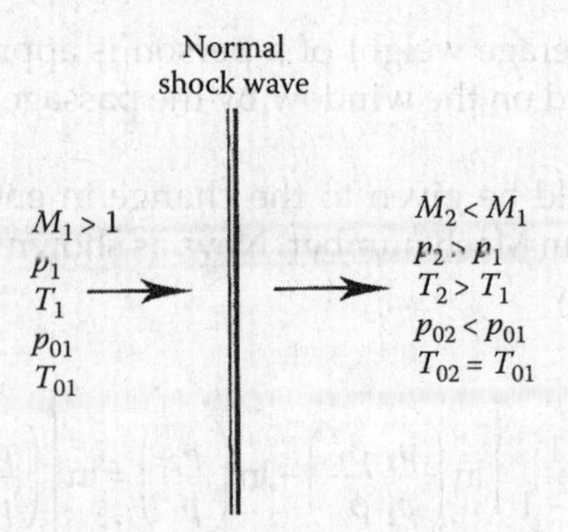

FIGURE 5.16
Values of variables before and after a normal shock.

i.e., p_2/p_1 must be greater than 1. It will also be noted that Equation 5.38 can be rearranged to give

$$M_2^2 = \frac{M_1^2 + \left(\frac{2}{\gamma - 1}\right)}{\left(\frac{2\gamma}{\gamma - 1}\right) M_1^2 - 1} \tag{5.49}$$

Hence, since γ will be between 1 and 2 and since M_1 is always greater than 1, it follows from this equation that M_2 will always be less than 1, i.e., the flow downstream of a normal shock wave will always be subsonic. These conclusions about a normal shock wave are summarized in Figure 5.16.

Limiting Cases of Normal Shock Wave Relations

It is instructive to consider the limiting case of a very strong normal shock, i.e., a normal shock wave for which M_1 is very large. Now, if M_1 is very large, the equations given above for a normal shock indicate that

For $M_1 >> 1$,

$$\frac{p_2}{p_1} = \frac{2\gamma M_1^2}{(\gamma + 1)}, \qquad \frac{\rho_2}{\rho_1} = \frac{(\gamma + 1)}{(\gamma - 1)}$$

$$\frac{T_2}{T_1} = \left\{\frac{2\gamma(\gamma - 1)M_1^2}{(\gamma + 1)^2}\right\}, \qquad M_2^2 = \frac{(\gamma - 1)}{2\gamma}$$

Thus, if M_1 tends to infinity, p_2/p_1 and T_2/T_1 tend to infinity, but ρ_2/ρ_1 tends to $(\gamma + 1)/(\gamma - 1)$ and M_2 tends to $\sqrt{(\gamma - 1)/2\gamma}$. In fact, the assumptions on which the above analysis of the changes across a normal shock wave are based, i.e., that the gas remains thermally and calorically perfect, will cease to be valid when the shock is very strong because very high temperatures will then usually exist behind the shock.

The above discussion concerned the flow across a very strong shock wave. Another limiting case is that of a very *weak* normal shock wave. Now the discussion of the entropy change across a normal shock indicates that in this weak shock case the flow is isentropic, i.e., that the relations for the pressure and density ratios derived for isentropic flow in the previous chapter apply across such weak shocks. Now, the continuity equation applies across the shock whether or not the flow is isentropic. Therefore, even in the case of the weak shock limit, Equation 5.31 applies, i.e., the following applies

$$\frac{\rho_2}{\rho_1}\frac{a_2}{a_1} = \frac{M_1}{M_2} \tag{5.50}$$

However, in isentropic flow,

$$\frac{\rho_2}{\rho_1} = \left(\frac{T_2}{T_1}\right)^{1/\gamma-1}$$

Therefore, noting that

$$\frac{a_2}{a_1} = \left(\frac{T_2}{T_1}\right)^{1/2}$$

Equation 5.50 gives

$$\frac{M_2}{M_1} = \left(\frac{T_1}{T_2}\right)^{(\gamma+1)/2(\gamma-1)} \tag{5.51}$$

However, in adiabatic flow and therefore in isentropic flow, the energy equation gives

$$\frac{T_1}{T_2} = \frac{\left(1+\dfrac{(\gamma-1)}{2}M_2^2\right)}{\left(1+\dfrac{(\gamma-1)}{2}M_1^2\right)} \tag{5.52}$$

Substituting this into Equation 5.51 then gives

$$\frac{M_2}{M_1} = \left\{\frac{\left(1+\dfrac{(\gamma-1)}{2}M_2^2\right)}{\left(1+\dfrac{(\gamma-1)}{2}M_1^2\right)}\right\}^{(\gamma+1)/2(\gamma-1)} \tag{5.53}$$

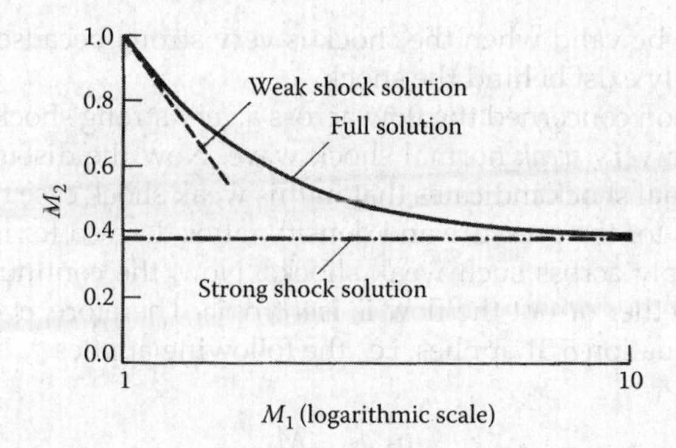

FIGURE 5.17
Relation between downstream Mach number values given by limiting normal shock wave solutions and the actual values given by the full normal shock relations. Results are for $\gamma = 1.4$.

This equation gives the value of the downstream Mach number, M_2, corresponding to any specified value of the upstream Mach number, M_1, for a very weak shock wave. Once M_2 is found, the changes in pressure, density, and temperature across the weak shock can be found using the isentropic relations in conjunction with Equation 5.52. This procedure gives, for example,

$$\frac{p_2}{p_1} = \left\{ \frac{\left(1 + \dfrac{(\gamma - 1)}{2} M_1^2\right)}{\left(1 + \dfrac{(\gamma - 1)}{2} M_2^2\right)} \right\}^{\gamma/2(\gamma-1)} \tag{5.54}$$

To illustrate the relation between the strong shock, the weak shock, and the actual normal shock relations, the variations of M_2 with M_1 given by these relations is shown in Figure 5.17 for the case of $\gamma = 1.4$. It will be seen from the results given in Figure 5.17 that the weak shock relations apply if $M_1 <$ about 1.1, whereas the strong shock relations only apply if M_1 is very large.

Normal Shock Wave Tables and Software

A number of sets of tables and graphs are available, which list the ratios of the various flow variables such as pressure, temperature, and density across a normal shock wave and the downstream Mach number as a function of the upstream Mach number for various gases, i.e., for various values of γ. Typical headings in such a set of normal shock tables are

$$M_1 \quad M_2 \quad \frac{p_2}{p_1} \quad \frac{T_2}{T_1} \quad \frac{\rho_2}{\rho_1} \quad \frac{a_2}{a_1} \quad \frac{p_{20}}{p_{10}} \quad \frac{p_{20}}{p_1}$$

The values in these tables and graphs are, of course, derived using the equations given in the previous section. Normal shock tables of this type are given in Appendix C for the case of $\gamma = 1.4$.

As with isentropic flow, instead of using tables, it is often more convenient to use software such as COMPROP that is available to support this book to find the changes across a shock wave. Alternatively, most calculators can be programmed to give results for a normal shock wave.

Example 5.4

Air is expanded from a large reservoir in which the pressure and temperature are 500 kPa and 35°C through a variable area duct. A normal shock occurs at a point in the duct where the Mach number is 2.5. Find the pressure and temperature in the flow just downstream of the shock wave. Downstream of the shock wave, the flow is brought to rest in another large reservoir. Find the pressure and temperature in this reservoir. Assume that the flow is one-dimensional and isentropic everywhere except through the shock wave.

Solution

The flow situation being considered is shown in Figure E5.4.

The flow upstream of the shock wave can be assumed to be isentropic so isentropic flow relations or tables or software give for the flow upstream of the shock

$$\frac{p_{01}}{p_1} = 17.085,$$

$$\frac{T_{01}}{T} = 2.25$$

Using the specified stagnation point conditions, the following are then obtained

$$p_1 = 500/17.085 = 29.3 \text{ kPa}, \quad T_1 = 308/2.25 = 136.9 \text{ K}$$

Next consider the changes across the shock wave. For air flow at Mach 2.5, normal shock relations, or tables or software gives

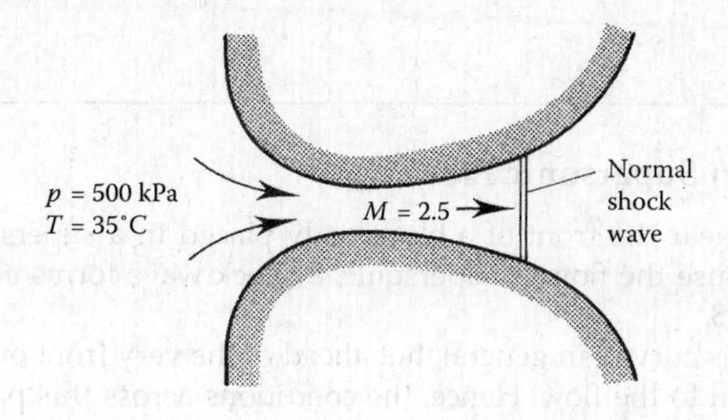

FIGURE E5.4
Flow situation considered.

$$M_2 = 0.5130,$$

$$\frac{p_2}{p_1} = 7.215,$$

$$\frac{T_2}{T_1} = 2.1375$$

From the above results, it follows that

$$p_2 = 7.215 \times 29.3 = 208.8 \text{ kPa}$$

and

$$T_2 = 2.1375 \times 136.9 = 292.6 \text{ K} = 19.6°\text{C}$$

Therefore, the pressure and temperature immediately downstream of the shock wave are 208.8 kPa and 19.6°C, respectively.

Lastly, consider the flow downstream of the shock. This flow is also assumed to be isentropic so isentropic flow relations or tables or the software give for the flow downstream of the shock where the Mach number is 0.513:

$$\frac{p_{02}}{p_2} = 1.194,$$

$$\frac{T_{02}}{T_2} = 1.052$$

Therefore, the downstream stagnation conditions are as follows:

$$p_{02} = 208.8 \times 1.194 = 249.4 \text{ kPa}, \quad T_{02} = 292.6 \times 1.052 = 308 \text{ K} = 35°\text{C}$$

Therefore, the pressure and temperature in the downstream reservoir are 249.4 kPa and 35°C, respectively.

It will be noted that, because the entire flow is assumed to be adiabatic, there is no change in the stagnation temperature. There is, however, as a result of the presence of the shock, an almost 50% loss of stagnation pressure.

The Pitot Tube in Supersonic Flow

Consider the flow near the front of a blunt body placed in a supersonic flow as shown in Figure 5.18. Because the flow is supersonic, a shock wave forms ahead of the body as shown in Figure 5.18.

This shock wave is curved in general, but ahead of the very front of the body, the shock is effectively normal to the flow. Hence, the conditions across this part of the shock, i.e., between points 1 and 2 in Figure 5.18, are related by the normal shock relations. Further, since the flow downstream of a normal shock wave is always subsonic, the deceleration

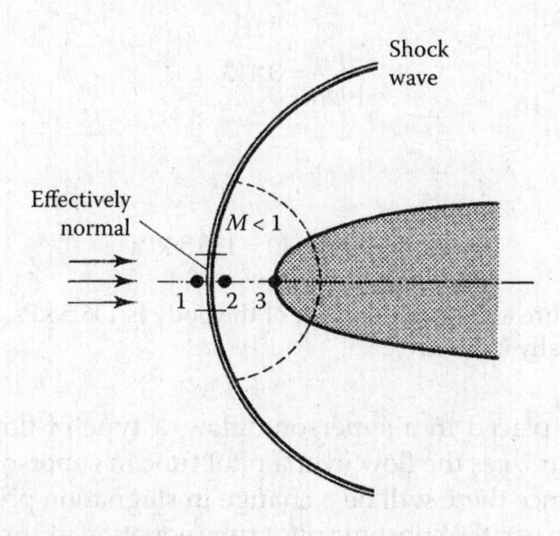

FIGURE 5.18
Supersonic flow over a blunt nosed body.

from point 2 in Figure 5.18 to point 3 in this figure where the velocity is effectively zero can, as discussed in the previous chapter, be assumed to be an isentropic process. Using this model of the flow, the pressure at the stagnation point can be calculated for any specified upstream conditions.

Example 5.5

Air flows over a blunt-nosed body. The air flow in the freestream ahead of the body has Mach 1.5 and a static pressure of 40 kPa. Find the pressure acting on the front of this body. Sketch the flow pattern near the nose.

Solution

The flow situation being considered in this example is shown in Figure E5.5.

Here $M_1 = 1.5$ and $p_1 = 40$ kPa. However, normal shock relations or tables or software give for this Mach number

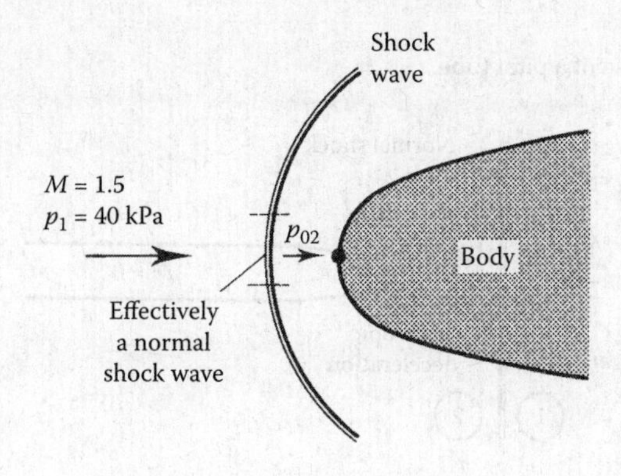

FIGURE E5.5
Flow situation considered.

$$\frac{p_{02}}{p_1} = 3.413$$

Hence,

$$p_{02} = 3.413 \times 40 = 136.5 \text{ kPa}$$

Therefore, the pressure acting on the front of the body is 136.5 kPa. The flow pattern is as shown schematically in Figure E5.5.

When a pitot tube is placed in a supersonic flow, a type of flow similar to that indicated in Figure 5.18 occurs, i.e., the flow over a pitot tube in supersonic flow resembles that shown in Figure 5.19. Since there will be a change in stagnation pressure across the shock wave, it is not possible to use the subsonic pitot tube equation in supersonic flow. However, as noted above, over the small area of the flow covered by the pressure tap in the nose of the pitot tube, the shock wave is effectively normal and the flow behind this portion of the shock wave is therefore subsonic and the deceleration downstream of this portion of the wave is isentropic, these assumptions being illustrated in Figure 5.20.

The flow can therefore be analyzed as follows:

1. The pressure ratio across the shock wave, p_2/p_1, can be found using normal shock wave relations.
2. The pressure at the stagnation point can be found by assuming that the isentropic relations apply between the flow behind the shock and the stagnation point.

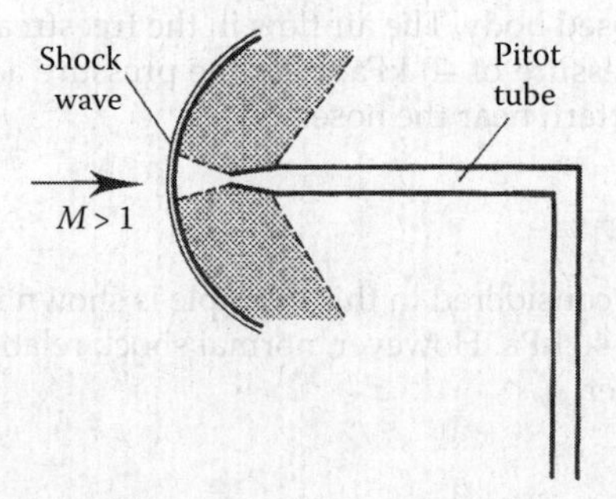

FIGURE 5.19
Supersonic flow near the nose of a pitot tube.

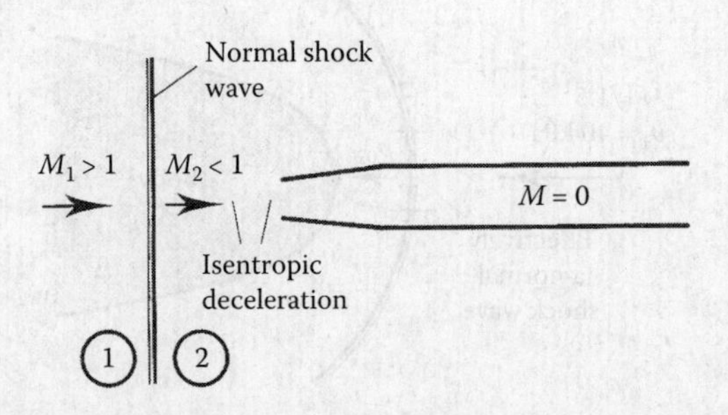

FIGURE 5.20
Assumed flow near the nose of a pitot tube.

Hence, since

$$\frac{p_{02}}{p_1} = \frac{p_{02}}{p_2}\frac{p_2}{p_1} \tag{5.55}$$

where the subscripts 1 and 2 denote the conditions upstream and downstream of the shock wave, respectively, using the relations previously given, this equation becomes

$$\frac{p_{02}}{p_1} = \left[1 + \left(\frac{\gamma-1}{2}\right)M_2^2\right]^{\frac{\gamma}{\gamma-1}}\left[\frac{2\gamma M_1^2 - (\gamma-1)}{(\gamma+1)}\right] \tag{5.56}$$

Therefore, using the expression for the downstream Mach number, it follows that

$$\frac{p_{02}}{p_1} = \left\{1 + \left(\frac{\gamma-1}{2}\left[\frac{(\gamma-1)M_1^2 + 2}{2\gamma M_1^2 - (\gamma-1)}\right]\right)\right\}\left[\frac{2\gamma M_1^2 - (\gamma-1)}{(\gamma+1)}\right]$$

$$= \frac{[(\gamma+1)M_1^2/2]^{\frac{\gamma}{(\gamma-1)}}}{\left[\left(\frac{2\gamma M_1^2}{\gamma+1}\right) - \left(\frac{\gamma-1}{\gamma+1}\right)\right]^{\frac{1}{(\gamma-1)}}} \tag{5.57}$$

This equation is known as the Rayleigh supersonic pitot tube equation. If p_{02} and p_1 are measured, this equation allows M_1 to be found. The value of p_{02}/p_1 is usually listed in shock tables or given by software such as COMPROP. This fact was utilized in solving Example 5.5.

It should be noted that the static pressure ahead of the shock wave, i.e., p_1, must be measured. If the flow is very nearly parallel to a plane wall, there will be essentially no static pressure changes normal to the flow direction and p_1 can then be found using a static hole in the wall as indicated in Figure 5.21.

However, it has also been found that a pitot–static tube can be used in supersonic flow since the shock wave interacts with the expansion waves (see later) that occur near the nose of the pitot–static tube causing the shock to decay rapidly to a Mach wave and the pressure downstream of the vicinity of the nose of the pitot tube is thus essentially equal to p_1 again as indicated in Figure 5.22.

Example 5.6

A pitot–static tube is placed in a supersonic air flow. The static pressure and temperature in the flow are 45 kPa and –20°C, respectively. The difference between the pitot and static pressures is measured and found to be 350 kPa. Find the Mach number and the air velocity.

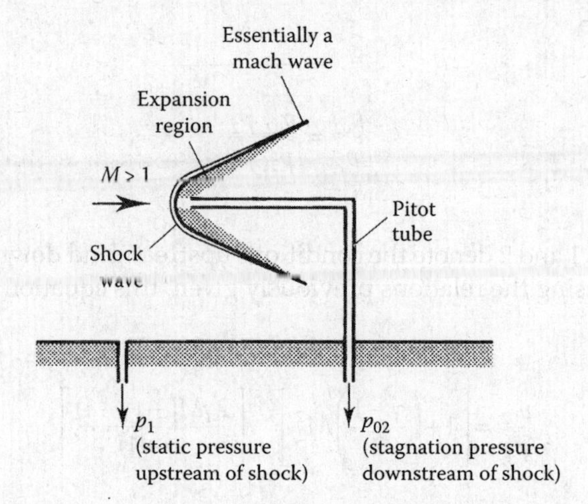

FIGURE 5.21
Use of a wall static pressure tap with a pitot tube in a supersonic flow.

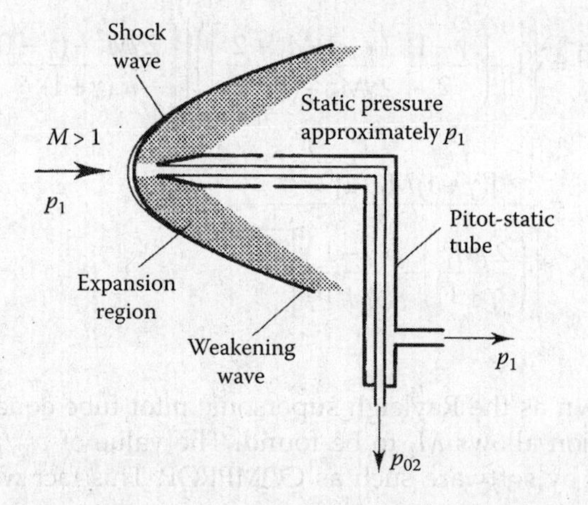

FIGURE 5.22
Pitot–static tube in a supersonic flow.

Solution

Because

$$\frac{p_{02}}{p_1} = \frac{(p_{02} - p_1) + p_1}{p_1}$$

it follows using the given information that

$$p_{02}/p_1 = (350 + 45)/45 = 8.778$$

However, for $p_{02}/p_1 = 8.778$, normal shock relations or tables or software give

$$M_1 = 2.536$$

Therefore,

$$V_1 = M_1 \times a_1 = 2.536 \times \sqrt{1.4 \times 287 \times 253} = 809 \text{ m/s}$$

Hence, the Mach number and velocity in the flow are 2.536 and 809 m/s, respectively.

Moving Normal Shock Waves

In the above discussion of normal shock waves, the coordinate system was so chosen that the shock wave was at rest. In many cases, however, it is necessary to derive results for the case where the shock wave is moving relative to the coordinate system used to describe the flow geometry. Consider the case where the gas ahead of the shock wave is stationary with respect to the coordinate system chosen and where the normal shock wave is moving into this stationary gas inducing a velocity in the direction of shock motion as indicated in Figure 5.23.

Such moving shock waves occur, for example, in the inlet and exhaust systems of internal combustion engines, in air compressors, as the result of explosions, and in pipelines following the opening or closing of a valve.

The required results for a moving normal shock wave can be obtained from those that were derived above for a stationary normal shock wave by noting that the velocities relative to a coordinate system fixed to the shock wave are as indicated in Figure 5.24.

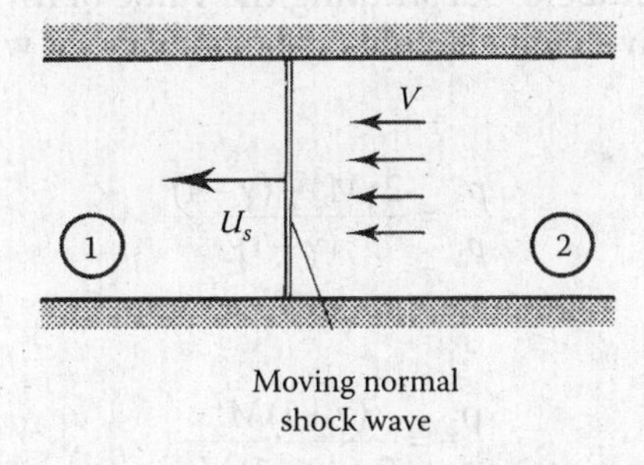

FIGURE 5.23
Moving normal shock wave.

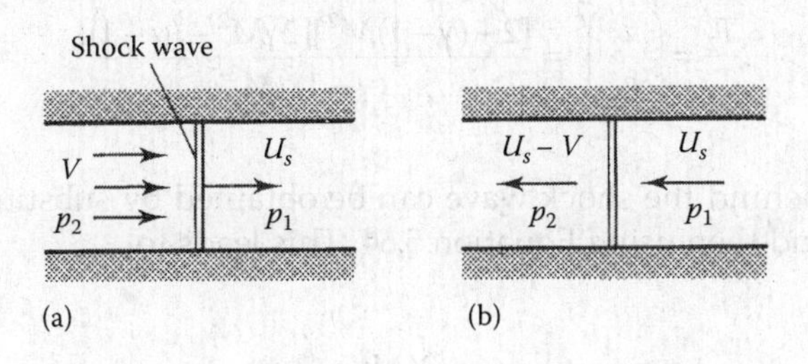

FIGURE 5.24
Relation between flow relative to walls (a) and to moving normal shock wave (b).

Hence, it follows that

$$V_1 = U_s, \qquad V_2 = U_s - V \tag{5.58}$$

Since the direction of the flow is obvious, only the magnitudes of the velocities will be considered here. The Mach numbers upstream and downstream of the shock wave relative to the shock wave are given by

$$M_1 = \frac{U_s}{a_1} = M_s \tag{5.59}$$

and

$$M_2 = \frac{U_s}{a_2} - \frac{V}{a_2} = \frac{U_s}{a_1}\frac{a_1}{a_2} - \frac{V}{a_2} = M_s\frac{a_1}{a_2} - M_2' \tag{5.60}$$

where

$$M_s = \frac{U_s}{a_1} \quad \text{and} \quad M_2' = \frac{V}{a_2} \tag{5.61}$$

M_s is the "shock Mach number." Substituting the value of M_1 for the moving shock into the equations previously given for a stationary normal shock wave then gives

$$\frac{p_2}{p_1} = \frac{2\gamma M_s^2 - (\gamma - 1)}{(\gamma + 1)} \tag{5.62}$$

$$\frac{\rho_2}{\rho_1} = \frac{(\gamma + 1)M_s^2}{2 + (\gamma - 1)M_s^2} \tag{5.63}$$

$$\frac{T_2}{T_1} = \left(\frac{a_2}{a_1}\right)^2 = \frac{[2 + (\gamma - 1)M_s^2][2\gamma M_s^2 - (\gamma - 1)]}{(\gamma + 1)^2 M_s^2} \tag{5.64}$$

The gas velocity behind the shock wave can be obtained by substituting Equation 5.60 into Equation 5.38 and then using Equation 5.64. This leads to

$$M_2' = \frac{2(M_s^2 - 1)}{[2\gamma M_s^2 - (\gamma - 1)]^{0.5}[2 + (\gamma - 1)M_s^2]^{0.5}} \tag{5.65}$$

It will be noted that for an infinitely strong moving normal shock wave, i.e., for $M_s \to \infty$, the above equation shows that there is a limiting value for the Mach number downstream of the shock wave, M_2', which is given by

$$M_2' \to \sqrt{\frac{2}{\gamma(\gamma-1)}} \tag{5.66}$$

Thus, for example, for air, a moving normal shock wave, no matter how strong, cannot generate a flow that has a Mach number that is greater than 1.89.

The actual velocity behind a moving normal shock wave, V, is given by

$$\begin{aligned} V = M_2' a_2 &= M_2' \left(\frac{a_2}{a_1}\right) a_1 \\ &= \frac{2(M_s^2-1)a_1[2+(\gamma-1)M_s^2]^{0.5}[2\gamma M_s^2-(\gamma-1)]^{0.5}}{[2\gamma M_s^2-(\gamma-1)]^{0.5}[2+(\gamma-1)M_s^2]^{0.5}(\gamma+1)M_s} \\ &= \frac{2(M_s^2-1)}{(\gamma+1)M_s} a_1 \end{aligned} \tag{5.67}$$

Normal shock wave software (COMPROP) or tables can be used to evaluate the properties of a moving normal shock wave. To do this, M_1 is set equal to M_s and the tables or software are then used to find the pressure, density, and temperature ratios across the moving shock wave. Further, since

$$M_s' = M_s \frac{a_1}{a_2} - M_2$$

and since M_2 is given by the normal shock tables or software, M_2' can be found. Also, since

$$\frac{V}{a_1} = M_2' \frac{a_2}{a_1} = M_s - M_2 \left(\frac{a_2}{a_1}\right)$$

V can also be deduced using normal shock tables or the software.

Example 5.7

A shock wave across which the pressure ratio is 1.25 is moving into still air at a pressure of 100 kPa and a temperature of 15°C. Find the velocity, pressure, and temperature of the air behind the shock wave.

Solution

The flow situation here being considered is shown in Figure E5.7.

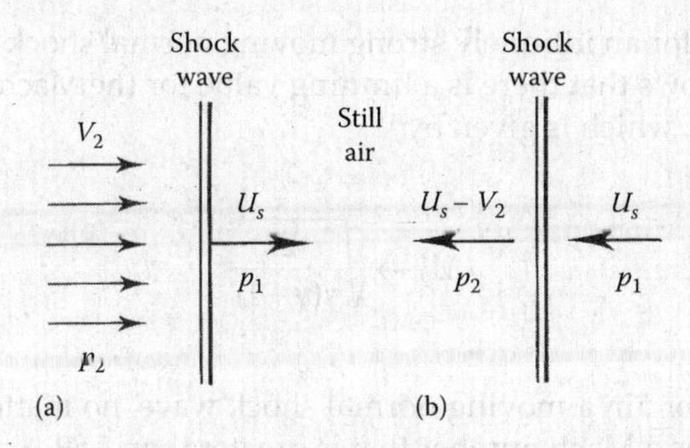

FIGURE E5.7
(a) Flow relative to undisturbed air. (b) Flow relative to wave.

Here

$$p_1 = 100 \text{ kPa}, \quad T_1 = 15°\text{C}, \quad \frac{p_2}{p_1} = 1.25$$

For the given $p_2/p_1 = 1.25$, M_1, M_2, T_2/T_1 are obtained using the normal shock relations or tables or the software. This gives

$$M_1 = 1.102, \quad M_2 = 0.9103, \quad T_2/T_1 = 1.0662$$

Thus, the pressure downstream of the shock is given by

$$p_2 = 1.25 \times 100 = 125 \text{ kPa}$$

whereas the temperature downstream of the shock is given by

$$T_2 = 1.0662 \times 288 = 307 \text{ K}$$

Now, considering the flow relative to the wave, the following apply

$$M_1 = \frac{U_s}{a_1},$$

$$M_2 = \frac{U_s - V}{a_2}$$

Therefore, the velocity downstream of the wave is given by

$$\begin{aligned} V = U_s - M_2 \times a_2 &= M_1 \times a_1 - M_2 \times a_2 \\ &= 1.102 \times \sqrt{1.4 \times 287 \times 288} - 0.9103 \\ &\quad \times \sqrt{1.4 \times 287 \times 1.0662 \times 288} \\ &= 374.87 - 319.71 = 55.2 \text{ m/s} \end{aligned}$$

Therefore, the velocity, pressure, and temperature behind the shock wave are 55.2 m/s, 125 kPa, and 307 K (24°C), respectively.

Example 5.8

A normal shock wave across which the pressure ratio is 1.17 moves down a duct into still air at a pressure of 105 kPa and a temperature of 30°C. Find the pressure, temperature, and velocity of the air behind the shock wave. This shock wave passes over a small circular cylinder as shown in Figure E5.8a. Assuming that the shock is unaffected by the small cylinder, find the pressure acting at the stagnation point on the cylinder after the shock has passed over it.

Solution

The flow relative to the shock wave as shown in Figure E5.8b is considered.

Since the pressure ratio across the shock wave is 1.17, normal shock relations or tables or the COMPROP software give

$$M_1 = 1.07, \quad M_2 = 0.936, \quad \frac{T_2}{T_1} = 1.046$$

Using these pressure and temperature ratio values then gives

$$p_2 = 1.17 \times 105 = 122.9 \text{ kPa}, \quad T_2 = 1.046 \times 303 = 316.9 \text{ K}$$

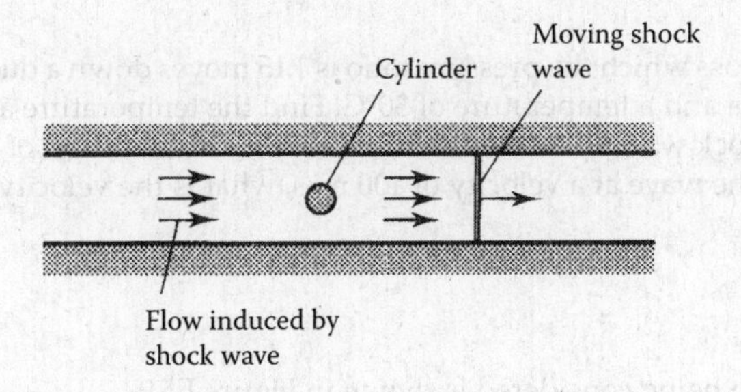

FIGURE E5.8a
Flow situation considered.

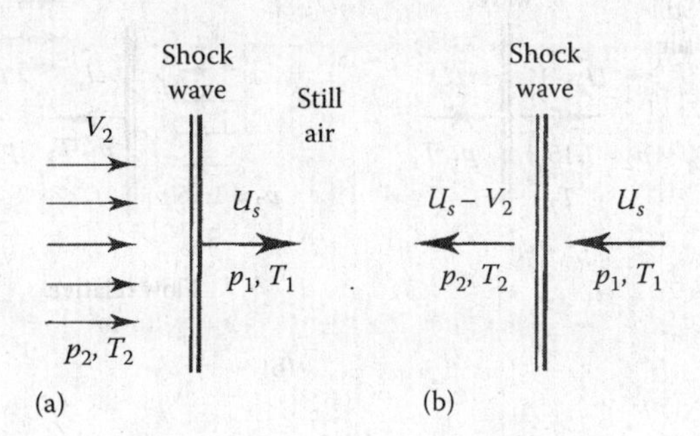

FIGURE E5.8b
(a) Flow relative to undisturbed air. (b) Flow relative to wave.

It is also noted that the speed of sound in the air ahead of the wave is given by

$$a_1 = \sqrt{\gamma R T_1} = \sqrt{1.4 \times 287 \times 303} = 348.9 \text{ m/s}$$

and the speed of sound behind the wave is given by

$$a_2 = \left[\frac{T_2}{T_1}\right]^{0.5} a_1 = 1.046^{0.5} \times 348.9 = 356.8 \text{ m/s}$$

Now,

$$M_2 = \frac{U_s - V_2}{a_2}, \quad \text{i.e.,} \quad V_2 = U_s - M_2 a_2 = M_s a_1 - M_2 a_2$$

However, $M_s = M_1$ so the above equation gives

$$V_2 = 1.07 \times 348.9 - 0.936 \times 356.8 = 39.38 \text{ m/s}$$

The cylinder is therefore exposed to a flow with a velocity of 39.37 m/s at a temperature of 316.9 K and a pressure of 122.9 kPa. The Mach number in this flow is equal to 39.37/356.8 = 0.11. Now, isentropic relations or tables or software give, for Mach 0.11, $p_0/p = 1.0085$. Therefore, the pressure at the stagnation point on the cylinder is 1.0085 × 122.85 = 123.9 kPa.

Example 5.9

A shock wave across which the pressure ratio is 1.15 moves down a duct into still air at a pressure of 50 kPa and a temperature of 30°C. Find the temperature and velocity of the air behind the shock wave. If instead of being at rest, the air ahead of the shock wave is moving toward the wave at a velocity of 100 m/s, what is the velocity of the air behind the shock wave?

Solution

The flow situation being considered is shown in Figure E5.9.

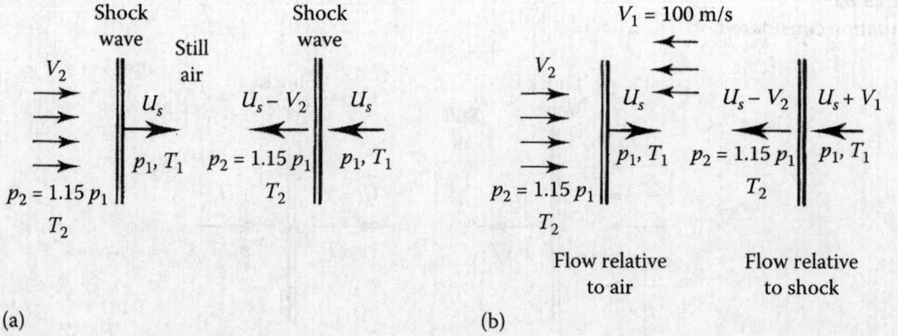

FIGURE E5.9
Flow relative to undisturbed air and flow relative to shock wave for (a) case where undisturbed air is at rest (b) case where undisturbed air has a velocity of 100 m/s.

For the normal shock wave moving into still air which is shown in Figure E5.9, the following are given

$$p_2/p_1 = 1.15, \quad p_1 = 50 \text{ kPa}, \quad T_1 = 30°\text{C}$$

Considering the flow relative to the shock, for $p_2/p_1 = 1.15$, normal shock relations or tables or software give for $\gamma = 1.4$

$$M_1 = 1.062, \quad M_2 = 0.943, \quad T_2/T_2 = 1.041$$

Therefore, using the known initial conditions

$$T_2 = (273 + 30) \times 1.041 = 315.4 \text{ K } (42.4°\text{C}) \quad \text{and} \quad p_2 = 1.15 \times 50 = 57.5 \text{ kPa}$$

Also, since

$$M_1 = U_s/a_1 \quad \text{and} \quad M_2 = (U_s - V)/a_2$$

it follows that

$$V = M_1 a_1 - M_2 a_2 = 1.062 \times \sqrt{1.4 \times 287 \times 303} - 0.943 \times \sqrt{1.4 \times 287 \times 315.4} = 35.1 \text{ m/s}$$

Hence, when the shock is moving into still air, the temperature and velocity behind the shock are 42.4°C and 35.1 m/s, respectively.

Next, consider the case where the air ahead of the shock is not at rest, this situation also being shown in Figure E5.9. Because the pressure ratio across the shock wave is still 1.15, the following still apply

$$M_1 = 1.062, \quad M_2 = 0.943, \quad T_2/T_2 = 1.041$$

So, again

$$T_2 = (273 + 30) \times 1.041 = 315.4 \text{ K } (42.4°\text{C})$$

However, since the flow relative to the wave is being considered, it follows that

$$M_1 = \frac{(U_s + V_1)}{a_1} \quad \text{and} \quad M_2 = \frac{(U_s - V_2)}{a_2}$$

Using these relations then gives

$$V_2 = U_s - M_2 \times a_2 = (M_1 \times a_1 - V_1) - M_2 a_2 = (1.062 \times \sqrt{1.4 \times 287 \times 303} - 100) - 0.943 \times \sqrt{1.4 \times 287 \times 315.4} = -64.9 \text{ m/s}$$

Therefore, the velocity behind the shock is −64.9 m/s. The negative sign indicates that the velocity behind the wave is in the opposite direction to that in which the wave is moving. Of course, the result for the second part of the question could have been directly deduced from that for the first part of the question by noting that, since the pressure ratio across the shock is the same in both cases, if the flow changes relative to the upstream flow is considered, the same situation as dealt with in the first part of the question is obtained. Therefore, the velocity behind the wave relative to the upstream flow will be 35.1 m/s. Therefore, the velocity behind the wave relative to the walls of the duct will be 35.1 − 100 = −64.9 m/s, as obtained before.

Brief consideration will now be given to the "reflection" of a moving shock wave off the closed end of a duct. Consider a moving normal shock wave propagating into a gas at rest in a duct. The shock, as discussed above, induces a flow behind it in the direction of shock motion. If the end of the duct is closed, however, there can be no flow out of the duct, i.e., the velocity of the gas in contact with the closed end must always be zero. Therefore, a normal shock wave must be "reflected" off the closed end, the strength of this "reflected" shock wave being just sufficient to reduce the velocity to zero. This is illustrated in Figure 5.25.

Consider a set of coordinates attached to the reflected normal shock wave. The gas velocities relative to this reflected shock wave are therefore as shown in Figure 5.26.

Hence, since

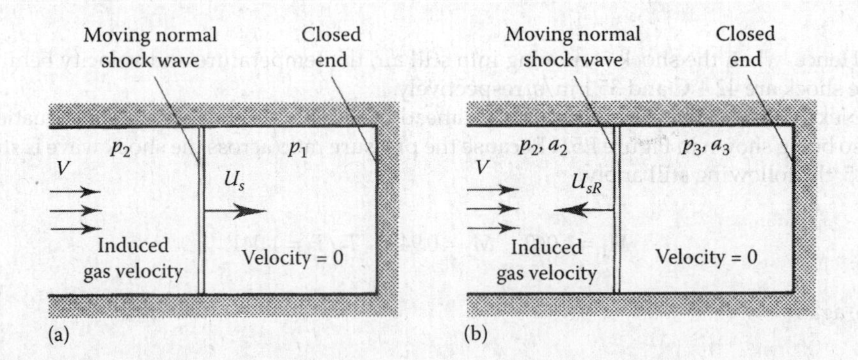

FIGURE 5.25
Reflection of a moving normal shock wave from the closed end of a duct. (a) Flow relative to duct, (b) Flow relative to shock wave.

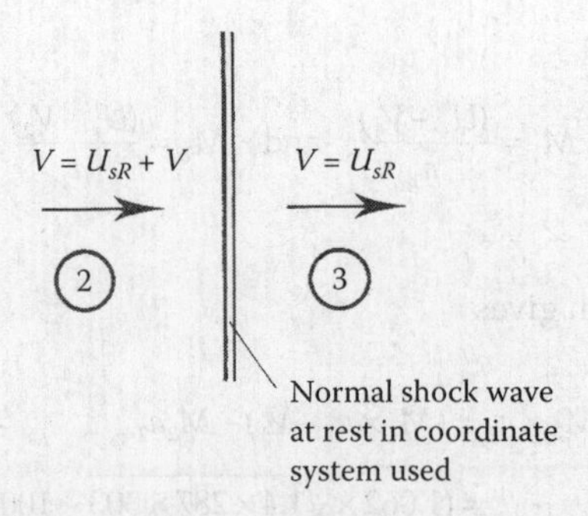

FIGURE 5.26
Gas velocities relative to reflected moving normal shock wave.

$$M_{R1} = \frac{(U_{sR} + V)}{a_2} = M_{sR} + M_2$$

and

$$M_{R2} = \frac{U_{sR}}{a_3} = \left(\frac{U_{sR}}{a_2}\right)\left(\frac{a_2}{a_3}\right) = M_{sR}\left(\frac{a_2}{a_3}\right)$$

These equations can be used in conjunction with the normal shock relations previously given or with the shock tables or with the software COMPROP to find the properties of the reflected shock. The procedure is illustrated in the following example.

Example 5.10

A normal shock wave across which the pressure ratio is 1.45 moves down a duct into still air at a pressure of 100 kPa and a temperature of 20°C. Find the pressure, temperature, and velocity of the air behind the shock wave. If the end of the duct is closed, find the pressure acting on the end of the duct after the shock is "reflected" from it.

Solution

The flow situation before and after the shock reflection is shown in Figure E5.10.

First, consider the shock wave before the reflection. For this wave,

$$p_2/p_1 = 1.45, \quad p_1 = 100 \text{ kPa}, \quad T_1 = 20°\text{C} = 293 \text{ K}$$

Consider the flow relative to the wave as shown in Figure E5.10. For $p_2/p_1 = 1.45$, normal shock relations or tables or software give

$$M_1 = 1.1772, \quad M_2 = 0.8567, \quad T_2/T_1 = 1.1137$$

Therefore,

$$p_2 = 1.45 \times 100 = 145 \text{ kPa} \quad \text{and} \quad T_2 = 1.1137 \times 293 = 326.3 \text{ K } (= 53.3°\text{C})$$

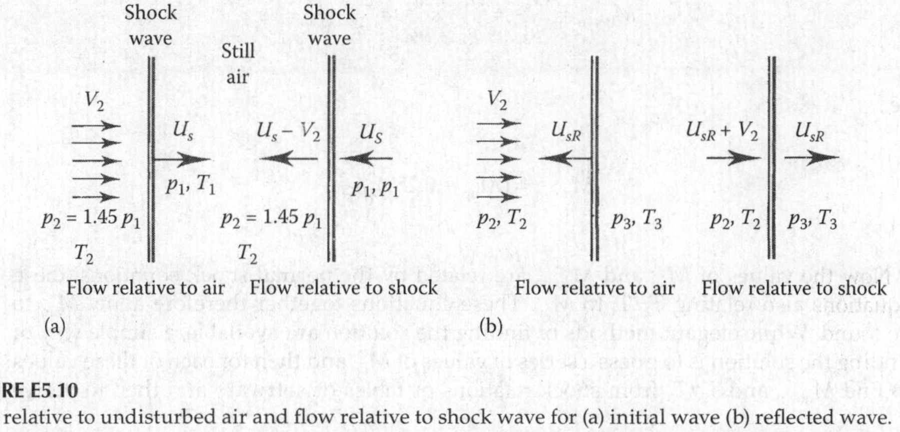

FIGURE E5.10
Flow relative to undisturbed air and flow relative to shock wave for (a) initial wave (b) reflected wave.

Also, since

$$M_1 = U_s/a_1 \quad \text{and} \quad M_2 = (U_s - V_2)/a_2$$

It follows that

$$V_2 = M_1 a_1 - M_2 a_2$$

Hence,

$$V_2 = 1.1772 \times \sqrt{1.4 \times 287 \times 293} - 0.8567 \times \sqrt{1.4 \times 287 \times 326.3} = 93.8 \text{ m/s}$$

Therefore, the pressure, temperature, and velocity behind the initial shock wave are 145 kPa, 53.3°C, and 93.8 m/s.

Next consider the wave that is "reflected" off the closed end. The strength of this wave must be such that it brings the flow to rest. Hence, the Mach numbers of the air flow upstream and downstream of the reflected wave relative to this wave are

$$M_{\text{up}} = \frac{V_2 + U_{sR}}{a_2} \quad \text{and} \quad M_{\text{down}} = \frac{U_{sR}}{a_3}$$

where U_{sR} is the velocity of the reflected wave and a_3 is the speed of sound in the flow downstream of the reflected wave.

However,

$$a_2 = \sqrt{\gamma R T_2} = \sqrt{1.4 \times 287 \times 326.3} = 362.1 \text{ m/s}$$

Hence,

$$M_{\text{up}} = \frac{93.8}{362.1} + \frac{U_{sR}}{a_2} = 0.259 + \frac{U_{sR}}{a_2}$$

and

$$M_{\text{down}} = \frac{U_{sR}}{a_3} = \frac{U_{sR}}{a_2} \times \frac{a_2}{a_3} = \frac{U_{sR}}{a_2}\sqrt{\frac{T_2}{T_3}}$$

i.e.,

$$M_{\text{down}} = (M_{\text{up}} - 0.259)\sqrt{\frac{T_2}{T_3}}$$

Now the values of M_{up} and M_{down} are related by the normal shock equations, these equations also relating T_2/T_3 to M_{up}. These equations together therefore allow M_{up} to be found. While elegant methods of finding the solution are available, a simple way of finding the solution is to guess a series of values of M_{up} and then for each of these values to find M_{down} and T_2/T_3 from shock relations or tables or software and then to derive

TABLE E5.10

Iterative Results Used in Obtaining Solution

M_{up} (Guessed)	T_3/T_2 (Shock Relations)	M_{down} (Shock Relations)	$(M_{up}-0.259)\sqrt{T_2/T_3}$ (Calculated)
1.000	1.000	1.000	0.741
1.100	1.065	0.912	0.802
1.200	1.128	0.842	0.886
1.500	1.320	0.701	1.080
1.250	1.159	0.813	0.920

the value of $M_{down} = (M_{up} - 0.259)\sqrt{T_2/T_3}$. The correct value of M_{up} is that which has the value of M_{down} as given directly by the normal shock relations equal to that given by the above equation. This correct value can be deduced from the results for various M_{up}. The results for various values of M_{up} are shown in Table E5.10.

From these results, it can be deduced that when $M_{up} = 1.17$, the values of M_{down} given by the shock relations and by the above equation are the same (approximately 0.863). Now for an upstream Mach number of 1.17, normal shock relations or tables or software give $p_3/p_2 = 1.4304$. Hence, the pressure behind the reflected shock, i.e., the pressure acting on the closed end is given by

$$p_3 = 1.4304 \times 145 = 207 \text{ kPa}$$

High pressures can therefore be generated when a moving shock wave is reflected off a closed end of a duct.

Lastly, consider what happens if a gas is flowing out of a duct at a steady rate when the end of the duct is suddenly closed. Since the velocity of the gas in contact with the closed end must again be zero, a shock wave is generated that moves into the moving gas bringing it

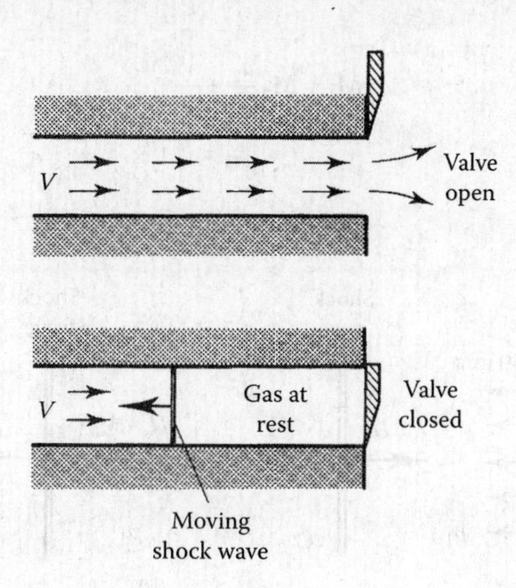

FIGURE 5.27
Moving normal shock wave generated by closure of valve.

to rest, i.e., the strength of the shock wave must be such that the velocity is reduced to zero behind it. This is illustrated in Figure 5.27.

A shock wave generated in this way can be analyzed using the same procedure as used to analyze a normal shock wave reflected from a closed end of a duct. This is illustrated by the following example.

Example 5.11

Air is flowing out of a duct at a velocity of 250 m/s with a temperature of 0°C and a pressure of 70 kPa. A valve at the end of the duct is suddenly closed. Find the pressure acting on the valve immediately after the valve closure.

Solution

The flow situation considered in this example is shown in Figure E5.11. The flow relative to the air behind the shock is shown in Figure E5.11a and the flow relative to the shock is shown in Figure E5.11b.

Consider the flow relative to the shock wave. Because the wave must bring the air to rest, it will be seen that if U_s is the velocity of the shock wave,

$$M_1 = \frac{U_s + V}{a_1}$$

and

$$M_2 = \frac{U_s}{a_2} = \frac{U_s}{a_1} \times \frac{a_1}{a_2}$$

Substituting this into the equation for M_1 gives

$$M_1 = M_2 \frac{a_2}{a_1} + \frac{V}{a_1}$$

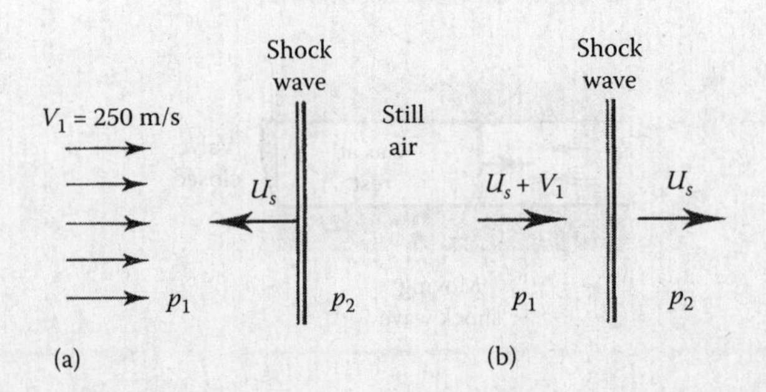

FIGURE E5.11
(a) Flow relative to undisturbed air. (b) Flow relative to wave.

TABLE E5.11

Iterative Results Used in Obtaining Solution

M_1	M_2	a_2/a_1	RHS
1.0	1.0	1.0	1.755
1.4	0.740	1.120	1.584
1.5	0.701	1.149	1.56
1.6	0.668	1.178	1.54

However, the speed of sound ahead of the wave is

$$a_1 = \sqrt{\gamma RT} = \sqrt{1.4 \times 287 \times 273} = 331.2 \text{ m/s}$$

and $V = 250$ m/s so the above equation gives

$$M_1 = M_2 \frac{a_2}{a_1} + \frac{250}{331.2} = M_2 \frac{a_2}{a_1} + 0.7548$$

Because the normal shock relations determine M_2 and a_2/a_1 as functions of M_1, the above equation together with the normal shock relations can be used to determine M_1. Although there are more elegant methods of obtaining the solution, the brute force approach is to choose a series of values of M_1 and then use normal shock relations to find the right-hand side, i.e., to find the value of

$$M_2 \frac{a_2}{a_1} + 0.7548$$

for each of these values of M_1. The value of M_1 that makes the right-hand side equal to M_1 can then be deduced from these results. A set of results is shown in Table E5.11, the values of M_2 and a_2/a_1 being deduced for $\gamma = 1.4$ from the normal shock relations or from normal shock tables or using the software. "RHS" is the right-hand side of the equation derived earlier and given by the above equation.

From these results, it will be seen that $M_1 = 1.55$. Normal shock relations or tables or software then gives $p_2/p_1 = 2.636$, so $p_2 = 2.636 \times 70 = 184.5$ kPa.

Concluding Remarks

A normal shock wave is an extremely thin region at right angles to the flow across which large changes in the flow variables can occur. Although the flow within the shock wave is complex, it was shown that expressions for the overall changes across the shock can be relatively easily derived. It was also shown that entropy considerations indicate that only compressive shock waves, i.e., shock waves across which the pressure increases, can occur and that the flow ahead of the shock must be supersonic. It was also shown that the flow downstream of a normal shock wave is always subsonic. The analysis of normal shock waves that are moving through a gas was also discussed.

PROBLEMS

1. A normal shock wave occurs in an air flow at a point where the velocity is 680 m/s, the static pressure is 80 kPa, and the static temperature is 60°C. Find the velocity, static pressure, and static temperature downstream of the shock. Also, find the stagnation temperature and stagnation pressure upstream and downstream of the shock.
2. The pressure ratio across a normal shock wave that occurs in air is 1.25. Ahead of the shock wave, the pressure is 100 kPa and the temperature is 15°C. Find the velocity, pressure, and temperature of the air behind the shock wave.
3. A perfect gas flows through a stationary normal shock. The gas velocity decreases from 480 to 160 m/s through the shock. If the pressure and the density upstream of the shock are 62 kPa and 1.5 kg/m^3, find the pressure and density downstream of the shock and the specific heat ratio of the gas.
4. A normal shock wave occurs in air at a point where the velocity is 600 m/s and the stagnation temperature and pressure are 200°C and 600 kPa, respectively. Find the Mach numbers, pressures, and temperatures upstream and downstream of the shock wave.
5. Show that the downstream Mach number of a normal shock approaches a minimum value as the upstream Mach number increases toward infinity. What is this minimum Mach number for a gas with a specific heat ratio of 1.67?
6. Air is expanded isentropically from a reservoir in which the pressure is 1000 kPa to a pressure of 150 kPa. A normal shock occurs at this point in flow. Find the pressure downstream of the shock wave.
7. Air is expanded isentropically from a reservoir in which the pressure and temperature are 150 psia and 60°F to a static pressure of 20 psia. A normal shock occurs at this point in the flow. Find the static pressure, static temperature, and the air velocity behind the shock.
8. The exhaust gases from a rocket engine have a molar mass of 14. They can be assumed to behave as a perfect gas with a specific heat ratio of 1.25. These gases are accelerated through a convergent–divergent nozzle. A normal shock wave occurs in the nozzle at a point in the flow where the Mach number is 2. Find the pressure, temperature, density, and stagnation pressure ratio across this shock wave.
9. Air is expanded isentropically from a reservoir in which the pressure is 1000 kPa and the temperature is 30°C until the pressure has dropped to 25 kPa. A normal shock wave occurs at this point. Find the static pressure, the static temperature, the air velocity, and the stagnation pressure after the shock wave.
10. A gas with a molar mass of 4 and a specific heat ratio of 1.67 is expanded from a large reservoir in which the pressure and temperature are 600 kPa and 35°C, respectively, through a nozzle system until the Mach number is 1.5. A normal shock wave then occurs in the flow. Find the pressure and velocity behind the shock wave.
11. Air is expanded from a large chamber through a variable area duct. The pressure and temperature in the large chamber are 115 psia and 100°F, respectively. At some point in the flow where the Mach number is 2.5, a normal shock wave occurs. Find the pressure, temperature, stagnation pressure, and velocity behind the shock wave.

12. A normal shock wave occurs in an air flow at a point where the velocity is 750 m/s, the pressure is 50 kPa, and the temperature is 10°C. Find the velocity, pressure, and static temperature downstream of the shock wave.
13. Air is isentropically expanded from a large chamber in which the pressure is 10,000 kPa and the temperature is 50°C until the Mach number reaches a value of 2. A normal shock wave then occurs in the flow. Following the shock wave, the air is isentropically decelerated until the velocity is again essentially zero. Find the pressure and temperature that then exist.
14. Air is expanded from a large reservoir in which the pressure and temperature are 500 kPa and 35°C through a variable area duct. A normal shock occurs at a point in the duct where the Mach number is 2.5. Find the pressure and temperature in the flow just downstream of the shock wave. Downstream of the shock wave, the flow is brought to rest in another large reservoir. Find the pressure and temperature in this reservoir. Assume that the flow is one-dimensional and isentropic everywhere except through the shock wave.
15. Air is expanded from a reservoir in which the pressure and temperature are maintained at 1000 kPa and 30°C. At a point in the flow at which the static pressure is 150 kPa, a normal shock wave occurs. Find the static pressure, the static temperature, and the air velocity behind the shock wave. Assume the flow to be isentropic everywhere except through the shock wave.
16. Air is expanded through a convergent–divergent nozzle from a large chamber in which the pressure and temperature are 200 kPa and 310 K, respectively. A normal shock wave occurs at a point in the nozzle where the Mach number is 2.5. The air is then brought to rest in a second large chamber. Find the pressure and temperature in this second chamber. Clearly state the assumptions you have made in arriving at the solution.
17. Air at a temperature of 10°C and a pressure of 50 kPa flows over a blunt nosed body at a velocity of 500 m/s. Estimate the pressure acting on the front of the body.
18. A pitot–static tube is placed in a supersonic air flow at Mach 2.0. The static pressure and static temperature in the flow are 101 kPa and 30°C, respectively. Estimate the difference between the pitot and static pressures.
19. A pitot–static tube is placed in a supersonic flow in which the static pressure and temperature are 60 kPa and –20°C, respectively. The difference between the pitot and static pressures is measured and found to be 449 kPa. Find the Mach number and velocity in the flow. Discuss the assumptions used in deriving the answers.
20. A pitot–static tube is placed in an air flow in which the Mach number is 1.7. The static pressure in the flow is 55 kPa and the static temperature is –5°C. What will be the measured difference between the pitot and the static pressures?
21. A pitot–static tube is placed in a supersonic flow in which the static temperature is 0°C. Measurements indicate that the static pressure is 80 kPa and that the ratio of the pitot to the static pressure is 4.1. Find the Mach number and the velocity in the flow.
22. A pitot tube is placed in a stream of carbon dioxide in which the pressure is 60 kPa and the Mach number is 3.0. What will the pitot pressure be?
23. A thermocouple placed in the mouth of a pitot tube can be used to measure the stagnation temperature of a flow. Such an arrangement placed in an air flow gives the stagnation pressure as 180 kPa, the static pressure as 55 kPa, and the stagnation

temperature as 95°C. Estimate the velocity of the stream assuming that the flow is supersonic.

24. A shock wave propagates down a constant area duct into stagnant air at a pressure of 101.3 kPa and a temperature of 25°C. If the pressure ratio across the shock wave is 3, find the shock speed and the velocity of the air downstream of the shock.
25. A normal shock wave propagates down a constant area tube containing stagnant air at a temperature of 300 K. Find the velocity of the shock wave if the air behind the wave is accelerated to Mach number 1.2.
26. A shock wave is moving down a constant area duct containing air. The air ahead of the shock wave is at rest and at a pressure and temperature of 100 kPa and 20°C, respectively. If the pressure ratio across the shock wave is 2.5, find the velocity, pressure, and the temperature in the air behind the shock wave.
27. A normal shock wave propagates at a speed of 2600 m/s down a pipe that is filled with hydrogen. The hydrogen is at rest and at a pressure and temperature of 101.3 kPa and 25°C, respectively, upstream of the wave. Assuming hydrogen to behave as a perfect gas with constant specific heats, find the temperature, pressure, and velocity downstream of the wave.
28. A normal shock wave across which the pressure ratio is 1.2 moves down a duct into still air at a pressure of 100 kPa and a temperature of 20°C. Find the pressure, temperature, and velocity of the air behind the shock wave. This shock wave passes over a small circular cylinder. Assuming that the shock is unaffected by the small cylinder, find the pressure acting at the stagnation point on the cylinder after the shock has passed over it.
29. As a result of a rapid chemical reaction, a normal shock wave is generated, which propagates down a duct in which there is air at a pressure of 100 kPa and a temperature of 30°C. The pressure behind this shock wave is 130 kPa. Half a second after the generation of this shock wave, a second normal shock wave is generated by another chemical reaction. This second shock wave follows the first one down the duct, the pressure behind this second wave being 190 kPa. Find the velocity of the air and the temperature behind the second shock wave. Also, find the distance between the two waves at a time of 0.7 s after the generation of the first shock wave.
30. A normal shock wave across which the pressure ratio is 1.25 is propagating down a duct containing still air at a pressure of 120 kPa and a temperature of 35°C. This shock wave is reflected off the closed end of the duct. Find the pressure and temperature behind the reflected shock wave.
31. A normal shock wave is propagating down a duct in which p = 110 kPa and T = 30°C. The pressure ratio across the shock is 1.8. Find the velocity of the shock wave and the air velocity behind the shock. This moving shock strikes a closed end to the duct. Find the pressure on the closed end after the shock reflection.
32. A normal shock wave is moving down a duct into still air in which the pressure is 100 kPa and the temperature is 20°C. The pressure ratio across the shock is 1.8. Find the velocity of the shock and the velocity of the air behind the shock. If this shock strikes the closed end of the duct, find the pressure on this closed end after the shock reflection.
33. A shock across which the pressure ratio is 1.18 moves down a duct into still air at a pressure of 100 kPa and a temperature of 30°C. Find the temperature and velocity

of the air behind the shock wave. If instead of being at rest, the air ahead of the shock wave is moving toward the wave at a velocity of 75 m/s, what would be the velocity of the air behind the shock wave?

34. Air at a pressure of 105 kPa and a temperature of 25°C is flowing out of a duct at a velocity of 250 m/s. A valve at the end of the duct is suddenly closed. Find the pressure acting on the valve.
35. Air is flowing out of a duct at a velocity of 250 m/s. The static temperature and pressure in the flow are 0°C and 70 kPa. A valve at the end of the duct is suddenly closed. Estimate the pressure acting on this valve immediately after it is closed.
36. Air is flowing down a duct at a velocity of 200 m/s. The pressure and the temperature in the flow are 85 kPa and 10°C, respectively. If a valve at the end of the duct is suddenly closed, find the pressure acting on the valve immediately after closure.
37. A piston in a pipe containing stagnant air at a pressure of 101 kPa and a temperature of 25°C is suddenly given a velocity of 100 m/s into the pipe, causing a normal shock wave to propagate through the air down the pipe. Find the velocity of the shock wave and the pressure acting on the piston.
38. Air flows at a velocity of 90 m/s down a 20 cm diameter pipe. The air is at a pressure of 120 kPa and a temperature of 30°C. A valve at the end of the pipe is suddenly closed. This valve is held in place by eight mild steel bolts each with a diameter of 12 mm. Will the bolts hold the valve without yielding? Assume that the pipe is discharging the air to the atmosphere and that the ambient pressure is 101 kPa. It will be necessary to look up the yield strength of the steel to answer this question.
39. A cannon fires a shell that causes a projectile to move down the barrel at a velocity of 740 m/s. What is the speed of the normal shock proceeding down the barrel in front of the projectile if the undisturbed air in the barrel is at 101 kPa and 20°C? How fast would the projectile have to be moving down the barrel if the velocity of the shock ahead of the projectile was 2.5 times the velocity of the projectile?

6

Oblique Shock Waves

Introduction

Attention is now turned from normal shock waves, which are straight and in which the flow before and after the wave is normal to the shock, to oblique shock waves. Such shock waves are, by definition, also straight but they are at an angle to the upstream flow and, in general, they produce a change in flow direction as indicated in Figure 6.1.

The oblique shock relations can be deduced from the normal shock relations by noting that the oblique shock can produce no momentum change parallel to the plane in which it lies. To show this, consider the control volume shown in Figure 6.2. In this figure, the components of the velocity parallel to the wave are L_1 and L_2 while the components normal to the wave are N_1 and N_2 as shown.

Because there are no changes in the flow variables in the direction parallel to the wave there is no net force on the control volume parallel to the wave and there is, consequently, no momentum change parallel to the wave. Because there is no momentum change parallel to the shock, L_1 must equal L_2. Hence, if the coordinate system moving parallel to the wave front at a velocity $L = L_1 = L_2$ is considered, the flow in this coordinate system through the wave is as shown in Figure 6.3.

In this coordinate system, the oblique shock has been reduced to a normal shock and the normal shock relations must therefore apply to the velocity components N_1 and N_2. Further, since the scalar flow properties p, ρ, and T are unaffected by the coordinate system used, the Rankine–Hugoniot relations must apply without any modification to oblique shocks. Thus, all the properties of oblique shocks can be obtained by the modification and manipulation of the normal shock relations provided that the angle of the shock relative to the upstream flow is known. However, it is more instructive and, in some respects, simpler to deduce these oblique shock relations from the fundamental conservation of mass, momentum, and energy laws, using the normal shock relations when a parity is formally established.

Oblique Shock Wave Relations

Consider again flow through a control volume that spans the shock wave and which, without any loss of generality, can be assumed to have unit area parallel to the oblique shock wave. This control volume is shown in Figure 6.4. As shown in this figure, β is defined as the shock wave angle and δ is the change in flow direction induced by the shock wave.

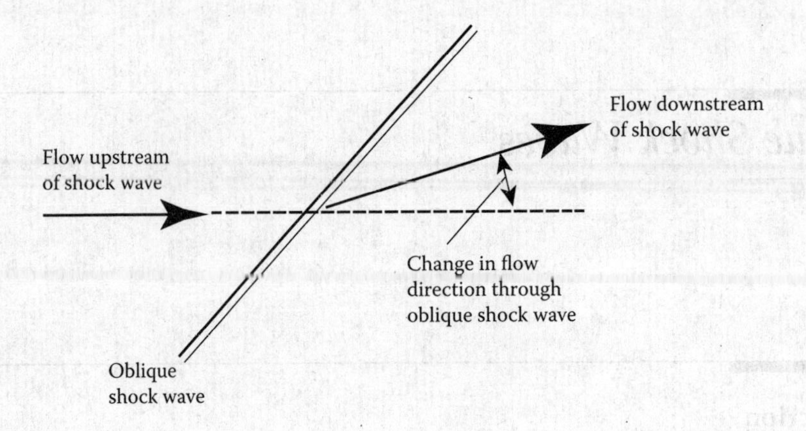

FIGURE 6.1
An oblique shock wave.

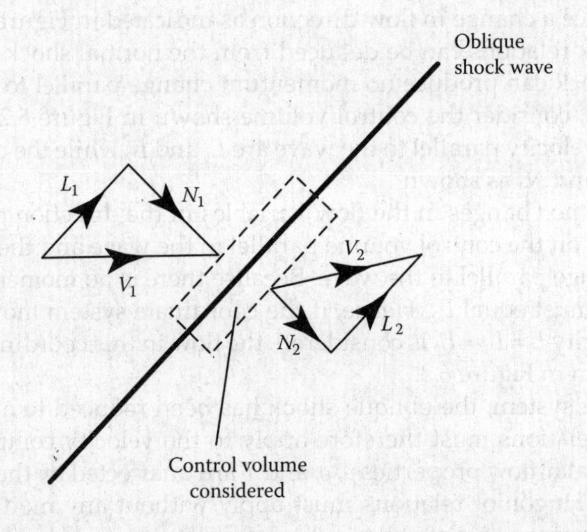

FIGURE 6.2
Control volume considered.

The conservation of mass, momentum, and energy principles are now applied to the control volume shown in Figure 6.4. Since there is no change in velocity parallel to the wave, the L velocity components can carry no net mass into the control volume. Conservation of mass therefore gives

$$\rho_1 N_1 = \rho_2 N_2 \tag{6.1}$$

The conservation of momentum equation is applied in the direction normal to the shock, giving

$$p_1 - p_2 = \rho_2 N_2^2 - \rho_1 N_1^2 \tag{6.2}$$

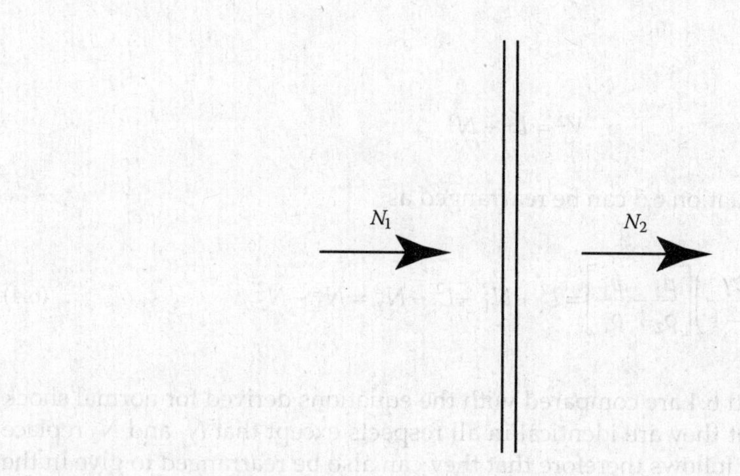

FIGURE 6.3
Flow normal to an oblique shock wave.

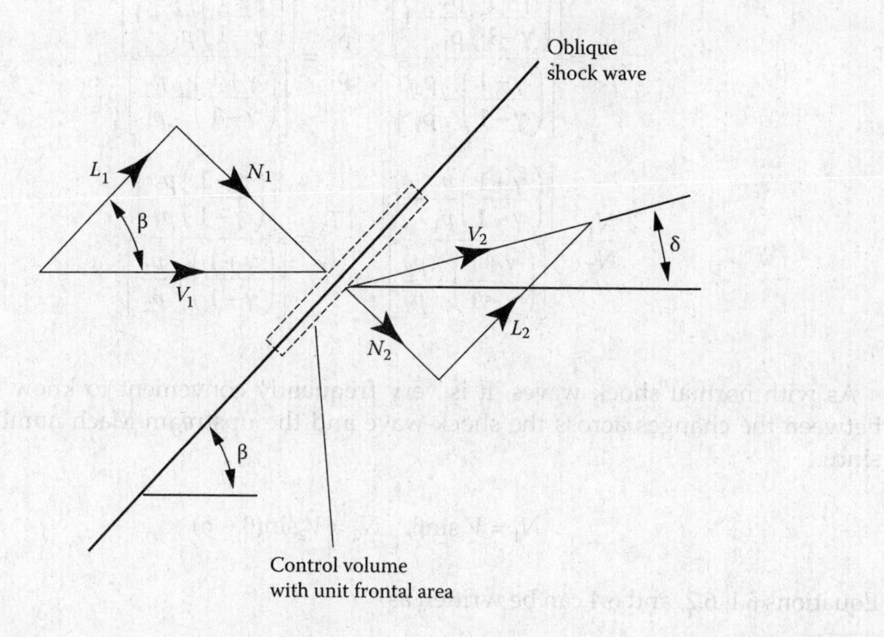

FIGURE 6.4
Control volume used in the analysis of flow through an oblique shock wave.

Because there are no gradients of temperature upstream and downstream of the shock wave, the flow through the control volume must, as in the case of the normal shock, be adiabatic. The energy equation therefore gives

$$\frac{2\gamma}{\gamma-1}\frac{p_1}{\rho_1}+V_1^2=\frac{2\gamma}{\gamma-1}\frac{p_2}{\rho_2}+V_2^2 \tag{6.3}$$

However,

$$V^2 = L^2 + N^2$$

Thus, because $L_1 = L_2$, Equation 6.3 can be rearranged as

$$\left(\frac{2\gamma}{\gamma-1}\right)\left[\frac{p_2}{\rho_2}-\frac{p_1}{\rho_1}\right] = L_1^2 + N_1^2 - L_2^2 - N_2^2 = N_1^2 - N_2^2 \tag{6.4}$$

If Equations 6.1, 6.2, and 6.4 are compared with the equations derived for normal shock waves, it will be seen that they are identical in all respects except that N_1 and N_2 replace V_1 and V_2, respectively. It follows therefore that they can also be rearranged to give in the same way as for normal shocks

$$\frac{p_2}{p_1} = \frac{\left[\left(\frac{\gamma+1}{\gamma-1}\right)\frac{\rho_2}{\rho_1}-1\right]}{\left[\left(\frac{\gamma+1}{\gamma-1}\right)-\frac{\rho_2}{\rho_1}\right]}, \quad \frac{\rho_2}{\rho_1} = \frac{\left[\left(\frac{\gamma+1}{\gamma-1}\right)\frac{p_2}{p_1}+1\right]}{\left[\left(\frac{\gamma+1}{\gamma-1}\right)+\frac{p_2}{p_1}\right]}$$

$$\frac{N_1}{N_2} = \frac{\left[\left(\frac{\gamma+1}{\gamma-1}\right)\frac{p_2}{p_1}+1\right]}{\left[\left(\frac{\gamma+1}{\gamma-1}\right)+\frac{p_2}{p_1}\right]}, \quad \frac{T_2}{T_1} = \frac{\left[\left(\frac{\gamma+1}{\gamma-1}\right)\frac{p_2}{p_1}\right]}{\left[\left(\frac{\gamma+1}{\gamma-1}\right)+\frac{p_1}{p_2}\right]} \tag{6.5}$$

As with normal shock waves, it is very frequently convenient to know the relations between the changes across the shock wave and the upstream Mach number, M_1. Now since

$$N_1 = V_1 \sin\beta, \quad N_2 = V_2 \sin(\beta - \delta) \tag{6.6}$$

Equations 6.1, 6.2, and 6.4 can be written as

$$\rho_1 V_1 \sin\beta = \rho_2 V_2 \sin(\beta - \delta) \tag{6.7}$$

$$p_1 - p_2 = \rho_2 (V_1 \sin\beta)^2 - \rho_1 [V_2 \sin(\beta - \delta)]^2 \tag{6.8}$$

$$\frac{2\gamma}{(\gamma-1)}\left[\frac{p_2}{\rho_2}-\frac{p_1}{\rho_1}\right] = (V_1 \sin\beta)^2 - [V_2 \sin(\beta-\delta)]^2 \tag{6.9}$$

These are, of course, again identical to those used to study normal shocks, except that $V_1 \sin\beta$ occurs in place of V_1 and $V_2 \sin(\beta - \delta)$ occurs in place of V_2. Hence, if in the normal shock relations M_1 is replaced by $M_1 \sin\beta$ and M_2 by $M_2 \sin(\beta - \delta)$, the following relations for oblique shocks are obtained using equations given in the previous chapter:

$$\frac{p_2}{p_1} = \frac{2\gamma M_1^2 \sin^2\beta - (\gamma - 1)}{\gamma + 1} \tag{6.10}$$

$$\frac{\rho_2}{\rho_1} = \frac{(\gamma+1)M_1^2 \sin^2\beta}{2 + (\gamma-1)M_1^2 \sin^2\beta} \tag{6.11}$$

$$\frac{T_2}{T_1} = \frac{[2+(\gamma-1)M_1^2\sin^2\beta][2\gamma M_1^2\sin^2\beta - (\gamma-1)]}{(\gamma+1)^2 M_1^2 \sin^2\beta} \tag{6.12}$$

$$M_2^2 \sin^2(\beta - \delta) = \frac{M_1^2 \sin^2\beta + 2/(\gamma-1)}{2\gamma M_1^2 \sin^2\beta/(\gamma-1) - 1} \tag{6.13}$$

It should be noted that since it was proved using entropy considerations that for normal shocks M_1 had to be greater than 1, i.e., the flow ahead of the shock had to be supersonic it therefore follows from the above discussion that for oblique shocks it is necessary that

$$M_1 \sin\beta \geq 1 \tag{6.14}$$

The minimum value that $\sin\beta$ can have is therefore $1/M_1$, i.e., the minimum shock angle is the Mach angle. When the shock has this angle, Equation 6.10 shows that (p_2/p_1) is equal to 1, i.e., the shock wave is a Mach wave.

The maximum value that β can have is, of course, 90°, the wave then being a normal shock wave. Hence, the limits on β are

$$\sin^{-1}\left(\frac{1}{M_1}\right) \leq \beta \leq 90° \tag{6.15}$$

It should further be noted that since it was proved that the flow behind a normal shock wave is subsonic, i.e., M_2 was less than 1, it follows that for an oblique shock wave

$$M_2 \sin(\beta - \delta) \leq 1 \tag{6.16}$$

Hence, for an oblique shock wave, M_2 can be greater than or less than 1.

To utilize the above relationships to find the properties of oblique shocks the relation among δ, β, and M_1 has to be known. Now, it will be seen from Figure 6.4 that

$$\tan\beta = \frac{N_1}{L_1}, \qquad \tan(\beta - \delta) = \frac{N_2}{L_2} \tag{6.17}$$

However, it was previously noted that

$$L_1 = L_2$$

and from the continuity equation, it follows that

$$\frac{N_1}{N_2} = \frac{\rho_1}{\rho_2}$$

Equation 6.17 therefore can be used to give

$$\frac{\tan(\beta - \delta)}{\tan\beta} = \frac{\rho_1}{\rho_2} = \frac{2 + (\gamma - 1)M_1^2 \sin^2\beta}{(\gamma + 1)M_1^2 \sin^2\beta} = X, \text{ say} \tag{6.18}$$

Equation 6.11 having been used. However, since

$$\tan(\beta - \delta) = \frac{\tan\beta - \tan\delta}{1 + \tan\beta\tan\delta}$$

it follows that

$$\frac{\tan(\beta - \delta)}{\tan\beta} = \frac{1 - \tan\delta/\tan\beta}{1 + \tan\beta\tan\delta}$$

Substituting this into Equation 6.18 then gives

$$1 - \frac{\tan\delta}{\tan\beta} = X + X\tan\beta\tan\delta$$

which becomes, on rearrangement,

$$\tan\delta = \frac{\tan\beta(1 - X)}{X\tan^2\beta + 1}$$

Substituting for X from Equation 6.18 then gives

$$\tan\delta = \frac{(\gamma + 1)M_1^2 \sin^2\beta - 2 - (\gamma - 1)M_1^2 \sin^2\beta}{2\tan^2\beta + (\gamma - 1)M_1^2 \sin^2\beta\tan^2\beta + (\gamma + 1)M_1^2 \sin^2\beta}\tan\beta$$

i.e.,

$$\tan\delta = \frac{2\cot\beta(M_1^2 \sin^2\beta - 1)}{2 + M_1^2(\gamma + \cos 2\beta)} \tag{6.19}$$

This equation gives the variation of δ with M_1 and β. It will be noted that the turning angle, δ, is zero when cot β = 0 and also when M_1 sin β is equal to 1, i.e.,

$$\delta = 0 \quad \text{when} \quad \beta = 90° \quad \text{and when} \quad \beta = \sin^{-1}(1/M_1)$$

these two limits being, of course, a normal shock and an infinitely weak Mach wave. Thus, as discussed before an oblique shock lies between a normal shock and a Mach wave as indicated in Figure 6.5. In both of these two limiting cases, there is no turning of the flow. Between these two limits, δ reaches a maximum.

The relation among δ, M_1, and β, as given by Equation 6.19, is frequently presented graphically and resembles that shown in Figure 6.6. A larger and more accurate graph is given in Appendix G. The software COMPROP also allows the relationship to be found.

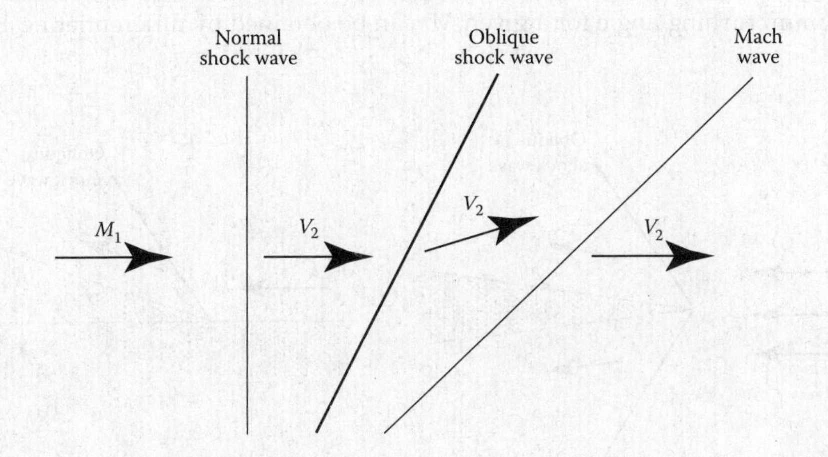

FIGURE 6.5
Limiting cases of an oblique shock wave.

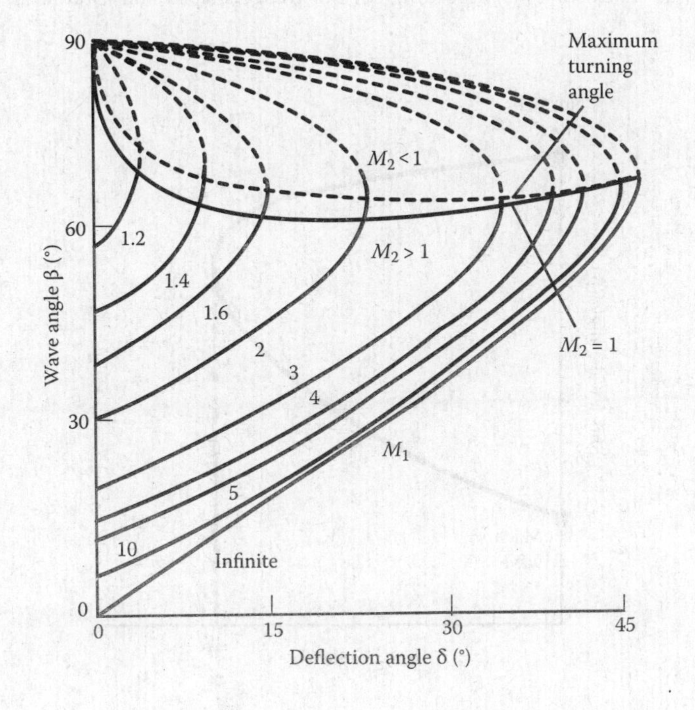

FIGURE 6.6
Oblique shock chart.

For flow over a body that has a surface set at an angle to the oncoming flow, this graph or Equation 6.19 from which it is derived or the software, allows the shock angle to be found for any value of M_1. This flow situation is illustrated in Figure 6.7.

The normal shock limit and the Mach wave limit on the oblique shock at a given value of M_1 are given by the intercepts of the curves in the oblique shock angle graph with the vertical axis at $\delta = 0$. This is illustrated in Figure 6.8.

It will also be seen from the diagram given in Figure 6.6 that, as already mentioned, there is a maximum angle through which a gas can be turned at a given M_1. The value of this maximum turning angle for a given M_1 can be obtained by differentiating Equation

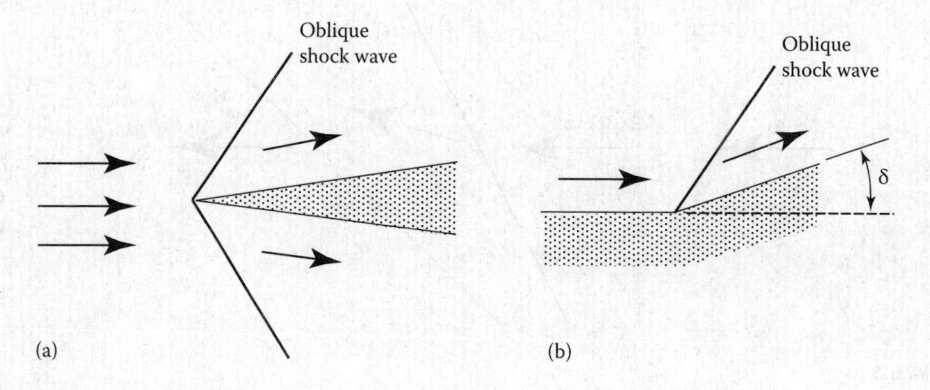

FIGURE 6.7
Generation of oblique shock waves (a) at leading edge of wedge shaped body and (b) at a change in wall direction.

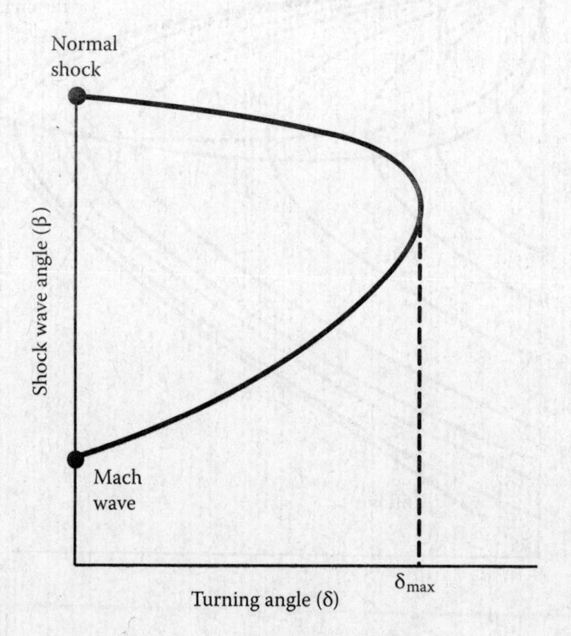

FIGURE 6.8
Limiting cases on oblique shock chart.

6.19 with respect to δ for a fixed M_1 and setting $d\delta/d\beta$ equal to zero. This leads to the following expression for the maximum turning angle

$$\sin^2\beta_{max} = \frac{\gamma+1}{4\gamma} - \frac{1}{\gamma M_1^2}\left[1 - \sqrt{(\gamma+1)\left(1+\frac{\gamma-1}{2}M_1^2+\frac{\gamma+1}{16}M_1^4\right)}\right]$$

where β_{max} is the shock angle that exists when δ has its maximum value for a given M_1. Once β_{max} has been found using this equation, Equation 6.19 can be used to find the value of δ_{max}. The variation of δ_{max} with M_1 so obtained for $\gamma = 1.4$ is shown in Figure 6.9.

For flow over bodies involving greater angles than this, a detached shock occurs as illustrated in Figure 6.10. A detached shock is curved and, in general, not amenable to simple analytical treatment.

It should also be noted that as M_1 increases, δ_{max} increases so that if a body involving a given turning angle, accelerates from a low to a high Mach number, the shock can be detached at the low Mach numbers and become attached at the higher Mach numbers. The solution giving the larger β is termed the 'strong shock' solution and is indicated by the dotted line in Figure 6.6. The figure also shows the $M_2 = 1$ locus and shows that M_2 is always less than 1 in the strong shock case.

It will further be noted from Figure 6.6 that if δ is less than δ_{max}, there are two possible solutions, i.e., two possible values for β, for a given M_1 and δ as indicated in Figure 6.11.

Experimentally, it is found that for a given M_1 and δ in external flows the shock angle, β, is usually that corresponding to the "weak" or nonstrong shock solutions. Under some circumstances, the conditions downstream of the shock may cause the strong shock solution to exist in part of the flow. In the event of no other information being available, the nonstrong shock solution should be used. The software COMPROP gives both the weak and strong oblique shock solutions.

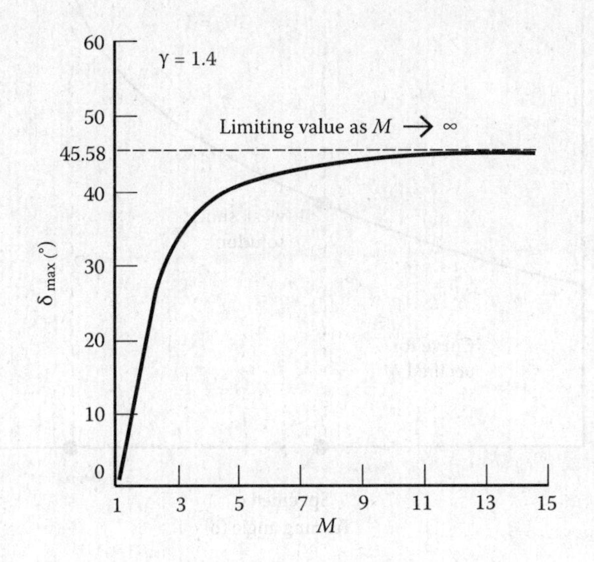

FIGURE 6.9
Variation of maximum turning angle with upstream Mach number, M_1.

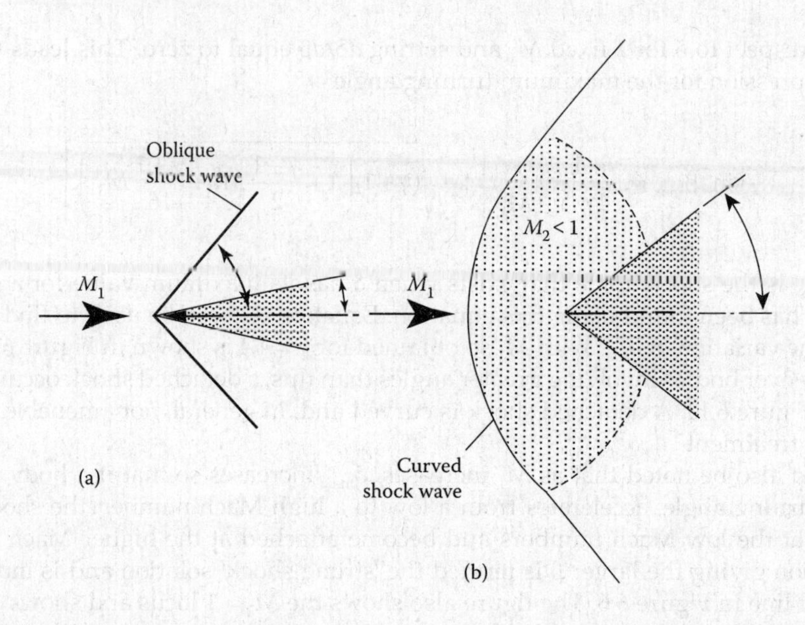

FIGURE 6.10
(a) Attached and (b) detached oblique shock wave.

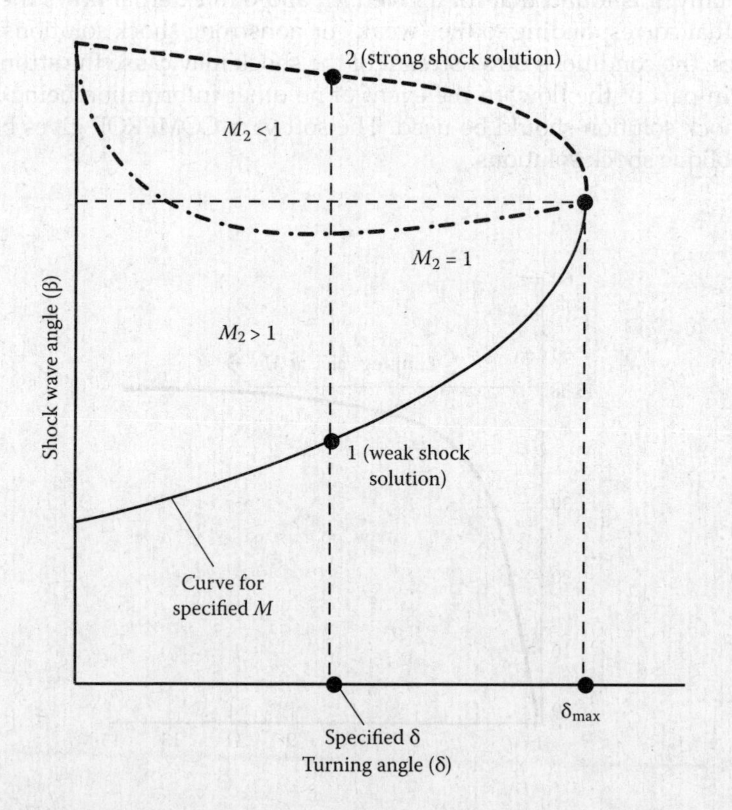

FIGURE 6.11
Strong and weak oblique shock wave solutions.

A curve that determines when the Mach number downstream of the shock, i.e., M_2, is equal to 1 is also given in Figure 6.6. It will be noted, as mentioned before, that the Mach number behind the shock in the 'strong' shock case is always less than 1.

The oblique shock waves results are often presented in the form of a series of diagrams. However, all calculations can be carried out using a single graph giving the relation among δ, β, and M_1 such as that shown in Figure 6.6, together with normal shock tables. This procedure is, for a given M_1 and δ, to use the diagram to give β and then to calculate $M_{N1} = M_1 \sin\beta$ and to use the normal shock relations or the normal shock tables to find p_2/p_1, ρ_2/ρ_1, T_2/T_1, and M_{N2} corresponding to this value of Mach number. M_2 is then obtained by setting it equal to $M_{N2}/\sin(\beta - \delta)$. Alternatively, the software COMPROP can be used to carry out this procedure.

Example 6.1

Air flowing at Mach 2 with a pressure of 80 kPa and a temperature of 30°C passes over a component of an aircraft that can be modeled as a wedge with an included angle of 8° that is aligned with the flow, i.e., the flow is turned by both the upper and lower surfaces of the wedge through an angle of 4°, leading to the generation of a oblique shock waves. Find the pressure acting on the surfaces of the wedge.

Solution

The flow situation being considered is shown in Figure E6.1a.

Here the flow is turned through an angle δ = 4° and the Mach number upstream of the shock wave is $M_1 = 2.0$. Hence, using the oblique shock chart or the software gives β = 33.4° (see Figure E6.1b).

Using this value of β then gives

$$M_{N1} = M_1 \sin\beta = 2 \times \sin 33.4° = 1.10$$

This value is directly given by the software.

Normal shock relations or tables or the software give for an upstream Mach number of 1.10 (M_{N1})

$$\frac{p_2}{p_1} = 1.245$$

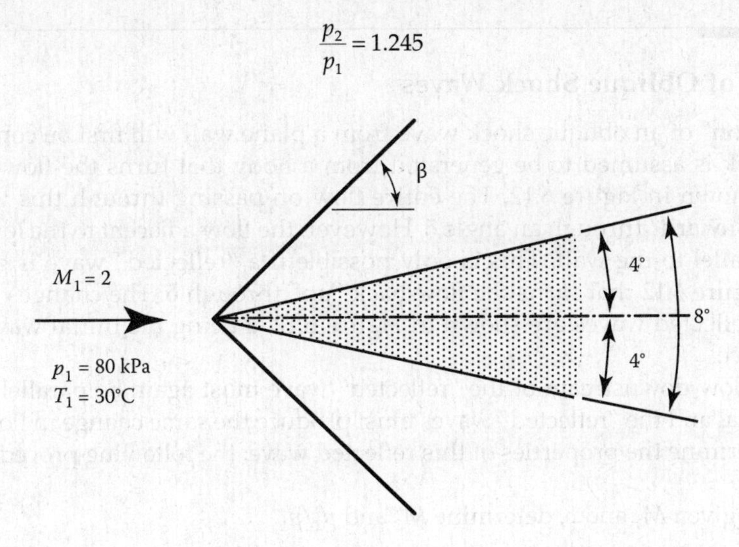

FIGURE E6.1a
Flow situation considered.

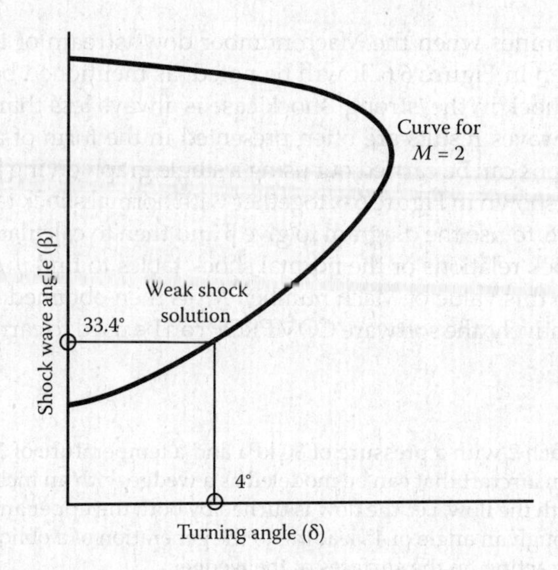

FIGURE E6.1b
Conditions considered on oblique shock wave chart.

Therefore,

$$p_2 = 1.245 \times 80 = 99.6 \text{ kPa}$$

i.e., the pressure downstream of the shock wave is 99.6 kPa. This will be the pressure acting on the surfaces of the wedge because the flow is uniform downstream of the shock waves.

Reflection of Oblique Shock Waves

The "reflection" of an oblique shock wave from a plane wall will first be considered. An oblique shock is assumed to be generated from a body that turns the flow through an angle δ as shown in Figure 6.12. The entire flow on passing through this wave is then turned "downward" through an angle δ. However, the flow adjacent to the lower flat wall must be parallel to the wall. This is only possible if a "reflected" wave is generated as shown in Figure 6.12 that turns the flow back "up" through δ. The changes through the initial and reflected waves are shown in Figure 6.12, **a** being the initial wave and **b** the reflected wave.

Since the flow downstream of the "reflected" wave must again be parallel to the wall, both the initial and the "reflected" waves must produce the same change in flow direction. Thus, to determine the properties of this reflected wave, the following procedure is used:

1. For the given M_1 and δ, determine M_2 and p_2/p_1.
2. For this value of M_2 and since the turning angle of the second wave is also δ, determine M_3 and p_3/p_2.

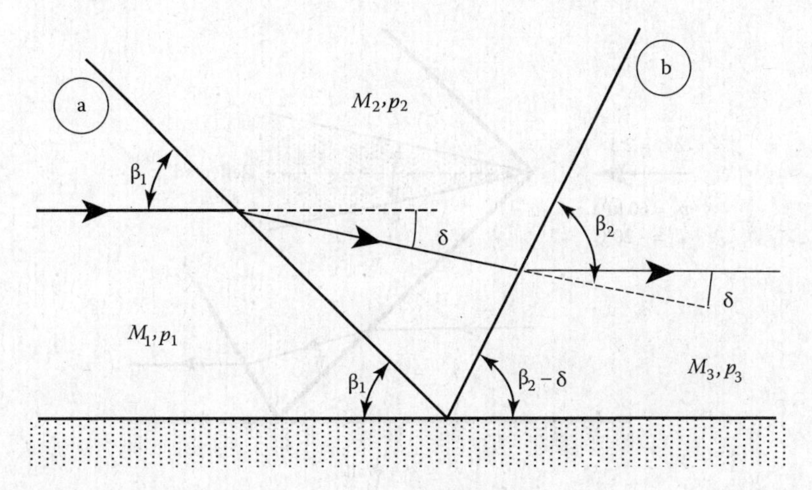

FIGURE 6.12
Reflected oblique shock wave.

3. The overall pressure ratio, p_3/p_1, is then found from

$$\frac{p_3}{p_1} = \frac{p_3}{p_2}\frac{p_2}{p_1}$$

4. The angle that the reflected wave makes with the wall is $\beta_2 - \delta$ and since β_2 was found in step 2, this angle can be determined.

Example 6.2

Air flowing at Mach 2.5 with a pressure of 60 kPa and a temperature of −20°C passes over a wedge which turns the flow through an angle of 4° leading to the generation of oblique shock waves. One of the oblique shock waves impinges on a flat wall, which is parallel to the flow upstream of the wedge and is "reflected" from it. Find the pressure and velocity behind the reflected shock wave.

Solution

The situation under consideration is shown in Figure E6.2a.

Upstream of the initial wave, the following conditions exist:

$$p_1 = 60 \text{ kPa}, \quad T_1 = 253 \text{ K}, \quad M_1 = 2.5$$

The conditions downstream of the initial wave, i.e., in region 2, are first obtained. Now, for

$$M_1 = 2.5 \quad \text{and} \quad \delta = 4°$$

the oblique shock chart or the software gives

$$\beta = 26.6°$$

Hence,

$$M_{N1} = M_1 \sin\beta = 2.5 \times \sin 26.6 = 1.12$$

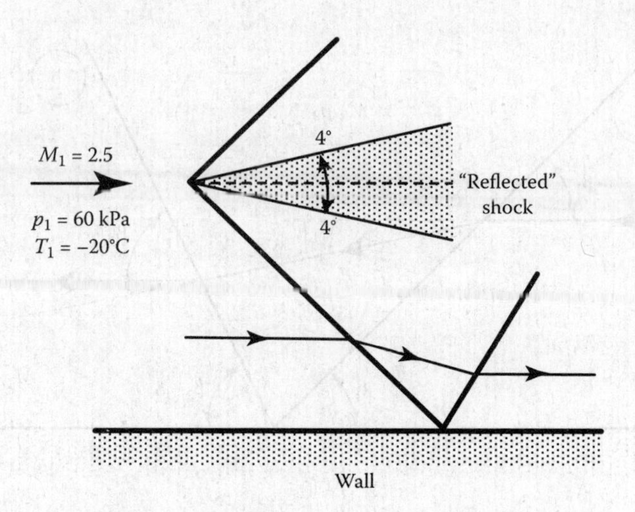

FIGURE E6.2a
Flow situation considered.

This value is also directly given by the software.

Next, using normal shock relations or tables or software for an upstream Mach number of 1.12 (M_{N1}), the following are obtained:

$$M_{N2} = 0.897, \quad \frac{p_2}{p_1} = 1.336, \quad \frac{T_2}{T_1} = 1.087$$

However,

$$M_{N2} = M_2 \sin(\beta - \delta)$$

Hence,

$$M_2 = \frac{0.897}{\sin(26.6 - 4)} = 2.334$$

Conditions behind the reflected wave, i.e., in region 3, will next be derived by considering the changes from region 2 to region 3 as shown in Figure E6.2b.

Now, for

$$M_2 = 2.334, \quad \delta = 4°$$

the oblique shock chart or the software gives

$$\beta_2 = 28.5°$$

Hence,

$$M_{N2} = M_2 \sin\beta_2 = 2.334 \quad \sin 28.5 = 1.113$$

This value is also directly given by the software.

Next, using normal shock relations or tables or software for an upstream Mach number of 1.113 (M_{N2}), the following are obtained:

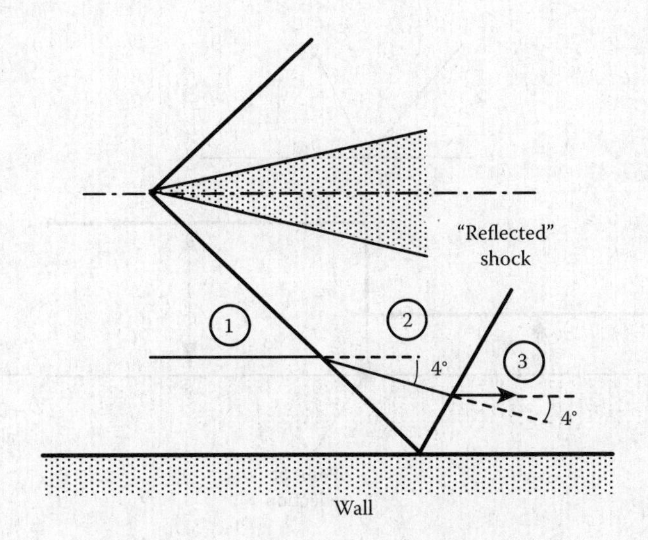

FIGURE E6.2b
Flow angles in situation considered.

$$M_{N3} = 0.9018, \quad \frac{p_3}{p_2} = 1.297, \quad \frac{T_3}{T_2} = 1.078$$

However,

$$M_{N3} = M_3 \sin(\beta - \delta)$$

Hence,

$$M_3 = \frac{0.9018}{\sin(28.5 - 4°)} = 2.17$$

Also,

$$p_3 = \left(\frac{p_3}{p_2} \times \frac{p_2}{p_1}\right) p_1 = 1.297 \times 1.336 \times 60 = 104.0 \text{ kPa}$$

$$T_3 = \left(\frac{T_3}{T_2} \times \frac{T_2}{T_1}\right) T_1 = 1.078 \times 1.087 \times 253 \text{ K} = 296 \text{ K}$$

$$a_3 = \sqrt{\gamma R T_3} = \sqrt{1.4 \times 287 \times 296} = 345 \text{ m/s}$$

From these results, it follows that

$$V_3 = M_3 a_3 = 2.17 \times 345 = 749 \text{ m/s}$$

Therefore, after the reflection the pressure and velocity are 104 kPa and 749 m/s, respectively.

The form of the variation of the pressure acting on the wall near an oblique shock reflection is, if viscous effects are ignored, as shown in Figure 6.13.

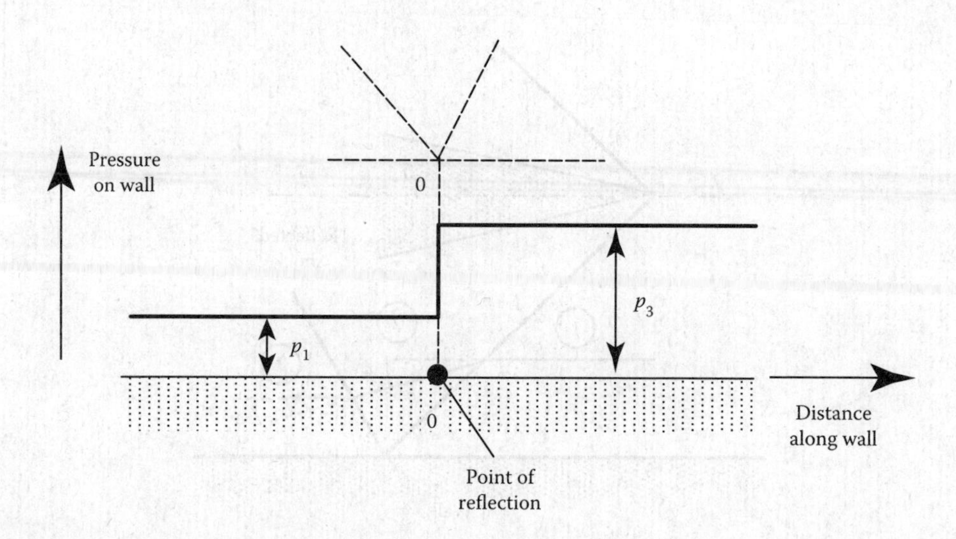

FIGURE 6.13
Wall pressure distribution near point of oblique shock wave reflection in ideal fluid case.

In real fluids, a boundary layer exists on the wall in which the velocity drops from its freestream value to zero at the wall. This means that the flow adjacent to the wall is subsonic and cannot sustain the pressure discontinuities associated with shock waves. Due to the presence of the boundary layer, there is therefore a "spreading out" of the pressure distribution along the wall, which may therefore resemble that shown in Figure 6.14.

The actual form of the pressure distribution will depend on the type of boundary layer flow, i.e., laminar or turbulent, the thickness of the boundary layer, and the shock strength. Because of the large positive pressure gradients induced by the shock wave impingement on the boundary layer, the interaction can also cause a local separation bubble in the boundary layer as shown schematically in Figure 6.15. The complexity of the reflection that can occur is also shown in Figure 6.16.

Another point concerning the reflection of oblique shock waves should be noted. It was previously indicated that for a given initial Mach number, there is a maximum angle through which an oblique shock wave can turn a flow and that this angle decreases with

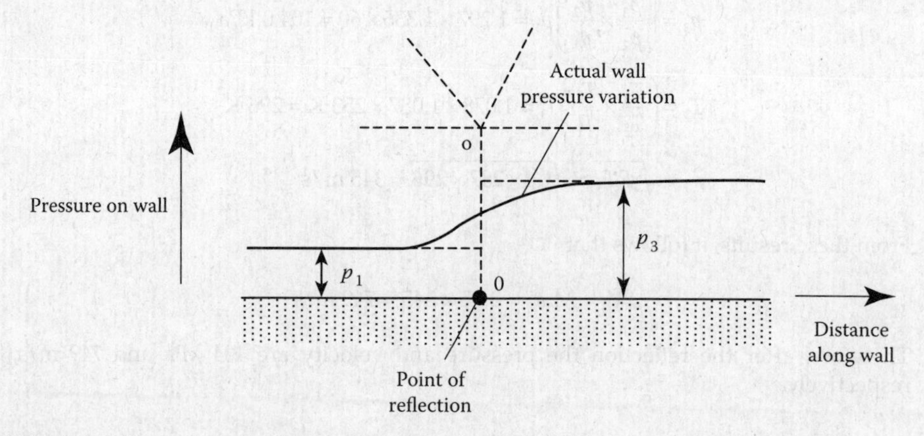

FIGURE 6.14
Wall pressure distribution near point of oblique shock wave reflection in real case.

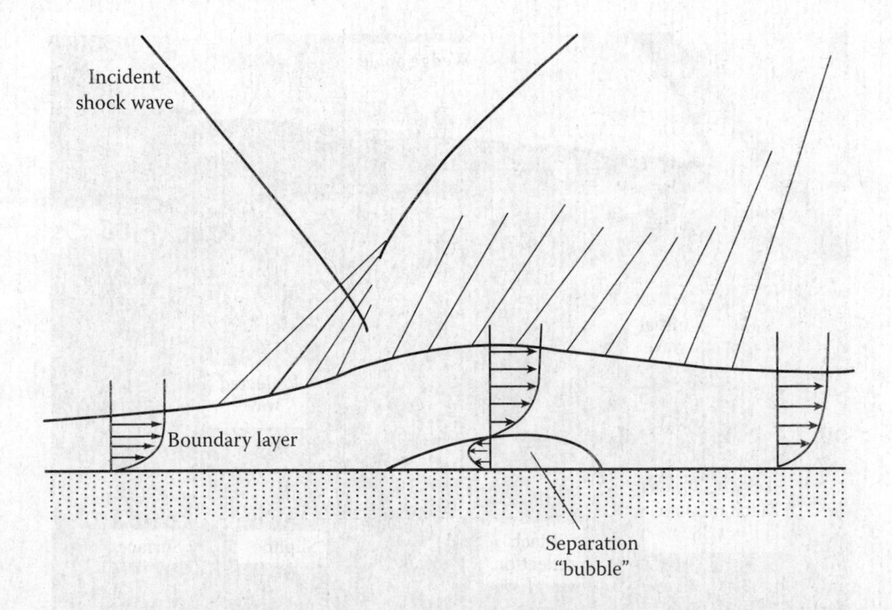

FIGURE 6.15
Boundary layer separation during shock wave–boundary layer interaction.

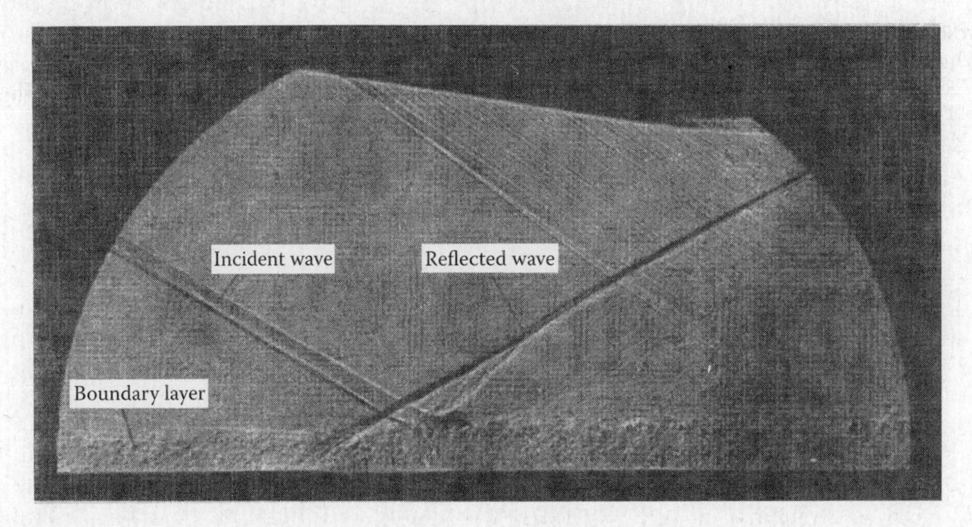

FIGURE 6.16
Reflection of oblique shock wave from wall with boundary layer. (From J. E. Green, *Journal of Fluid Mechanics*, 40(01), 81–95, Jan 1970, image number (l) from Figure 4 in plate 3. Copyright © 1970 Cambridge University Press. Reprinted with the permission of Cambridge University Press.)

decreasing Mach number. Therefore, considering the oblique shock wave reflection previously shown, it is possible for a situation to arise in which the maximum possible turning angle corresponding to M_2, i.e., to the flow behind the initial shock waves, is less than δ, the angle required to bring the flow parallel to the wall. In this case, a so-called Mach reflection occurs, this being illustrated in Figure 6.17 and shown schematically in Figure 6.18.

Here, a curved strong shock, behind which the flow is subsonic, forms near the wall. Since this subsonic flow need not all be parallel, the flow above the wall shock layer does not have to be parallel to the wall. The flow behind the curved wall shock is divided from the flow

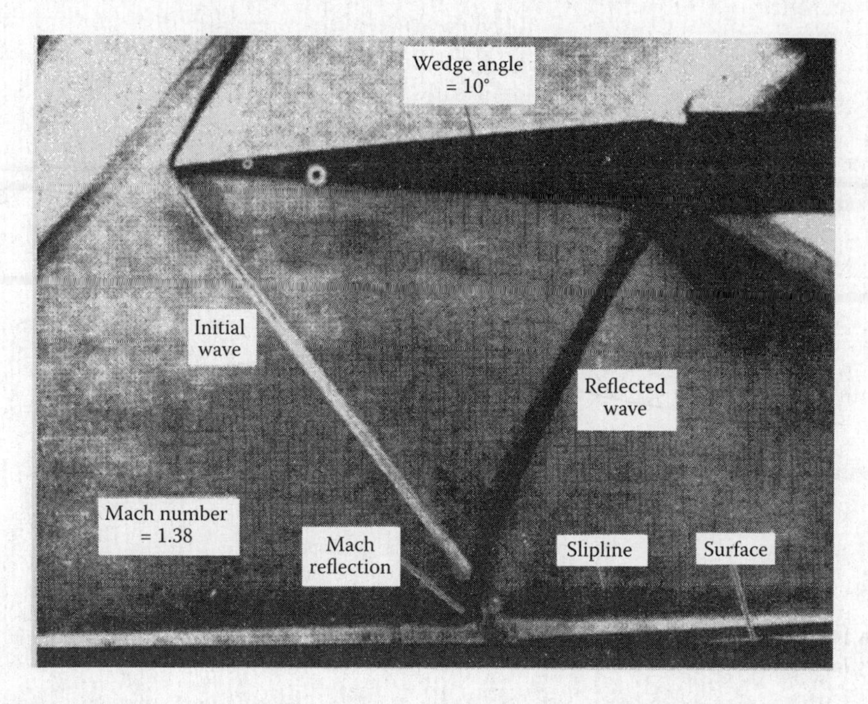

FIGURE 6.17
Mach reflection of an incident oblique shock wave. (H.W. Liepmann and A. Roshko: *Elements of Gas Dynamics*. 1957. Copyright Wiley-VCH Verlag GmbH & Co. KGaA. Reproduced with permission. By permission of Dover Publications, Inc.)

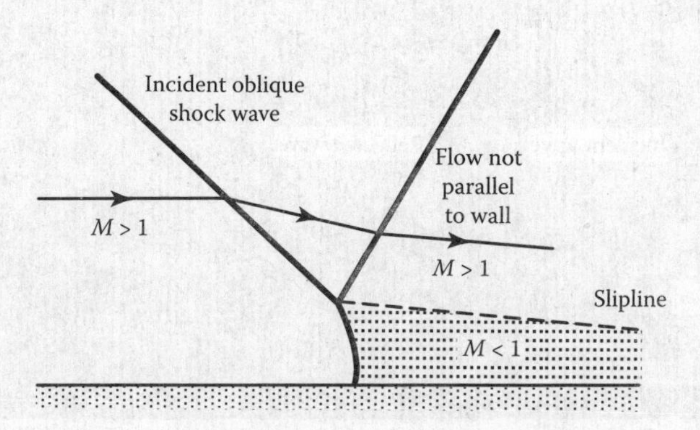

FIGURE 6.18
Schematic representation of a Mach reflection.

behind the "reflected" oblique shock by a slipline across which there are changes in velocity, temperature, and entropy. These sliplines grow in the downstream direction into thin regions across which the changes in flow properties occur, these regions being called slipstreams.

The details of the flow are influenced by downstream conditions due to the presence of the subsonic region, and the Mach reflection is difficult to treat analytically.

Example 6.3

Air flows at Mach 2.5 over a wedge, leading to the generation of an oblique shock wave. This oblique shock wave impinges on a flat wall, which is parallel to the flow upstream of the wedge, and is "reflected" from it. If various wedge angles have to be considered, what is the largest turning angle that a wedge could produce without a Mach reflection being produced at the wall?

Solution

The situation being considered is shown in Figure E6.3.

A Mach reflection occurs when the maximum turning angle corresponding to the Mach number downstream of the incident wave, i.e., M_2, is not enough to bring the flow in region 3 parallel to the wall, i.e., is less than the turning angle produced by the wedge.

A simple iterative approach will be adopted here. A turning angle will be assumed. The maximum turning angle that can be produced by the reflected wave will be calculated, and the difference between this maximum value and the required turning angle will be calculated. This is termed $\Delta\delta$ in Table E6.3.

When $\Delta\delta$ is negative, a Mach reflection will exist. The turning angle that makes $\Delta\delta = 0$ can then be deduced.

To illustrate how the results in Table E6.3 are derived, consider the case where δ is assumed to be 12°. Now, for

$$M_1 = 2.5, \quad \delta = 12°$$

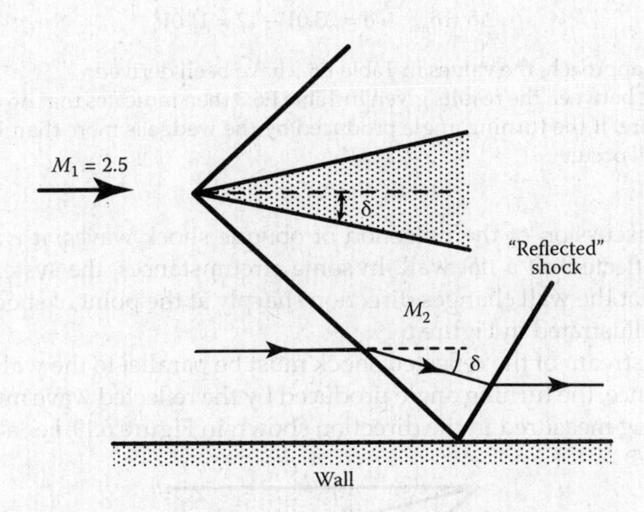

FIGURE E6.3
Flow situation considered.

TABLE E6.3
Iterative Results Used in Obtaining Solution

δ (Guess)	β	M_{N1}	M_{N2}	M_2	δ_{max}	$\Delta\delta$
4°	26.6°	1.120	0.8967	2.333	27.87°	23.87°
12°	33.8°	1.391	0.7436	2.002	23.01°	11.01°
15°	36.9°	1.503	0.7002	1.877	20.73°	5.73°
18°	40.4°	1.620	0.6625	1.739	17.88°	−0.12°
20°	42.9°	1.701	0.6402	1.645	15.74°	−4.26°

the oblique shock chart or the software gives

$$\beta = 33.8°$$

Hence,

$$M_{N1} = M_1 \sin\beta = 2.5 \quad \sin 33.8 = 1.391$$

This value is also directly given by the software.

Next, using normal shock relations or tables or software for an upstream Mach number of 1.391 (M_{N1}), the following is obtained:

$$M_{N2} = 0.7436$$

However,

$$M_{N2} = M_2 \sin(\beta - \delta)$$

Hence,

$$M_2 = \frac{0.7436}{\sin(33.8 - 12)} = 2.002$$

Now for $M = 2.002$ the oblique shock chart or the software gives $\delta_{max} = 23.01°$. Hence,

$$\Delta\delta = \delta_{max} - \delta = 23.01 - 12 = 11.01°$$

Using this approach, the values in Table E6.3 have been derived.

Interpolating between the results given in Table E6.3 then indicates that $\Delta\delta = 0$ when $\delta = 17.9°$. Therefore, if the turning angle produced by the wedge is more than 17.9°, a Mach reflection will occur.

In the above discussion of the reflection of oblique shock waves, it was assumed that the shock was reflected off a flat wall. In some circumstances, the system involved may be designed so that the wall changes direction sharply at the point of shock impingement. This situation is illustrated in Figure 6.19.

The flow downstream of the reflected shock must be parallel to the wall downstream of the reflection. Hence, the turning angle produced by the reflected wave must be $\delta_1 - \delta_w$, the wall angle δ_w being measured in the direction shown in Figure 6.19, i.e., a convex corner is

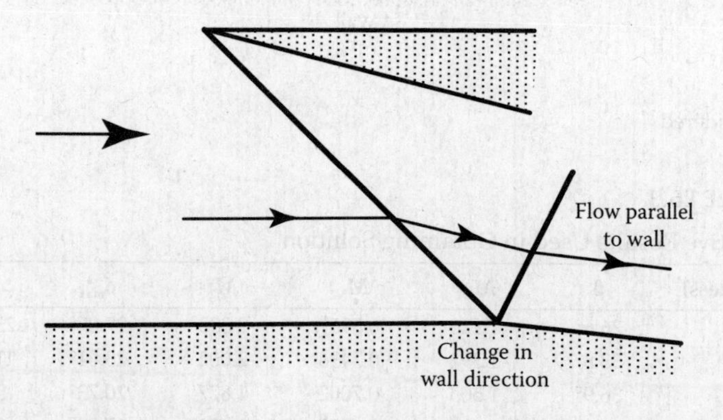

FIGURE 6.19
Change of wall direction at point of shock impingement.

being taken as positive. For such a convex corner, then, the reflected wave will be weaker than the wave that would be reflected from a flat wall. If $\delta_w = \delta_1$, no reflected wave will occur, i.e., the incident oblique shock wave will be "cancelled" by the turning of the wall. If $\delta_w > \delta_1$, an expansion wave of the type discussed in the next chapter will be generated.

Example 6.4

Air at a pressure of 60 kPa and a temperature of –20°C flows at Mach 2.5 over a wedge, which turns the flow through 4° leading to the generation of an oblique shock wave. This oblique shock wave impinges on a wall that "turns away from the flow" by 4° exactly at the point where the oblique shock wave impinges on it, i.e., the wall has a sharp convex 4° corner at the point where the incident shock impinges on it. Sketch the flow pattern. If the leading edge of the wedge is 1 m above the wall, how far behind this leading edge would the change in wall angle have to occur?

Solution

Since the change in wall direction is exactly equal to the turning angle produced by the incident shock, the flow behind this incident shock is parallel to the wall. There will therefore be no reflected wave, i.e., the shock is cancelled by the corner. The flow pattern will therefore be as shown in Figure E6.4.

Now, as indicated in the previous two examples, for

$$M_1 = 2.5, \quad \delta = 4^\circ$$

the oblique shock chart or the software gives

$$\beta = 26.6^\circ$$

Hence, if x is the distance of the point of impingement behind the leading edge of the wedge, it can be seen from Figure E6.4 that

$$\frac{h}{x} = \tan 26.6^\circ$$

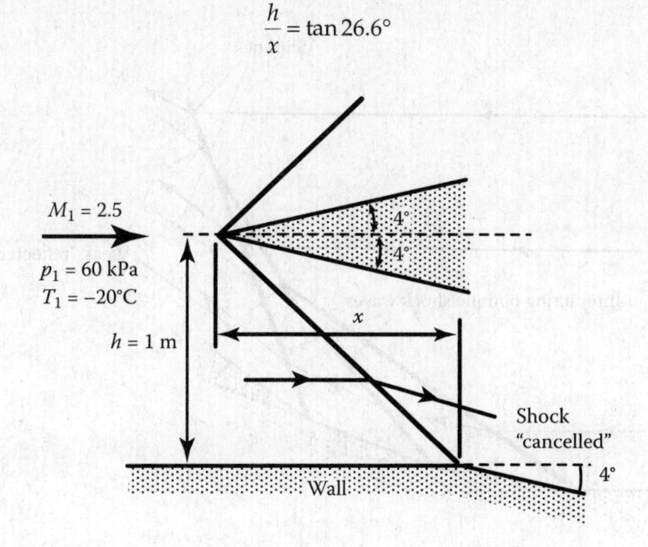

FIGURE E6.4
Flow situation considered.

Since $h = 1$ m, this equation gives

$$x = 1.99 \text{ m}$$

Therefore, the sharp corner in the wall would have to be 1.99 m behind the leading edge of the wedge.

Interaction of Oblique Shock Waves

It will be noted from the results and discussion given above about the characteristics of oblique shock waves that

- An oblique shock wave always decreases the Mach number, i.e., $M_2 \leq M_1$.
- Considering only the nonstrong shock solution, the shock angle, β, for a given turning angle, δ, increases with decreasing Mach number.

Hence, if the flow around a concave wall consisting of several angular changes of equal magnitude is considered, the oblique shock waves generated at each step will tend to converge and coalesce into a single oblique shock wave that is stronger than any of the initial waves. This is illustrated in Figure 6.20.

Now, the pressure and flow direction must be the same for all streamlines downstream of the last wave. However, two or more weaker waves cannot produce the same changes as a single stronger wave, and for this reason, the "reflected" shocks shown must be generated. These waves are much weaker than the initial waves. While these "reflected" waves equalize the pressure and flow direction they cannot equalize the velocity, density, and entropy. For this reason, the sliplines shown in the diagram exist

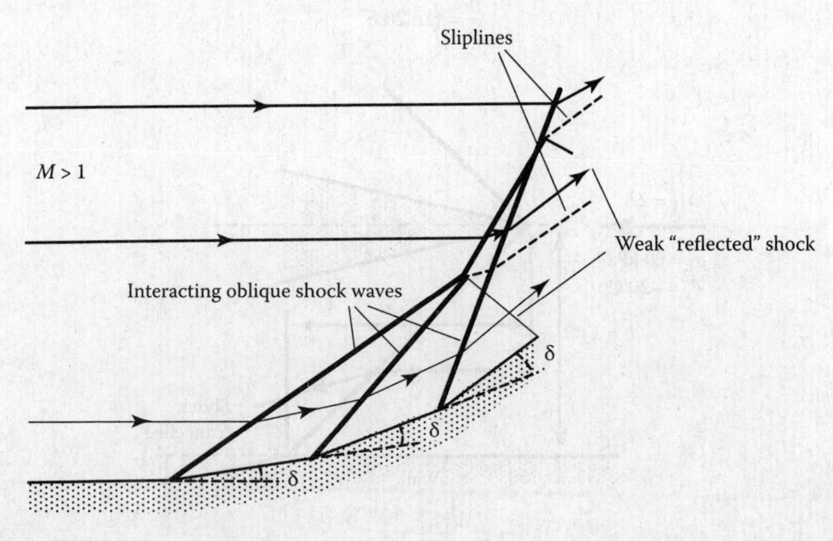

FIGURE 6.20
Interaction of oblique shock waves.

across which there is a jump in these properties. In theory, these sliplines are planes of discontinuity, but in reality, they grow into thin regions over which the changes in the properties occur.

A curved wall can be thought of as consisting of a series of small segments each producing a small fraction of the total change in flow direction. Each of these segments produces a weak wave and these weak waves interact in the manner described above to form a single oblique shock wave as shown in Figures 6.21 and 6.22.

In this case, instead of individual sliplines existing, there is a region of variable density, velocity, and entropy. The reflected waves discussed above will, in this case, be negligible

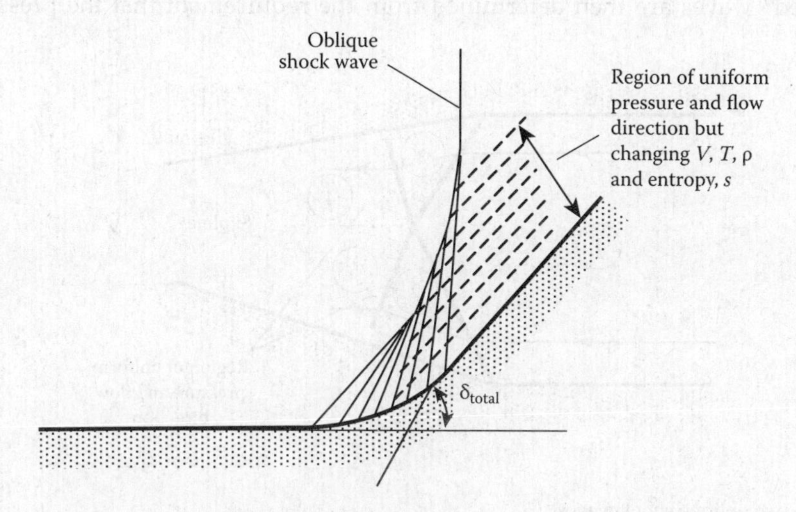

FIGURE 6.21
Interaction of oblique shock waves on a curved wall.

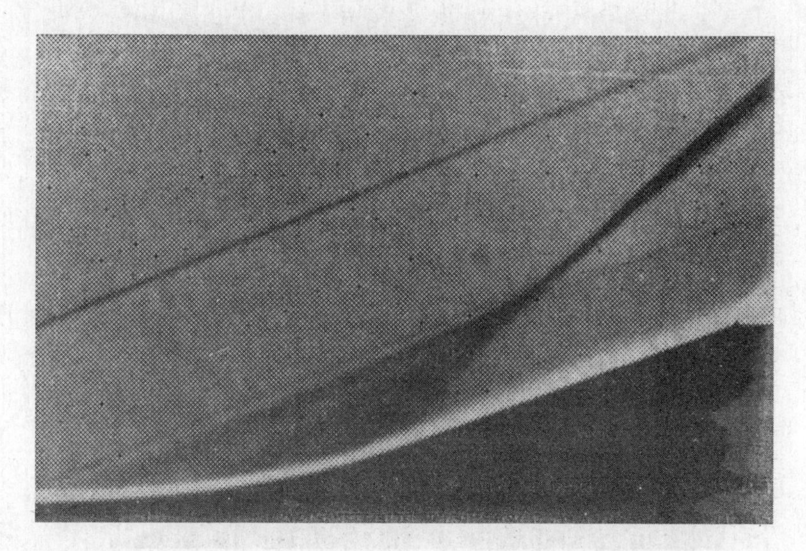

FIGURE 6.22
Development of an oblique shock wave in flow over a curved wall. (A.H. Shapiro: *The Dynamics and Thermodynamics of Compressible Fluid Flow, vol. 1,* The Ronald Press Company, 1953, New York. Copyright Wiley-VCH Verlag GmbH & Co. KGaA. Reproduced with permission.)

for most purposes because the initial waves are so weak. The final oblique shock resulting from the interaction must be that corresponding to the initial Mach number and the total turning angle, δ_{total}.

Lastly, consider what happens when oblique shock waves of differing strength generated by different surfaces interact as shown in Figures 6.23 and 6.24.

The flows in regions 4 and 5 shown in Figure 6.23 must, of course, be parallel to each other. Therefore, conservation of momentum applied in a direction normal to the flows in these two regions indicates that the pressures in regions 4 and 5 must be the same. The initial waves separating regions 1 and 2 and regions 1 and 3 are, of course, determined by the Mach number in region 1 and the turning angles, θ and φ. The properties of the "transmitted" waves are then determined from the requirement that the pressures and

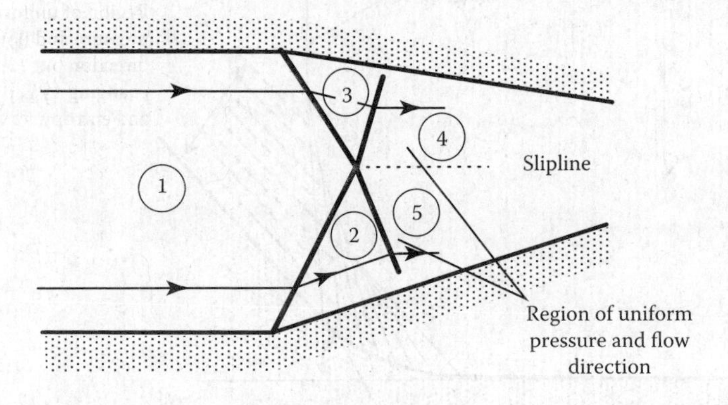

FIGURE 6.23
Interaction of two oblique shock waves.

FIGURE 6.24
Interaction of two oblique shock waves. (Courtesy of I.I. Glass. University of Toronto Institute for Aerospace Studies. With permission.)

flow directions in regions 4 and 5 must be the same. The density, velocity, and entropy will then be different in these two regions, and the slipstream shown must therefore exist. Of course, when $\varphi = \theta$, the initial waves are both of the same strength as are the transmitted waves. No slipline then exists.

Example 6.5

Air is flowing at Mach 3 with a pressure and a temperature of 30 kPa and −10°C, respectively, down a wide channel. The upper wall of this channel turns through an angle of 4° "toward the flow," whereas the lower wall turns through an angle of 3° "toward the flow" leading to the generation of two oblique shock waves. These two oblique shock waves intersect each other. Find the pressure and flow direction downstream of the shock intersection.

Solution

The flow situation being considered is shown in Figure E6.5a.

The conditions behind the two initial shock waves will first be considered, i.e., the conditions in regions 2 and 3 shown in Figure E6.5a will first be derived.

First, consider region 2. Using the software or the oblique shock wave chart for $M = 3$ and $\delta = 4°$ gives $\beta = 22.3°$, and therefore

$$M_{N1} = M_1 \sin\beta = 3 \quad \sin 22.3° = 1.140$$

This is directly given by the software. Using the normal shock tables for $M = 1.14$ gives

$$M_{N2} = 0.882$$

which is also directly given by the software, and

$$\frac{p_2}{p_1} = 1.350$$

Hence,

$$M_2 = \frac{M_{N2}}{\sin(\beta - \delta)} = \frac{0.882}{\sin(22.3 - 4)} = 2.799$$

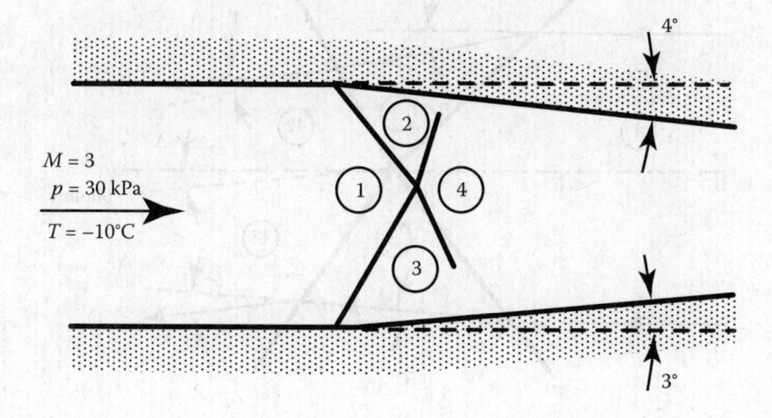

FIGURE E6.5a
Flow situation considered.

Next, consider region 3. Using the software or the oblique shock wave chart for $M = 3$ and $\delta = 3°$ gives $\beta = 21.6°$, and therefore,

$$M_{N1} = M_1 \sin\beta = 3 \quad \sin 21.6° = 1.104$$

This is directly given by the software. Using the normal shock tables for $M = 1.104$ gives

$$M_{N3} = 0.908$$

which is also directly given by the software, and

$$\frac{p_3}{p_1} = 1.255$$

Hence,

$$M_3 = \frac{M_{N3}}{\sin(\beta - \delta)} = \frac{0.908}{\sin(21.6 - 3)} = 2.848$$

Next, consider region 4. The strengths of the shock waves after the intersection must be such that the pressure and the flow direction is the same throughout this region. It is convenient to consider two parts of region 4: Region 42, which is downstream of region 2, and region 43, which is downstream of region 3. The solution must be such that the flow directions and pressures in regions 42 and 43 are the same. Although there are elegant ways of obtaining the solution, a very simple approach will be adopted here. A flow direction that is the same in both regions 42 and 43 will be assumed. This direction will be specified by the angle Δ, which is the flow direction relative to the flow upstream of the initial shock waves as defined in Figure E6.5b.

The turning angle produced by the oblique shock wave between regions 2 and 42 is given by

$$\delta = 4° - \Delta$$

Similarly, the turning angle produced by the oblique shock wave between regions 3 and 43 is given by

$$\delta = 3° + \Delta$$

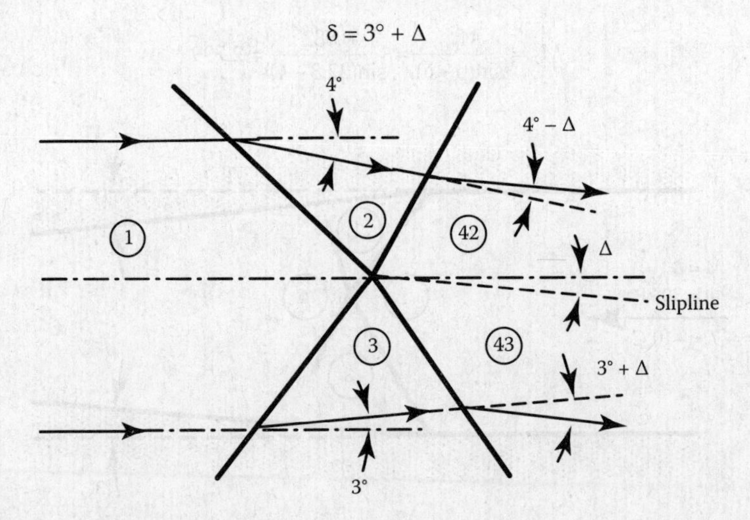

FIGURE E6.5b
Flow angles and flow regions in situation considered.

For this chosen flow direction, the pressures in the two regions, i.e., p_{42} and p_{43}, can then be calculated. These two pressures will not in general be the same because the value of Δ has been guessed. The procedure is repeated for several different values of Δ and the value of Δ that makes $p_{42} = p_{43}$ can then be deduced.

To illustrate the procedure, consider the case where the guessed value of Δ is 1°. In this case, the turning angle between region 2 and region 42 is $\delta = 3°$, and it will be recalled, $M_2 = 2.799$. Using the software or the oblique shock wave chart for this value of M and δ gives $\beta = 23.1°$, and therefore,

$$M_{N2} = M_2 \sin\beta = 2.799 \quad \sin 23.1 = 1.098$$

This is directly given by the software. Using the normal shock tables for $M = 1.098$ gives

$$\frac{p_{42}}{p_2} = 1.242$$

Next, consider the change between region 3 and region 43. For $\Delta = 1°$, the turning angle between region 3 and region 43 is $\delta = 4°$, and it will be recalled, $M_2 = 2.848$. Using the software or the oblique shock wave chart for this value of M and δ gives $\beta = 23.5°$, and therefore,

$$M_{N3} = M_3 \sin\beta = 2.848 \quad \sin 23.5 = 1.13$$

This is directly given by the software. Using the normal shock tables for $M = 1.13$ gives

$$\frac{p_{43}}{p_3} = 1.335$$

Using these results gives

$$p_{42} = \frac{p_{42}}{p_2}\frac{p_2}{p_1}p_1 = 1.242 \times 1.350 \times 30 = 50.2 \text{ kPa}$$

and

$$p_{43} = \frac{p_{43}}{p_3}\frac{p_3}{p_1}p_1 = 1.335 \times 1.255 \times 30 = 50.3 \text{ kPa}$$

This procedure is repeated for several other values of Δ giving the results shown in Table E6.5.

Interpolating between the results given in Table E6.5 then indicates that $p_{42} = p_{43}$ approximately when $\Delta = +0.5°$ and $p_4 = 52$ kPa. The flow direction is therefore as indicated in Figure E6.5c.

TABLE E6.5

Iterative Results Used in Obtaining Solution

Δ (Chosen)	p_{42}/p_1	p_{43}/p_1	p_{42} (kPa)	p_{43} (kPa)
+1°	1.677	1.675	50.2	50.3
–1°	1.920	1.453	57.6	43.6
0°	1.793	1.561	53.8	46.8

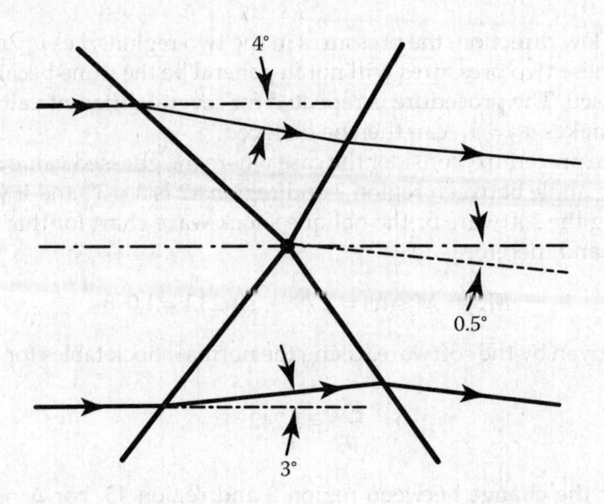

FIGURE E6.5c
Flow angles in situation considered.

Conical Shock Waves

The discussion given thus far in this chapter has been concerned with two-dimensional plane flows of the type shown in Figure 6.25. In such flows, the flow before and after the oblique shock wave is uniform as shown in Figure 6.26.

A related but more complex flow is that associated with axisymmetric supersonic flow over a cone as shown in Figure 6.27.

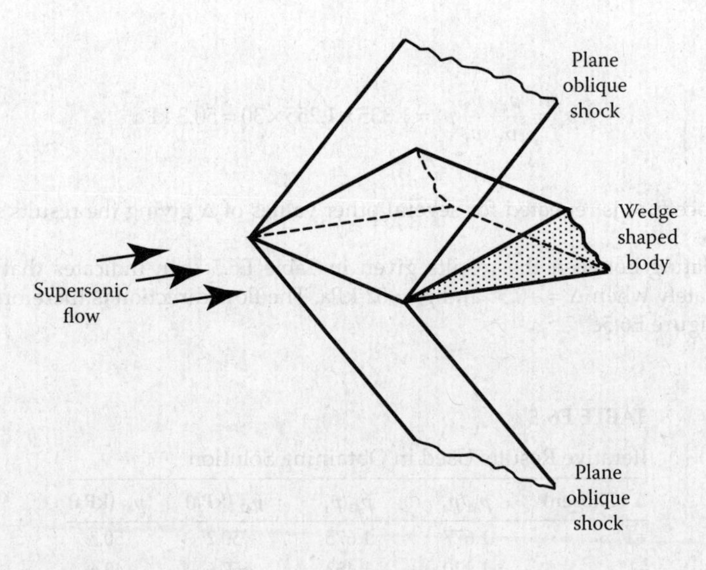

FIGURE 6.25
Oblique shock waves on wedge.

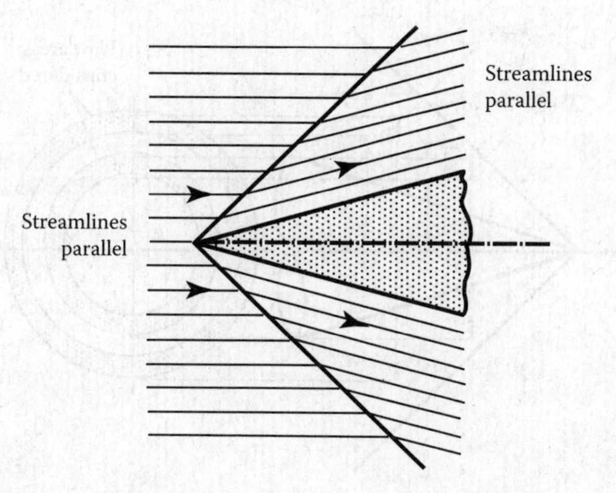

FIGURE 6.26
Flow direction upstream and downstream of oblique shock wave on wedge.

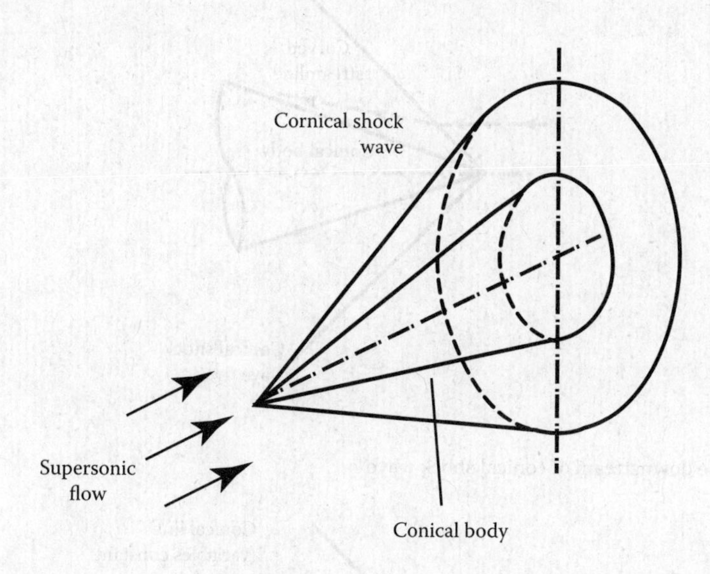

FIGURE 6.27
Conical shock wave.

In this situation, because the flow area increases with increasing distance from the centerline, as shown in Figure 6.28, conservation of mass will require that the flow downstream of a conical shock wave be curved, as indicated in Figure 6.29.

The flow between the cone surface and the conical shock wave is therefore two-dimensional and cannot be analyzed using the simple procedures adopted in dealing with plane oblique shock waves. However, as long as the conical shock is attached to the vertex of the cone, the flow behind a conical shock wave will be "conical," i.e., the flow variables are constant along conical surfaces originating at the vertex of the cone as shown in Figure 6.30. Using this fact, a relatively simple ordinary differential equation can be derived to

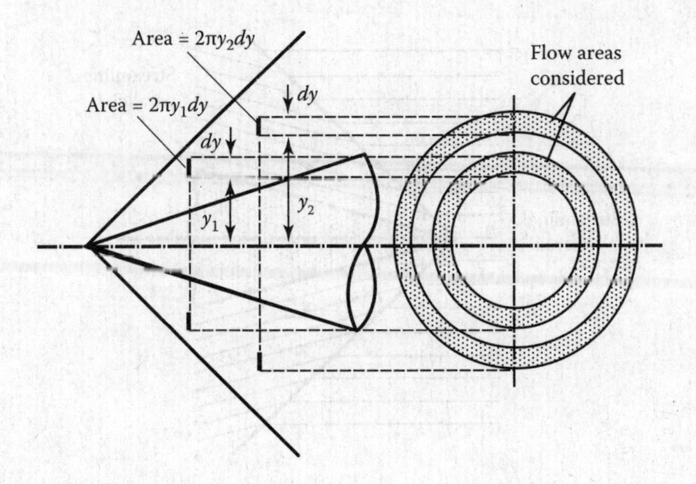

FIGURE 6.28
Flow area between cone surface and conical shock wave.

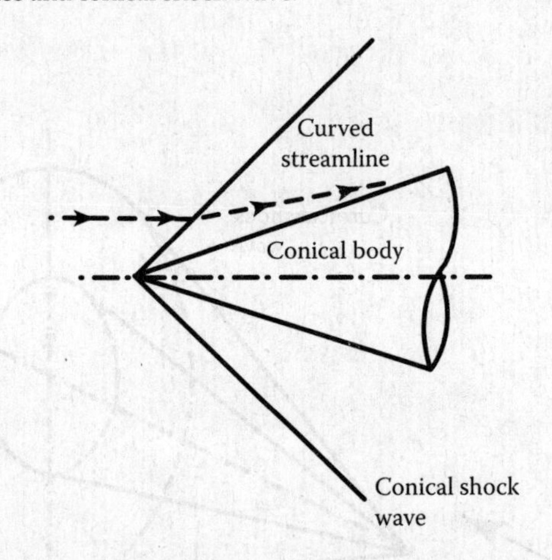

FIGURE 6.29
Curved streamline downstream of conical shock wave.

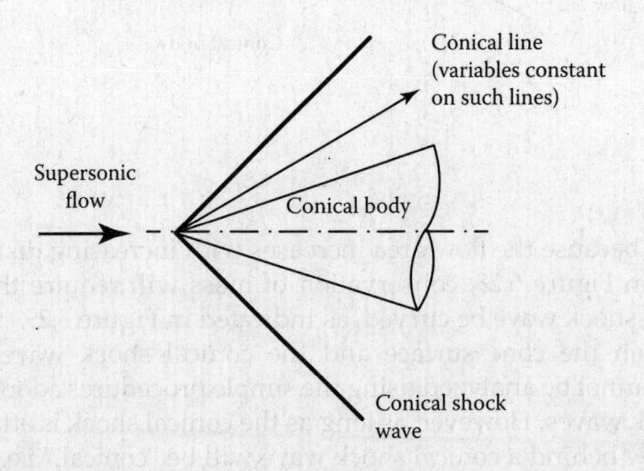

FIGURE 6.30
Conical flow downstream of conical shock wave.

describe the flow behind an attached conical shock wave. This equation has to be numerically integrated.

Concluding Remarks

When a supersonic flow is turned through a positive angle, i.e., is "toward itself," an oblique shock wave develops. The relation among the initial Mach number, the turning angle, and the oblique shock angle has been considered in this chapter. The interaction and reflection of oblique shock waves has also been considered. The difference between conical shock waves and oblique shock waves has also been discussed.

PROBLEMS

1. Air is flowing over a flat wall. The Mach number, pressure, and temperature in the air stream are 3, 50 kPa, and –20°C, respectively. If the wall turns through an angle of 4° leading to the formation of an oblique shock wave, find the Mach number, the pressure, and the temperature in the flow behind the shock wave.
2. An air flow in which the Mach number is 2.5 passes over a wedge with a half-angle of 10°, the wedge being symmetrically placed in the flow. Find the ratio of the stagnation pressures before and after the oblique shock wave generated at the leading edge of the wedge.
3. A symmetrical wedge with a 12° included angle is placed in an air flow in which the Mach number is 2.3 and the pressure is 60 kPa. If the centerline of the wedge is at an angle of 4° to the direction of flow, find the pressure difference between the two surfaces of the wedge.
4. Air flows over a wall at a supersonic velocity. The wall turns toward the flow generating an oblique shock wave. This wave is found to be at an angle of 50° to the initial flow direction. A scratch on the wall upstream of the shock wave is found to generate a very weak wave that is at an angle of 30° to the flow. Find the angle through which the wall turned.
5. Air flows over a plane wall at Mach 3.5, the pressure in the flow being 100 kPa. The wall turns through an angle leading to the generation of an oblique shock wave whose strength is such that the pressure downstream of the corner is 548 kPa. Find the turning angle of the corner.
6. Air, flowing down a plane-walled duct at Mach 3, passes over a wedge. What is the largest included angle that this wedge can have if the oblique shock wave that is generated is attached to the wedge? Sketch the flow pattern that will exist if the wedge angle is greater than this maximum value.
7. A uniform air flow at Mach 2.5 passes around a sharp concave corner in the wall that turns the flow through an angle of 10° and leads to the generation of an oblique shock wave. The pressure and temperature in the flow upstream of the corner are 70 kPa and 10°C, respectively. Find the Mach number, the pressure, the temperature, and the stagnation pressure downstream of the oblique shock wave. How large would the corner angle have to be before the shock became detached from the corner?

8. Find the minimum values of the Mach number for which the oblique shock generated at the leading edge of a wedge placed in a supersonic air flow remains attached to the wedge for deflection angles of 15°, 25°, and 40°.
9. A wedge symmetrically placed in a supersonic air stream is to be used to determine the Mach number in the flow. This will be done using optical methods to measure the angle that the oblique shock wave attached to the leading edge of the wedge makes to the upstream flow. If the total included angle of the wedge that is to be used is 45°, find the Mach number range over which this method can be used.
10. Air flowing at Mach 2.5 passes over a wedge that turns the flow through an angle of 5°. Find the pressure ratio across the oblique shock that is generated. If this oblique shock is reflected off a plane surface, find the overall pressure ratio.
11. An oblique shock wave with wave angle of 26° in an air stream in which the Mach number is 2.7, the pressure is 100 kPa, and the temperature is 30° impinges on a straight wall. Find the Mach number, pressure, temperature, and stagnation pressure downstream of the reflected wave.
12. Air flows down a duct at Mach 1.5. The top wall of the duct turns toward the flow leading to the generation of an oblique shock wave that strikes the flat lower wall of the duct and is reflected from it. What is the smallest turning angle that would give a Mach reflection off the lower wall?
13. A two-dimensional wedge with an included angle of 10° is placed in a wind tunnel that has parallel walls. If the Mach number in the freestream ahead of the wedge is 2 and if the axis of the wedge is inclined at an angle of 2° to the direction of the air flow, find the Mach numbers upstream and downstream of the oblique shock waves after they are reflected off the upper and lower walls of the tunnel. Also, sketch the flow pattern marking the angles the waves make to the tunnel walls.
14. Air in which the pressure is 60 kPa is flowing down a plane-walled duct at Mach 2.5. The air stream passes over a wedge with an included angle of 10°. The oblique shock wave that is generated by the wedge is reflected off the flat wall of the duct. Find the pressure and Mach number after the reflection.
15. Air at a pressure and temperature of 40 kPa and –30°C flows at Mach 3 down a wide duct. The upper wall of the duct turns sharply through an angle of 5° leading

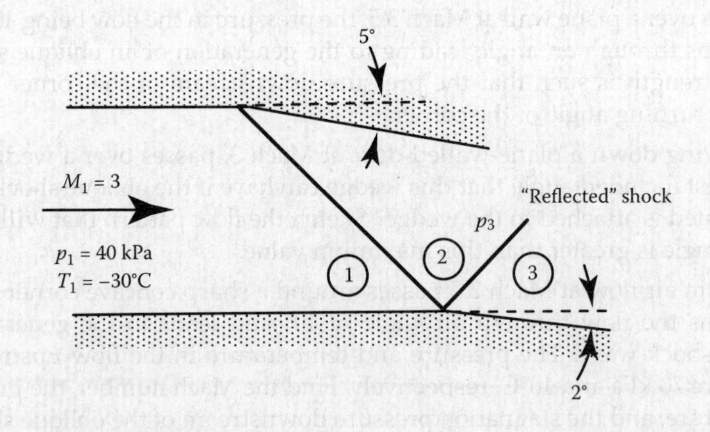

FIGURE P6.15
Flow situation considered.

to the formation of an oblique shock wave as shown in Figure P6.15. Find the Mach number, temperature, and pressure behind this shock wave. As shown in Figure P6.15, this shock wave strikes the lower wall of the duct exactly at a point where the lower wall turns away from the flow through an angle of 2°. Find the Mach number, pressure, and temperature behind the "reflected" wave.

16. Find the pressure ratio p_3/p_1 for the flow situation shown in Figure P6.16.
17. Find the pressure ratio p_4/p_1 for the flow situation shown in Figure P6.17.
18. If the Mach number and pressure ahead of the oblique shock wave system shown in Figure P6.18 are 3 and 50 kPa, respectively, find the pressure in the region 4 downstream of the wave intersection.
19. Air is flowing at Mach 2 and a pressure of 70 kPa in a two-dimensional channel. The upper wall turns toward the flow through an angle of 5° and the lower wall turns toward the flow through an angle of 3°, two oblique shock waves thus being generated. These two shock waves intersect each other. Find the pressure in the region just downstream of the shock intersection.
20. An air stream in which the Mach number is 3 and the pressure is 80 kPa flows between two parallel walls. The upper wall turns sharply through an angle of 18°

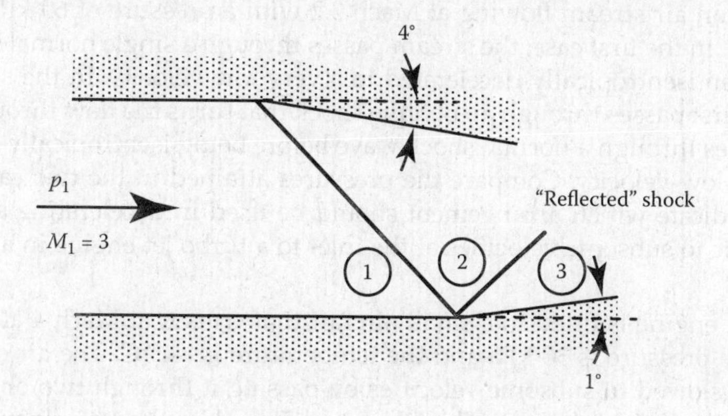

FIGURE P6.16
Flow situation considered.

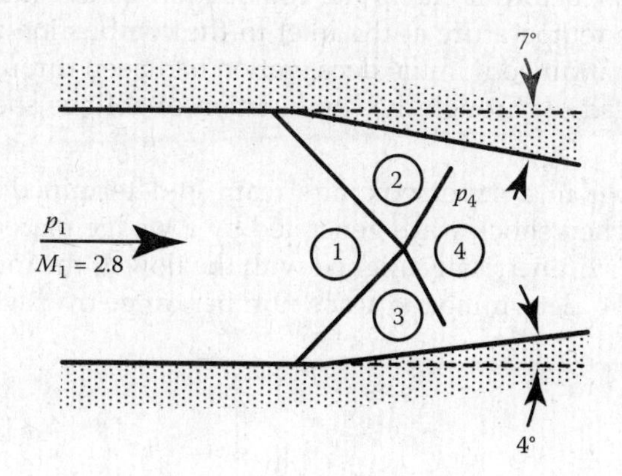

FIGURE P6.17
Flow situation considered.

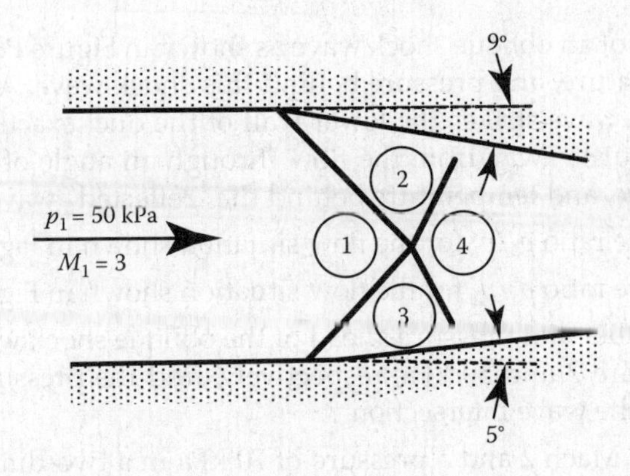

FIGURE P6.18
Flow situation considered.

and the lower wall turns sharply through an angle of 12° leading to the generation of two oblique shock waves that intersect each other. Sketch the flow pattern and find the flow direction, the Mach number, and the pressure immediately downstream of the shock intersection.

21. Consider an air stream flowing at Mach 3.2 with a pressure of 60 kPa. Consider two cases. In the first case, the stream passes through a single normal shock wave and is then isentropically decelerated to a very low velocity. In the second case, the flow first passes through an oblique shock that turns the flow through 25° and then passes through a normal shock wave before being isentropically decelerated to a very low velocity. Compare the pressures attained in the two cases. Do the results indicate which arrangement should be used in decelerating a flow from supersonic to subsonic velocities at the inlet to a turbo jet engine in a supersonic aircraft?
22. A ram-jet engine is fitted to a small aircraft that cruises at Mach 4 at an altitude where the pressure is 30 kPa and the temperature is –45°C. The air entering the engine is slowed to subsonic velocities by passing it through two oblique shock waves each of which turn the flow through 15° and by then passing it through a normal shock wave. Following the normal shock, the flow is isentropically decelerated to Mach 0.1 before it enters the combustion zone. Find the values of the pressure and the temperature at the inlet to the combustion zone. What values would have been attained if initial deceleration had been through a single normal shock wave instead of through the combination of oblique shocks and a normal shock?
23. The Mach number in a supersonic airstream is determined by measuring the angle of the attached shock wave generated by a wedge placed in the airstream, the wedge being symmetrically aligned with the flow. If the included angle of the wedge probe is 34° determine the Mach number range over which the probe can be effectively used.

7

Expansion Waves: Prandtl–Meyer Flow

Introduction

The discussion given in Chapters 5 and 6 was concerned with waves that involved an increase in pressure, i.e., with shock waves. In this chapter, attention will be given to the types of waves that are generated when there is a decrease in pressure. For example, the type of wave that is generated when a supersonic flow passes over a convex corner and the type of wave that is generated when the end of a duct containing a gas at a pressure that is higher than that in the surrounding air is suddenly opened will be discussed in this chapter. These two situations are illustrated in Figure 7.1.

Steady supersonic flows around convex corners will first be addressed in this chapter. Attention will then be given to unsteady flows.

Prandtl–Meyer Flow

In the previous chapter, supersonic flow around a concave corner, i.e., a corner involving a positive angular change in flow direction, was considered. It was indicated there that the flow over such a corner was associated with an oblique shock wave, this shock wave originating at the corner when it is sharp. Consider, now, the flow around a convex corner as shown in Figure 7.2. To determine whether an oblique shock wave also occurs in this case, it is assumed that it does occur, a sharp corner being considered for simplicity as shown in Figure 7.2.

Consider the velocity components indicated in Figure 7.2. For the reasons given in the previous chapter, $L_1 = L_2$, and since V_2 must be parallel to the downstream wall, geometrical considerations show that $N_2 > N_1$. However, N_2 and N_1 must be related by the normal shock wave relations, and in dealing with normal shock waves, it was shown that an expansive shock was not possible since it would violate the second law of thermodynamics. It is therefore not possible with an oblique shock wave for N_2 to be greater than N_1 and the flow over a convex corner cannot therefore take place through an oblique shock.

To understand the actual flow that occurs when a supersonic flow passes around a convex corner, consider what happens, in general, when the flow is turned through a differentially small angle, $d\theta$, this producing differentially small changes dp, $d\rho$, and dT in the pressure, density, and temperature, respectively. The present analysis applies whether $d\theta$ is positive or negative, i.e., whether the corner is concave or convex, the changes through the differentially weak Mach wave produced being isentropic (see later). By the reasoning

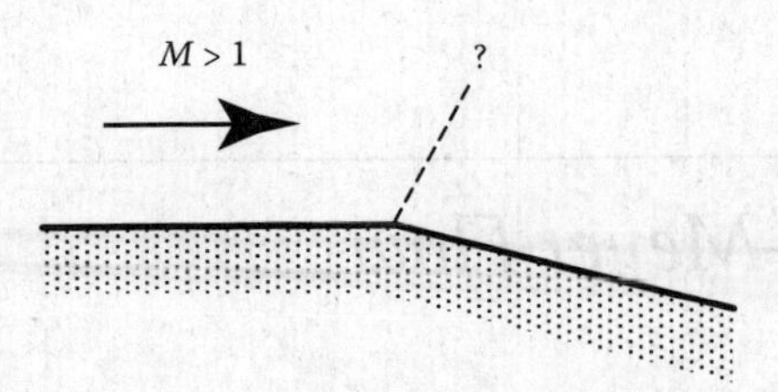

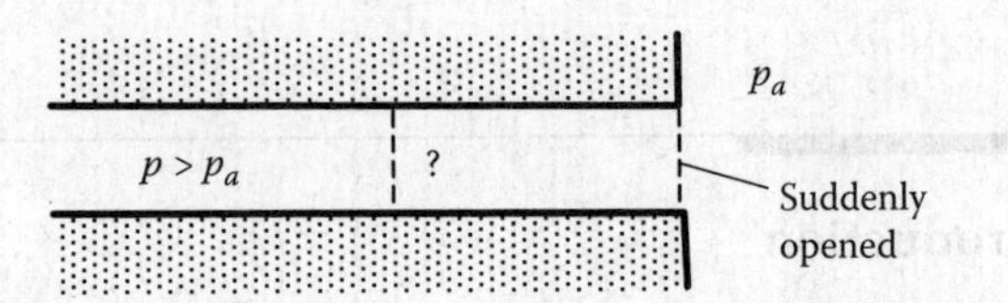

FIGURE 7.1
Flows involving a pressure decrease.

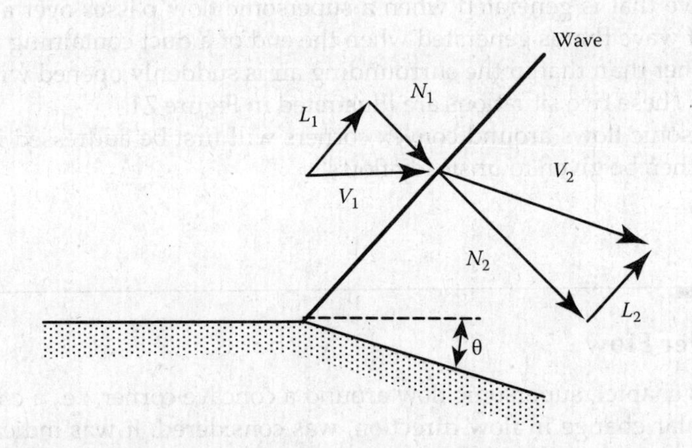

FIGURE 7.2
Assumed flow around convex corner.

previously given, the velocity component parallel to the wave, L, is unchanged by the wave. Hence, considering unit area of the wave shown in Figure 7.3, the equations of continuity and momentum give

$$\rho N = (\rho + d\rho)(N + dN)$$

i.e.,

$$\rho\, dN + N\, d\rho = 0 \tag{7.1}$$

higher-order terms having been neglected, and

$$p - (p + dp) = \rho N[(N + dN) - N]$$

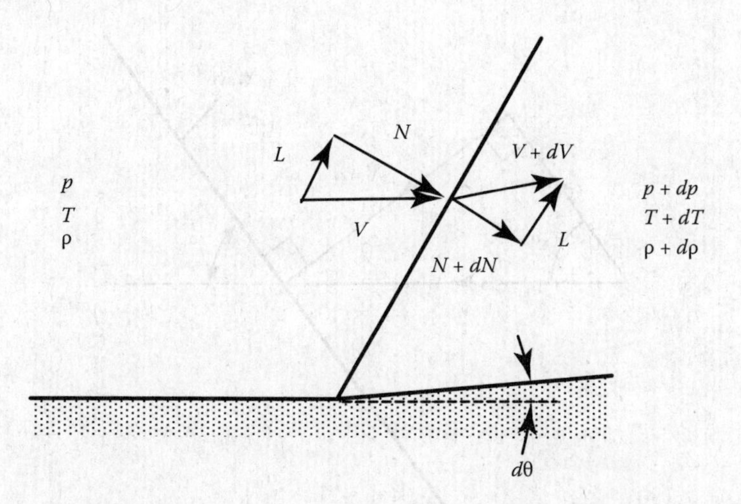

FIGURE 7.3
Changes produced by a weak wave.

i.e.,

$$-dp = \rho N\, dN \tag{7.2}$$

Substituting for dN from Equation 7.2 into Equation 7.1 gives

$$N^2 = \frac{dp}{d\rho} \tag{7.3}$$

Now, in the limiting case of a very weak wave that is being considered here, $dp/d\rho$ will be equal to the square of the upstream speed of sound, i.e.,

$$N^2 = a^2 \quad \text{or} \quad N = a \tag{7.4}$$

This is indicated in Figure 7.4. Further, since L is unchanged by the presence of the disturbance, it follows that

$$(V + dV)\cos(\alpha - d\theta) = V\cos\alpha$$

i.e.,

$$(V + dV)(\cos\alpha\cos d\theta + \sin\alpha\sin d\theta) = V\cos\alpha$$

Expanding this equation and ignoring higher-order terms then gives

$$V\cos\alpha + V\sin\alpha\, d\theta + dV\cos\alpha = V\cos\alpha$$

Therefore,

$$\frac{dV}{V} = -\tan\alpha\, d\theta = \frac{-d\theta}{\sqrt{M^2 - 1}} \tag{7.5}$$

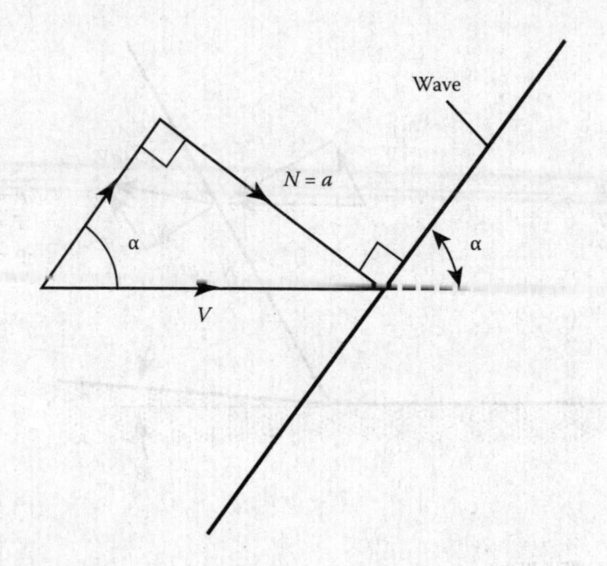

FIGURE 7.4
Upstream velocity components near a weak wave.

Further, since the energy equation gives

$$\left(\frac{2\gamma}{\gamma-1}\right)\left(\frac{p}{\rho}\right)+V^2=\left(\frac{2\gamma}{\gamma-1}\right)\left(\frac{p+dp}{\rho+d\rho}\right)+(V+dV)^2$$

i.e., ignoring higher-order terms,

$$\left(\frac{2\gamma}{\gamma-1}\right)\left(\frac{p}{\rho}\right)+V^2=\left(\frac{2\gamma}{\gamma-1}\right)\left(\frac{p}{\rho}\right)\left[1+\frac{dp}{p}-\frac{d\rho}{\rho}\right]+V^2+2V\,dV$$

The following applies

$$\left(\frac{2\gamma}{\gamma-1}\right)\left(\frac{p}{\rho}\right)\left[\frac{dp}{p}-\frac{dp}{p}\frac{p}{\rho}\frac{d\rho}{dp}\right]=-2V\,dV \tag{7.6}$$

However, by the previously made assumptions

$$\frac{\gamma p}{\rho}=\frac{dp}{d\rho}=a^2$$

Thus, Equation 7.6 becomes

$$\left(\frac{2a^2}{\gamma-1}\right)\frac{dp}{p}\left[1-\frac{1}{\gamma}\right]=-2V\,dV$$

i.e.,

$$\frac{dp}{p} = -\frac{\gamma V\, dV}{a^2} = -\gamma M^2 \frac{dV}{V} \tag{7.7}$$

or using Equation 7.5

$$\frac{dp}{p} = \frac{\gamma M^2}{\sqrt{M^2 - 1}} d\theta \tag{7.8}$$

Further, since

$$\frac{d\rho}{\rho} = \frac{d\rho}{dp}\frac{dp}{p}\frac{p}{\rho} = \frac{1}{a^2}\frac{dp}{p}\frac{a^2}{\gamma} = \frac{1}{\gamma}\frac{dp}{p}$$

it follows, using Equation 7.8, that

$$\frac{d\rho}{\rho} = \frac{M^2}{\sqrt{M^2 - 1}} d\theta \tag{7.9}$$

Similarly,

$$\begin{aligned}
\frac{ds}{R} &= \left(\frac{1}{\gamma - 1}\right)\ln\left(\frac{p_2}{p_1}\right) - \left(\frac{\gamma}{\gamma - 1}\right)\ln\left(\frac{\rho_2}{\rho_1}\right) \\
&= \left(\frac{1}{\gamma - 1}\right)\ln\left(1 + \frac{dp}{p}\right) - \left(\frac{\gamma}{\gamma - 1}\right)\ln\left(1 + \frac{d\rho}{\rho}\right) \\
&= \left(\frac{1}{\gamma - 1}\right)\frac{\gamma M^2 d\theta}{\sqrt{M^2 - 1}} - \left(\frac{\gamma}{\gamma - 1}\right)\frac{M^2 d\theta}{\sqrt{M^2 - 1}} \\
&= 0
\end{aligned} \tag{7.10}$$

which shows that there is no change in entropy across the weak wave being considered.

Lastly, since

$$M^2 = \frac{V^2}{a^2} = \frac{V^2 \rho}{\gamma p}$$

the differential change in M, i.e., dM, is given by

$$2M\, dM = 2V\, dV\left(\frac{\rho}{\gamma p}\right) + \left(\frac{V}{\gamma p}\right)d\rho - \left(\frac{V\rho}{\gamma p^2}\right)dp$$

Thus,

$$2\frac{dM}{M} = 2\frac{dV}{V} + \frac{d\rho}{\rho} - \frac{dp}{p} = [-2 + M^2 - \gamma M^2]\frac{d\theta}{\sqrt{M^2-1}}$$

Hence,

$$\frac{dM}{M} = \left[1 + \frac{\gamma - 1}{2}M^2\right]\frac{(-d\theta)}{\sqrt{M^2-1}} \tag{7.11}$$

Thus, a differentially small change in flow direction produces an isentropic disturbance, which is such that

$$dV \propto -d\theta$$

$$dp \propto d\theta$$

$$d\rho \propto d\theta$$

$$dM \propto -d\theta$$

$$ds = 0$$

these changes also being indicated in Figure 7.5.

Thus, if flow around a corner, which may be considered to consist of an infinite number of differentially small angular changes as shown in Figure 7.6, is considered, it follows from the preceding results that for positive angular changes, the Mach number decreases, and the waves converge to form an oblique shock wave, whereas for a negative angular change, the waves diverge. Thus, for negative changes in wall angle a region consisting of Mach waves is generated and the flow remains isentropic throughout. Such flows are called Prandtl–Meyer flows and their form is as shown in Figure 7.7.

To analytically determine the changes produced by such a flow, it is noted that Equation 7.11 must apply locally at all points within the expansion fan such as the point shown in Figure 7.8. Therefore, if this equation is integrated, it will give the relation between the changes in flow properties and the change in flow direction across the wave.

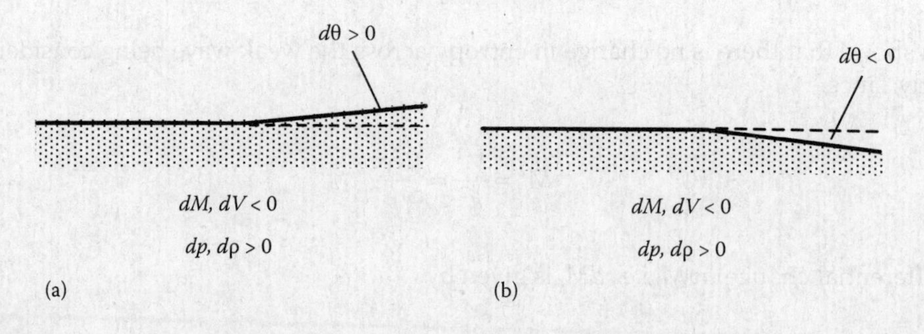

FIGURE 7.5
Changes in values of flow variables produced by a weak wave: (a) disturbance compressive and (b) disturbance expansive.

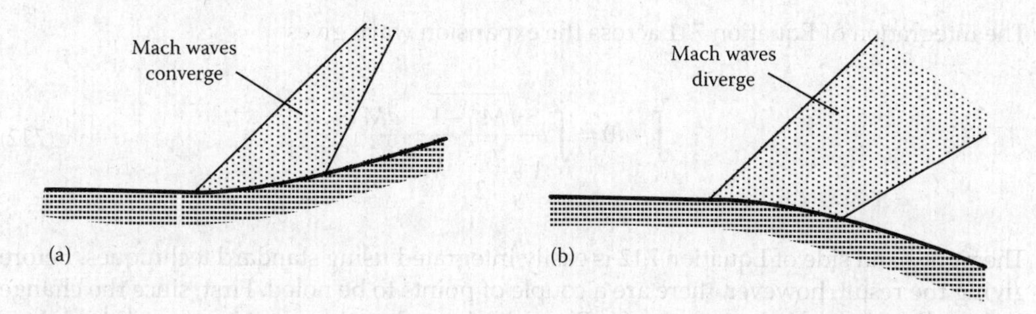

FIGURE 7.6
Mach waves produced in flow around a corner with finite angle: (a) pressure increases through wave system and (b) pressure decreases through wave system.

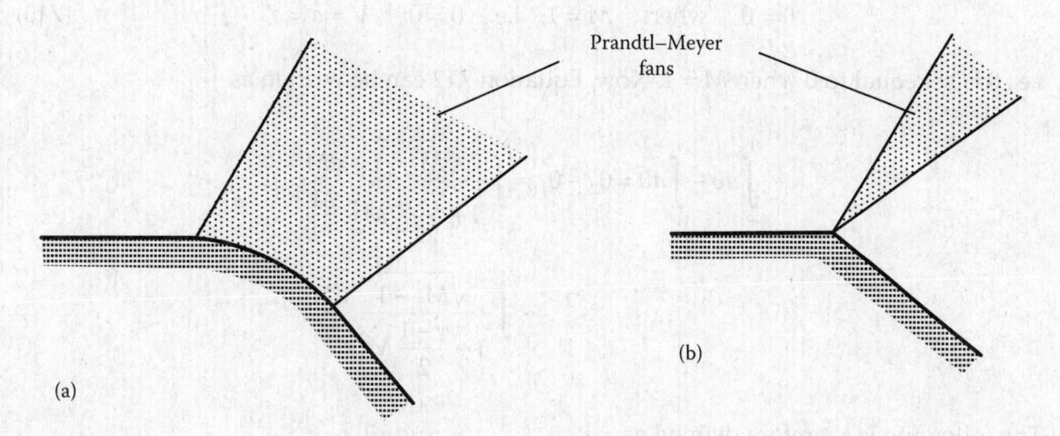

FIGURE 7.7
Prandtl–Meyer flow: (a) centered fan and (b) noncentered fan.

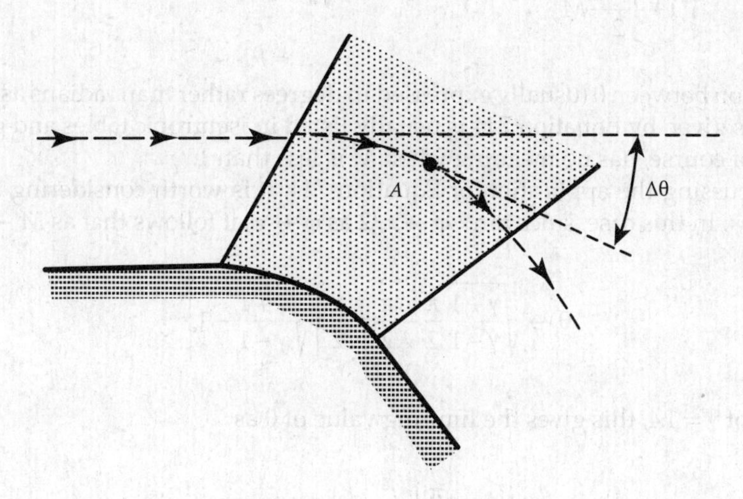

FIGURE 7.8
Prandtl–Meyer flow.

The integration of Equation 7.11 across the expansion wave gives

$$\int_{\theta_1}^{\theta_2} -d\theta = \int_{M_1}^{M_2} \frac{\sqrt{M^2-1}}{1+\frac{\gamma-1}{2}M^2}\frac{dM}{M} \tag{7.12}$$

The right-hand side of Equation 7.12 is easily integrated using standard techniques. Before giving the result, however, there are a couple of points to be noted. First, since the change in flow direction must be negative for Prandtl–Meyer flow to exist, it is convenient to drop the negative sign on the left-hand side of this equation. Second, to express the results in as convenient a form as possible, some standard base condition is used in evaluating the integral. This initial boundary condition is arbitrarily taken as

$$\theta = 0 \quad \text{when} \quad M = 1, \text{ i.e., } \theta = 0: \quad V = a = a^* \tag{7.13}$$

i.e., θ is set equal to 0 when $M = 1$. Now, Equation 7.12 can be written as

$$\int_0^{\theta_2} d\theta - \int_0^{\theta_1} d\theta = \theta_2 - \theta_1 = \int_1^{M_2} \frac{\sqrt{M^2-1}}{1+\frac{\gamma-1}{2}M^2}\frac{dM}{M} - \int_1^{M_1} \frac{\sqrt{M^2-1}}{1+\frac{\gamma-1}{2}M^2}\frac{dM}{M}$$

The following is therefore defined as

$$\theta = \int_1^{M} \frac{\sqrt{M^2-1}}{1+\frac{\gamma-1}{2}M^2}\frac{dM}{M} = \sqrt{\frac{\gamma+1}{\gamma-1}}\tan^{-1}\sqrt{\frac{\gamma-1}{\gamma+1}(M^2-1)} - \tan^{-1}\sqrt{M^2-1} \tag{7.14}$$

and the relation between θ (usually expressed in degrees rather than radians as in Equation 7.14) and M as given by Equation 7.14 is usually listed in isentropic tables and given by the software. θ, of course, has no meaning when M is less than 1.

Before discussing the application of Equation 7.14, it is worth considering the limiting case of $M \to \infty$. In this case, since $\tan^{-1}\varphi \to \pi/2$ as $\varphi \to \infty$, it follows that as $M \to \infty$,

$$\theta \to \sqrt{\frac{\gamma+1}{\gamma-1}}\frac{\pi}{2} - \frac{\pi}{2} = \frac{\pi}{2}\left\{\sqrt{\frac{\gamma+1}{\gamma-1}} - 1\right\} \tag{7.15}$$

For the case of $\gamma = 1.4$, this gives the limiting value of θ as

$$\theta_{max} = \frac{\pi}{2}\left\{\sqrt{6}-1\right\} = 130.5° \tag{7.16}$$

Thus, if a flow at Mach 1 is turned through an angle of 130.5°, an infinite Mach number is generated and the pressure falls to zero. Expansion through a greater angle would, according to the present theory, lead to a vacuum adjacent to the wall as indicated in Figure 7.9. Of course, in reality, the continuum and ideal gas assumptions cease to be valid before this situation is reached.

Next, consider the application of Equation 7.14 to the calculation of the flow changes produced by a Prandtl–Meyer expansion. Referring to Figure 7.10, the procedure is as follows.

1. From tables or graphs or from the equation find the value of θ corresponding to M_1, i.e., θ_1. This is equivalent to assuming that the initial flow was generated by an expansion around a hypothetical corner from Mach 1, the reference Mach number in the tables, to Mach M_1.

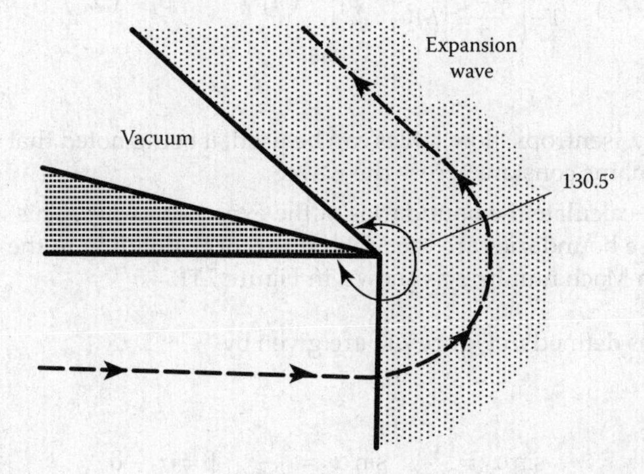

FIGURE 7.9
Expansion to zero pressure.

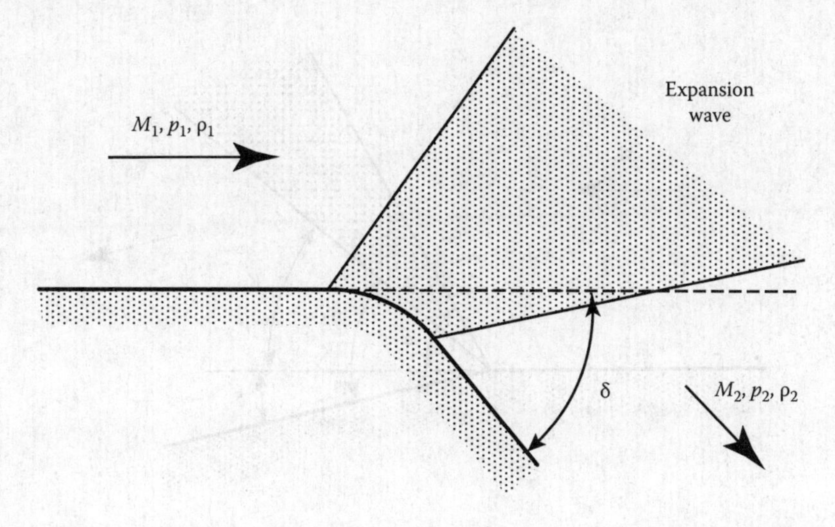

FIGURE 7.10
Flow changes through a Prandtl–Meyer wave.

2. Calculate the θ for the flow downstream of the corner. This will be given by

$$\theta_2 = \theta_1 + \delta$$

3. Find, using tables or graphs or the equation, the downstream Mach number, M_2, corresponding to this value of θ_2.
4. Any other required property of the downstream flow is obtained by noting that the expansion is isentropic and that the following relations therefore apply

$$\frac{T_2}{T_1} = \frac{1 + \left(\frac{\gamma - 1}{2}\right) M_1^2}{1 + \left(\frac{\gamma - 1}{2}\right) M_2^2}, \quad \frac{p_2}{p_1} = \left(\frac{T_2}{T_1}\right)^{\frac{\gamma}{\gamma - 1}}, \quad \frac{\rho_2}{\rho_1} = \left(\frac{T_2}{T_1}\right)^{\frac{1}{\gamma - 1}}$$

   Alternatively, isentropic flow tables can be used, it being noted that the stagnation pressure remains constant across the wave.
5. If necessary, calculate the boundaries of the expansion wave. This is done by noting that these boundaries are the Mach lines corresponding to the upstream and downstream Mach numbers as shown in Figure 7.11.

The various angles defined in Figure 7.11 are given by

$$\sin \alpha_1 = \frac{1}{M_1}, \quad \sin \alpha_2 = \frac{1}{M_2}, \quad \beta = \alpha_2 - \delta$$

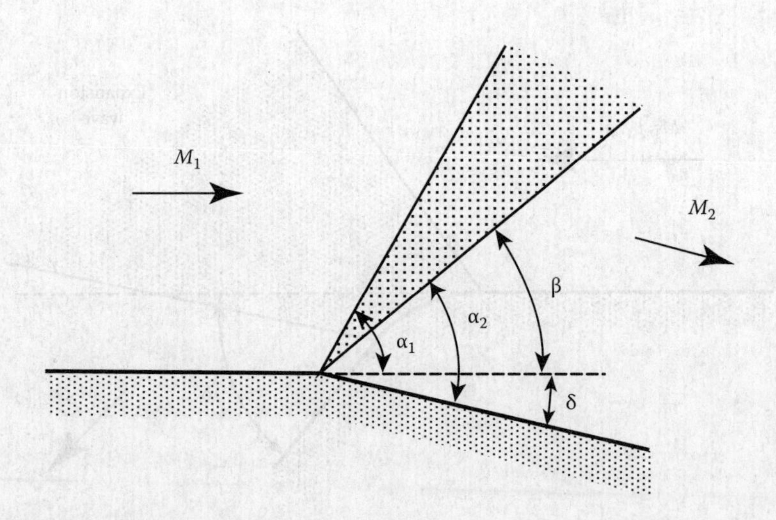

FIGURE 7.11
Angles associated with a centered expansion wave.

Example 7.1

Air flows at Mach 1.8 with a pressure of 90 kPa and a temperature of 15°C down a wide channel. The upper wall of this channel turns through an angle of 5° "away from the flow" leading to the generation of an expansion wave. Find the pressure, Mach number, and temperature behind this expansion wave.

Solution

The flow situation being considered is shown in Figure E7.1.

For $M_1 = 1.8$ isentropic flow relations or tables or the software give

$$\theta_1 = 20.73°, \quad \frac{p_{01}}{p_1} = 5.746, \quad \frac{T_{01}}{T_1} = 1.648$$

After, i.e., downstream of, the expansion wave

$$\theta_2 = \theta_1 + 5 = 20.73 + 5 = 25.73°$$

For this value of θ, isentropic flow relations or tables or the software give

$$M_2 = 1.98, \quad \frac{p_{02}}{p_2} = 7.585, \quad \frac{T_{02}}{T_2} = 1.784$$

Therefore, because the flow through the expansion wave is isentropic so that $p_{02} = p_{01}$ and $T_{02} = T_{01}$, it follows that

$$T_2 = \frac{T_{01}}{T_1}\frac{T_2}{T_{02}}T_1 = \frac{1.648}{1.784} \times (273 + 15) = 266.0 \text{ K} = -7°\text{C}$$

and

$$p_2 = \frac{p_{01}}{p_1}\frac{p_2}{p_{02}}p_1 = \frac{5.746}{7.585} \times 90 = 68.2 \text{ kPa}$$

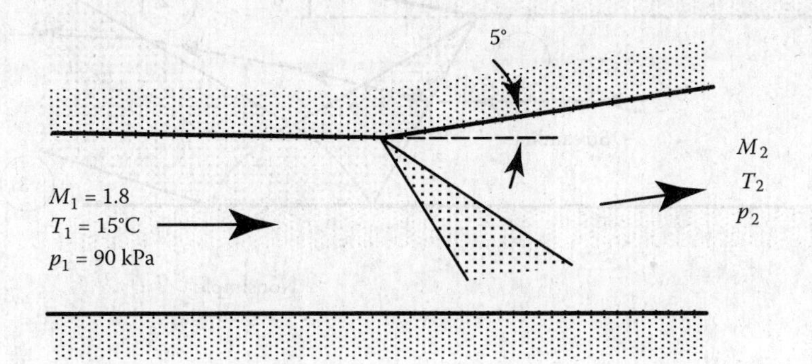

FIGURE E7.1
Flow situation considered.

Hence, the pressure, Mach number, and temperature downstream of the expansion wave are 68.2 kPa, 1.98, and −7°C, respectively.

Reflection and Interaction of Expansion Waves

Just as with oblique shock waves, expansion waves can undergo reflection and can interact with each other. Consider, first, the reflection of an expansion wave from a straight wall as shown in Figure 7.12.

If θ_1 is the Prandtl–Meyer angle corresponding to the initial flow conditions, i.e., to conditions in region 1, then the Prandtl–Meyer angle for the flow in the intermediate region 2 is, of course, given by

$$\theta_2 = \theta_1 + \delta$$

Since the end flow, i.e., the flow in region 3, must again be parallel to the wall, the reflected wave must also turn the flow through an angle of δ so that

$$\theta_3 = \theta_2 + \delta = \theta_1 + 2\delta$$

Once this angle is determined, the Mach number in region 3 can be found, and since the whole flow is isentropic, the conditions in region 3 can then be found in terms of those in region 1 using isentropic relations.

Inside the region of interaction of the incident and reflected waves, the relation between θ and M previously derived cannot be directly applied. This region is known as a nonsimple region. However, the flow is, of course, isentropic throughout.

Example 7.2

Air is flowing down a duct at Mach 2.0 with a pressure of 90 kPa. The upper wall of the duct turns "away" from the flow through an angle of 10°, leading to the formation of an

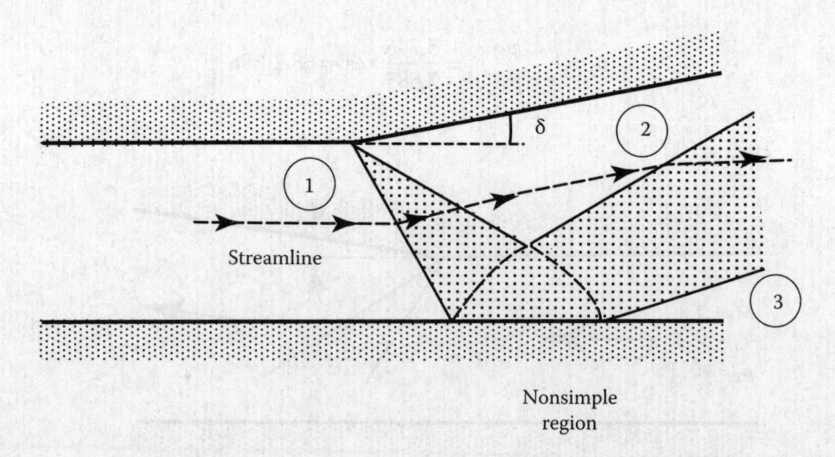

FIGURE 7.12
Reflection of an expansion wave off a flat wall. The regions 1, 2, and 3 used in determining the changes in such a flow are shown in this figure.

expansion wave. This wave is reflected off the flat lower wall of the channel. Find the pressure after the reflection.

Solution

The flow situation here being considered is shown in Figure E7.2

The initial wave and the reflected wave both turn the flow through 10°. Now, in the initial flow at $M_1 = 2$, isentropic relations or tables or the software give

$$\frac{p_{01}}{p_1} = 7.83, \qquad \theta_1 = 26.38°$$

Hence,

$$\theta_3 = \theta_1 + 10 + 10 = 46.38°$$

Isentropic relations or tables or the software then give for this value of θ

$$M_3 = 2.83, \quad \frac{p_{03}}{p_3} = 28.41$$

However, the entire flow is isentropic; thus,

$$p_{03} = p_{01}$$

and thus,

$$p_3 = \frac{p_3}{p_{03}}\frac{p_{01}}{p_1}p_1 = \frac{7.83}{28.41} \times 90 = 24.81 \text{ kPa}$$

Therefore, the pressure and Mach number behind the reflected wave are 24.81 kPa and 2.83, respectively.

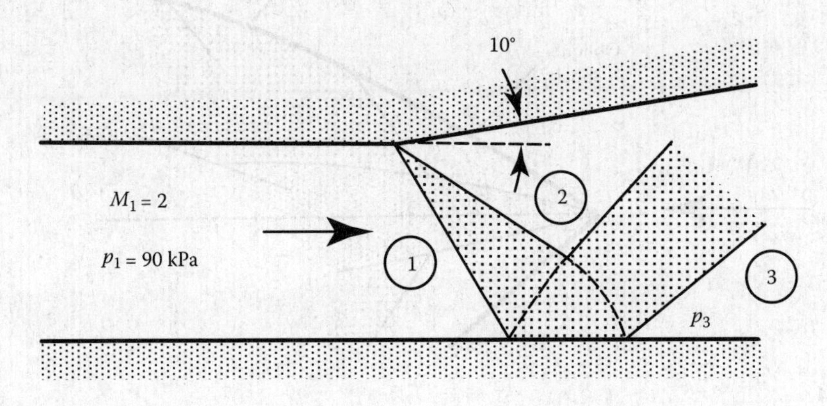

FIGURE E7.2
Flow situation considered.

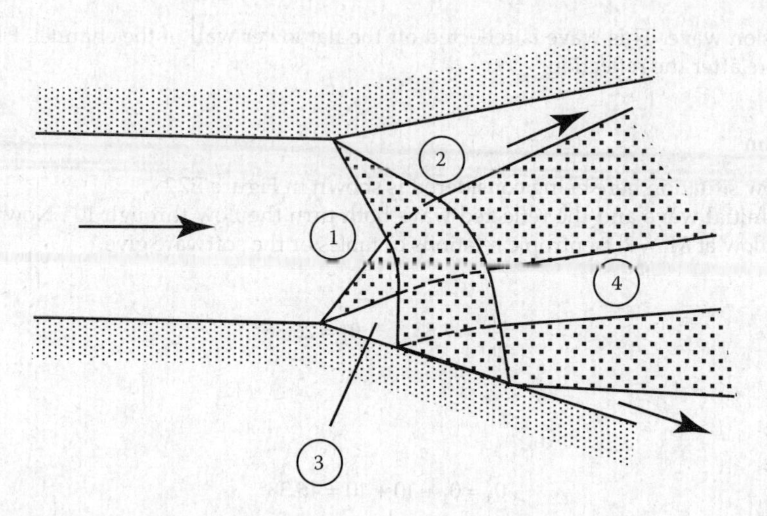

FIGURE 7.13
Interaction of expansion waves.

When expansion waves interact, the flow resembles that shown in Figure 7.13. Since the entire flow is isentropic, region 4 must be a region of uniform properties, no slipstreams, such as those that are generated when shock waves interact, being generated when expansion waves interact.

In flows over bodies, expansion waves often interact with a shock wave, the shock being attenuated (weakened) by the interaction. An example of such an interaction is shown in Figure 7.14. The interaction is complicated by the generation of a series of reflected waves.

An expansion wave can also be generated by the "reflection" of an oblique shock wave off a constant pressure boundary. To see how this can happen, consider a wedge-shaped body placed in a two-dimensional supersonic jet flow as shown in Figure 7.15.

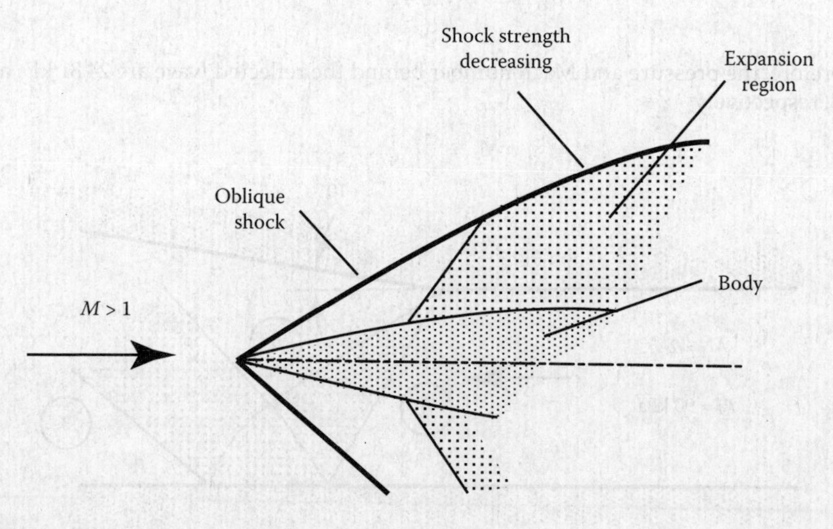

FIGURE 7.14
Interaction of an expansion wave with an oblique shock wave.

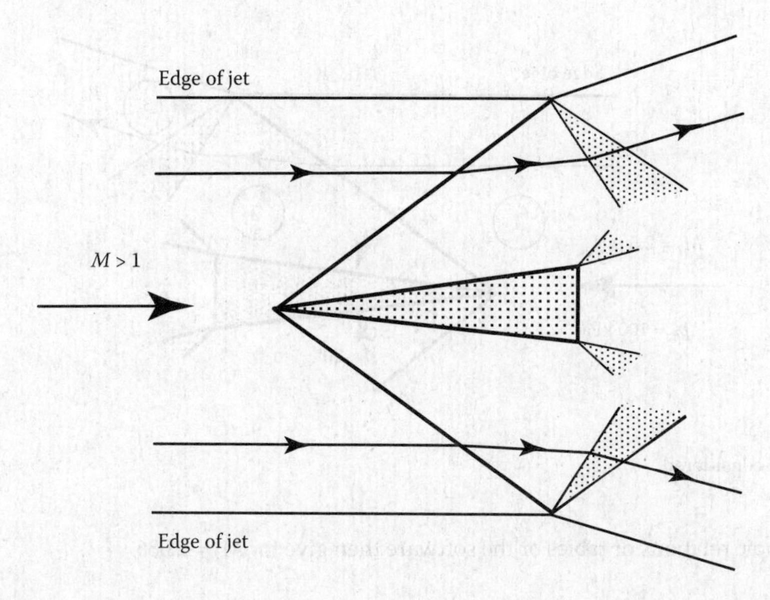

FIGURE 7.15
Reflection of an oblique shock wave from the edge of a jet flow.

The edges of the jet are exposed to the stagnant surrounding gas and must therefore remain at the pressure that exists in this ambient gas. The flow ahead of the body, because it is a parallel flow, must also all be at the ambient pressure as indicated in Figure 7.15. Oblique shock waves are generated at the leading edge of the body, these shock waves increasing the pressure. Expansion waves are therefore generated at the edges of the jet at the points at which the shock impinges on these edges, the expansion waves being of such a strength that they decrease the pressure back to that in the ambient fluid as shown in Figure 7.15.

Example 7.3

A wedge-shaped body with an included angle of 20° is placed symmetrically in a plane jet of air, the Mach number in the jet being 2.5. The ambient air pressure is 100 kPa. The oblique shock waves generated at the leading edge of the wedge are reflected off the edges of the jet. Find the Mach number and pressure in the flow downstream of the reflected waves.

Solution

The flow situation considered in this example is shown in Figure E7.3.

First consider the oblique shock wave. For $M = 2.5$ and $\delta = 10°$, using the oblique shock relations or the oblique shock chart and the normal shock tables or the software give, as discussed in the previous chapter,

$$M_2 = 2.086, \qquad \frac{p_2}{p_1} = 1.866$$

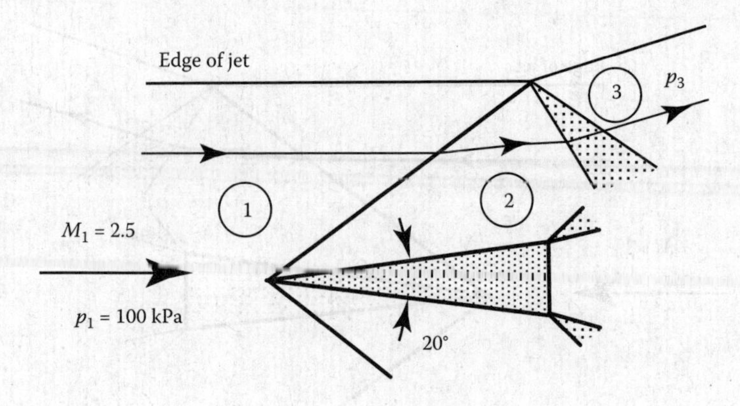

FIGURE E7.3
Flow situation considered.

Isentropic relations or tables or the software then give for $M_2 = 2.086$,

$$\frac{p_{20}}{p_2} = 8.945, \qquad \theta_2 = 28.72°$$

The strength of the "reflected" expansion waves must be such that $p_3 = p_1$, i.e., such that

$$\frac{p_{03}}{p_3} = \frac{p_{03}}{p_{02}}\frac{p_{02}}{p_2}\frac{p_2}{p_1}$$

i.e., because the flow through the expansion wave is isentropic, and therefore, $p_{03} = p_{02}$,

$$\frac{p_{03}}{p_3} = 8.95 \times 1.866 = 16.70$$

For this value of p_{03}/p_3, isentropic relations or tables or the software give $M_3 = 2.485$.

The Mach number downstream of the "reflected" expansion waves is therefore 2.485. The pressure behind these waves is, as assumed in the above calculations, the same as that in the initial flow, i.e., 100 kPa.

Boundary Layer Effects on Expansion Waves

For the same reasons that the presence of a boundary layer causes a "spreading out" of the pressure change when a shock wave impinges on a wall, the presence of a boundary layer modifies an expansion wave near a wall. This is illustrated in Figure 7.16. The extent of the interaction again depends on the thickness and type, i.e., laminar or turbulent, of boundary layer.

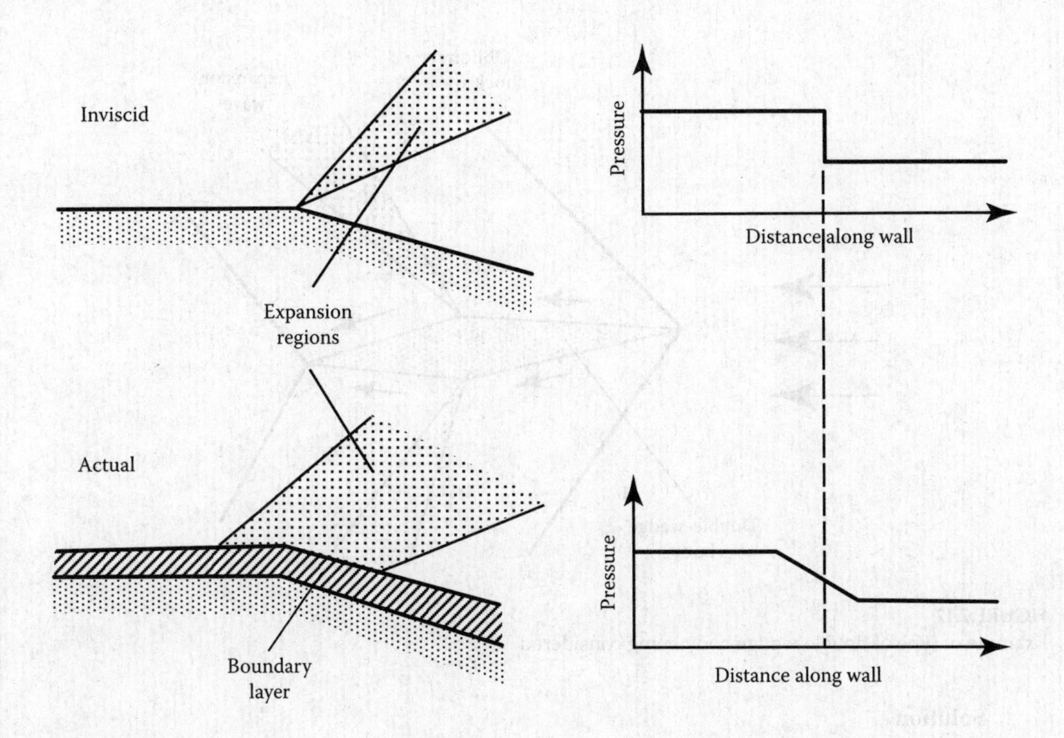

FIGURE 7.16
Interaction of an expansion wave with a wall boundary layer.

Flow over Bodies Involving Shock and Expansion Waves

Many bodies over which an effectively two-dimensional supersonic flow occurs in practice can be assumed to consist of a series of flat surfaces. For example, the type of body shown in Figure 7.17 is similar to the cross-sectional shape of the control surfaces used on some supersonic vehicles.

The flow over such a body can be calculated by noting that a series of oblique shock waves and expansion waves occur that cause the flow to be locally parallel to each of the surfaces as illustrated in Figure 7.17. Provided that the body is "slender," any secondary waves generated as a result of the interaction of the shock waves and expansion waves generated by the body will not impinge on the body and their presence will not effect the flow over the body. The flow over the body, and the pressure acting on the surfaces over the body can then be calculated by separately using the oblique shock wave and the expansion wave results. Once the pressures on the surfaces of the body are found, the net force on the body can be found. This procedure is illustrated in the following examples.

Example 7.4

A simple wing may be modeled as a 0.25 m wide flat plate set at an angle of 3° to an air flow at Mach 2.5, the pressure in this flow being 60 kPa. Assuming that the flow over the wing is two-dimensional, estimate the lift and drag force per meter span due to the wave formation on the wing. What other factor causes drag on the wing?

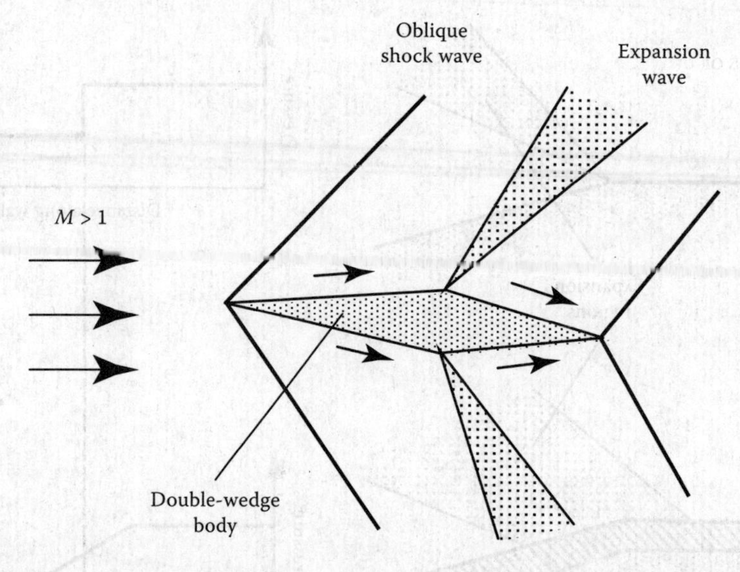

FIGURE 7.17
Example of type of double-wedge body being considered.

Solution

The flow situation being considered is shown in Figure E7.4.

An expansion wave forms on the upper surface at the leading edge. This wave turns the flow parallel to the upper surface of the plate. Similarly, an oblique shock wave forms on the lower surface at the leading edge. This wave turns the flow parallel to the lower surface of the plate. Waves also form at the trailing edge of the plate but these waves have no effect on the pressures on the surfaces of the plate, and they will not be considered here.

First consider the expansion wave, which turns the flow parallel to the upper surface, the region adjacent to the upper surface being designated as 2 as indicated in Figure E7.4.

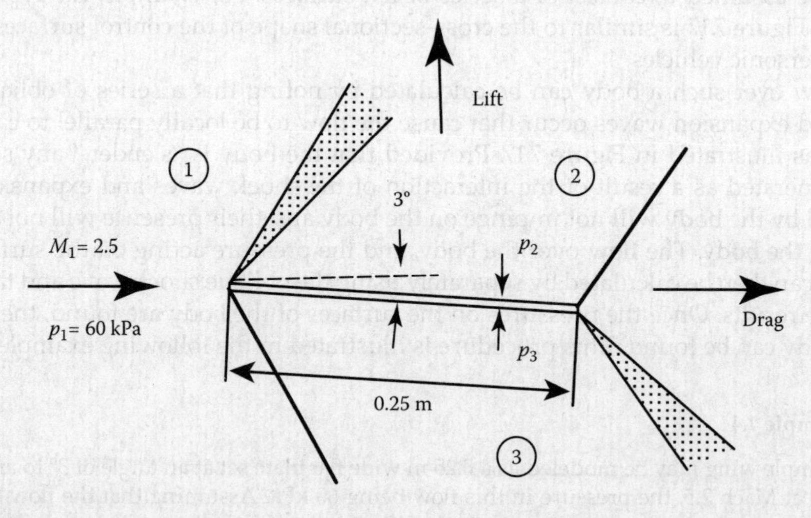

FIGURE E7.4
Flow situation considered.

Now, in the freestream, i.e., in region 1, where the Mach number, M_1, is 2.5, isentropic relations or tables or the software give

$$\frac{p_{01}}{p_1} = 17.09, \qquad \theta_1 = 39.13°$$

Hence, since the flow is turned through 3° by the expansion wave, it follows that

$$\theta_2 = 39.13 + 3° = 42.13°$$

Using this value of θ_2, isentropic relations or tables or the software give

$$M_2 = 2.63, \qquad \frac{p_{02}}{p_2} = 20.92$$

It therefore follows that, since the flow through the expansion wave is isentropic, which means that $p_{02} = p_{01}$,

$$p_2 = \frac{p_2}{p_{02}} \frac{p_{01}}{p_1} p_1 = \frac{17.09}{20.92} \times 60 = 49.02 \text{ kPa}$$

Therefore, the pressure acting on the upper surface of the plate is 49 kPa.

Next, consider the oblique shock wave that turns the flow parallel to the lower surface, the region adjacent to the lower surface being designated as 3 as indicated in Figure E7.4. Now, since M_1 is 2.5 and the turning angle δ produced by the oblique shock wave is 3°, oblique shock relations or charts or the software give

$$\beta = 26°, \quad M_{N1} = M_1 \sin\beta = 2.5 \sin 26 = 1.096$$

with the latter values being given directly by the software.

Normal shock relations or tables or the software then give for Mach 1.096

$$\frac{p_3}{p_1} = 1.23$$

From this, it follows that

$$p_3 = \frac{p_3}{p_1} p_1 = 1.23 \times 60 = 74 \text{ kPa}$$

Therefore, the pressure acting on the lower surface of the plate is 74 kPa.

The lift is the net force acting on the plate normal to the direction of initial flow, whereas the drag is the net force parallel to the direction of initial flow. Therefore, since the plate area per meter span is 0.25 m^2, it follows that

$$\text{Lift per meter span} = (p_3 - p_2)\, A\cos 3 = (74 - 49) \times 0.25 \times 0.999 = 6.23 \text{ kN/m span}$$

$$\text{Drag per meter span} = (p_3 - p_2)\, A\sin 3 = (74 - 49) \times 0.25 \times 0.0523 = 0.33 \text{ kN/m span}$$

Therefore, the lift and drag per meter span are 6.23 and 0.33 N, respectively. This drag is that due to the pressure variation about the plate. It is termed the "wave drag." The skin friction drag, i.e., the force on the plate due to viscous forces, will also contribute to the drag.

Example 7.5

Find the lift per meter span for the wedge shaped airfoil shown in Figure E7.5a. Also, sketch the flow pattern about the airfoil. The Mach number and the pressure ahead of the airfoil are 2.6 and 40 kPa, respectively.

Solution

Consider the angles shown in Figure E7.5b.

It will be seen that

$$0.4\tan 2° = 0.3\tan\psi$$

This gives $\psi = 2.67°$. It then follows that

$$\phi = 2 + \psi = 2 + 2.67 = 4.67°$$

The wave pattern about the airfoil is then as shown in Figure E7.5c.

The waves that occur at the trailing edge do not effect the pressures on the surfaces of the airfoil and will not be analyzed here. The angles of turning produced by the waves are as follows:

Shock wave A – Angle of turn = 5°

Expansion wave B – Angle of turn = 1°

Expansion wave C – Angle of turn = 4.67°

Expansion wave D – Angle of turn = 4.67°

First consider shock wave A, which separates regions 1 and 2. Since M_1 is 2.6 and δ is 5°, oblique shock relations or charts or the software give

$$\beta = 26.5°, \quad M_{N1} = M_1\sin\beta = 2.6\sin 26.5 = 1.16$$

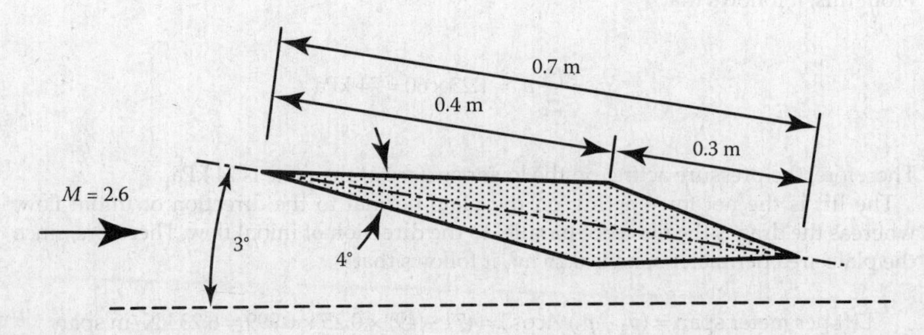

FIGURE E7.5a
Flow situation considered.

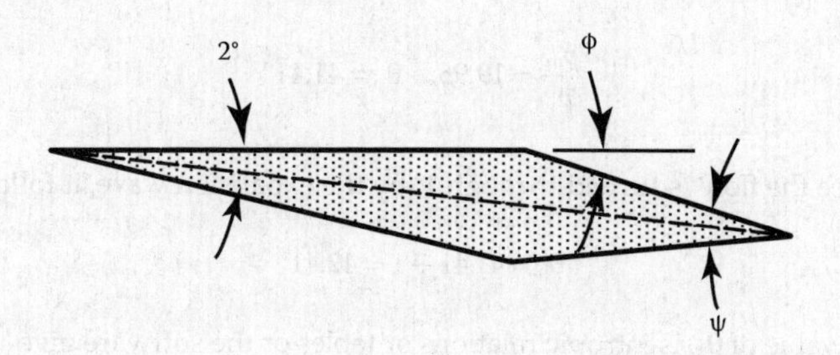

FIGURE E7.5b
Angles considered.

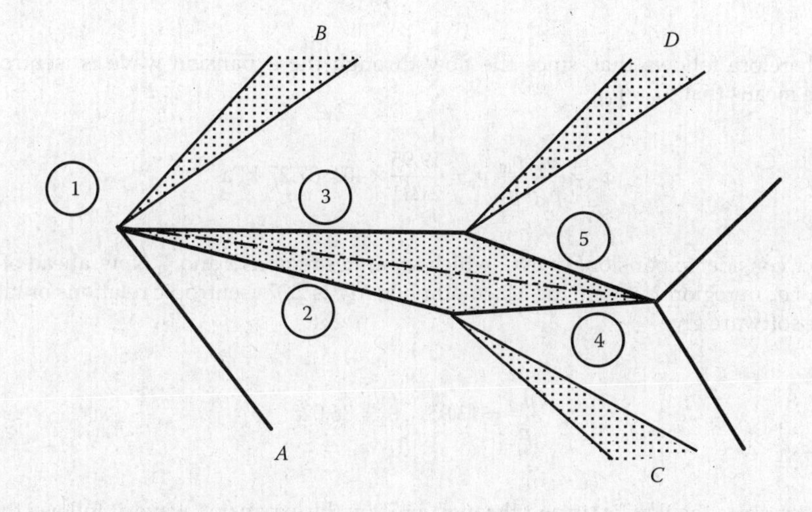

FIGURE E7.5c
Waves generated and flow regions considered.

The latter value is given directly by the software.

Normal shock relations or tables or the software then give for Mach 1.16

$$\frac{p_2}{p_1} = 1.403, \quad M_{N2} = 0.868$$

From this, it follows that

$$p_2 = \frac{p_2}{p_1} p_1 = 1.403 \times 40 = 56.1 \text{ kPa}$$

and

$$M_2 = \frac{M_{N2}}{\tan(26.5 - 5)} = 2.37$$

Next, consider expansion wave *B*, which separates regions 1 and 3. Now, in the freestream, i.e., in region 1, where the Mach number, M_1 is 2.6, isentropic relations or tables or the software give

$$\frac{p_{01}}{p_1} = 19.95, \quad \theta_1 = 41.41°$$

Hence, since the flow is turned through 1° by the expansion wave, it follows that

$$\theta_3 = 41.41 + 1 = 42.41°$$

Using this value of θ_3, isentropic relations or tables or the software give

$$M_3 = 2.64, \qquad \frac{p_{03}}{p_3} = 21.41$$

It therefore follows that, since the flow through the expansion wave is isentropic, which means that $p_{03} = p_{01}$,

$$p_3 = \frac{p_3}{p_{03}}\frac{p_{01}}{p_1}p_1 = \frac{19.95}{21.41} \times 40 = 37.27 \text{ kPa}$$

Next, consider expansion wave *C*, which separates regions 2 and 4. Now, ahead of the wave, i.e., in region 2, where the Mach number, M_2, is 2.37, isentropic relations or tables or the software give

$$\frac{p_{02}}{p_1} = 13.95, \quad \theta_2 = 36.02°$$

Hence, since the flow is turned through 4.67° by the expansion wave, it follows that

$$\theta_4 = 36.02 + 4.67 = 40.69°$$

Using this value of θ_4, isentropic relations or tables or the software give

$$M_4 = 2.57, \quad \frac{p_{04}}{p_4} = 19.05$$

It therefore follows that, since the flow through the expansion wave is isentropic, which means that $p_{04} = p_{02}$,

$$p_4 = \frac{p_4}{p_{04}}\frac{p_{02}}{p_2}p_2 = \frac{13.95}{19.05} \times 56.1 = 41.08 \text{ kPa}$$

Next consider expansion wave *D*, which separates regions 3 and 5. Now, ahead of the wave, i.e., in region 3, where the Mach number, M_3, is 2.64, isentropic relations or tables or the software give

$$\frac{p_{03}}{p_3} = 21.23, \quad \theta_3 = 42.41°$$

Hence, since the flow is turned through 4.67° by the expansion wave, it follows that

$$\theta_5 = 42.41 + 4.67 = 47.08°$$

Using this value of θ_5, isentropic relations or tables or the software give

$$M_5 = 2.87, \quad \frac{p_{05}}{p_5} = 30.19$$

It therefore follows that, since the flow through the expansion wave is isentropic, which means that $p_{05} = p_{03}$,

$$p_5 = \frac{p_5}{p_{05}}\frac{p_{03}}{p_3}p_3 = \frac{21.23}{30.19} \times 37.27 = 26.2 \text{ kPa}$$

Hence,

$$p_2 = 56.1 \text{ kPa}, \quad p_3 = 37.3 \text{ kPa}, \quad p_4 = 41.1 \text{ kPa}, \quad p_5 = 26.2 \text{ kPa}$$

and therefore, since the areas of the various surfaces per meter span are

$$A_2 = 0.4/\cos 2 = 0.400 \text{ m}^2 = A_3$$

$$A_2 = 0.3/\cos 2.67 = 0.300 \text{ m}^2 = A_5$$

the lift and drag are given by

$$\text{Lift} = 56.1 \times 0.4 \times \cos 5 - 37.3 \times 0.4 \times \cos 1$$

$$+ 41.1 \times 0.3 \times \cos 0.33 \times 26.2 \times 0.3 \times \cos 5.67 = 11.95 \text{ kN}$$

Hence, the lift per meter span is approximately 12 kN.

The waves that form at the trailing edge of the airfoil were not considered in the above two examples. Their strengths are, however, easily found by noting that the direction of flow and pressure must be the same in the entire region downstream of the airfoil.

Unsteady Expansion Waves

A type of flow that is related to the Prandtl–Meyer flow discussed above is flow through an unsteady expansion wave. To understand the basic characteristics of this flow consider a piston at the end of a long duct as shown in Figure 7.18.

If the piston is suddenly given a velocity, dV, in the direction of withdrawing it from the duct, the gas adjacent to the piston must have a velocity, dV, in the same direction as the piston and a weak wave (a sound wave) must propagate into the stationary gas at the local speed of sound, a. Across this wave, there will be decreases in the speed of sound and pressure of magnitudes da and dp, respectively. If the piston is then given another sudden

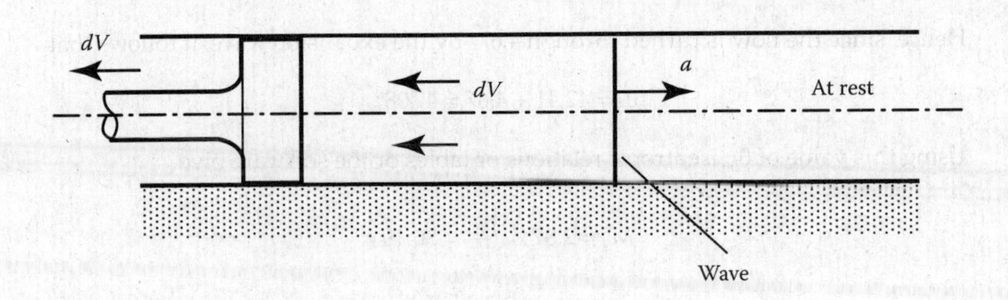

FIGURE 7.18
Generation of first weak wave.

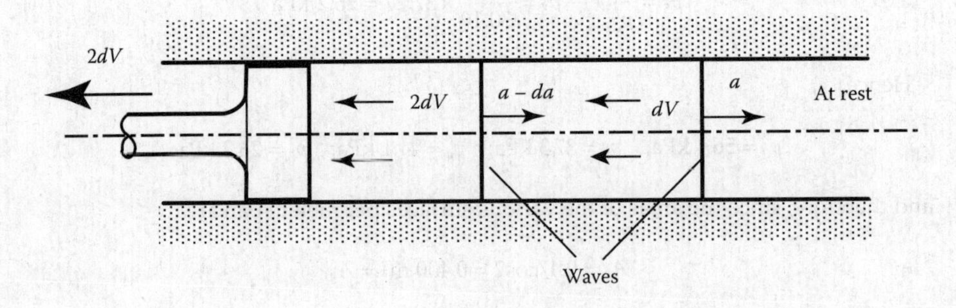

FIGURE 7.19
Generation of second weak wave.

acceleration to a new velocity, 2 dV, as shown in Figure 7.19, then another weak wave will be generated, which will move relative to the fluid at velocity $(a - da)$, i.e., the local velocity of sound. However, since the fluid ahead of this second wave already has velocity dV in the opposite direction to that of wave propagation, the actual speed at which the wave propagates relative to the duct is $(a - da - dV)$. This is shown in Figure 7.19.

If the piston had, in fact, been smoothly accelerated up to a velocity V_p, then the acceleration can be thought of as consisting of a series of differentially small jumps in velocity each of which produces an expansion wave that propagates down the duct at a lower velocity than its predecessor, thereby leading to an ever-widening expansion region. The process is conveniently shown on an x–t diagram, where x is the distance along the duct and t is the time. Such a diagram for the process being considered is shown in Figure 7.20.

The initial wave propagates at a speed, a_1, into the fluid so that the position of the head of the expansion is given by

$$\text{Head:} \quad x = a_1 t \tag{7.17}$$

The tail of the wave propagates at speed, a_2, relative to the fluid in region 2, and therefore, since the fluid in this region has velocity, u_p, in the direction of piston motion, the velocity of the tail of the wave relative to the walls of the duct is

$$\text{Tail:} \quad x - x_f = (a_2 - u_p)(t - t_f) \tag{7.18}$$

If the piston is instantly accelerated from zero velocity to velocity u_p, the x–t diagram takes the form shown in Figure 7.21.

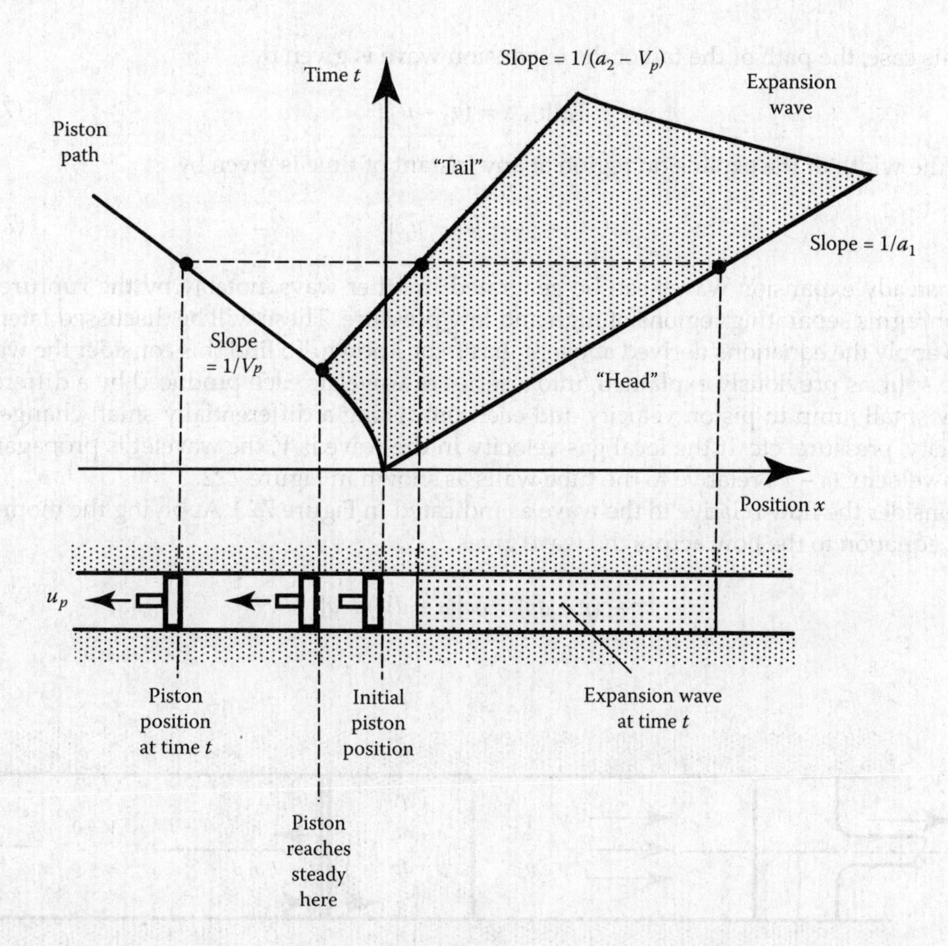

FIGURE 7.20
x–t diagram for wave system generated by accelerating piston.

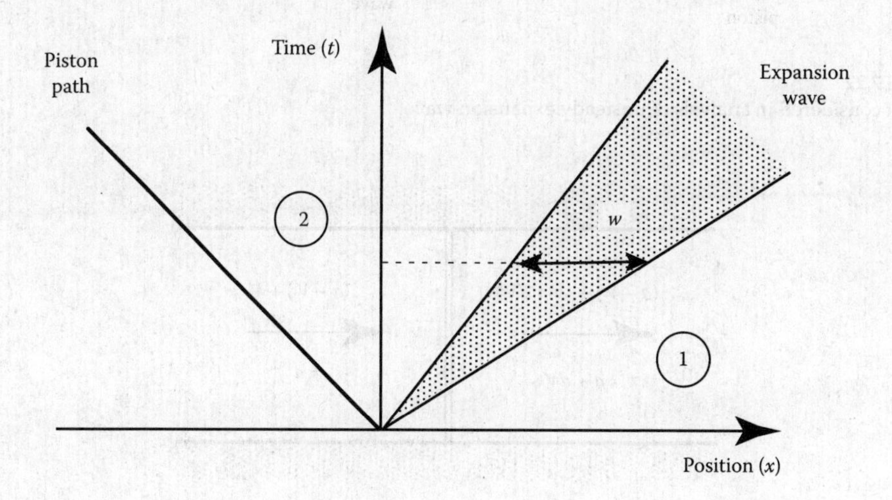

FIGURE 7.21
x–t diagram for instantly accelerated piston case.

In this case, the path of the tail of the expansion wave is given by

$$\text{Tail:} \quad x = (a_2 - u_p)t \tag{7.19}$$

and the width of the expansion region at any instant of time is given by

$$w = [a_1 - (a_2 - u_p)]t \tag{7.20}$$

Unsteady expansion waves can be generated in other ways, notably by the rupture of diaphragms separating regions of high and low pressure. These will be discussed later.

To apply the equations derived above, a_2 has to be known. To find this consider the wave to be split, as previously explained, into a series of wavelets each produced by a differentially small jump in piston velocity and each producing a differentially small change in velocity, pressure, etc. If the local gas velocity in the wave is V, the wavelet is propagated with velocity $(a - V)$ relative to the tube walls as shown in Figure 7.22.

Consider the flow relative to the wave as indicated in Figure 7.23. Applying the momentum equation to the flow across the wave gives

$$p - (p + dp) = \rho a[(a + dV) - a]$$

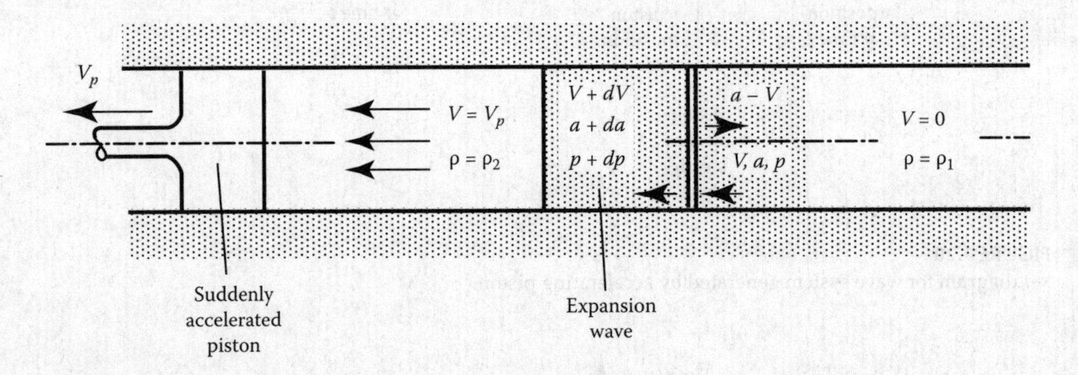

FIGURE 7.22
Wavelet considered in analysis of unsteady expansion wave.

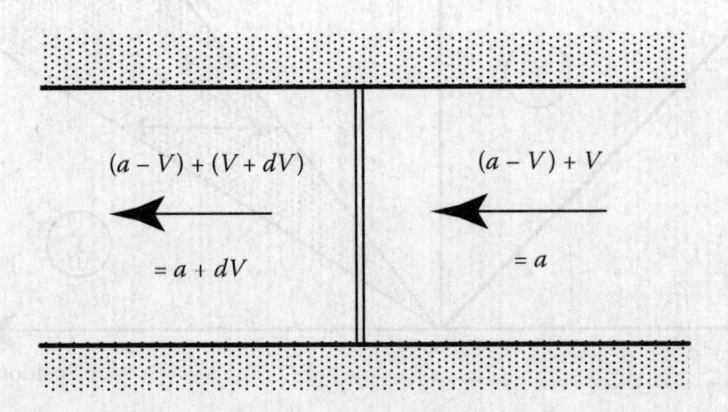

FIGURE 7.23
Flow relative to wavelet considered.

i.e.,

$$-dp = \rho a\, dV \tag{7.21}$$

However,

$$\frac{dp}{d\rho} = a^2$$

Thus, Equation 7.21 can be written as

$$-\frac{d\rho}{\rho} = \frac{dV}{a} \tag{7.22}$$

Integrating this equation across the complete wave gives

$$\int_0^{u_p} dV = -\int_{\rho_1}^{\rho_2} a \frac{d\rho}{\rho} \tag{7.23}$$

However, since the entire flow is isentropic, a is related to ρ by

$$\frac{a}{a_1} = \left(\frac{\rho}{\rho_1}\right)^{\frac{\gamma-1}{2}}$$

Thus, the right-hand side of Equation 7.23 becomes

$$\int_{\rho_1}^{\rho_2} a \frac{d\rho}{\rho} = \int_{\rho_1}^{\rho_2} a_1 \left(\frac{\rho}{\rho_1}\right)^{\frac{\gamma-1}{2}} \frac{d\rho}{\rho}$$

$$= \left(\frac{2}{\gamma-1}\right) \frac{a_1}{\rho_1^{\frac{\gamma-1}{2}}} \left[\rho_2^{\frac{\gamma-1}{2}} - \rho_1^{\frac{\gamma-1}{2}}\right]$$

$$= \left(\frac{2}{\gamma-1}\right) a_1 \left[\left(\frac{\rho_2}{\rho_1}\right)^{\frac{\gamma-1}{2}} - 1\right] \tag{7.24}$$

$$= \left(\frac{2}{\gamma-1}\right) a_1 \left[\left(\frac{a_2}{a_1}\right) - 1\right]$$

$$= \frac{2a_2}{\gamma-1} - \frac{2a_1}{\gamma-1}$$

Equation 7.23 can therefore be written as

$$u_p = -\frac{2a_2}{\gamma - 1} + \frac{2a_1}{\gamma - 1}$$

i.e.,

$$\frac{a_2}{a_1} = 1 - \left(\frac{\gamma - 1}{2}\right)\left(\frac{u_p}{a_1}\right) \tag{7.25}$$

This is the basic equation for unsteady expansion wave flow. Once (a_2/a_1) is known, the other changes across the wave can be calculated by noting that the flow through the wave is isentropic so that

$$\frac{p_2}{p_1} = \left(\frac{a_2}{a_1}\right)^{\frac{2\gamma}{\gamma-1}}, \quad \frac{\rho_2}{\rho_1} = \left(\frac{a_2}{a_1}\right)^{\frac{2}{\gamma-1}}$$

In the above discussion, unsteady expansion waves associated with the motion of pistons were considered. As was noted earlier, however, such waves can be generated in other ways. As an example, consider a tube that is initially sealed at both ends and contains a gas at a pressure above atmospheric pressure. If one end of the tube is suddenly opened, the pressure at this end of the tube will drop to atmospheric and an unsteady expansion wave will propagate down the tube inducing flow out of the tube. This situation is shown in Figure 7.24.

The flow is equivalent to that which would have been generated by the instantaneous acceleration of a piston to velocity, V_2. From the previous work, it follows that across the expansion wave

$$\frac{a_2}{a_1} = 1 - \left(\frac{\gamma - 1}{2}\right)\frac{V_2}{a_1} \tag{7.26}$$

However, the expansion is isentropic, so that

$$\left(\frac{p_2}{p_1}\right) = \left(\frac{a_2}{a_1}\right)^{\frac{2\gamma}{\gamma-1}}$$

Combining these equations and noting that $p_2 = p_a$, the atmospheric pressure gives

$$\left(\frac{p_a}{p_1}\right)^{\frac{\gamma-1}{2\gamma}} = 1 - \left(\frac{\gamma - 1}{2}\right)\frac{V_2}{a_1}$$

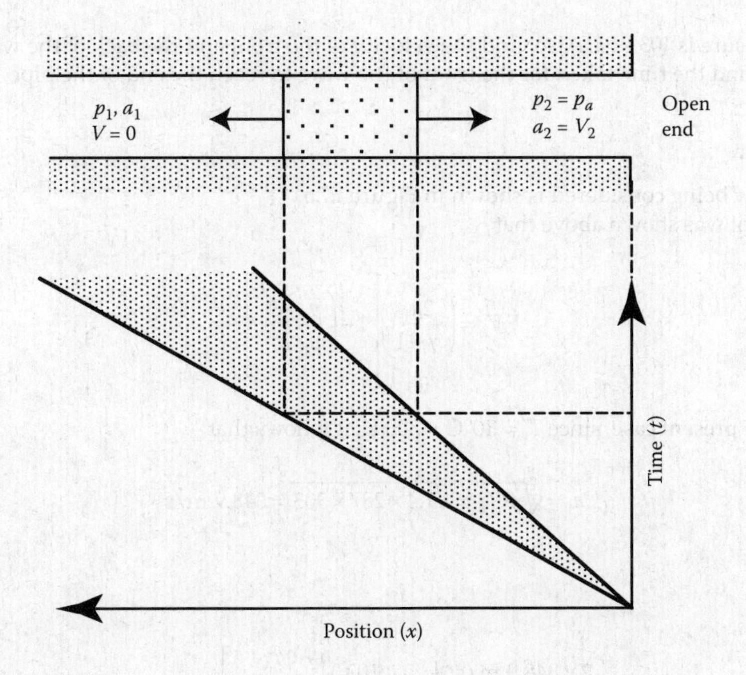

FIGURE 7.24
Unsteady expansion wave produced by sudden opening of the end of a tube.

i.e.,

$$V_2 = \left(\frac{2a_1}{\gamma - 1}\right)\left\{1 - \left(\frac{p_a}{p_1}\right)^{\frac{\gamma-1}{2\gamma}}\right\} \tag{7.27}$$

which gives the velocity at which the gas will be discharged from the tube.

It is interesting to note that no matter how large the initial pressure, p_1, is, there is a maximum velocity at which the gas will be discharged, this being

$$V_{2\max} = \frac{2a_1}{\gamma - 1} \tag{7.28}$$

This is, then, the maximum velocity that can be generated by an unsteady expansion wave propagating into a gas at rest. It should not be confused with the maximum escape velocity previously discussed, which was the maximum velocity that could be generated by a steady isentropic expansion.

Example 7.6

A diaphragm at the end of 4-m-long pipe containing air at a pressure of 200 kPa and a temperature of 30°C suddenly ruptures causing an expansion wave to propagate down the pipe. Find the velocity at which the air is discharged from the pipe if the ambient air pressure is 103 kPa. Also, find the velocity of the front and the back of the wave and hence, find the time taken for the front of the wave to reach the end of the pipe.

Solution

The flow being considered is shown in Figure E7.6.

Now, it was shown above that

$$V_2 = \left(\frac{2a_1}{\gamma - 1}\right)\left[1 - \left(\frac{p_a}{p_1}\right)^{\frac{\gamma-1}{2\gamma}}\right]$$

In the present case, since $T_1 = 30°\text{C} = 303$ K, it follows that

$$a_1 = \sqrt{\gamma R T_1} = \sqrt{1.4 \times 287 \times 303} = 348.9 \text{ m/s}$$

Hence,

$$V_2 = \left(\frac{2 \times 348.9 \text{ m/s}}{1.4 - 1}\right)\left[1 - \left(\frac{103}{200}\right)^{(1.4-1)/(2\times 1.4)}\right] = 157.8 \text{ m/s}$$

Therefore, the velocity at which the air is discharged from the pipe is 157.8 m/s.

The front of the wave propagates at the local speed of sound in the undisturbed air, a_1, i.e., at 348.9 m/s. The tail of wave propagates at the local speed of sound behind the wave, a_2, relative to gas behind the wave, i.e., at $a_2 - V_2$ relative to the pipe. However, as shown above,

$$\frac{a_2}{a_1} = 1 - \left(\frac{\gamma - 1}{2}\right)\left(\frac{V_2}{a_1}\right)$$

i.e.,

$$\frac{a_2}{a_1} = 1 - \left(\frac{1.4 - 1}{2}\right)\left(\frac{157.8}{348.9}\right) = 0.91$$

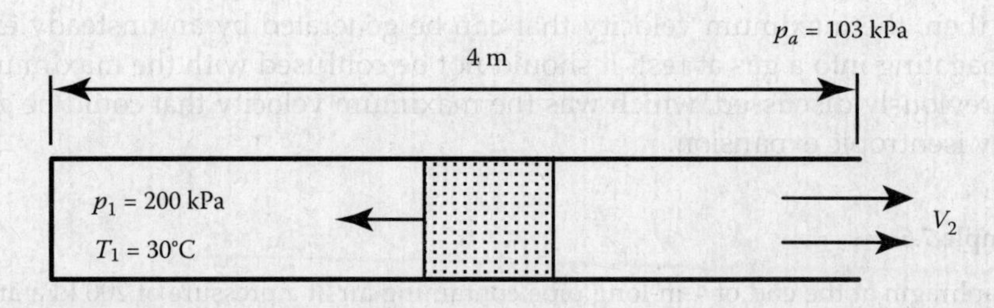

FIGURE E7.6
Flow situation considered.

Hence,

$$a_2 = 0.91a_1 = 0.91 \times 348.9 = 317.5 \text{ m/s}$$

Therefore, the velocity of the tail of the wave relative to the walls of the pipe is 348.9 – 157.8 = 191.1 m/s.

Because the front of the wave is moving at a velocity of 348.9 m/s, the time taken for the front of the wave to reach the end of the pipe is given by

$$t = \frac{4}{348.9} = 0.0115 \text{ s}$$

i.e., the time for the head of the wave to reach the end of the pipe is 0.0115 s.

An unsteady expansion wave is also generated when a moving shock wave reaches the end of an open duct as indicated in Figures 7.25 and 7.26.

Situations arise in a number of practical situations in which unsteady shock waves and expansion waves are simultaneously generated. Perhaps the simplest example of this is the flow that occurs in a so-called shock tube. In its simplest form, this consists of a long tube of constant area divided into two sections by a diaphragm that is typically made from a thin sheet of metal that often has grooves cut into it to ensure that it can be easily and cleanly broken. The tube contains a high-pressure gas on one side of the diaphragm and a low-pressure gas on the other side of the diaphragm, as shown in Figure 7.27.

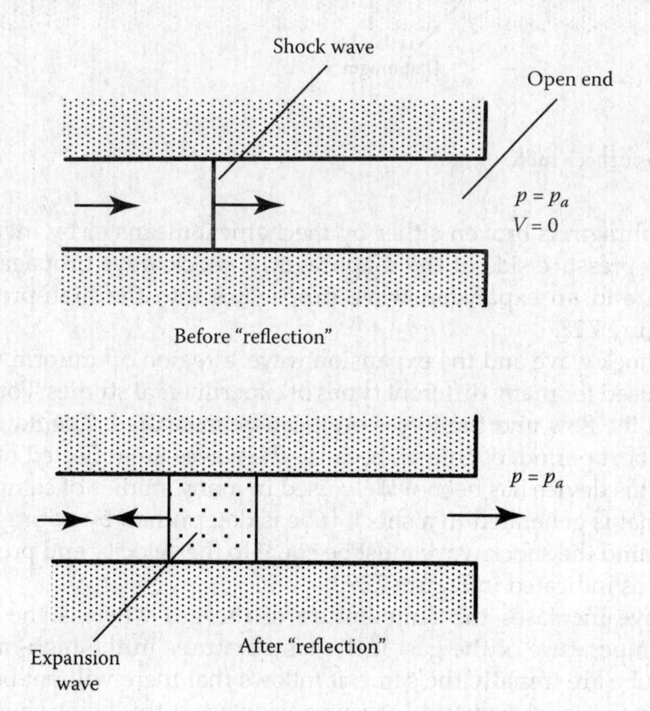

FIGURE 7.25
Generation of expansion wave by reflection of a moving shock wave from the open end of a duct.

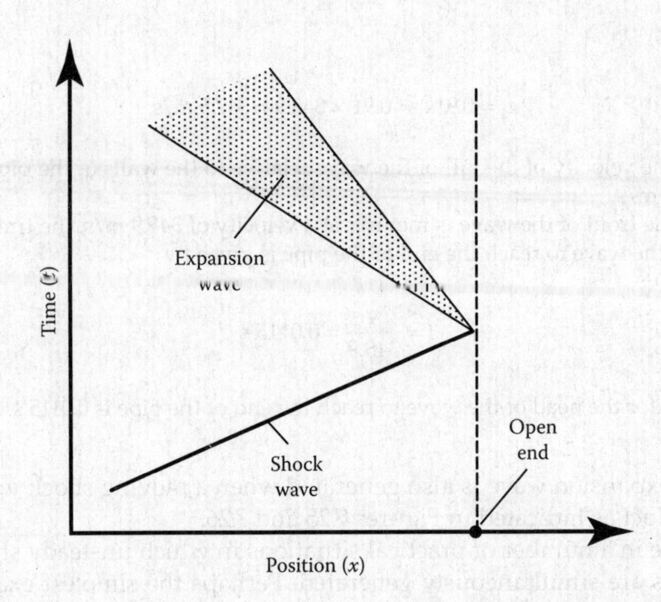

FIGURE 7.26
x–t diagram for flow associated with the reflection of a moving shock wave from the open end of a duct.

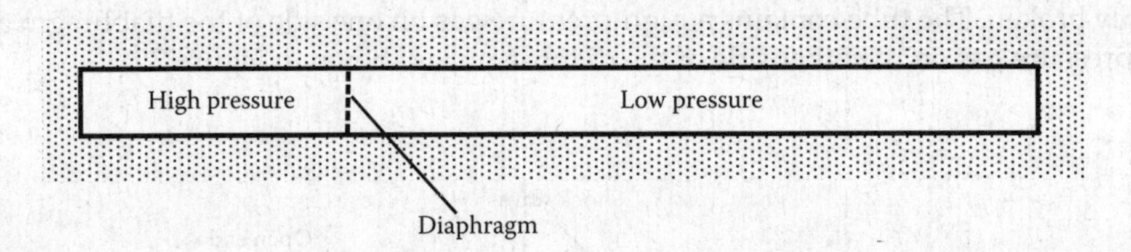

FIGURE 7.27
Arrangement of a basic shock tube.

When the diaphragm is broken either by mechanical means or by increasing the pressure on the high pressure side of the diaphragm, a shock wave propagates into the low-pressure section and an expansion wave propagates into the high-pressure section as illustrated in Figure 7.28.

Between the shock wave and the expansion wave, a region of uniform velocity is generated that can be used for many different types of experimental studies. For example, a body can be placed in the flow and the forces on it can be measured. The flow in a shock tube only lasts for a short period, of course, because the waves are reflected off the ends of the tube. However, this device has been widely used in many studies of compressible flows.

The velocity that is generated in a shock tube is determined by noting that the velocity and pressure behind the shock wave must be equal to the velocity and pressure behind the expansion wave as indicated in Figure 7.29.

The shock wave increases the temperature of the gas whereas the expansion wave decreases the temperature of the gas. If the temperatures in the high- and low-pressure sections of the tube are initially the same, it follows that there will not be a uniform temperature between the shock wave and the expansion wave, this being shown in Figure 7.29.

The way in which the flow generated in a shock tube can be analyzed is illustrated in a very basic way in the following example.

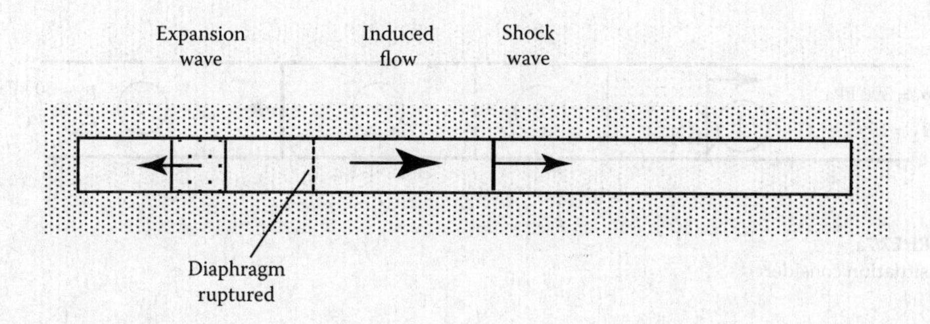

FIGURE 7.28
Waves generated in a basic shock tube following rupture of the diaphragm.

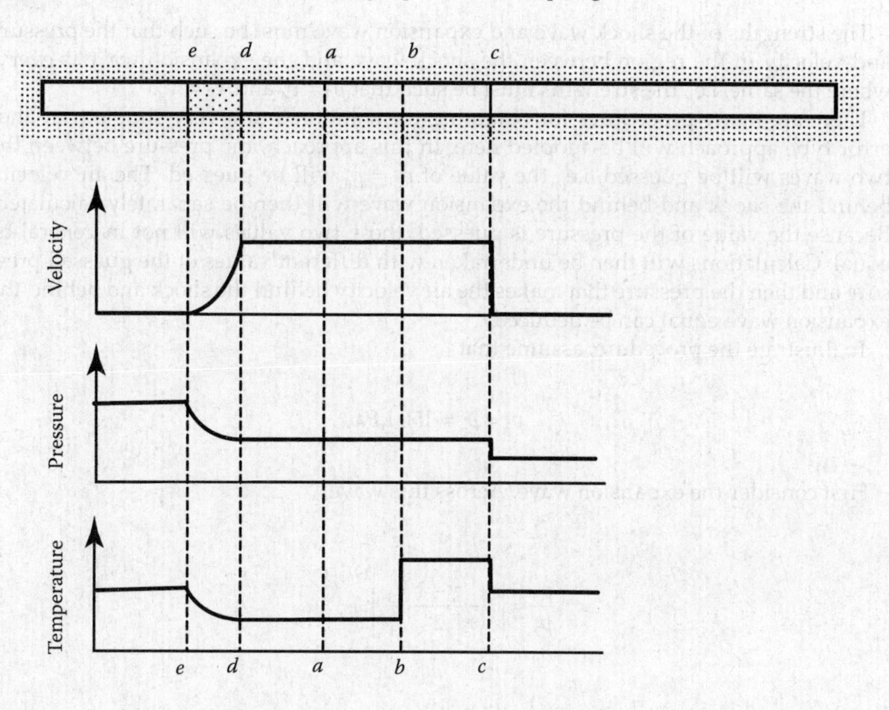

FIGURE 7.29
Velocity, pressure, and temperature variations in a shock tube.

Example 7.7

A shock tube essentially consists of a long tube containing air and separated into two sections by a diaphragm. The pressures on the two sides of the diaphragm are 300 and 30 kPa, and the temperature is 15°C in both sections. If the diaphragm is suddenly ruptured, find the velocity of the air between the moving shock wave and the moving expansion wave that is generated.

Solution

The flow that is generated by the rupturing of the diaphragm is shown in Figure E7.7a. The speed of sound in the undisturbed air, i.e., in sections 1 and 2 is given by

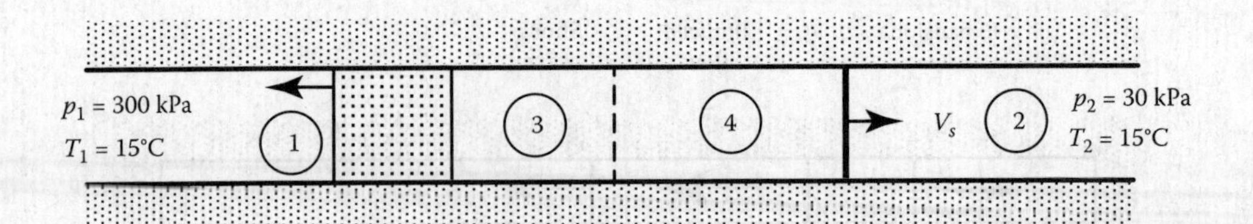

FIGURE E7.7a
Flow situation considered.

$$a_1 = a_2 = \sqrt{\gamma RT} = \sqrt{1.4 \times 287 \times 288} = 340.2 \text{ m/s}$$

The strengths of the shock wave and expansion wave must be such that the pressure and velocity in the region between the shock wave and the expansion wave is everywhere the same, i.e., the strengths must be such that $p_3 = p_4$ and $V_3 = V_4$.

There are many procedures for obtaining the solution, but a very simple trial-and-error type approach will be adopted here. In this approach, the pressure between the two waves will be guessed, i.e., the value of $p_3 = p_4$ will be guessed. The air velocity behind the shock and behind the expansion wave will then be separately calculated. Because the value of the pressure is guessed, these two values will not in general be equal. Calculations will then be undertaken with different values of the guessed pressure and then the pressure that makes the air velocity behind the shock and behind the expansion wave equal can be deduced.

To illustrate the procedure, assume that

$$p_3 = p_4 = 150 \text{ kPa}$$

First consider the expansion wave. Across this wave,

$$\frac{p_3}{p_1} = \left[1 - \frac{\gamma - 1}{2}\frac{V_3}{a_1}\right]^{2\gamma/\gamma-1}$$

For the specified conditions in region 1,

$$p_1 = 300 \text{ kPa}, \quad a_1 = 340.2 \text{ m/s}$$

Hence, for $p_3 = 150$ kPa, i.e., for

$$\frac{p_3}{p_1} = \frac{150}{300} = 0.5$$

this equation gives

$$0.5 = \left[1 - \frac{0.4}{0.2}\frac{V_3}{340.2}\right]^{2.8/0.4}$$

Solving for V_3 gives

$$V_3 = 160.4 \text{ m/s}$$

Next consider the flow across the shock wave. The flow relative to the shock is shown in Figure E7.7b.

Since for $p_4 = 150$ kPa,

$$\frac{p_4}{p_2} = \frac{150}{30} = 5$$

For this pressure ratio, normal shock relations or tables or the software gives

$$M_2 = 2.11, \quad M_4 = 0.5598, \quad \frac{T_4}{T_2} = 1.7789$$

However, M_2 is the shock Mach number, i.e.,

$$M_2 = \frac{V_s}{a_2} = \frac{V_s}{340.2 \text{ m/s}}$$

Hence, for the guessed pressure,

$$V_s = 340.2 \times 2.11 = 717.8 \text{ m/s}$$

Also,

$$T_4 = 1.7789 \times 288 = 512 \text{ K}$$

Thus,

$$a_4 = \sqrt{1.4 \times 287 \times 512} = 454 \text{ m/s}$$

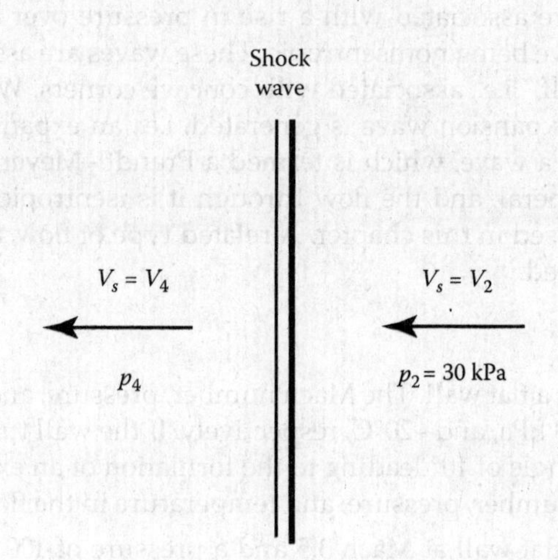

FIGURE E7.7b
Flow relative to shock wave.

TABLE E7.7

Iterative Results Used in Obtaining Solution

Guessed $p_3 = p_4$ (kPa)	p_4/p_2	M_2	M_4	a_4/a_1	a_4 (m/s)	V_s (m/s)	V_4 (m/s)	V_3 (m/s)	$V_4 - V_3$ (m/s)
80	2.67	1.56	0.6809	1.1664	397	531	261	293	−32
85	2.83	1.60	0.6684	1.1781	401	544	276	280	−4
90	3	1.65	0.6540	1.1928	405	561	296	269	27
100	3.3	1.73	0.6330	1.2166	414	589	327	247	80

However, as discussed in the previous chapter,

$$M_4 = \frac{V_s - V_4}{a_4}$$

from which it follows that

$$V_4 = V_s - M_4 a_4 = 717.8 - 0.5598 \times 454 = 463.7 \text{ m/s}$$

Hence, when it is guessed that $p_3 = p_4 = 150$ kPa, it is found that $V_3 = 160.4$ m/s and $V_4 =$ 463.7 m/s.

Calculations of this type have been carried out for a number of other values of the guessed pressures, the results of some of these calculations being given in Table E7.7.

By interpolation between these results, it can be deduced that $V_3 = V_4$ when $p_3 = p_4 =$ 86 kPa and that at this pressure, $V_3 = V_4 = 279$ m/s, i.e., the velocity of air between the moving shock wave and the moving expansion wave is 279 m/s.

Concluding Remarks

Oblique shock waves are associated with a rise in pressure over a very thin region, the flow through such a wave being nonisentropic. These waves are associated with a turning of the flow "toward itself," i.e., associated with concave corners. When the flow is turned "away from itself," an expansion wave is generated, i.e., an expansion wave is generated at convex corners. Such a wave, which is termed a Prandtl–Meyer wave or an expansion wave, is not thin, in general, and the flow through it is isentropic. The characteristics of such waves were discussed in this chapter. A related type of flow, an unsteady expansion wave, was also considered.

PROBLEMS

1. Air is flowing over a flat wall. The Mach number, pressure, and temperature in the air stream are 3, 50 kPa, and −20°C, respectively. If the wall turns "away" from the flow through an angle of 10° leading to the formation of an expansion wave, what will be the Mach number, pressure, and temperature in the flow behind the wave?
2. Air flows along a flat wall at Mach 3.5 and a pressure of 100 kPa. The wall turns toward the flow through an angle of 25° leading to the formation of an oblique shock wave. A short distance downstream of this, the wall turns away from the

flow through an angle of 25°, leading to the generation of an expansion wave causing the flow to be parallel to its original direction. Find the Mach number and pressure downstream of the expansion wave.

3. Air is flowing at Mach 2 at a temperature and pressure of 100 kPa and 0°C down a duct. One wall of this duct turns through an angle of 5° away from the flow leading to the formation of an expansion wave. This expansion wave is reflected off the flat opposite wall of the duct. Find the pressure and temperature behind the reflected wave.
4. Air is flowing through a wide channel at Mach 1.5, the pressure being 120 kPa. The upper wall of the channel turns through an angle of 4° "away" from the flow, leading to the generation of an expansion wave. This expansion wave "reflects" off the flat lower surface of the channel. Find the Mach number and pressure after this reflection.
5. Air flowing at Mach 3 is turned through an angle that leads to the generation of an expansion wave across which the pressure decreases by 60%. Find the angle that the upstream and downstream ends of the expansion wave make to the initial flow direction.
6. An air stream flowing at Mach 4 is expanded around a concave corner with an angle of 15° leading to the generation of an expansion wave. Some distance downstream of this the air flows around a concave corner leading to the generation of an oblique shock wave and returning the flow to its original direction. If the pressure in the initial flow is 80 kPa, find the pressure downstream of the oblique shock wave.
7. An oblique shock wave occurs in an air flow in which the Mach number is 2.5, this shock wave turning the flow through 10°. The shock wave impinges on a free boundary along which the pressure is constant and equal to that existing upstream of the shock wave. The shock is "reflected" from this boundary as an expansion wave. Find the Mach number and flow direction downstream of this expansion wave.
8. Air is flowing through a wide channel at Mach 2 at a pressure of 140 kPa. The upper wall of the channel turns through an angle of 8° "away" from the flow leading to the generation of an expansion wave, whereas the lower wall of the channel turns through an angle of 6° "away" from the flow, also leading to the generation of an expansion wave. The two expansion waves interact and "pass through" each other. Find the Mach number, flow direction, and pressure just downstream of this interaction.
9. A symmetrical double-wedge–shaped body with an included angle of 15° is aligned with an air flow in which the Mach number is 3 and the pressure is 20 kPa. The flow situation is therefore as shown in Figure P7.9. Find the pressures acting on the surfaces of the body.

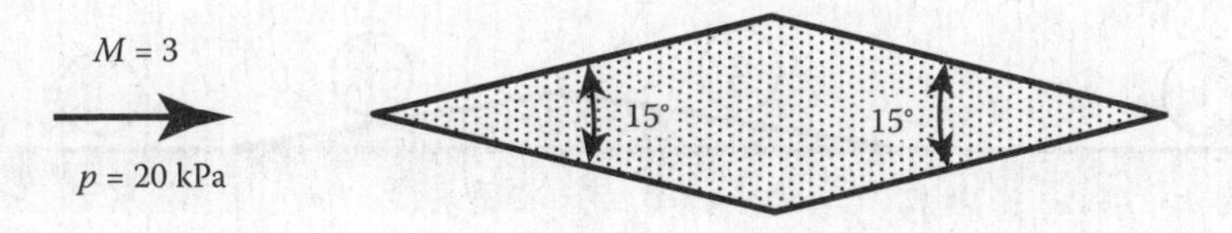

FIGURE P7.9
Flow situation considered.

10. A simple wing may be modeled as a 0.3 m wide flat plate set at an angle of 3° to an airflow at a Mach number of 2.5, the pressure in this flow being 40 kPa. Assuming that the flow over the wing is two-dimensional, estimate the lift and drag force per meter span due to the wave formation on the wing. What other factor causes drag on the wing?
11. Consider two-dimensional flow over the double-wedge airfoil shown in Figure P7.11. Find the lift and drag per meter span acting on the airfoil and sketch the flow pattern. How does the pressure vary over the surface of the airfoil?
12. For the double-wedge airfoil shown in Figure P7.12, find the lift per meter span if the Mach number and pressure in the uniform air flow ahead of the airfoil are 3 and 40 kPa, respectively.
13. A uniform air flow over a plane wall at Mach 2.2 and with a temperature of 325 K encounters a symmetrical triangular "bump" on the wall. The upstream and downstream faces of this bump are at an angle of 15° to the plane wall over which the air is flowing. The situation being considered is as shown in Figure P7.13. Find the Mach numbers and temperatures in the flows over the upstream and downstream faces of the "bump" and in the flow over the wall downstream of the bump, i.e., find the Mach numbers and temperatures in the flow regions 2, 3, and 4 shown in Figure P7.13.

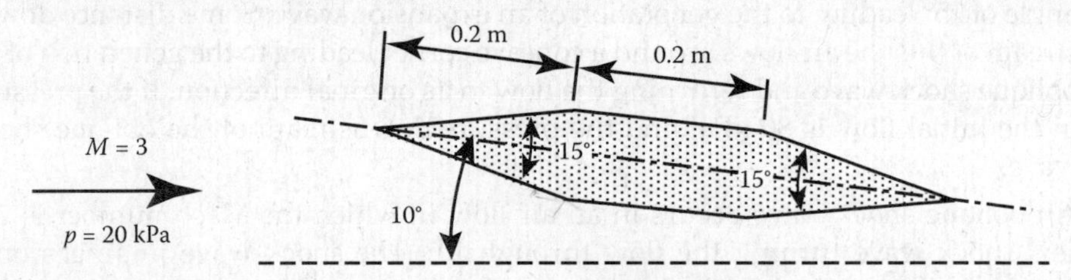

FIGURE P7.11
Flow situation considered.

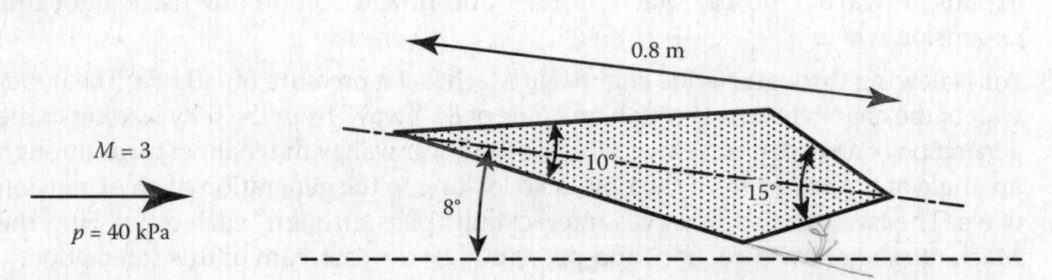

FIGURE P7.12
Flow situation considered.

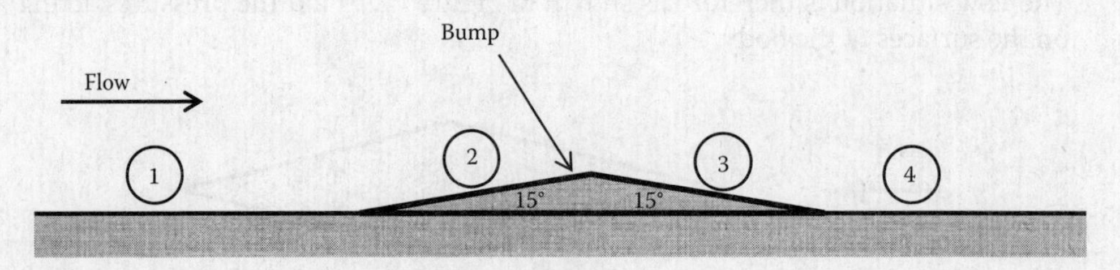

FIGURE P7.13
Flow situation considered.

14. A uniform air flow over a plane wall at Mach 2.2 and with a temperature of 270 K and a pressure of 90 kPa is deflected downward by the changes in the wall direction shown in Figure P7.14. Determine the Mach numbers, the pressures, and the stagnation pressures in regions 2 and 3 shown in Figure P7.14. Sketch the oblique shock waves generated by the changes in wall direction and on this sketch indicate the angles that the shock waves make to the initial flow direction. Determine the maximum turning angle that the second wall segment can have (15° in the above calculation) if the second shock wave is to remain attached to this wall segment.
15. A thin flat plate is placed in a uniform air flow in which the Mach number is 2.3 and the pressure is 10 kPa. The plate is set at an angle to the flow, and as a result, an oblique shock wave originating at the leading edge of the lower surface of the plate is generated, and an expansion wave originating at the leading edge of the upper surface of the plate is generated. The oblique shock wave makes an angle of 40° to the direction of the undisturbed flow upstream of the plate. Find the angle at which the plate is set relative to the undisturbed flow direction and the pressures acting on the upper and lower surfaces of the plate.
16. A safety diaphragm at the end of a 3 m long pipe containing air at a pressure of 200 kPa and a temperature of 10°C suddenly ruptures, causing an expansion wave to propagate down the pipe. Find the velocity at which the air is discharged from the pipe, the velocity of the front and the back of the wave, and the time taken for the front of the wave to reach the end of the pipe. Assume that the ambient pressure of the air surrounding the pipe is 100 kPa.
17. An unsteady expansion wave propagates down a duct containing air at rest at a pressure of 800 kPa and a temperature of 2000°C. The pressure behind the wave is 300 kPa. Find the velocity and the Mach number in the flow that is induced behind the wave in the duct.
18. Air is flowing through a long pipe at a velocity of 50 m/s at a pressure and temperature of 150 kPa and 40°C, respectively. Valves at the inlet and exit to this pipe are suddenly and simultaneously closed. Discuss the waves that are generated in the pipe following valve closure and find the pressures acting on each valve immediately following the valve closure.
19. A long pipe is conveying air at a pressure and temperature of 150 kPa and 100°C, respectively, at a velocity of 160 m/s. Valves are fitted at both the inlet and the exit to the pipe. Discuss what waves will be developed and what pressure will act on the valve if (1) the inlet valve is suddenly closed and (2) the exit valve is suddenly closed.

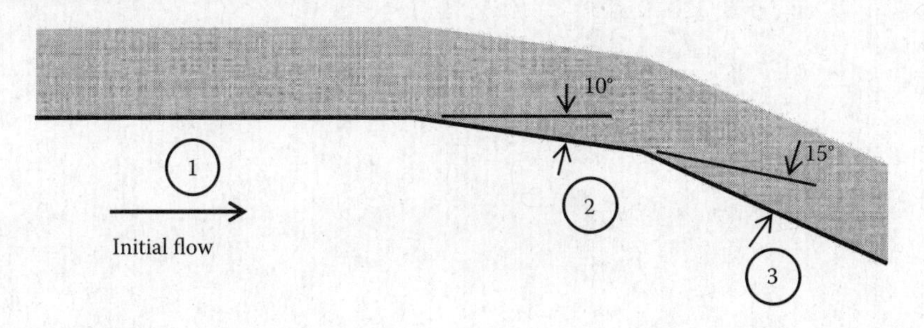

FIGURE P7.14
Flow situation considered.

20. A closed tube contains air at a pressure and temperature of 200 kPa and 30°C, respectively. One end of the tube is suddenly opened to the surrounding atmosphere. At what velocity does the air leave the open end of the tube if the ambient pressure is 100 kPa?
21. A long pipe containing air is separated into two sections by a diaphragm. The pressure on one side of the diaphragm is 500 kPa and the pressure on the other side of the diaphragm is 100 kPa, the air temperature being 20°C in both sections. If the diaphragm suddenly ruptures causing a shock wave to move into the low pressure section, and an expansion wave to move into the high pressure section, find the pressure and air velocity in the region between the two waves.
22. A shock tube containing air has initial pressures on the two sides of the diaphragm of 400 and 10 kPa, with the temperature of the air being 25°C in both sections. If the diaphragm separating the two sections is suddenly ruptured, find the velocity, pressure, and temperature of the air between the moving shock wave and the moving expansion wave that are generated.
23. The air pressure in the high and low-pressure sections of a constant diameter shock tube are 600 and 20 kPa, respectively. The temperatures in both sections are 30°C. After the diaphram that separates the two sections is ruptured, a shock wave propagates into the low pressure section and an expansion wave propagates into the high pressure section. Find the air velocity and temperatures between the two waves and the velocity of the shock wave.

8

Variable Area Flow

Introduction

The steady flow of a gas through a duct (or a streamtube) that has a varying cross-sectional area will be considered in this chapter. Such flows, i.e., compressible gas flows through a duct whose cross-sectional area is varying, occur in many engineering devices, e.g., in the nozzle of a rocket engine and in the blade passages in turbo machines.

It will be assumed throughout this chapter that the flow can be adequately modeled by assuming it to be one-dimensional at all sections of the duct, i.e., quasi-one-dimensional flow will be assumed in this chapter. This means, by virtue of the discussion given in Chapter 2, that the rate of change of cross-sectional area with distance along the duct is not very large. It will also be assumed in this chapter in studying the effects of changes in area on compressible gas flow that the flow is isentropic everywhere except through any shock waves that may occur in the flow. This assumption is usually quite adequate since the effects of friction and heat transfer are usually restricted to a thin boundary layer adjacent to the walls in the types of flows here being considered and their effects therefore can often be ignored or be adequately accounted for by introducing empirical constants. The presence of shock waves will have to be accounted for in the work of this chapter and the flow through these waves is, as discussed before, not isentropic.

Effects of Area Changes on Flow

Consider, first, the general effects of a change in area on isentropic flow through a channel. The situation considered is shown in Figure 8.1, i.e., the effects of a differentially small change in area, dA, on the other variables, i.e., V, p, T, and ρ, are considered. The effects of dA on the changes in pressure, density, velocity, etc., i.e., on dp, $d\rho$, dV, will be derived using the governing equations discussed earlier. The analysis presented here is an extension of some of the analyses given earlier and there is some overlap with earlier work.

First, it is recalled that the continuity equation gives

$$\rho AV = \text{mass flow rate} = \text{constant} \tag{8.1}$$

where A is the cross-sectional area of the duct at any point. Applying this to the flow being considered gives

$$\rho AV = (\rho + d\rho)(A + dA)(V + dV)$$

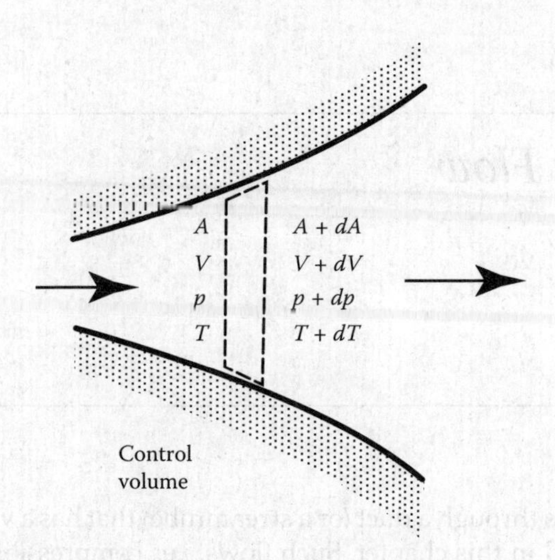

FIGURE 8.1
Flow changes considered in variable area channel.

Since dp, $d\rho$, dV, and dA are, by assumption, all small, this equation becomes, to first-order accuracy (i.e., if terms involving the products and squares of the differentially small quantities such as $d\rho \times dA$ are ignored)

$$\rho AV = AVd\rho + \rho VdA + \rho AdV$$

i.e., dividing through by ρAV

$$\frac{d\rho}{\rho} + \frac{dA}{A} + \frac{dV}{V} = 0 \tag{8.2}$$

Next, it is recalled that the energy equation gives

$$c_p T + \frac{V^2}{2} = \text{constant}$$

which gives for the situation being considered

$$c_p T + \frac{V^2}{2} = c_p(T + dT) + \frac{(V + dV)^2}{2}$$

i.e., to first-order accuracy

$$c_p dT + V\,dV = 0 \tag{8.3}$$

Further, the equation of state gives

$$p = \rho RT \quad \text{and} \quad p + dp = (\rho + d\rho)\,R(T + dT)$$

Subtracting these two equations and dividing the result by the first of the two equations give, to first-order accuracy,

$$\frac{dp}{p} = \frac{d\rho}{\rho} + \frac{dT}{T} \tag{8.4}$$

Lastly, since the flow being considered is, by assumption, isentropic, it follows that

$$\frac{p}{\rho^\gamma} = \text{constant} \quad \text{and} \quad \frac{p + dp}{(\rho + d\rho)^\gamma} = \text{constant} \tag{8.5}$$

Because dp/p and $d\rho/\rho$ are by assumption small, the second of the above two equations gives

$$\frac{p}{\rho^\gamma} \frac{\left[1 + \dfrac{dp}{p}\right]}{\left[1 + \dfrac{d\rho}{\rho}\right]^\gamma} = \text{constant}$$

i.e., to first-order accuracy

$$\frac{p}{\rho^\gamma} \frac{\left[1 + \dfrac{dp}{p}\right]}{\left[1 + \gamma\dfrac{d\rho}{\rho}\right]} = \text{constant}$$

i.e.,

$$\frac{p}{\rho^\gamma}\left[1 + \frac{dp}{p} - \gamma\frac{d\rho}{\rho}\right] = \text{constant}$$

Combining this with the first equation then gives to first-order accuracy

$$\frac{dp}{p} = \gamma\frac{d\rho}{\rho} \tag{8.6}$$

Equations 8.2, 8.3, 8.4, and 8.6 together are sufficient to determine the required results, i.e., to determine the relationship between the four variables dp/p, dV/V, dT/T, and $d\rho/\rho$, and the fractional area change dA/A. As discussed before, because isentropic flow is being considered, the momentum equation has not been used, it basically giving the same result as the energy equation. Combining Equations 8.4, and 8.6 gives

$$\frac{dT}{T} = (\gamma - 1)\frac{d\rho}{\rho} \tag{8.7}$$

which can be substituted into Equation 8.3 to give

$$(\gamma - 1)\frac{d\rho}{\rho} + \frac{V^2}{c_p T}\frac{dV}{V} = 0 \tag{8.8}$$

Now,

$$\frac{V^2}{c_p T} = \frac{V^2}{c_p a^2}\gamma R = \gamma\left(1 - \frac{1}{\gamma}\right)M^2 = (\gamma - 1)M^2$$

Thus, Equation 8.8 can be written as

$$\frac{d\rho}{\rho} = -M^2\frac{dV}{V} \tag{8.9}$$

Substituting this into Equation 8.2 then gives

$$\frac{dA}{A} = (M^2 - 1)\frac{dV}{V} \tag{8.10}$$

which may alternatively be written as

$$\frac{dA}{dV} = (M^2 - 1)\frac{A}{V} \tag{8.11}$$

Because A and V are positive, it may be concluded from the above two equations that

1. If $M < 1$, i.e., if the flow is subsonic, then dA has the opposite sign to dV, i.e., decreasing the area increases the velocity and vice versa.
2. If $M > 1$, i.e., if the flow is supersonic, then dA has the same sign as dV, i.e., decreasing the area decreases the velocity and vice versa.
3. If $M = 1$ then $dA/dV = 0$ and A reaches an extremum. From (1) and (2), it follows that when $M = 1$, A must be a minimum.

Further important conclusions regarding the effects of varying area on the flow variables can be obtained by writing Equation 8.10 as

$$\frac{dA}{A} = (M^2 - 1)\frac{dM}{M}\frac{1}{a}\frac{dV}{dM} \tag{8.12}$$

However, by noting that $V = Ma$, it follows that

$$\frac{dV}{V} = \frac{dM}{M} + \frac{da}{a} \tag{8.13}$$

Further,

$$a = \sqrt{\gamma RT} \quad \text{and} \quad a + da = \sqrt{\gamma R(T + dT)}$$

The second of these two equations gives to first-order accuracy

$$a + da = \sqrt{\gamma RT}\left(1 + \frac{dT}{2T}\right)$$

Hence, dividing this result by the equation for a gives

$$\frac{da}{a} = \frac{1}{2}\frac{dT}{T}$$

and since the energy equation (8.3) gives

$$\frac{dT}{T} = -\frac{V^2}{c_pT}\frac{dV}{V} = -(\gamma - 1)M^2\frac{dV}{V}$$

it follows that

$$\frac{da}{a} = -\left(\frac{\gamma - 1}{2}\right)M^2\frac{dV}{V} \tag{8.14}$$

Substituting this result into Equation 8.13 then gives

$$\frac{dM}{M} = \frac{dV}{V}\left[1 + \left(\frac{\gamma - 1}{2}\right)M^2\right] \tag{8.15}$$

Substituting this, in turn, into Equation 8.12 gives

$$\frac{dA}{A} = \frac{(M^2 - 1)\frac{1}{a}\frac{V}{M}}{1 + \left(\frac{\gamma - 1}{2}\right)M^2}\frac{dM}{M}$$

i.e.,

$$\frac{dA}{A} = \frac{(M^2 - 1)}{1 + \left(\frac{\gamma - 1}{2}\right) M^2} \frac{dM}{M} \tag{8.16}$$

From this equation, it follows that

1. When $M < 1$, dA has the opposite sign to dM, e.g., when A increases, M decreases.
2. When $M > 1$, dA has the same sign as dM, e.g., when A increases, M increases.
3. When $M = 1$, $dA = 0$, i.e., A is a minimum when $M = 1$. dA can also be zero when dM is zero, i.e., a minimum in the flow area can also be associated with a maximum or minimum in the Mach number.

The results derived above concerning the effect of area changes on the Mach number and the velocity are summarized in Figure 8.2.

From these results, it follows that if a subsonic flow is to be accelerated to a supersonic velocity it must be passed through a convergent–divergent passage or nozzle. The convergent portion accelerates the flow up to Mach 1 and the divergent portion then accelerates the flow to supersonic velocity. At the throat, since $dA = 0$, the Mach number is equal to 1. This is summarized in Figure 8.3.

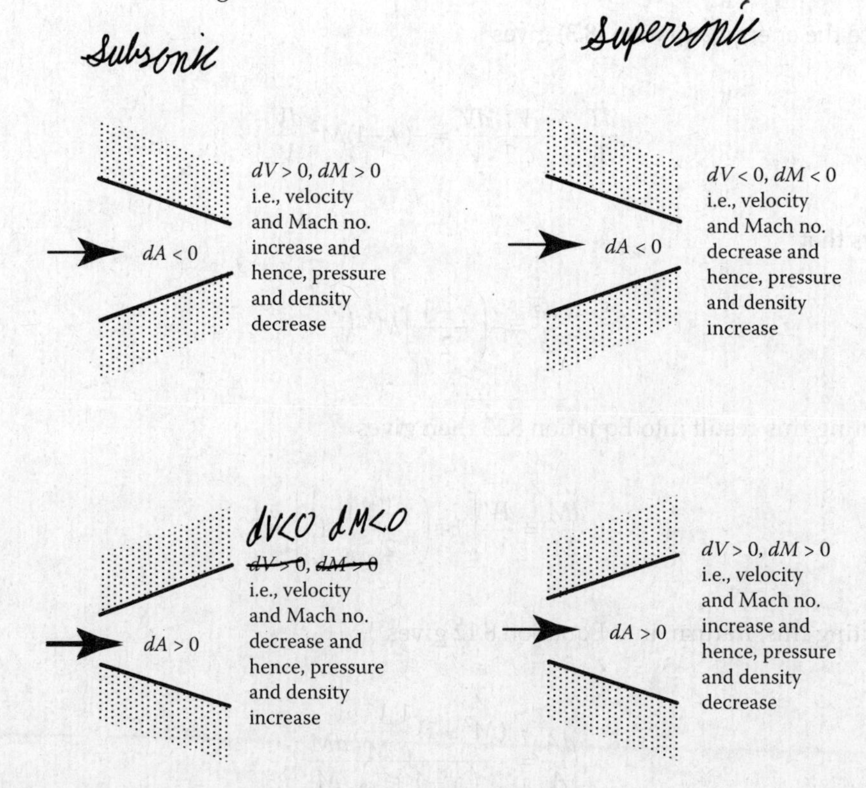

FIGURE 8.2
Effect of area change on Mach number and velocity.

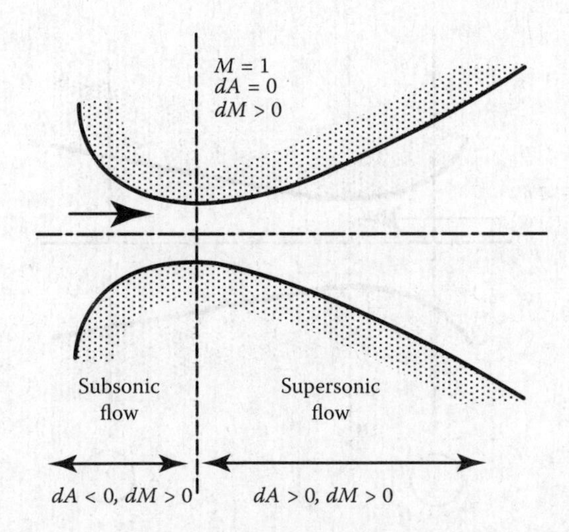

FIGURE 8.3
Generation of supersonic flow in a convergent–divergent duct.

Of course, in such a nozzle, the pressure will decrease continuously along the nozzle. If the end pressure is not low enough, the flow will remain subsonic throughout as in a Venturi meter, i.e., supersonic flow will not be generated. In this situation, since dA is still zero at the throat, dV and dM must be zero at the throat, i.e., the velocity and Mach number both reach a maximum at the throat (see later for a detailed discussion of the actual operating characteristics of such nozzles).

In a similar way, if the flow entering the nozzle is supersonic, two possibilities exist. Either the Mach number will decrease to 1 at the throat and then continue to decrease to subsonic values in the divergent portion of the nozzle or the flow will remain supersonic throughout the nozzle, the Mach number and velocity, in this case, decreasing in the convergent portion of the nozzle, reaching minimum although still supersonic value at the throat, and then increasing again in the divergent portion of the nozzle.

Equations for Variable Area Flow

Attention was given in the previous section to the changes in the flow variables produced by a differentially small change in area. Equations for the changes in the flow variables produced by finite changes in the flow area will be derived in the present section. The presence of shock waves in the flow will, for the present, be ignored. The flow is then, as previously discussed, analyzed using the assumption that it is one-dimensional at all sections and also that it is isentropic. The required relations could have been obtained by integrating the differential relations given in the previous section. It is easier, however, to derive the relations by directly applying the full energy, continuity, and state equations.

Consider the flow of a gas from a large reservoir through some duct system as shown in Figure 8.4.

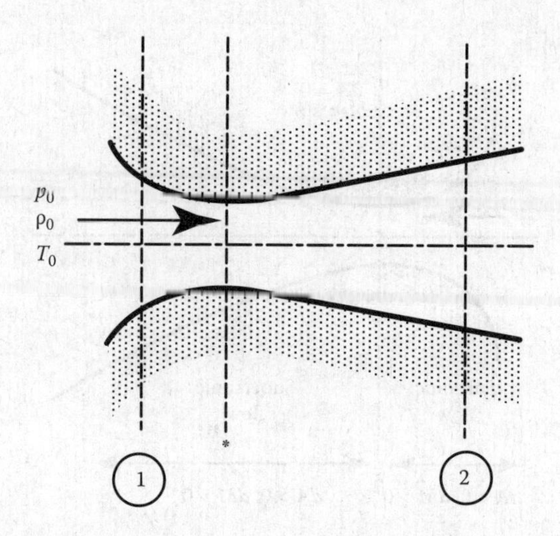

FIGURE 8.4
Flow through a convergent–divergent duct.

Because the reservoir is large, the velocity in it is essentially zero and the stagnation conditions therefore essentially exist in the reservoir, i.e., the pressure, density, and temperature in the reservoir are p_0, ρ_0, and T_0, respectively. The equations governing the flow at some arbitrary section 1 are the continuity equation

$$\rho_1 V_1 A_1 = \text{constant} = \text{mass flow rate, } \dot{m} \tag{8.17}$$

and the energy equation

$$V_1^2 + \frac{2}{(\gamma-1)} a_1^2 = \frac{2}{(\gamma-1)} a_0^2 \tag{8.18}$$

The momentum equation could have been used instead of the energy equation it giving, as discussed earlier because the flow is isentropic, the same result as that obtained using the energy equation.

In addition to Equations 8.17 and 8.18, the isentropic relations apply at all points, i.e.,

$$\frac{a_1}{a_0} = \left(\frac{T_1}{T_0}\right)^{\frac{1}{2}} = \left(\frac{p_1}{p_0}\right)^{\frac{\gamma-1}{2\gamma}} = \left(\frac{\rho_1}{\rho_0}\right)^{\frac{\gamma-1}{2}}$$

An attempt will first be made to relate the conditions existing at the arbitrary section 1 of the duct to the pressure ratio, p_1/p_0, existing at this section. To do this, it is noted that the energy equation (8.18) gives

$$V_1^2 = \left(\frac{2}{\gamma-1}\right) a_0^2 \left[1 - \left(\frac{a_1}{a_0}\right)^2\right]$$

which becomes, using the isentropic relations,

$$V_1 = \left\{ \left(\frac{2a_0^2}{\gamma - 1} \right) 1 - \left(\frac{p_1}{p_0} \right)^{\frac{\gamma-1}{\gamma}} \right\}^{\frac{1}{2}}$$

i.e.,

$$V_1 = \left\{ \left(\frac{2\gamma}{\gamma - 1} \right) \left(\frac{p_0}{\rho_0} \right) \left[1 - \left(\frac{p_1}{p_0} \right)^{\frac{\gamma-1}{\gamma}} \right] \right\}^{\frac{1}{2}} \tag{8.19}$$

This equation can also be written as

$$V_1 = \left\{ (2c_p T_0) \left[1 - \left(\frac{p_1}{p_0} \right)^{\frac{\gamma-1}{\gamma}} \right] \right\}^{\frac{1}{2}}$$

Thus, the velocity at any section can be determined if the pressure at this section and the stagnation conditions are known.

Substituting Equation 8.19 into the continuity equation (8.17) and then using the isentropic relations gives

$$\dot{m} = \rho_0 V_1 A_1 \frac{\rho_1}{\rho_0} = \rho_0 A_1 \left(\frac{p_1}{p_0} \right)^{\frac{1}{\gamma}} \left\{ \left(\frac{2\gamma}{\gamma - 1} \right) \left(\frac{p_0}{\rho_0} \right) \left[1 - \left(\frac{p_1}{p_0} \right)^{\frac{\gamma-1}{\gamma}} \right] \right\}^{\frac{1}{2}} \tag{8.20}$$

Since $\dot{m}$ is a constant, i.e., the mass flow rate along the duct does not change because the flow is assumed to be steady, this equation can be used to relate the pressure at any point in the duct to the area, i.e., if subscript 2 refers to conditions at some other section of the duct, then Equation 8.20 gives

$$\dot{m} = \rho_0 A_2 \left(\frac{p_2}{p_0} \right)^{\frac{1}{\gamma}} \left\{ \left(\frac{2\gamma}{\gamma - 1} \right) \left(\frac{p_0}{\rho_0} \right) \left[1 - \left(\frac{p_2}{p_1} \right)^{\frac{\gamma-1}{\gamma}} \right] \right\}^{\frac{1}{2}} \tag{8.21}$$

Dividing Equation 8.21 by Equation 8.20 then gives, on rearrangement,

$$\frac{A_2}{A_1} = \left(\frac{p_1}{p_2}\right)^{\frac{1}{\gamma}} \left[\frac{1 - \left(\frac{p_1}{p_0}\right)^{\frac{\gamma-1}{\gamma}}}{1 - \left(\frac{p_2}{p_0}\right)^{\frac{\gamma-1}{\gamma}}}\right]^{\frac{1}{2}} \tag{8.22}$$

This equation relates the pressures at any two sections of the duct to the areas of these sections.

In presenting the equations for flow in a variable area duct, it is convenient to choose, for reference, conditions at some specific point. A convenient point to use for this purpose is the point in the flow at which the Mach number is equal to 1. Of course, there may not actually be a real point in the flow at which the Mach number is equal to 1, but the conditions at such a point whether or not it really exists in the flow are convenient to use for reference purposes. As discussed before, the conditions at the point where $M = 1$ are known as the critical conditions and expressions for the pressure and velocity at such a point have previously been derived. They will, however, be repeated here for reference. The energy equation gives the critical velocity, V^*, as

$$V^{*2} = \frac{2}{\gamma+1} a_0^2 \tag{8.23}$$

Since $V^* = a^*$, another way of writing this equation is

$$\left(\frac{a^*}{a_0}\right)^2 = \frac{T^*}{T_0} = \frac{2}{\gamma+1} \tag{8.24}$$

Using the isentropic relations allows this to be written as

$$\frac{p^*}{p_0} = \left(\frac{2}{\gamma+1}\right)^{\frac{\gamma}{\gamma-1}} \tag{8.25}$$

Substituting this into Equation 8.20 gives the following equation that can be used to find the area of the duct where the critical conditions exist

$$\dot{m} = \rho_0 A^* \left(\frac{2}{\gamma+1}\right)^{\frac{1}{\gamma-1}} \left\{\left(\frac{2\gamma}{\gamma-1}\right) \frac{p_0}{\rho_0} \left(\frac{\gamma-1}{\gamma+1}\right)\right\}^{\frac{1}{2}}$$

This can be rearranged to give

$$A^* = \frac{\dot{m}}{\sqrt{\gamma p_0 \rho_0}}\left(\frac{2}{\gamma+1}\right)^{-\frac{\gamma+1}{2(\gamma-1)}} \tag{8.26}$$

It must be stressed, as already mentioned, that the critical conditions may not exist in the real flow, e.g., if the flow remains subsonic throughout, critical conditions will not exist at any point in the flow. Even in such cases, however, the critical conditions are used as reference conditions.

The area at any section of the duct, expressed in terms of the critical area, A^*, can be related to the pressure by substituting Equation 8.25 into Equation 8.22, giving

$$\frac{A}{A^*} = \left(\frac{2}{\gamma+1}\right)^{\frac{1}{\gamma-1}} \frac{1}{\left(\dfrac{p}{p_0}\right)^{\frac{1}{\gamma}}} \left\{\frac{1-\dfrac{2}{\gamma+1}}{1+\left(\dfrac{p}{p_0}\right)^{\frac{\gamma-1}{\gamma}}}\right\}^{\frac{1}{2}}$$

i.e.,

$$\frac{A}{A^*} = \frac{\left(\dfrac{2}{\gamma+1}\right)^{\frac{\gamma+1}{2(\gamma-1)}}\left(\dfrac{\gamma-1}{2}\right)^{\frac{1}{2}}}{\left\{\left(\dfrac{p}{p_0}\right)^{\frac{2}{\gamma}} - \left(\dfrac{p}{p_0}\right)^{\frac{\gamma+1}{\gamma}}\right\}^{\frac{1}{2}}} \tag{8.27}$$

If a nozzle is to be designed for a given mass flow rate, $\dot{m}$, with a given overall pressure ratio, (p_e/p_0) (see Figure 8.5), then applying Equations 8.19, 8.20, and 8.27 gives

$$V_e = \left\{\left(\frac{2\gamma}{\gamma-1}\right)\left(\frac{p_0}{\rho_0}\right)\left[1-\left(\frac{p_e}{p_0}\right)^{\frac{\gamma-1}{\gamma}}\right]\right\}^{\frac{1}{2}} \tag{8.28}$$

$$\dot{m} = \rho_0 A_e \left(\frac{p_e}{p_0}\right)^{\frac{1}{\gamma}} \left\{\left(\frac{2\gamma}{\gamma-1}\right)\left(\frac{p_0}{\rho_0}\right)\left[1-\left(\frac{p_e}{p_0}\right)^{\frac{\gamma-1}{\gamma}}\right]\right\}^{\frac{1}{2}} \tag{8.29}$$

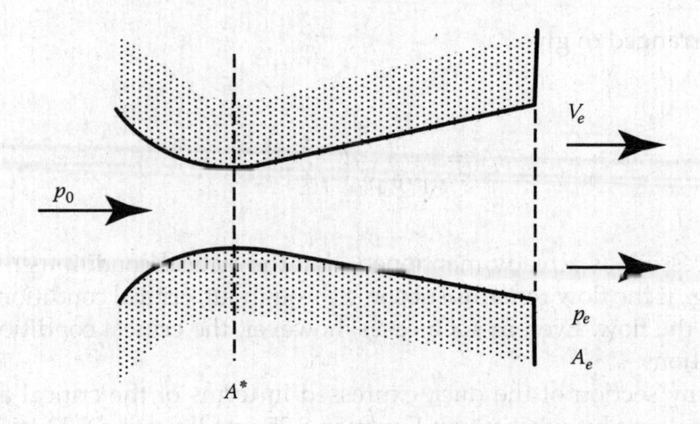

FIGURE 8.5
Exit plane conditions.

$$\frac{A_e}{A^*} = \frac{\left(\dfrac{2}{\gamma+1}\right)^{\frac{\gamma+1}{2(\gamma-1)}}\left(\dfrac{\gamma-1}{2}\right)^{\frac{1}{2}}}{\left\{\left(\dfrac{p_e}{p_0}\right)^{\frac{2}{\gamma}} - \left(\dfrac{p_e}{p_0}\right)^{\frac{\gamma+1}{\gamma}}\right\}^{\frac{1}{2}}} \tag{8.30}$$

For a given set of stagnation conditions and given values of $\dot{m}$ and (p_e/p_0), Equation 8.29 allows the exit area, A_e, to be found, whereas Equation 8.30 allows the throat area, A^*, to be found. Further, for the given pressure ratio, (p_e/p_0), Equation 8.28 allows the discharge velocity, V_e, to be found. The present analysis cannot be used to make any predictions concerning the optimum shape of the nozzle since it is based on the assumption that the flow is one-dimensional at all sections.

It is often convenient to express the variable area relations in terms of the Mach number, M, existing at any section. To do this, it is first noted, as before, that the energy equation gives

$$V^2 + \left(\frac{2}{\gamma-1}\right)a^2 = \left(\frac{2}{\gamma-1}\right)a_0^2$$

which can be used to give

$$V = Ma_0\left\{1+\left(\frac{\gamma-1}{2}\right)M^2\right\}^{-\frac{1}{2}} \tag{8.31}$$

The energy equation also gives

$$\left(\frac{a}{a_0}\right)^2 = \left\{1+\left(\frac{\gamma-1}{2}\right)M^2\right\}^{-1} \tag{8.32}$$

This in turn gives, using the isentropic relations,

$$\left(\frac{\rho}{\rho_0}\right) = \left\{1 + \left(\frac{\gamma - 1}{2}\right)M^2\right\}^{-\frac{1}{\gamma - 1}} \tag{8.33}$$

Hence, since

$$\dot{m} = \rho V A$$

it follows by using Equations 8.31 through 8.33 that

$$\frac{\dot{m}}{A} = \frac{\rho_0 a_0 M}{\left\{1 + \left(\frac{\gamma - 1}{2}\right)M^2\right\}^{\frac{\gamma+1}{2(\gamma-1)}}} \tag{8.34}$$

Since $\dot{m}$ must be the same at all sections, applying this equation between any two sections 1 and 2 of a duct then gives

$$\frac{A_2}{A_1} = \left(\frac{M_1}{M_2}\right)\left\{\frac{1 + \left(\frac{\gamma - 1}{2}\right)M_2^2}{1 + \left(\frac{\gamma - 1}{2}\right)M_1^2}\right\}^{\frac{\gamma+1}{2(\gamma-1)}} \tag{8.35}$$

Thus, if the ratio of the areas at the two sections is known, the Mach numbers at these sections can be related by this equation. It is again convenient to write this equation in terms of the duct area, A^*, at which the critical conditions exist. Since M^* is by definition equal to 1 at this section, the Mach number, M, at some other section where the area is A is then given by

$$\frac{A}{A^*} = \left(\frac{1}{M}\right)\left\{\frac{1 + \left(\frac{\gamma - 1}{2}\right)M^2}{1 + \left(\frac{\gamma - 1}{2}\right)}\right\}^{\frac{\gamma+1}{2(\gamma-1)}} = \frac{1}{M}\left\{\left(\frac{2}{\gamma + 1}\right)\left[1 + \left(\frac{\gamma - 1}{2}\right)M^2\right]\right\}^{\frac{\gamma+1}{2(\gamma-1)}} \tag{8.36}$$

Typical variations of A/A^* with M as given by this equation are shown in Figure 8.6. This relationship is usually listed in isentropic tables (see Appendix E) and is given by the software COMPROP.

It should be clearly understood that Equation 8.36 indicates that the Mach number is uniquely related to the cross-sectional area for a given value of A^*. This point will be considered further in the discussion of nozzle characteristics given below.

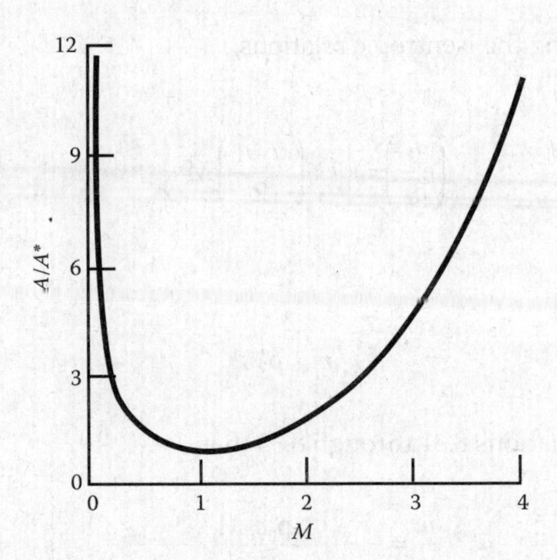

FIGURE 8.6
Variation of area ratio with Mach number.

Example 8.1

Hydrogen flows from a large reservoir through a convergent–divergent nozzle, the pressure and temperature in the reservoir being 600 kPa and 40°C, respectively. The throat area of the nozzle is 10^{-4} m^2 and the pressure on the nozzle exit plane is 130 kPa. Assuming that the flow throughout the nozzle is isentropic and that the flow is steady and one-dimensional, find the mass flow rate through the nozzle and the exit area of the nozzle.

Solution

The flow situation being considered is shown in Figure E8.1.

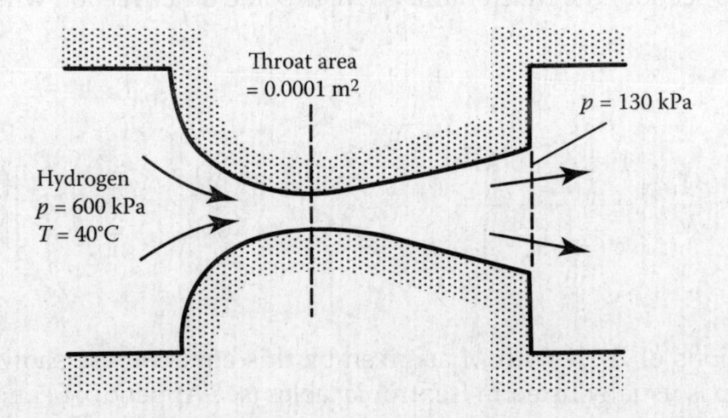

FIGURE E8.1
Flow situation considered.

Because

$$\frac{p^*}{p_0} = \left(\frac{2}{\gamma+1}\right)^{\frac{\gamma}{\gamma-1}}$$

it follows that, since for hydrogen $\gamma = 1.407$,

$$\frac{p^*}{p_0} = \left(\frac{2}{2.407}\right)^{\frac{1.407}{0.407}} = 0.5271$$

This could also have been obtained using the software COMPROP.

Now, in the situation being considered, $p_e/p_0 = 130/600 = 0.2167$ (< 0.5271). Hence, in this situation, the flow is being accelerated to a supersonic velocity and at the throat $M = 1$; thus,

$$\dot{m} = \rho_0 A^* \left(\frac{2}{\gamma+1}\right)^{\frac{1}{\gamma-1}} \left\{ \left(\frac{2\gamma}{\gamma-1}\right) \frac{p_0}{\rho_0} \left(\frac{\gamma-1}{\gamma+1}\right) \right\}^{\frac{1}{2}}$$

Hence,

$$\dot{m} = \rho_0 A^* \left(\frac{2}{2.407}\right)^{\frac{1}{0.407}} \left\{ \left(\frac{2.814}{0.407}\right) \frac{p_0}{\rho_0} \left(\frac{0.407}{2.407}\right) \right\}^{\frac{1}{2}}$$

However, $\rho_0 = p_0/RT_0$; thus, since $R = 8314.3/2.016 = 4124.2$, it follows that $\rho_0 = 600{,}000/(4124.2 \times 313) = 0.4648$ kg/m^3. Therefore,

$$\dot{m} = 0.4648 \times 0.0001 \times \left(\frac{2}{2.407}\right)^{\frac{1}{0.407}} \left\{ \left(\frac{2.814}{0.407}\right) \frac{600{,}000}{0.4648} \left(\frac{0.407}{2.407}\right) \right\}^{\frac{1}{2}} = 0.03622 \text{ kg/s}$$

Hence, the mass flow rate through the nozzle is 0.03622 kg/s.

The nozzle exit area can be found using

$$\frac{A_e}{A^*} = \frac{\left(\frac{2}{\gamma+1}\right)^{\frac{\gamma+1}{2(\gamma-1)}} \left(\frac{\gamma-1}{2}\right)^{\frac{1}{2}}}{\left\{ \left(\frac{p_e}{p_0}\right)^{\frac{2}{\gamma}} - \left(\frac{p_e}{p_0}\right)^{\frac{\gamma+1}{\gamma}} \right\}^{\frac{1}{2}}}$$

which gives

$$\frac{A_e}{0.0001} = \frac{\left(\dfrac{2}{2.407}\right)^{\frac{2.407}{2\times 0.407}}\left(\dfrac{0.407}{2}\right)^{\frac{1}{2}}}{\left\{\left(\dfrac{130,000}{600,000}\right)^{\frac{2}{1.407}} - \left(\dfrac{130,000}{600,000}\right)^{\frac{2.407}{1.407}}\right\}^{\frac{1}{2}}} = 0.0006416\ \text{m}^2$$

Therefore, the nozzle exit area is 6.416×10^{-4} m^2.

Operating Characteristics of Nozzles

The concern here is with the effect of changes in the upstream and downstream pressures on the nature of the flow in and on the mass flow rate through a nozzle, i.e., through a variable area passage designed to accelerate a gas flow. In the present discussion of the operating characteristics of nozzles, it will be assumed that the nozzle is connected to an upstream chamber in which the conditions, i.e., the upstream stagnation conditions, are kept constant while the conditions in the downstream chamber into which the nozzle discharges are varied. The pressure in the downstream chamber is termed the back-pressure. The situation considered is therefore as shown in Figure 8.7. The characteristics of convergent and convergent–divergent nozzles will be separately discussed.

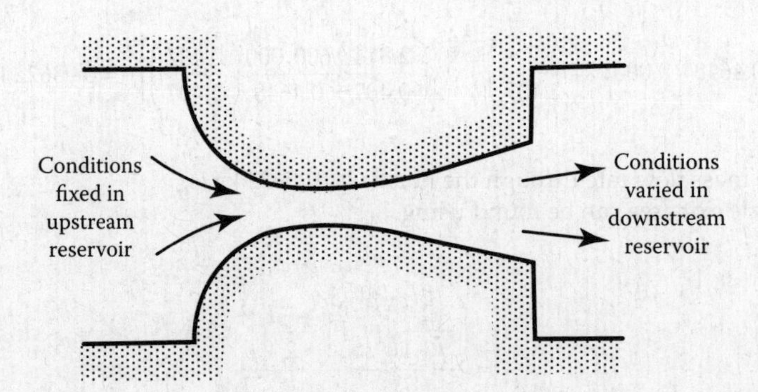

FIGURE 8.7
Flow situation considered in discussing nozzle operating characteristics.

Convergent Nozzle

In this section, a nozzle of the type shown in Figure 8.8 will be considered.

If the back-pressure, p_b, is initially equal to the supply pressure, p_0, there will be no flow through the nozzle. As the back-pressure, p_b, is decreased, flow commences, this flow initially being subsonic throughout the nozzle. Under these circumstances, the pressure on the exit plane of the nozzle, p_e, remains equal to the back-pressure, p_b, and the Mach number on the exit plane is less than 1. In this region of operation, a reduction in p_b produces an increase in the mass flow rate, $\dot{m}$. This type of flow continues to exist until p_b is reduced to the critical pressure corresponding to p_0, i.e., until

$$p_b = p^* = p_0\left(\frac{2}{\gamma+1}\right)^{\frac{\gamma}{\gamma-1}}$$

When p_b has been decreased to this value, the Mach number on the exit plane becomes equal to 1. Further reductions in p_b have no effect on the flow in the nozzle, i.e., p_e remains equal to p^*, the mass flow rate remains constant and the Mach number on the exit plane remains equal to 1. Since $p_b < p_e$ in this state, the expansion from p_e to p_b takes place outside the nozzle through a series of expansion waves, the flow then resembling that shown in Figure 8.9.

The reason that reductions in the back-pressure have no effect on the flow in the nozzle once the Mach number at the nozzle exit plane has reached 1 can be understood by realizing that, as previously discussed, the effect of changes in the back-pressure are propagated into the nozzle at the speed of sound relative to the fluid. The speed of propagation of the effects of the changes in the back-pressure up the nozzle relative to the nozzle will therefore be equal to the speed of sound minus the local velocity of the gas. Hence, once the gas velocity on the exit plane becomes equal to the local speed of sound, i.e., once the exit plane Mach number becomes equal to 1, the effect of changes in the back-pressure cannot be propagated up the nozzle. As a result, once the exit plane Mach number has reached

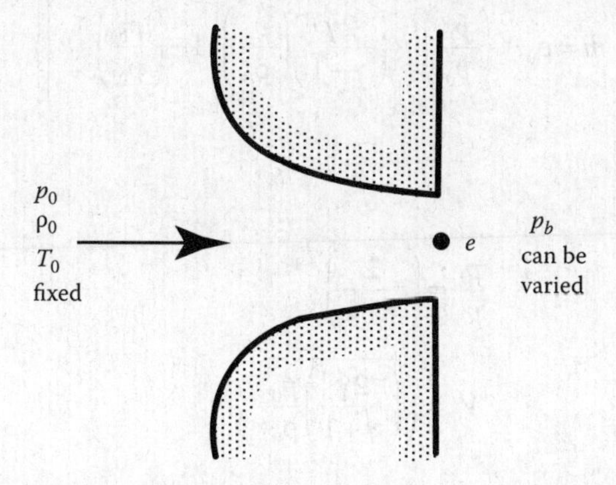

FIGURE 8.8
Convergent nozzle situation considered.

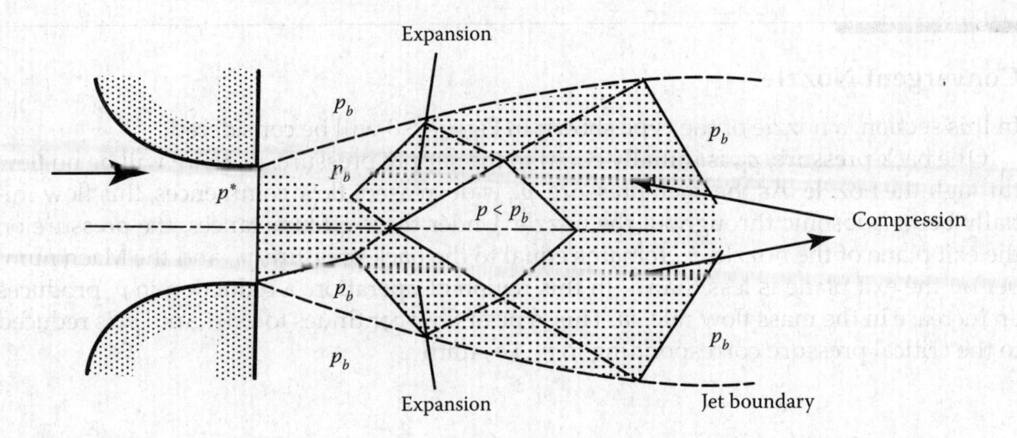

FIGURE 8.9
Flow near the exit to a convergent nozzle when p_b is less than the critical pressure.

a value of 1, further reductions in back-pressure can have no influence on the flow in the nozzle and cannot therefore effect the mass flow rate through the nozzle.

From the above discussion, it follows that the following relations apply to the flow through a convergent nozzle.

When $p_b > p^*$,

$$p_e = p_b$$

$$V_e = \left\{\left(\frac{2\gamma}{\gamma-1}\right)\left(\frac{p_0}{\rho_0}\right)\left[1-\left(\frac{p_b}{p_0}\right)^{\frac{\gamma-1}{\gamma}}\right]\right\}^{\frac{1}{2}}$$

$$\dot{m} = \rho_0 A_e \left(\frac{p_b}{p_0}\right)^{\frac{1}{\gamma}}\left\{\left(\frac{2\gamma}{\gamma-1}\right)\left(\frac{p_0}{\rho_0}\right)\left[1-\left(\frac{p_b}{p_0}\right)^{\frac{\gamma-1}{\gamma}}\right]\right\}^{\frac{1}{2}}$$

When $p_b \le p^*$,

$$\frac{p_e}{p_0} = \left(\frac{2}{\gamma+1}\right)^{\frac{\gamma}{\gamma-1}}$$

$$V_e = \sqrt{\left(\frac{2\gamma}{\gamma+1}\right)\frac{p_0}{\rho_0}}$$

$$\dot{m} = \sqrt{\gamma p_0 \rho_0}\, A_e \left(\frac{2}{\gamma+1}\right)^{\frac{\gamma+1}{2(\gamma-1)}}$$

All of these quantities, i.e., p_e/p_0, V_e, $\dot{m}$, being independent of p_b, i.e., being constant.

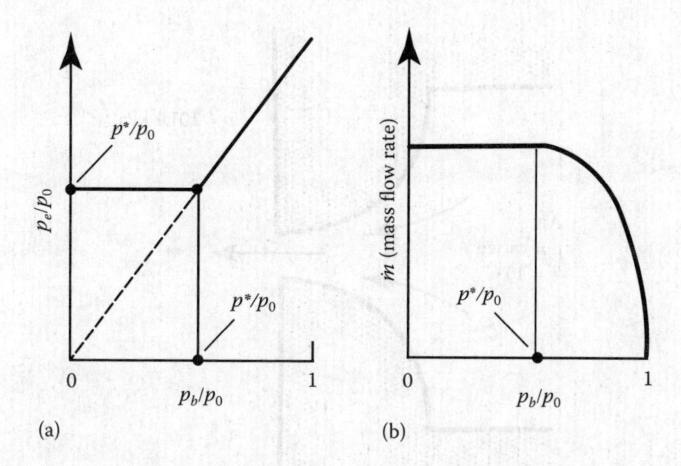

FIGURE 8.10
Effect of back-pressure on the exit plane pressure (a) and on the mass flow rate (b) with flow through a convergent nozzle.

The variation of p_e and $\dot{m}$ with p_b for a convergent nozzle is therefore as shown in Figure 8.10. While $p_b/p_0 > p^*/p_0$, $p_e/p_0 = p_b/p_0$ and $\dot{m}$ increase with decreasing p_b/p_0. However, when $p_b/p_0 < p^*/p_0$, $p_e/p_0 = p^*/p_0$ and $\dot{m}$ are unaffected by changes in p_b/p_0. Under these circumstances, when changes in the back-pressure have no effect on the mass flow rate through the nozzle, the nozzle is said to be "choked."

Example 8.2

Consider the flow of air out of a large vessel through a convergent nozzle to the atmosphere, the atmospheric pressure being 101.1 kPa. The temperature of the air in the vessel is 40°C and the pressure in this vessel is varied. Show how the mass flow rate out of the nozzle per unit exit section area varies with the pressure in the vessel. Assume steady, one-dimensional, isentropic flow in the nozzle.

Solution

The flow situation here being considered is shown in Figure E8.2a.

The velocity on the exit plane is given by

$$V_e = M_e a_e = M_e a_0 \frac{a_e}{a_0}$$

The mass flow rate is given by

$$\dot{m} = \rho_e V_e A_e = \frac{\rho_e}{\rho_0} \rho_0 V_e A_e$$

Hence,

$$\frac{\dot{m}}{A_e} = \frac{\rho_e}{\rho_0} \rho_0 M_e a_0 \frac{a_e}{a_0} = M_e a_0 \rho_0 \frac{\rho_e}{\rho_0} \frac{a_e}{a_0}$$

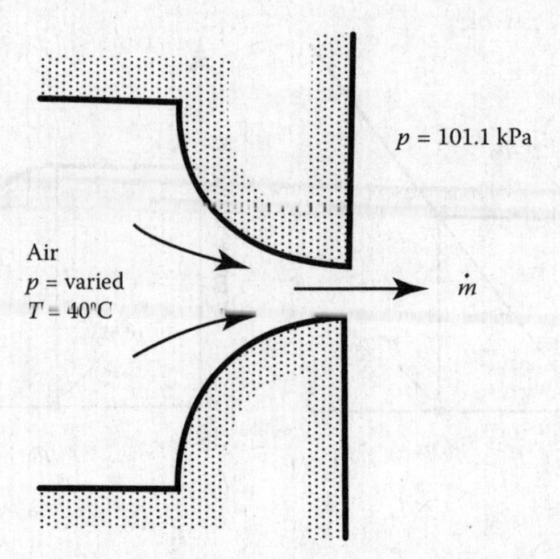

FIGURE E8.2a
Flow situation considered.

However,

$$\rho_0 = \frac{p_0}{RT_0} = \frac{p_0}{287 \times 313} = \frac{p_0}{89{,}831}$$

and

$$a_0 = \sqrt{\gamma RT_0} = \sqrt{1.4 \times 287 \times 313} = 354.6 \text{ m/s}$$

Hence,

$$\frac{\dot{m}}{A_e} = M_e \times 354.6 \times \frac{p_0}{89{,}831} \frac{\rho_e}{\rho_0} \frac{a_e}{a_0}$$

Because the flow in the nozzle is isentropic, isentropic relations or tables or the software COMPROP give the values of ρ_e/ρ_0, M_e, and a_e/a_0 for any value of p_e/p_0 and therefore for any value of $101{,}100/p_0$, with p_0 in pascals. This will apply until p_0 has risen to a value that causes M_e to reach a value of 1 which, according to isentropic relations or tables or the software, occurs when $p_0/p_e = 1.8929$. Once p_0 has risen to this value, i.e., 191.37 kPa, p_0/p_e remains equal to 1.8929 and

$$\frac{\dot{m}}{A_e} = 1 \times 354.6 \times \frac{p_0}{89{,}831} \frac{\rho^*}{\rho_0} \frac{a^*}{a_0}$$

$$= 1 \times 354.6 \times \frac{p_0}{89{,}831} \times 0.63394 \times 0.91287$$

$$= 0.0022844\, p_0$$

Using these relations and isentropic relations or tables or the COMPROP software, Table E8.2 can be constructed.

The variation is shown in Figure E8.2b.

TABLE E8.2

Calculated Variation of Mass Flow Rate per Unit Flow Area with p_0

p_0 (kPa)	p_0/p_e	M_e	a_0/a_e	ρ_0/ρ_e	$\dot{m}/A_e$
101.1	1.000	0.0000	1.000	1.000	0.000
120	1.187	0.5009	1.130	1.025	204.9
140	1.385	0.6981	1.262	1.047	292.0
160	1.583	0.8371	1.388	1.068	356.7
180	1.780	0.9465	1.510	1.086	410.1
191.4	1.893	1.0000	1.577	1.096	437.1
200	1.893	1.0000	1.577	1.096	456.8
250	1.893	1.0000	1.577	1.096	571.0
300	1.893	1.0000	1.577	1.096	685.2
400	1.893	1.0000	1.577	1.096	913.6
500	1.893	1.0000	1.577	1.096	1141.9

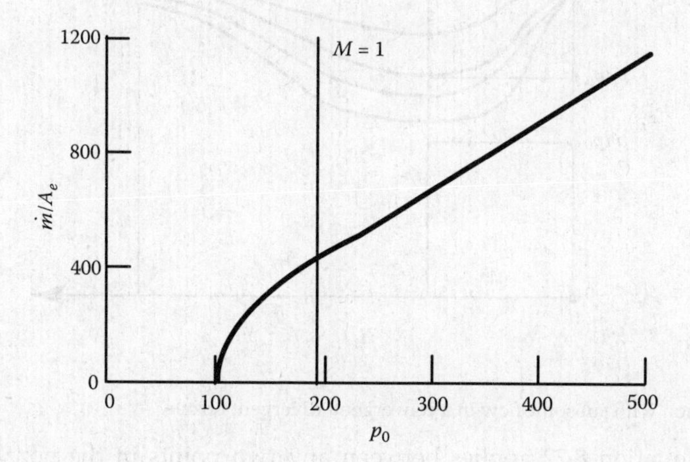

FIGURE E8.2b
Variation of mass flow rate per unit area with supply pressure.

Convergent–Divergent Nozzle

In this section, a nozzle of the type shown in Figure 8.11 will be considered.

As p_b varies, four more or less separate flow regimes can be identified. The nature of the flow that exists in these four regimes can again be explained by considering how the flow in the nozzle changes as the back-pressure p_b, is decreased. The four flow regimes are then as follows:

1. When p_b is very nearly the same as p_0 the flow remains subsonic throughout. The flow in the nozzle is then similar to that through a venturi, the pressure dropping from p_0 to a minimum value that is greater than p^* at the throat and then increasing again to $p_e = p_b$ at the exit. In this flow regime, the pressure distribution in the nozzle therefore resembles that shown in Figure 8.12. With this type of flow, the mass flow rate through the nozzle increases as p_b, is decreased.

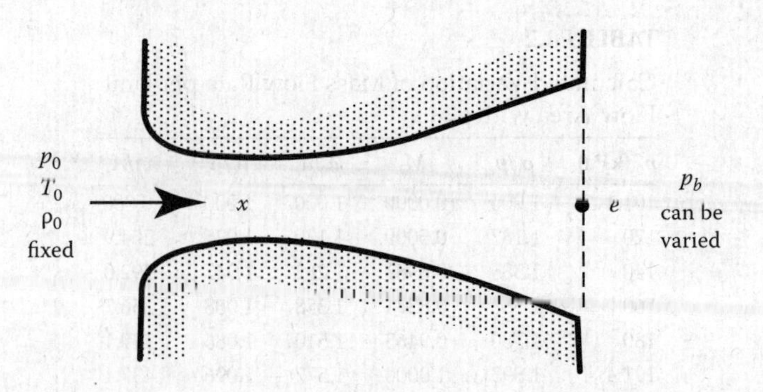

FIGURE 8.11
Convergent–divergent nozzle.

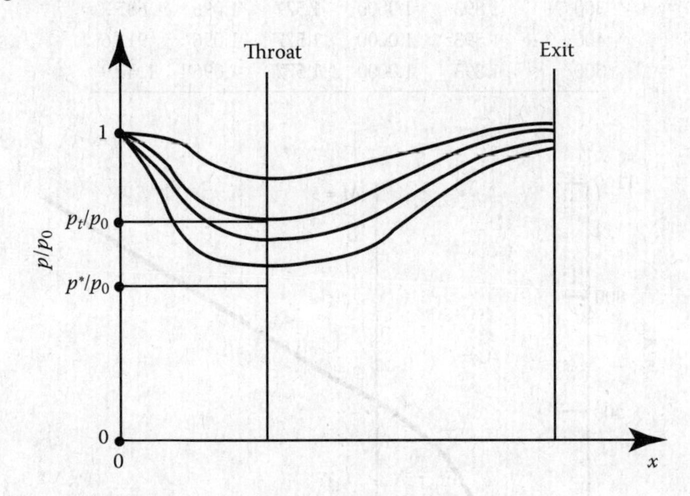

FIGURE 8.12
Pressure distribution with subsonic flow in a convergent–divergent nozzle.

Since Equation 8.22 applies between any two points in the nozzle, it can be applied between the throat and the nozzle exit to relate the pressure at the throat, p_t, to the back-pressure, p_b, which, with subsonic flow throughout the nozzle, is equal to p_e. This procedure gives

$$\frac{A_e}{A^*} = \left(\frac{p_t}{p_b}\right)^{\frac{1}{\gamma}} \left[\frac{1-\left(\frac{p_t}{p_0}\right)^{\frac{\gamma-1}{\gamma}}}{1-\left(\frac{p_b}{p_0}\right)^{\frac{\gamma-1}{\gamma}}}\right]^{\frac{1}{2}}$$

i.e.,

$$\left(\frac{p_t}{p_0}\right)^{\frac{1}{\gamma}} \left\{1-\left(\frac{p_t}{p_0}\right)^{\frac{\gamma-1}{\gamma}}\right\}^{\frac{1}{2}} = \left(\frac{A_e}{A^*}\right)\left(\frac{p_b}{p_0}\right)^{\frac{1}{\gamma}} \left\{1-\left(\frac{p_b}{p_0}\right)^{\frac{\gamma-1}{\gamma}}\right\}^{\frac{1}{2}} \tag{8.37}$$

2. In the first flow regime, discussed above, as the back-pressure decreases, the throat pressure, which is lower than the back-pressure, also decreases. This continues until p_b has dropped to a value at which the throat pressure becomes equal to the critical pressure and the Mach number at the throat becomes equal to 1. The back-pressure at which this occurs is obtained by substituting

$$p_t = p^* = p_0\left(\frac{2}{\gamma+1}\right)^{\frac{\gamma}{\gamma-1}}$$

into Equation 8.37, giving

$$\left(\frac{A_e}{A^*}\right)\left(\frac{p_{b\,\mathrm{crit}}}{p_0}\right)^{\frac{1}{\gamma}}\left\{1-\left(\frac{p_{b\,\mathrm{crit}}}{p_0}\right)^{\frac{\gamma-1}{\gamma}}\right\}=\left(\frac{\gamma-1}{\gamma+1}\right)\left(\frac{2}{\gamma+1}\right)^{\frac{1}{\gamma-1}} \tag{8.38}$$

Here $p_{b\,\mathrm{crit}}$ is the back-pressure at which the throat pressure first drops to the critical value.

Once Mach 1 has been reached at the throat, further reductions in the back-pressure cannot effect conditions upstream of the throat and therefore cannot alter the mass flow rate through the nozzle. The nozzle is therefore choked once the back-pressure is decreased to $p_{b\,\mathrm{crit}}$ and the second flow regime is entered when p_b has decreased to $p_{b\,\mathrm{crit}}$.

As the back-pressure is reduced below $p_{b\,\mathrm{crit}}$, a region of supersonic flow develops just downstream of the throat. This region of supersonic flow is terminated by what is effectively a normal shock wave. The shock wave increases the pressure and reduces the velocity to a subsonic value. The flow downstream of the shock wave then decelerates subsonically until the pressure on the exit plane p_e is equal to the back-pressure, p_b. The flow in the nozzle under these circumstances is as shown in Figure 8.13.

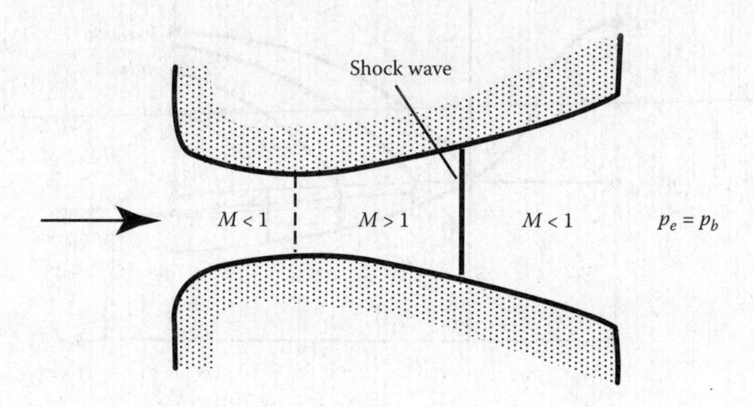

FIGURE 8.13
Flow in a convergent–divergent nozzle in Regime 2.

In the above discussion, the supersonic portion of the flow is assumed to be terminated by a normal shock wave. In real flows, if the nozzle is of relatively small size and the boundary layer consequently relatively thick, a complex wave system can actually occur near the end of the supersonic flow region as a result of the interaction of the shock wave with the boundary layer. However, even in such cases, the characteristics of the flow can often be adequately modeled by assuming a normal shock wave.

As the back-pressure is further reduced, the extent of the supersonic flow region increases, the shock wave moving further down the divergent portion of the nozzle. The nozzle pressure distribution in this regime therefore resembles that shown in Figure 8.14.

3. As noted above, in flow Regime 2, as the back-pressure decreases the shock wave moves down the divergent portion of the nozzle toward the exit plane. Eventually, p_b will drop to a value at which the shock wave is on the exit plane of the nozzle. It is when this occurs that flow Regime 3 is entered. The back-pressure at which this occurs can be estimated by again assuming that the shock wave, which at the beginning of this regime lies in the exit plane of the nozzle, can be adequately modeled as a normal shock. The flow at the exit plane of the nozzle is then as shown in Figure 8.15.

   Since the flow in the nozzle is now isentropic throughout, the pressure, p_e, and Mach number, M_e, on the exit plane ahead of the shock can be found using isentropic relations. They are therefore given by the following two equations:

$$\left(\frac{p_e}{p_0}\right)^{\frac{2}{\gamma}} - \left(\frac{p_e}{p_0}\right)^{\frac{\gamma+1}{\gamma}} = \left(\frac{2}{\gamma+1}\right)^{\frac{\gamma+1}{\gamma-1}} \left(\frac{\gamma-1}{2}\right)\left(\frac{A^*}{A_e}\right)^{\frac{1}{2}}$$

$$\frac{1}{M_e}\left\{\left(\frac{2}{\gamma+1}\right)\left[1+\left(\frac{\gamma-1}{2}\right)M_e^2\right]^{\frac{\gamma+1}{2(\gamma-1)}}\right\} = \frac{A_e}{A^*}$$

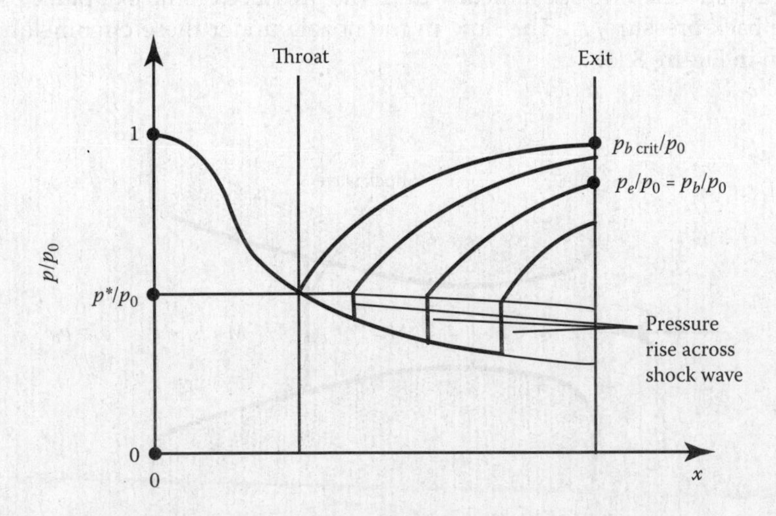

FIGURE 8.14
Pressure distribution in a convergent–divergent nozzle in Regime 2.

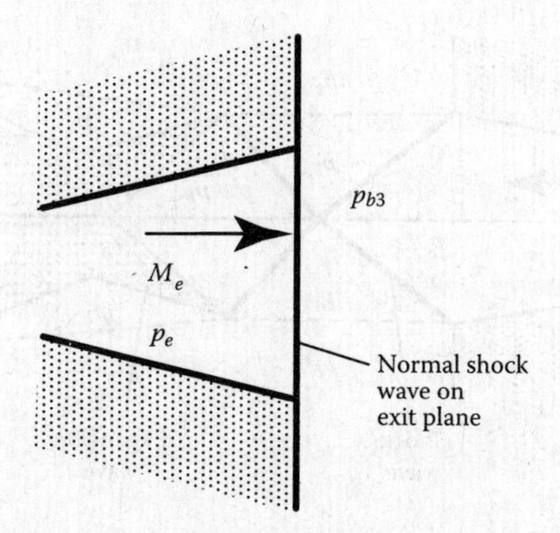

FIGURE 8.15
Flow at exit of convergent–divergent nozzle at start of Regime 3.

these being obtained using Equations 8.27 and 8.36, respectively. Thus, since (A_e/A^*) is known, (p_e/p_0) and M_e can be calculated. They will, of course, be equal to the design pressure ratio and the design exit Mach number because the flow in the nozzle is isentropic throughout. The design conditions are those that involve isentropic flow throughout the nozzle with the exit plane pressure equal to the back-pressure.

Next, it is noted that since the flow behind a normal shock wave is subsonic, the static pressure behind the shock wave must be the back-pressure, p_{b3}. Therefore, the normal shock wave relations given in Chapter 5 give

$$\frac{p_{b3}}{p_e} = \frac{2\gamma M_e^2 - (\gamma - 1)}{(\gamma + 1)}$$

Hence, since M_e is known, p_{b3}/p_e can be calculated. p_{b3}/p_0 is then given by noting that

$$\frac{p_{b3}}{p_0} = \left(\frac{p_{b3}}{p_e}\right)\left(\frac{p_e}{p_0}\right)$$

As p_b is decreased below p_{b3}, the conditions at all sections of the nozzle remain unchanged and the pressure on the exit plane, p_e, remains unchanged. With decreasing p_b, however, the shock wave moves outside the nozzle and the compression from p_e to p_b takes place through a series of oblique shock waves outside the nozzle, the flow pattern resembling that shown in Figure 8.16.

In this state, the nozzle is said to be "overexpanded" because the exit plane pressure is less than the back-pressure. The pressure distribution in the nozzle in this regime resembles that shown in Figure 8.17.

4. In flow Regime 3, as p_b is decreased the oblique shock waves in the discharge flow become weaker and weaker and the difference between p_e and p_b becomes smaller and smaller. Eventually, a point is reached at which p_b is just equal to the exit plane

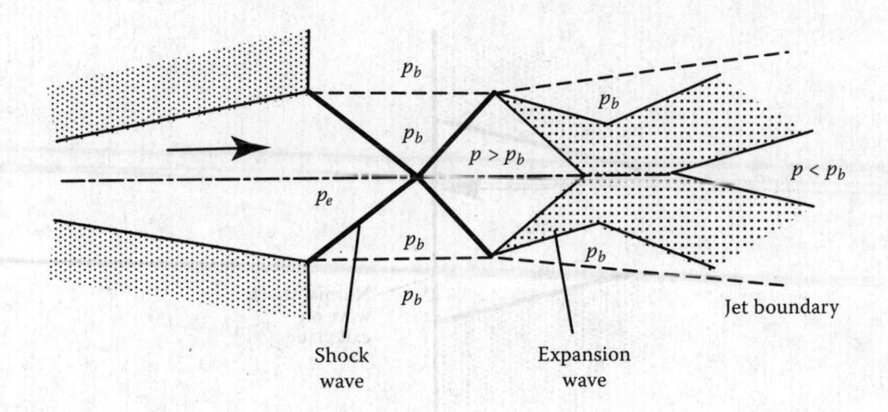

FIGURE 8.16
Flow near exit of overexpanded convergent–divergent nozzle.

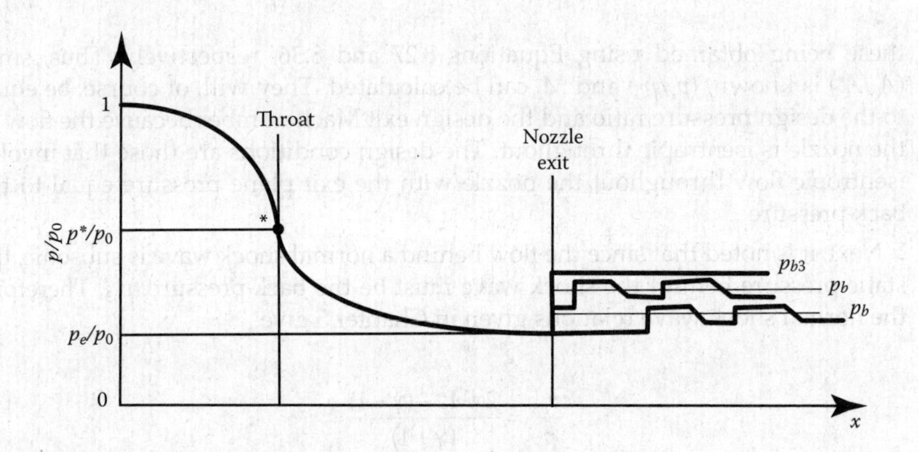

FIGURE 8.17
Pressure distribution in a convergent–divergent nozzle in Regime 3.

pressure, p_e. The nozzle is then operating at its "design" pressure ratio and there are no waves inside or outside the nozzle. A further reduction in p_b moves the flow into Regime 4. The design back-pressure, i.e., the back-pressure at which Regime 4 begins, is given by Equation 8.30 as

$$\left(\frac{p_{b4}}{p_0}\right)^{\frac{2}{\gamma}} - \left(\frac{p_{b4}}{p_0}\right)^{\frac{\gamma+1}{\gamma}} = \left(\frac{2}{\gamma+1}\right)^{\frac{\gamma+1}{(\gamma-1)}}\left(\frac{\gamma-1}{2}\right)\left(\frac{A^*}{A_e}\right)^{\frac{1}{2}}$$

If p_b is further reduced, it becomes less than the exit plane pressure, p_e, which remains constant, of course, from the beginning of Regime 3, and the expansion from p_e to p_b takes place through a series of expansion waves outside the nozzle as shown in Figure 8.18.

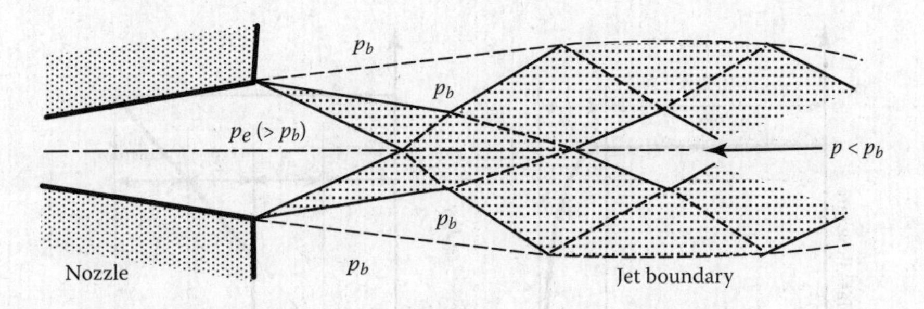

FIGURE 8.18
Flow near exit of underexpanded convergent–divergent nozzle.

In this state, the nozzle is said to be "underexpanded" because the pressure on the exit plane is greater than the back-pressure. The nozzle pressure distribution in this regime resembles that shown in Figure 8.19.

The characteristics of a convergent–divergent nozzle are summarized in Figures 8.20 and 8.21. Some of the features of flow through a convergent–divergent nozzle discussed above can be seen in the set of photographs of the flow near the exit of a two-dimensional nozzle shown in Figure 8.22.

Instead of using the equations given above to determine the limits of the various flow regimes, they can be determined using isentropic and normal shock tables or the COMPROP software. This is illustrated in the examples given below.

The above discussion was concerned with the situation where the supply pressure p_0 is kept constant and the back-pressure p_b is varied. The flow changes that occur in other situations, e.g., when there are variations in p_0 for a fixed value of p_b, can easily be deduced from the above discussion.

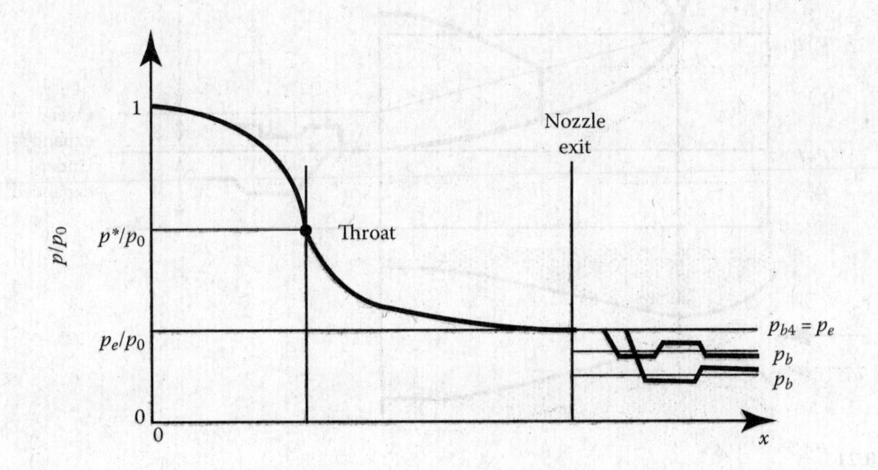

FIGURE 8.19
Pressure distribution in a convergent–divergent nozzle in Regime 4.

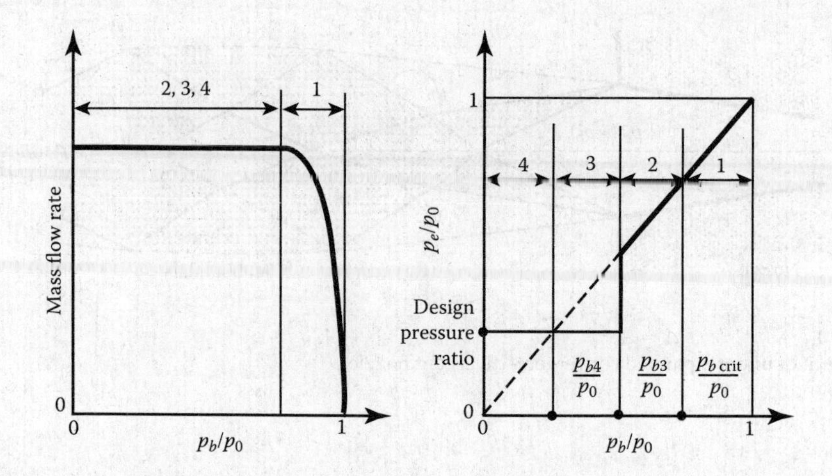

FIGURE 8.20
Effect of back-pressure on mass flow rate and exit plane pressure in a convergent–divergent nozzle.

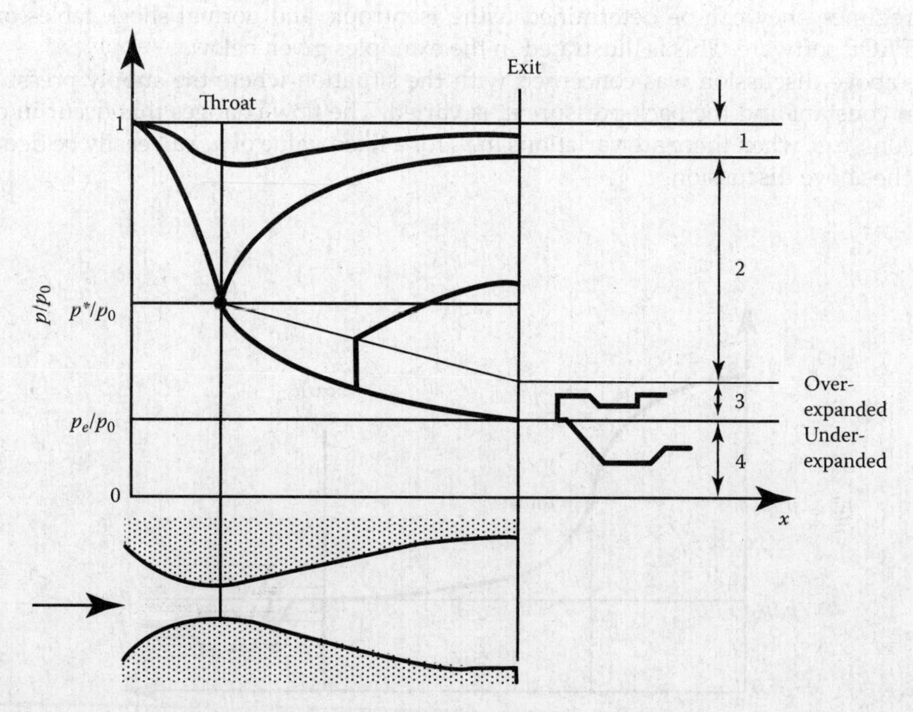

FIGURE 8.21
Pressure distribution in a convergent–divergent nozzle.

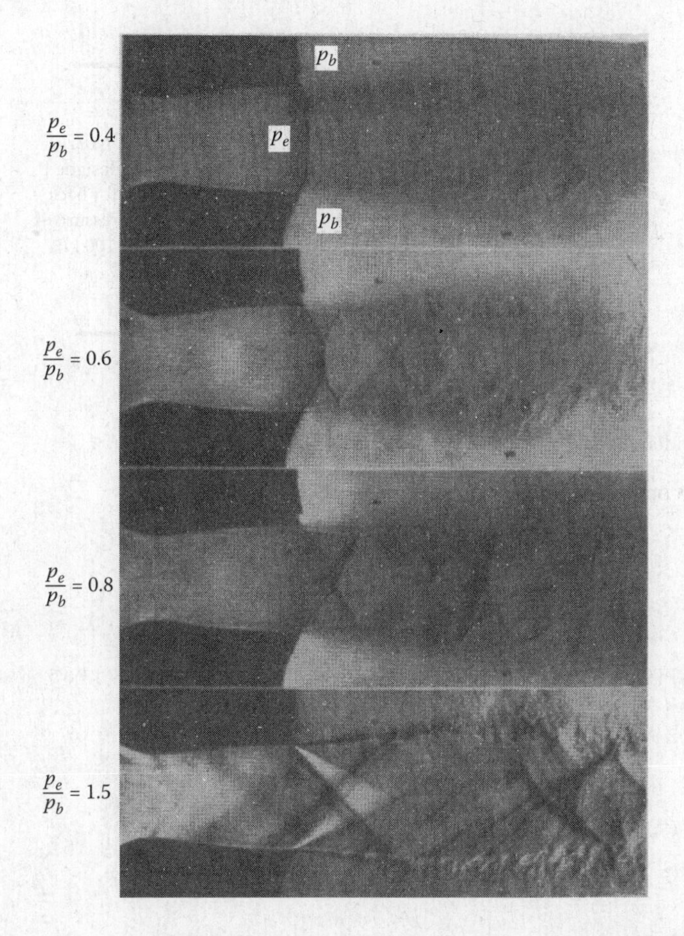

FIGURE 8.22
Photograph of flow near the exit of a convergent–divergent nozzle. (From L. Howarth (ed.), *Modern Developments in Fluid Dynamics*, vol. 1, plate 4, p. 203, 1956. Oxford University Press. With permission.)

Example 8.3

Air is expanded through a convergent–divergent nozzle from a large reservoir in which the pressure and temperature are 600 kPa and 40°C, respectively. The design back-pressure is 100 kPa.

Find

1. The ratio of the nozzle exit area to the nozzle throat area
2. The discharge velocity from the nozzle under design considerations
3. At what back-pressure will there be a normal shock at the exit plane of the nozzle?

Solution

The flow situation here being considered is shown in Figure E8.3.

Here, $p_0 = 600$ kPa and $T_0 = 40°C$ and the design back-pressure is $p_b = 100$ kPa.

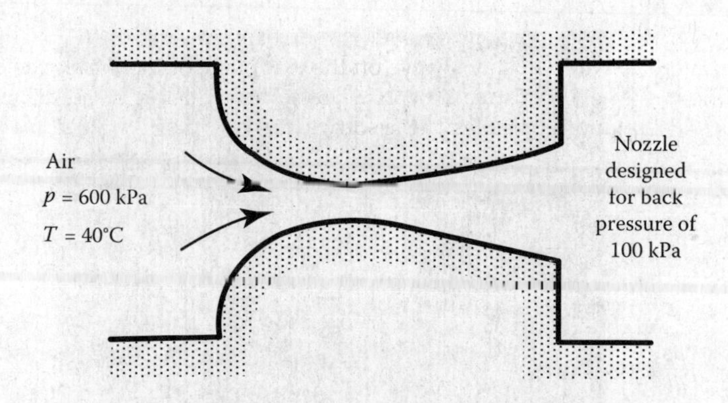

FIGURE E8.3
Flow situation considered.

1. When operating at the design conditions $p_e = p_b$, thus

$$\frac{p_e}{p_0} = \frac{100}{600} = 0.1667$$

Hence, using isentropic relations or tables or the software gives

$$M_e = 1.83$$

and

$$\frac{A^*}{A_e} = 0.6792$$

Hence, the ratio of the nozzle exit area to the nozzle throat area is

$$\frac{A_e}{A^*} = \frac{1}{0.6792} = 1.472$$

2. Since $M_e = 1.83$, using isentropic relations or tables or the software gives

$$\frac{a_e}{a_0} = 0.7739$$

However,

$$a_0 = \sqrt{\gamma R T_0} = \sqrt{1.4(286.8)(40 + 273)} = 345 \text{ m/s}$$

Hence,

$$V_e = \frac{V_e}{a_e}\frac{a_e}{a_0}a_0 = 1.83 \times 0.7739 \times 354.5 = 502.1 \text{ m/s}$$

Hence, the nozzle discharge velocity under design conditions is 502.1 m/s.

3. When there is a normal shock wave on the exit plane of the nozzle the design conditions will exist upstream of the shock. Hence, using $M_1 = 1.83$, normal shock wave relations or tables or the software give

$$\frac{p_2}{p_1} = 3.740$$

Hence,

$$p_2 = \frac{p_2}{p_1} p_1 = 3.740 \times 100 = 374.0 \text{ kPa}$$

Therefore, there will be a normal shock wave on the exit plane of the nozzle when $p_b = p_2 = 374.0$ kPa.

Example 8.4

A convergent–divergent nozzle is designed to expand air from a chamber in which the pressure is 800 kPa and the temperature is 40°C to give Mach 2.7. The throat area of the nozzle is 0.08 m^2. Find

1. The exit area of the nozzle
2. The mass flow rate through the nozzle when operating under design conditions
3. The design back-pressure
4. The lowest back-pressure for which there is only subsonic flow in the nozzle
5. The back-pressure at which there will be a normal shock wave on the exit plane of the nozzle
6. The back-pressure below which there are no shock waves in the nozzle
7. The range of back-pressures over which there are oblique shock waves in the exhaust from the nozzle
8. The range of back-pressures over which there are expansion waves in the exhaust from the nozzle

Solution

The flow situation being considered is shown in Figure E8.4.

1. For $M = 2.7$, isentropic flow relations or tables or software give

$$\frac{A}{A^*} = 3.183$$

Hence,

$$A_e = 3.183 \times 0.08 = 0.255 \text{ m}^2$$

2. Now $\dot{m} = \rho^* V^* A^*$. However,

$$\rho_0 = \frac{p_0}{RT_0} = \frac{800 \times 10^3}{287 \times 313} = 8.91 \text{ kg/m}^3$$

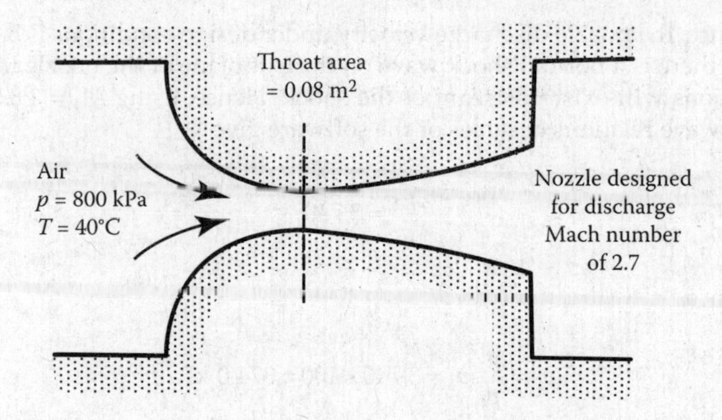

FIGURE E8.4
Flow situation considered.

Also, isentropic flow relations or tables or software give for $M = 1$

$$\frac{\rho^*}{\rho_0} = 0.63394, \qquad \frac{T^*}{T_0} = 0.83055$$

from which it follows that

$$T^* = 0.83055 \times 313 = 260 \text{ K}$$

Also,

$$V^* = a^* = \sqrt{\gamma R T^*} = \sqrt{1.4 \times 287 \times 260} = 323.2 \text{ m/s}$$

and

$$\rho^* = 0.63394 \times 8.91 = 5.65 \text{ kg/m}^3$$

Hence,

$$\dot{m} = \rho^* V^* A^* = 5.65 \times 323.2 \times 0.08 = 146 \text{ kg/s}$$

3. Isentropic flow relations or tables or the software give for $M = 2.7$

$$\frac{p_0}{p} = 23.283; \quad \text{hence,} \quad p_{\text{design}} = 800/23.283 = 34.36 \text{ kPa}$$

4. Subsonic isentropic flow relations or tables or software give $p_0/p = 1.025$ for $A/A^* = 3.183$. Hence, the Mach number at the throat will reach 1 when the back-pressure has dropped to $800/1.025 = 780.5$ kPa.

5. When there is a shock wave on the exit plane of the nozzle, the Mach number ahead of the shock is 2.7 and the pressure is 34.36 kPa. However, for Mach 2.7, normal shock relations or tables or software give

$$\frac{p_b}{p_{\text{design}}} = 8.33832$$

Hence, the back-pressure at which there is a shock wave on the nozzle exit plane is

$$p_b = 8.33832 \times 34.36 = 286.5 \text{ kPa}$$

6. When the back-pressure has dropped below that established in (5), the shock wave moves out of the nozzle. Hence, there are no shock waves in the nozzle when

$$p_b < 286.5 \text{ kPa}$$

7. There are oblique shocks in the exhaust when the back-pressure is below that established in (5). And is greater than the design back-pressure. Hence, there are oblique shock waves in the exhaust when

$$34.36 \text{ kPa} < p_b < 286.5 \text{ kPa}$$

8. Expansion waves will occur when $p_b < p_{\text{design}}$, i.e., when $p_b < 34.36$ kPa.

Convergent–Divergent Supersonic Diffusers

With most present day air-breathing aircraft engines, it is necessary to decelerate the air to subsonic velocity before passing it to the engine. This deceleration, which also increases the pressure of the air, is carried out in the diffuser. From the discussion given in the earlier part of this chapter, it follows that this deceleration can be accomplished by passing the air through a convergent–divergent diffuser, the deceleration potentially being shockless at the design flight Mach number of the aircraft. In such a case, the flow through the diffuser will be as shown in Figure 8.23.

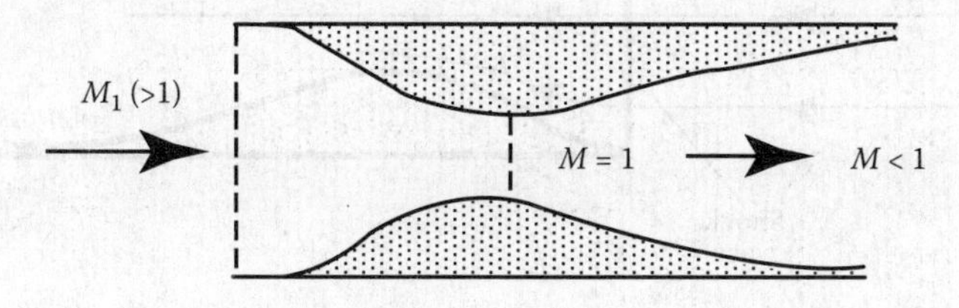

FIGURE 8.23
Ideal flow through a convergent–divergent supersonic diffuser.

The present discussion will be concerned with how this state is achieved and how the diffuser performs when the aircraft is flying at Mach numbers different from the design value. Initially, the concern will be with fixed diffusers, i.e., diffusers in which the throat and inlet areas are fixed.

Now, in the ideal case, the diffuser will, of course, be designed so that at the design Mach number, the ratio of throat to inlet area is equal to the value of A^*/A corresponding to the design Mach number. However, since A^*/A increases as M decreases, it is obvious that when the aircraft is accelerating up to the design Mach number the diffuser will not be able to "swallow" all the air flowing toward the intake. Under these conditions therefore a normal shock stands ahead of the diffuser and decreases the velocity to a subsonic velocity. The air is then able to "spill" over the intake. This is illustrated in Figure 8.24.

As the flight Mach number is increased and the throat area becomes closer to that required for the flight Mach number, the shock moves toward the intake. However, even when the design Mach number is reached, there is still a normal shock ahead of the diffuser and spillage still occurs. It is only when some Mach number that is greater than the design Mach number is reached that the shock moves up to the intake and all the air reaching the diffuser is ingested. This situation is illustrated in Figure 8.25.

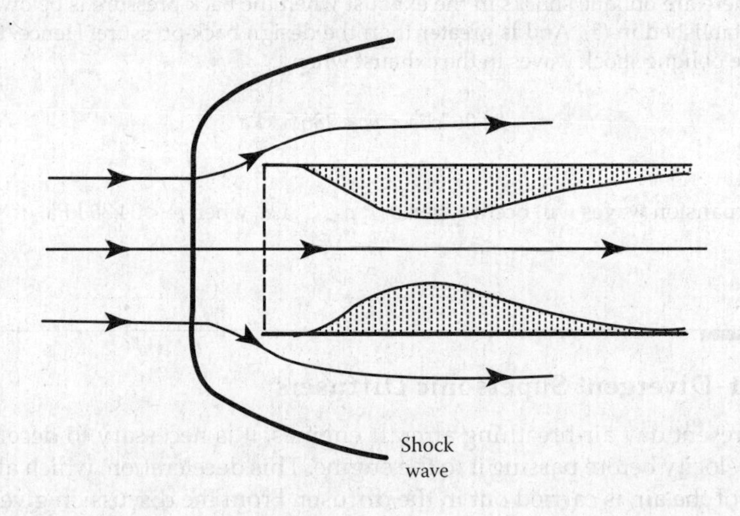

FIGURE 8.24
Flow near intake to fixed supersonic diffuser when $M < M_{design}$.

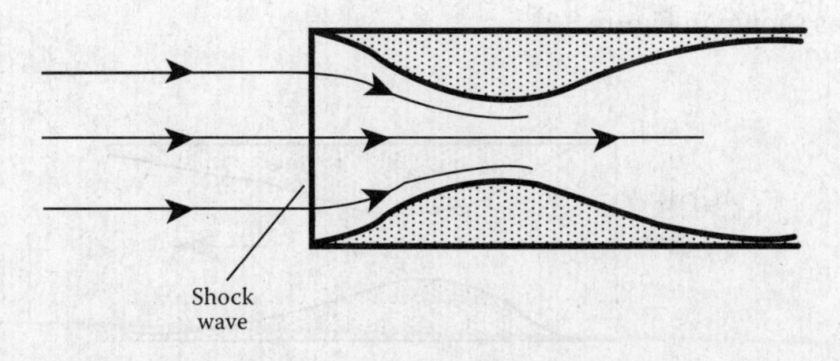

FIGURE 8.25
Flow near intake to fixed supersonic diffuser just before shock is swallowed.

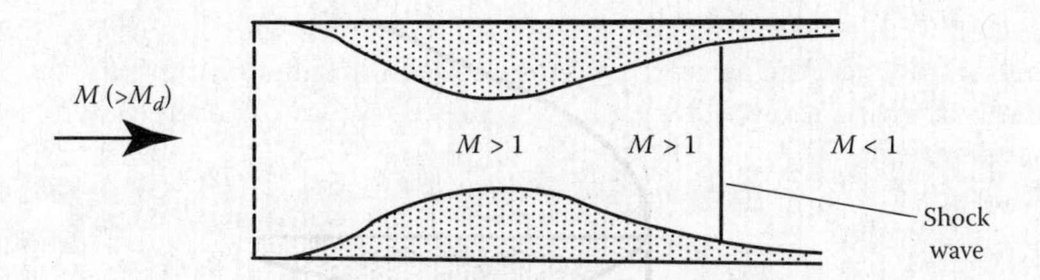

FIGURE 8.26
Flow in supersonic diffuser after shock wave is swallowed.

The Mach number, M_m, at which this occurs is found by noting that this set of circumstances is reached because the Mach number behind the shock wave is now sufficiently small for the throat to intake area ratio to be equal to the value A^*/A that corresponds to the Mach number in the subsonic flow immediately behind the shock wave. M_m is thus found using subsonic relations or tables or the COMPROP software to give the subsonic Mach number corresponding to the diffuser area ratio and then using normal shock relations or tables or software to obtain the supersonic Mach number, which gives this downstream Mach number value.

Any further slight increase in the Mach number will cause the shock wave to be "swallowed" by the diffuser and the shock wave then settles in the divergent portion of the diffuser as shown in Figure 8.26.

Thus, for the fixed area diffuser to swallow the shock, it is necessary to use "overspeeding," i.e., to accelerate the aircraft to a Mach number that is greater than the design Mach number. Once the shock has been swallowed, the flight Mach number can be decreased and as it is decreased the shock wave moves up the divergent portion of the diffuser toward the throat, decreasing in strength as it does so because the Mach number upstream of the shock decreases as the shock moves toward the throat. At the design Mach number, the shock wave reaches the throat and, because the upstream Mach number is then 1, it disappears. The originally discussed shockless design condition then exists. In reality, it would not be practical to operate under exactly these conditions because the slightest decrease in Mach number would cause the shock to be "disgorged," i.e., for the shock to move right out of the diffuser again and the whole "overspeeding" process required to swallow the shock would have to be repeated.

Example 8.5

A small jet aircraft that is designed to cruise at Mach 1.7 has a convergent–divergent intake diffuser with a fixed area ratio. Find the ideal area ratio for this diffuser and the Mach number to which the aircraft must be taken to swallow the normal shock wave if the diffuser has this ideal area ratio.

Solution

The flow just before the shock is swallowed is as shown in Figure E8.5.

From isentropic relations or tables or software at $M_1 = 1.7$, i.e., with no shock,

$$\frac{A_i}{A^*} = 1.338$$

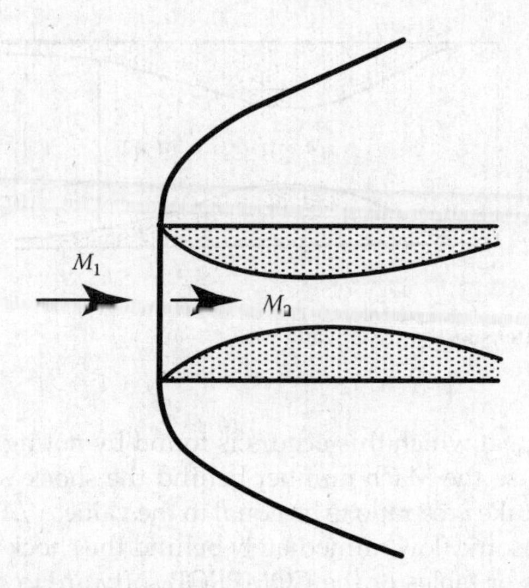

FIGURE E8.5
Flow just before shock wave is swallowed.

From isentropic relations or tables or software for $A/A^* = 1.338$ and subsonic flow

$$M_2 = 0.5$$

Hence, from normal shock relations or tables or software for $M_2 = 0.5$,

$$M_1 = 2.64$$

Therefore, the aircraft must be taken to a Mach number of 2.64 to swallow the normal shock wave formed upstream of the diffuser.

One criterion that can be used to describe the performance of a diffuser is the stagnation pressure recovery ratio, i.e., the ratio of the pressure that would be obtained if the air were brought to rest by the diffuser to the stagnation pressure corresponding to the freestream flow conditions. With the fixed area ratio diffuser discussed above under design conditions when no shock waves exist, this ratio is 1, but under other circumstances there is a stagnation pressure loss across the shock and the ratio is less than 1. Because of the poor performance at off-design conditions and because of the overspeeding required to swallow the shock, simple fixed convergent–divergent diffusers are seldom used. These disadvantages can, to some extent, be overcome using a diffuser with a variable area throat, but this, of course, adds mechanical complexities and increases the weight. The operation of some forms of variable area convergent–divergent diffuser are illustrated in a very schematic way in Figure 8.27.

With a variable throat area diffuser, the throat can be opened to allow the shock to be swallowed and then the throat area can be decreased until the shock disappears at all Mach numbers, i.e., the throat area is adjusted to match the actual Mach number and no overspeeding is required. This is illustrated in the following example.

Example 8.6

A small jet aircraft designed to cruise at Mach 2.5 has an intake diffuser with a variable area ratio. Find the ratio of the throat area under these cruise conditions to the throat area required when the aircraft is flying at Mach 1.3. Assume the diffuser intake area does not

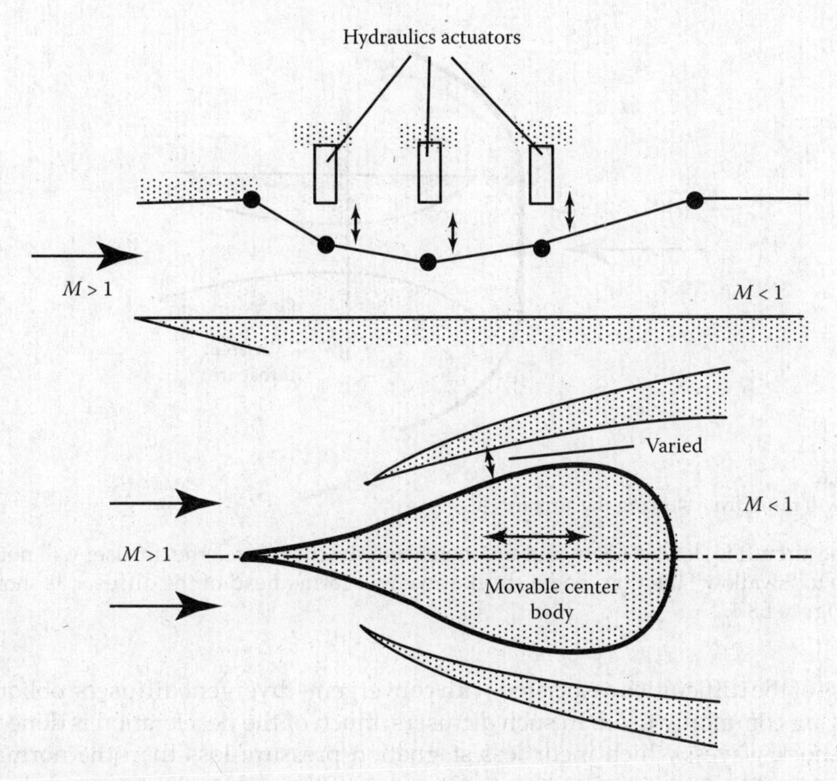

FIGURE 8.27
Simple variable area supersonic diffusers.

change. If the aircraft when flying at cruise conditions is suddenly slowed down without altering the diffuser area ratio, sketch the diffuser flow pattern that will then exist.

Solution

Using isentropic relations or tables or COMPROP software gives $A/A^* = 2.6367$ for Mach 2.5 and $A/A^* = 1.0663$ for Mach 1.3. Hence,

For

$$M = 2.5: \quad \frac{A_e}{A_{\text{throat}}} = 2.6367$$

For

$$M = 1.3: \quad \frac{A_e}{A_{\text{throat}}} = 1.0663$$

Since the intake area A_e is the same at the two Mach numbers, it follows that

$$\frac{A_{\text{throat}} \text{ at } M = 2.5}{A_{\text{throat}} \text{ at } M = 1.3} = \frac{1/2.6367}{1/1.0663} = 0.404$$

Hence, the throat area required under cruise conditions is 40.4% of that required at Mach 1.3.

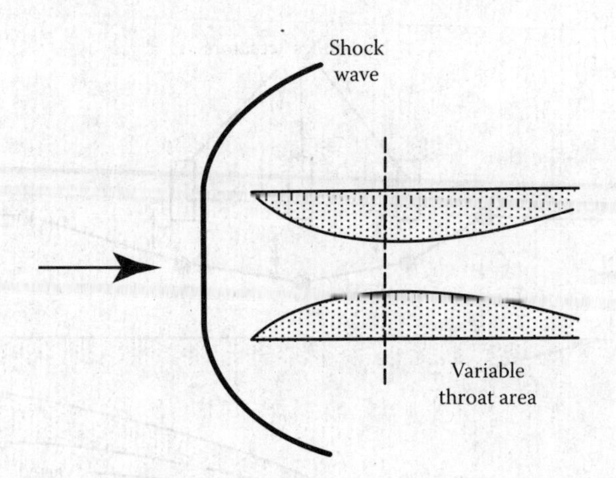

FIGURE E8.6
Flow that will result from slowing the aircraft.

If the aircraft is slowed down without increasing the throat area the diffuser will not be able to "swallow" the flow, and a shock wave will form ahead of the diffuser as shown in Figure E8.6.

Because of the difficulties associated with convergent–divergent diffusers, oblique shock diffusers are commonly used. In such diffusers, much of the deceleration is done through oblique shock waves, which incur less stagnation pressure loss than the normal shock waves that tend to occur in convergent–divergent diffusers. Usually a normal shock wave will exist at some point in the system, but because the Mach number upstream of this normal shock has been decreased through the oblique shocks, the losses are less than would exist if the normal shock occurred at the freestream Mach number. A typical oblique shock diffuser is shown very schematically in Figure 8.28.

In some cases, a curved compressive corner instead of a series of distinct steps each generating an oblique shock wave is used. The corner flow is still followed by a normal shock wave. This is shown schematically in Figure 8.29 and a photograph of the flow near

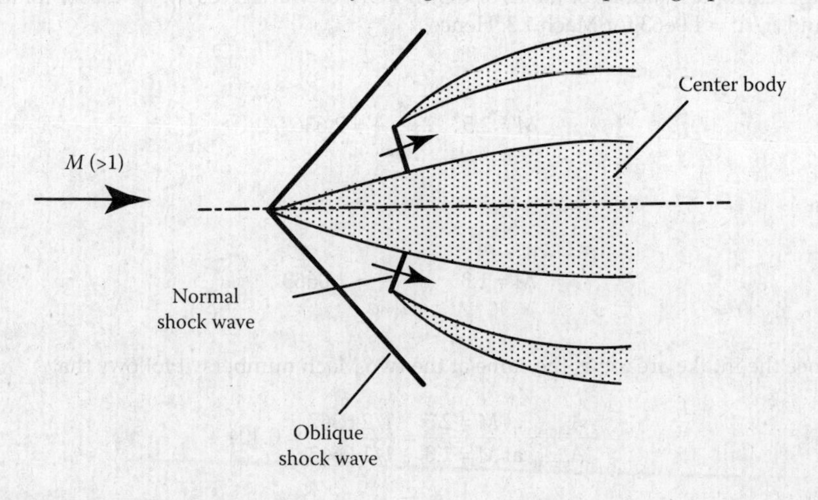

FIGURE 8.28
Simple oblique shock wave diffuser.

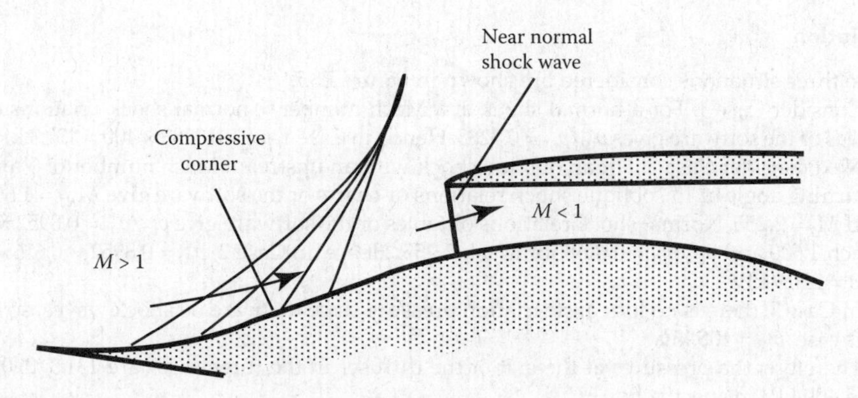

FIGURE 8.29
Diffuser with curved compressive corner.

FIGURE 8.30
Flow into diffuser with oblique and curved shock waves. (Courtesy of NASA.)

an intake that utilizes this type of system is shown in Figure 8.30. The situation shown actually involves an axisymmetric (conical) system.

Example 8.7

An aircraft is to cruise at Mach 3. The stagnation pressure in the flow ahead of the aircraft is 400 kPa. Consider three possible intake scenarios:

1. An intake that involves a normal shock in the freestream ahead of the intake followed by an isentropic deceleration of the subsonic flow behind the shock wave to an essentially zero velocity
2. An oblique shock wave diffuser in which the air flows through an oblique shock wave, which causes the flow to turn through 15°, followed by a normal shock and then an isentropic deceleration of the subsonic flow behind the normal shock wave to an essentially zero velocity
3. An ideal shockless convergent–divergent diffuser in which the air is isentropically brought to an essentially zero velocity

Find the pressures at the exits of each of these diffusers.

Solution

The three situations considered are shown in Figure E8.7.

Consider Case 1. For a normal shock at a Mach number 3, normal shock relations or tables or the software gives $p_{02}/p_{01} = 0.3283$. Hence, in Case 1, $p_{02} = 0.3283 \times 400 = 131.3$ kPa.

Next consider Case 2. For an oblique shock with an upstream Mach number of 3 and a turning angle of 15°, oblique shock relations or charts or the software give $M_{N1} = 1.600$ and $M_2 = 2.255$. Normal shock relations or tables or the software give $p_{02}/p_{01} = 0.8952$ for Mach 1.600 and $p_{03}/p_{02} = 0.6168$ for Mach 2.255. Hence, in Case 2, $p_{03} = 0.8952 \times 0.6168 \times 400 = 220.9$ kPa.

In Case 3, there is no loss in stagnation pressure since there are no shock waves, so in this case, $p_{02} = 400$ kPa.

Therefore, the pressures at the exit of the diffuser in the three cases are 131.3, 220.9, and 400 kPa, respectively.

The above discussion was concerned with supersonic diffusers for aircraft. Similar problems often arise in machines that involve compressible gas flow. As an illustration of this, consider a supersonic wind tunnel. In such a tunnel, the air is accelerated to the desired supersonic Mach number in the working section of the tunnel by passing it through a convergent–divergent nozzle. The air must then be decelerated back to a subsonic Mach number after it leaves the working section. This is usually done by fitting a convergent–divergent diffuser to the tunnel. Under ideal circumstances, the flow will be accelerated to a supersonic Mach number in the nozzle and then back to a subsonic Mach number in the diffuser, the throat area of the diffuser under these circumstances ideally then being equal to the throat area of the nozzle. This flow situation is shown in Figure 8.31.

The tunnel is started, however, by either increasing the pressure ahead of the nozzle or decreasing the pressure behind the diffuser. If operating characteristics of a nozzle discussed earlier are considered, it will be seen that a normal shock wave must exist in the nozzle

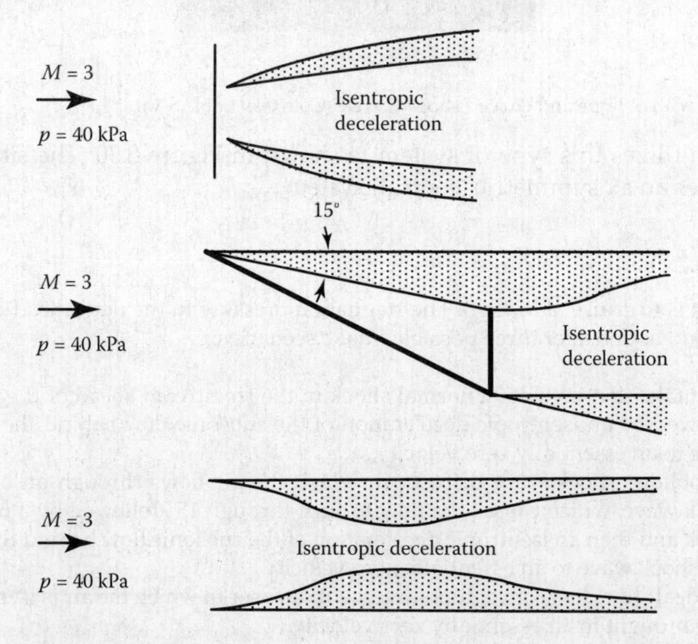

FIGURE E8.7
Three flow situations considered.

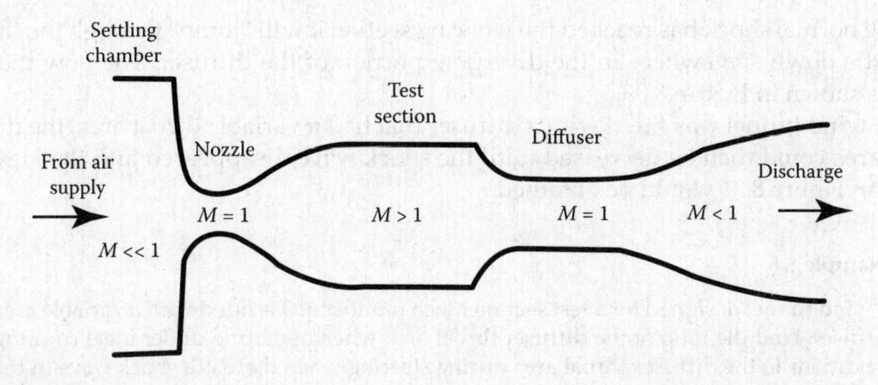

FIGURE 8.31
Ideal flow in a supersonic wind tunnel.

during the startup process. This shock will get stronger and stronger as it moves down the divergent section of the nozzle from the throat toward the working section and will be strongest when it lies at the end of the nozzle. This flow situation is shown in Figure 8.32.

Because there is a loss of stagnation pressure through the shock wave, the area of the throat of the diffuser will have to be bigger than that of the nozzle if the diffuser is to be able to "swallow" this starting shock wave. If the diffuser throat area is sufficiently large,

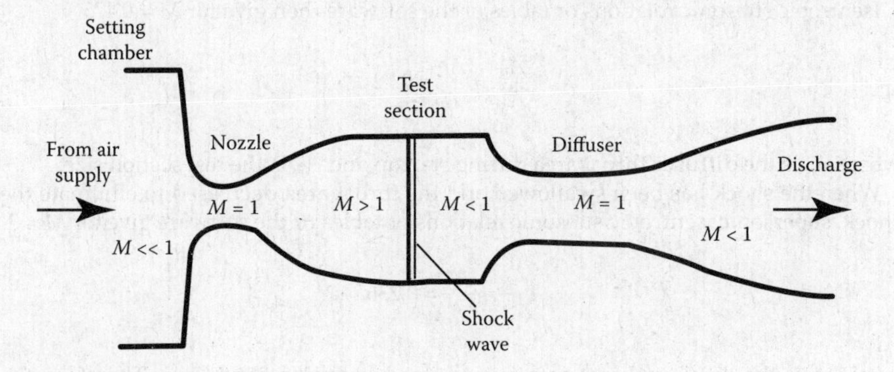

FIGURE 8.32
Flow in a supersonic wind tunnel during starting.

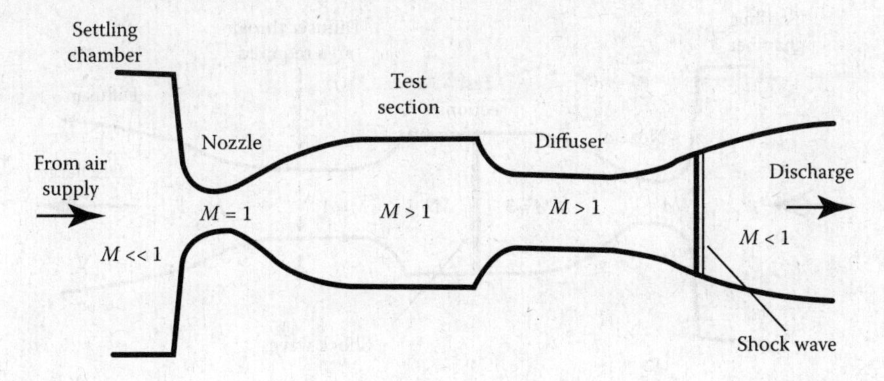

FIGURE 8.33
Flow in a supersonic wind tunnel after the shock wave has been "swallowed."

once the normal shock has reached the working section it will "jump" through the diffuser and settle down somewhere in the diverging portion of the diffuser. The flow that then exists is shown in Figure 8.33.

If the wind tunnel was fitted with a diffuser that had a variable throat area, the diffuser throat area could then be decreased until the shock wave disappeared and the ideal flow shown in Figure 8.31 would be obtained.

Example 8.8

A wind tunnel designed for a test section Mach number of 3 is fitted with a variable area diffuser. Find the ratio of the diffuser throat area when operating under ideal running conditions to the diffuser throat area during starting when there is a shock wave in the working section.

Solution

The flow situation just before the shock wave is swallowed is shown in Figure E8.8.

When the shock is about to be swallowed, normal shock relations or tables or the software give for $M_1 = 3$

$$M_2 = 0.475$$

Isentropic subsonic relations or tables or the software then give for $M = 0.475$

$$\frac{A_{ts}}{A_w} = \frac{A^*}{A} = 0.72$$

where A_{ts} is the diffuser throat area during startup and A_w is the test section size.

When the shock has been swallowed and the throat area decreased to eliminate the shock, supersonic isentropic subsonic relations or tables or the software give for $M = 3$

$$\frac{A_t}{A_w} = \frac{A^*}{A} = 0.236$$

where A_t is the diffuser throat area during steady running conditions. Therefore, the ratio of the diffuser throat area when operating under ideal running conditions to the diffuser throat area during startup is

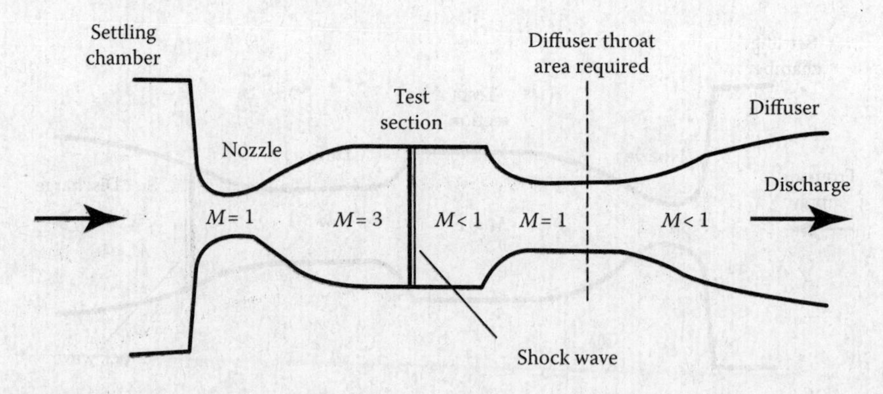

FIGURE E8.8
Flow situation considered when shock wave is in working section.

$$\frac{A_t}{A_{ts}} = \frac{A_t/A_i}{A_{ts}/A_i} = \frac{0.236}{0.72} = 0.328$$

Transonic Flow over a Body

In the flows discussed above, the flow was, in general, subsonic in part of the system and supersonic in other parts of the system, the supersonic flow region usually being terminated by a normal shock, e.g., consider the nozzle flow shown in Figure 8.13. Another situation in which the same general type of flow occurs will be discussed in the present section in a very qualitative manner. Consider flow over a body such as that shown in Figure 8.34.

The velocity and, hence, the Mach number is increased as the flow passes over the surface. As a result, as the freestream velocity is increased, a point will be reached at which, even when the flow in the freestream ahead of the body is subsonic, the Mach number at some point in the flow over the body reaches a value of 1. The freestream Mach number at which this occurs, i.e., at which the Mach number at some point in the flow over the body first reaches a value of 1, is called the critical Mach number, M_{crit}. As the freestream Mach number is increased above the critical value, regions of supersonic flow develop in the flow over the body, these regions being terminated by shock waves as indicated in Figure 8.35.

Because of the sharp pressure rise across the shock wave, the boundary layer on the body tends to separate from the surface at the shock wave as indicated in Figure 8.36. As a result of this boundary layer separation, the drag on the body rises as the freestream Mach number rises above the critical Mach number, the form of the variation being as

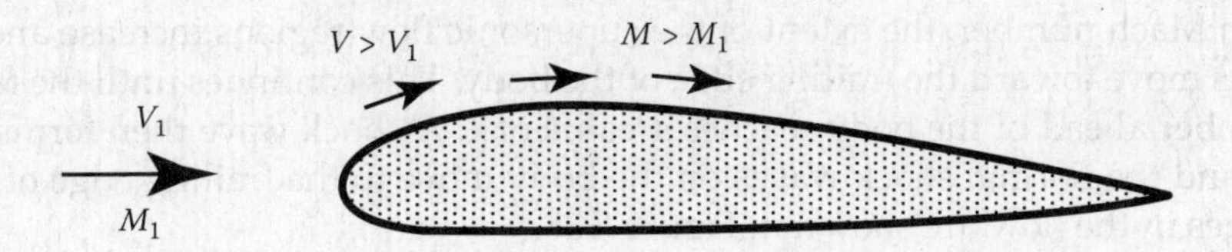

FIGURE 8.34
Velocity changes in external flow over a body.

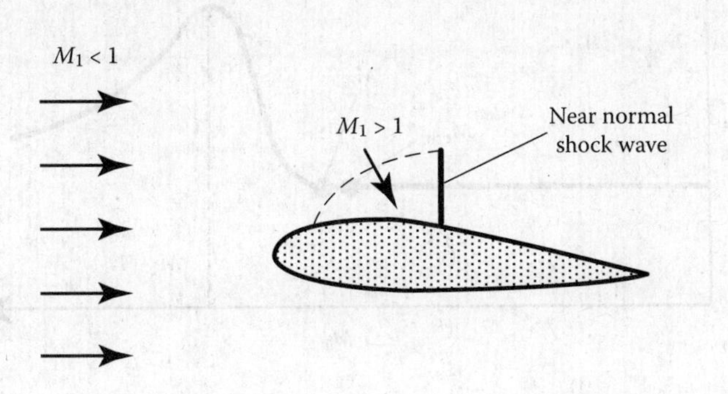

FIGURE 8.35
Transonic flow over a body.

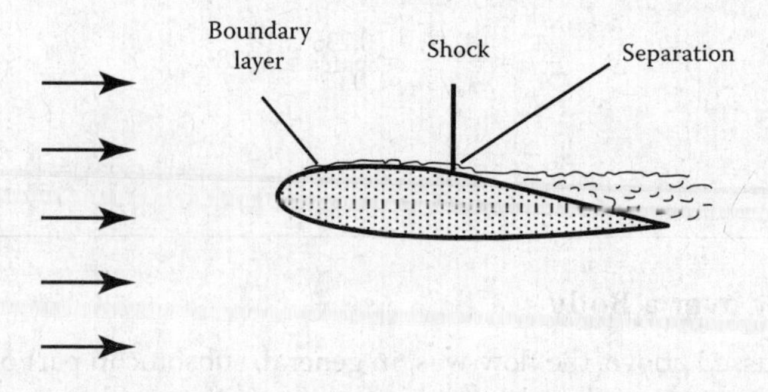

FIGURE 8.36
Shock induced boundary layer separation in transonic flow.

shown in Figure 8.37. The drag force on the body rises to a maximum near Mach 1 and then decreases with further increase in the freestream Mach number, as shown in Figure 8.37.

In Figure 8.37, the drag has been expressed in terms of the dimensionless drag coefficient, C_D, which is defined by

$$C_D = \frac{D}{\rho_1 V_1^2 A/2}$$

where D is the drag on the body, ρ_1 and V_1 are the density and velocity in the freestream ahead of the body, and A is the characteristic area of the body. The drag only starts to increase significantly at a freestream Mach number that is higher than the critical Mach number, the value of the freestream Mach number at which a significant drag rise first occurs being termed the divergence Mach number, M_{div}. With further increase in the freestream Mach number, the extent of the supersonic flow regions increase and the normal shocks move toward the trailing edge of the body. This continues until the freestream Mach number ahead of the body reaches a value of 1. A shock wave then forms ahead of the body and the normal shock waves on the body move to the trailing edge of the body. The changes in the flow are shown in Figure 8.38.

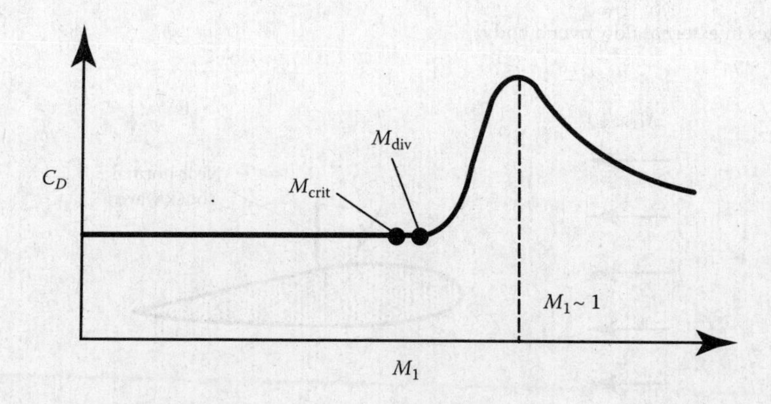

FIGURE 8.37
Variation of coefficient of drag with Mach number in transonic flow.

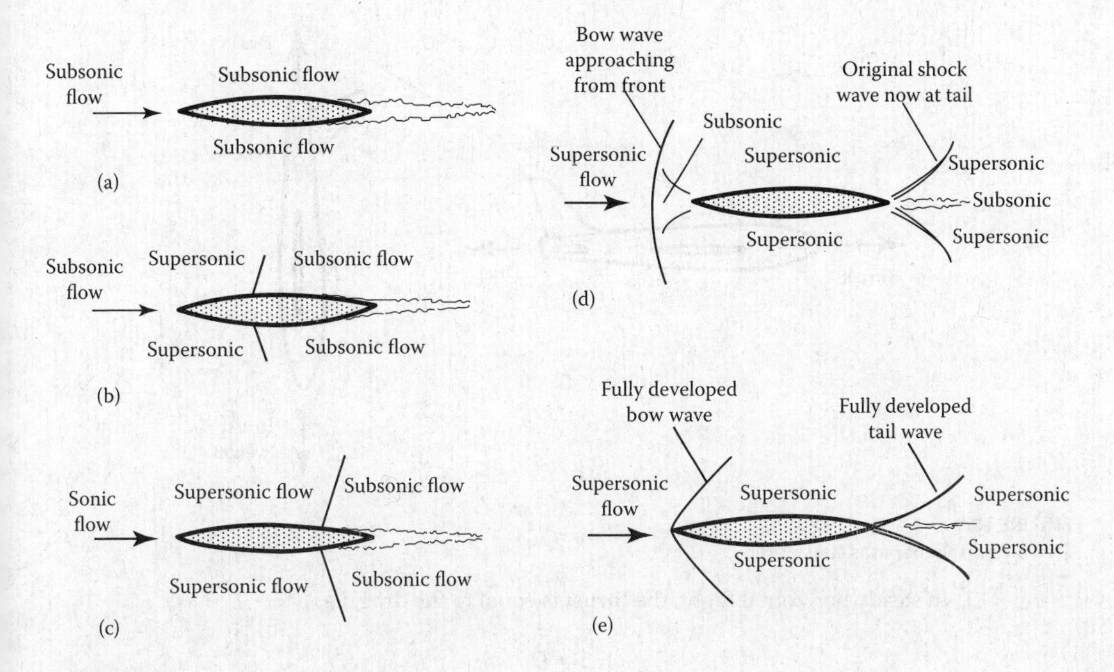

FIGURE 8.38
Variations in the flow over a body with increasing Mach number in (a) low subsonic Mach number flow, (b) and (c) transonic flow, (d) and (e) supersonic flow.

Because the thrust generated by the engines of an aircraft must balance the drag in steady horizontal flight, many aircraft will not have enough thrust to allow them to pass through the region of drag rise in the transonic region. For this reason, the drag rise near a freestream Mach number of 1 is often termed the "sound barrier." When flying in the transonic region, the position of the shock waves and the extent of the boundary layer separation induced by the shock waves is usually fluctuating with time, and this together with the unsteady forces produced by the impingement of the unsteady separated flow on other components of the aircraft gives rise to the buffeting often associated with "breaking the sound barrier."

Example 8.9

An aircraft has a wing area of 30 m^2 and a mass of 9000 kg. The turbojet engine that propels this aircraft has a maximum thrust of 50 kN. The drag coefficient, based on wing area for this aircraft has a value of 0.02 at low Mach numbers. Near Mach 1, this drag coefficient rises to a maximum value of 0.045. Determine

1. What is the highest speed at which this aircraft could fly horizontally at an altitude of 5000 m if there was no increase in drag near Mach 1 due to compressibility?
2. Whether this aircraft can "break the sound barrier" in steady horizontal flight and in a steady vertical dive at an altitude of 5000 m.

Solution

The flow situations considered are shown in Figure E8.9.

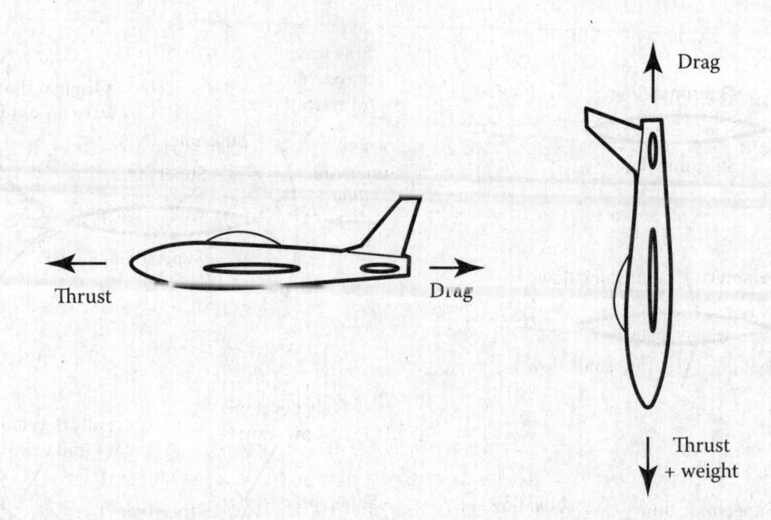

FIGURE E8.9
Two aircraft flight paths considered.

1. In steady horizontal flight, the thrust is equal to the drag, i.e.,

$$T = D$$

However,

$$D = C_D \frac{1}{2}\rho V^2 A_w$$

At an altitude of 5000 m in the standard atmosphere, the density is 0.737 kg/m³; thus, if $C_D = 0.02$, the maximum speed at which the aircraft can fly is determined by

$$50,000 = 0.02 \times \frac{1}{2} \times 0.737 \times V^2 \times 30$$

Solving this gives $V = 475$ m/s. However, the speed of sound at an altitude of 5000 m in the standard atmosphere is 320 m/s, so if there was no drag rise due to compressibility, the aircraft would be able to fly at a maximum Mach number of 475/320 = 1.48.

2. To determine whether the aircraft can "break the sound barrier," the drag on the aircraft when flying at Mach 1 with a drag coefficient of 0.045 is calculated. Since when flying at $M = 1$ at the specified altitude, $V = a = 320$ m/s, the drag on the aircraft is

$$D = 0.045 \times \frac{1}{2} \times 0.737 \times 320^2 \times 30 = 50,941 \text{ N}$$

This is greater than the maximum thrust of the engine, so the aircraft will not be able to reach Mach 1 in horizontal flight. However, in a vertical dive, the force available to overcome the drag is the sum of the thrust and the gravitational force on the aircraft, i.e., 50,000 + (9000 × 9.81), i.e., 138,290 N, so the aircraft will be able to reach supersonic velocities in a dive.

Concluding Remarks

To accelerate a gas flow from subsonic to supersonic velocities in a nozzle, it is necessary to use a convergent–divergent nozzle, the Mach number being 1 at the throat. If conditions upstream of the nozzle are kept constant, there is a limit to the mass flow rate through the nozzle that can be achieved by lowering the back-pressure. If the back-pressure is low enough for this maximum mass flow rate to be reached, the nozzle is said to be choked. Convergent–divergent diffuser systems designed to decelerate a supersonic flow to a subsonic velocity have also been discussed.

PROBLEMS

1. Air is discharged from a large reservoir, in which the pressure and temperature are 0.8 MPa and 25°C, respectively, through a convergent nozzle with an exit diameter of 5 cm. The nozzle discharges to the atmosphere. Find the mass flow rate through the nozzle and the pressure and temperature on the nozzle exit plane.
2. A supersonic nozzle possessing an area ratio of 3.0 is supplied from a large reservoir and is allowed to exhaust to atmospheric pressure. Determine the range of reservoir pressures over which a normal shock will appear in the nozzle. For what value of reservoir pressure will the nozzle be perfectly expanded, with supersonic flow at the exit plane? Find the minimum reservoir pressure to produce sonic flow at the nozzle throat. Assume isentropic flow except for shocks with $\gamma = 1.4$.
3. A converging–diverging nozzle is designed to operate with an exit Mach number of 1.75. The nozzle is supplied from an air reservoir at 1000 psia. Assuming one-dimensional flow, calculate
   a. Maximum back-pressure to choke the nozzle
   b. Range of back-pressures over which a normal shock will appear in the nozzle
   c. Back-pressure for the nozzle to be perfectly expanded to the design Mach number
   d. Range of back-pressures for supersonic flow at the nozzle exit plane
4. A convergent–divergent nozzle is designed to expand air from a chamber in which the pressure is 700 kPa and the temperature is 35°C to give Mach 1.6. The mass flow rate through the nozzle under design conditions is 0.012 kg/s. Find
   a. The throat and exit areas of the nozzle
   b. The design back-pressure and the temperature of the air leaving the nozzle with this back-pressure
   c. The lowest back-pressure for which there will be no supersonic flow in the nozzle
   d. The back-pressure below which there are no shock waves in the nozzle
5. A converging–diverging nozzle is designed to generate an exit Mach number of 2. The nozzle is supplied with air from a large reservoir in which the pressure is kept at 6.5 MPa. Assuming one-dimensional isentropic flow, find
   a. The maximum back-pressure at which the nozzle will be choked
   b. The range of back-pressures over which there will be a shock in the nozzle

c. The design back-pressure

d. The range of back-pressures over which there is supersonic flow on the nozzle exit plane

6. A convergent–divergent nozzle is designed to expand air from a chamber in which the pressure is 800 kPa and the temperature is 40°C to give Mach 2.5. The throat area of the nozzle is 0.0025 m^2. Find

a. The flow rate through the nozzle under design conditions

b. The exit area of the nozzle

c. The design back-pressure and the temperature of the air leaving the nozzle with this back-pressure

d. The lowest back-pressure for which there is only subsonic flow in the nozzle

e. The back-pressure at which there is a normal shock wave on the exit plane of the nozzle

f. The back-pressure below which there are no shock waves in the nozzle

g. The range of back-pressures over which there are oblique shock waves in the exhaust from the nozzle

h. The range of back-pressures over which there are expansion waves in the exhaust from the nozzle

i. The back-pressure at which a normal shock wave occurs in the divergent section of the nozzle at a point where the nozzle area is half way between the throat and the exit plane areas

7. A variable area diffuser is fitted to an aircraft designed to operate at Mach 3.5. If the shock wave is "swallowed" at this Mach number, find the ratio of the throat area for shockless operation to the throat area at which the shock wave is swallowed.

8. A nozzle is designed to expand air from a chamber in which the pressure and temperature are 800 kPa and 40°C, respectively, to Mach 2.5. The throat area of this nozzle is to be 0.1 m^2. Find

a. The exit area of the nozzle

b. The mass flow rate through the nozzle when operating at design conditions

c. The back-pressure at which there will be a normal shock wave on the exit plane of the nozzle

d. The range of back-pressures over which expansion waves will occur outside the nozzle

9. A rocket nozzle is designed to operate supersonically with a chamber pressure of 500 psia and an ambient pressure of 14.7 psia. Find the ratio between the thrust at sea level to the thrust in space (0 psia). Assume a constant chamber pressure, with a chamber temperature of 2500°R. Assume the rocket exhaust gases behave as a perfect gas with $\gamma = 1.4$ and R = 20 ft-lbf/lbm R.

10. Air flows through a convergent–divergent nozzle that has an inlet area of 0.0025 m^2. The inlet temperature and pressure are 50°C and 550 kPa, respectively, and the velocity at the inlet is 80 m/s. If the flow is assumed to be isentropic, and if the exit pressure is 120 kPa, find the throat and exit areas and the exit velocity.

11. Air flows through a convergent–divergent passage. The passage inlet area is 5 cm^2, the minimum area is 3 cm^2 and the exit area is 4 cm^2. The air velocity at the inlet

to the passage is 120 m/s. The pressure is 700 kPa and the temperature is 40°C. Assuming that the flow is isentropic, find the mass flow rate through the passage, the Mach number at the minimum area section, and the velocity and pressure at the exit section.

12. Air flows through a convergent–divergent nozzle from a large reservoir in which the pressure and temperature are maintained at 700 kPa and 60°C, respectively. The rate of air flow through the nozzle is 1 kg/s. On the exit plane of the nozzle the stagnation pressure is 550 kPa and the static pressure is 500 kPa. A shock wave occurs in the nozzle and the flow can be assumed to be isentropic everywhere except through the shock wave. Find the nozzle throat area, the Mach numbers before and after the shock, the nozzle areas at the point where the shock occurs and on the exit plane, and the air density on the exit plane of the nozzle.
13. Air flows through a convergent–divergent nozzle. The air has Mach 0.50 and a pressure and a temperature of 280 kPa and 10°C, respectively, at the inlet to the nozzle. The nozzle throat area is 6.5×10^{-4} m^2 and the ratio of the exit area to the throat area is 4. If the pressure on the exit plane of the nozzle is 170 kPa, find the Mach number and the temperature on the nozzle exit plane and the nozzle area at the point in the nozzle at which the normal shock wave occurs.
14. Air is supplied to a convergent–divergent nozzle from a large tank in which the pressure and temperature are kept at 700 kPa and 40°C, respectively. If the nozzle has an exit area that is 1.6 times the throat area and if a normal shock occurs in the nozzle at a section where the area is 1.2 times the throat area, find the pressure, temperature, and Mach number at the nozzle exit. Assume one-dimensional, isentropic flow.
15. Air enters a convergent–divergent nozzle at Mach number 0.2. The stagnation pressure is 700 kPa and the stagnation temperature is 5°C. The throat area of the nozzle is 46×10^{-4} m^2 and the exit area is 230×10^{-4} m^2. If the pressure at the exit to the nozzle is 500 kPa, determine if there is a shock in the divergent portion of the nozzle. If there is a shock wave, determine the nozzle area at which the shock occurs and the Mach number and pressure just before and just after the shock wave.
16. Air at a temperature of 20°C and a pressure of 101 kPa flows through a convergent–divergent nozzle at the rate of 0.5 kg/s. The exit area of the nozzle is 1.355 times the inlet area. If the air leaves the nozzle at a static temperature of 20°C and a stagnation temperature of 30°C, calculate the inlet and exit Mach numbers, the increase in entropy (if any) between the inlet and the outlet, the area at which the shock (if any) occurs and the stagnation pressure at the exit.
17. Air flows from a large reservoir in which the temperature and pressure are 80°C and 780 kPa through a convergent–divergent nozzle that has a throat diameter of 2.5 cm. When the back-pressure is 560 kPa, a shock wave is found to occur at a location in the nozzle where the static pressure is 210 kPa. Find the exit area, exit temperature, the exit Mach number, the area at which the shock wave occurs, and the pressure ratio across the shock.
18. Air flows from a reservoir in which the pressure is kept at 124 kPa through a convergent–divergent nozzle and exhausts to the atmosphere where the pressure is 101.3 kPa. Under these conditions, the nozzle is choked and the flow is subsonic on both sides of the throat. To what value must the pressure in the reservoir be increased so that there is a normal shock on the nozzle exit plane?

19. Air flows through a convergent–divergent nozzle. The nozzle exit and throat areas are 0.5 and 0.25 m^2, respectively. If the inlet stagnation pressure is 200 kPa and the back-pressure is 120 kPa, determine the nozzle area at which the normal shock wave is located. What is the increase in entropy across the shock? At what back-pressure will the shock wave be located on the nozzle exit plane?
20. Air at a pressure of 350 kPa, a temperature of 80°C, and a velocity of 180 m/s enters a convergent–divergent nozzle. A normal shock occurs in the nozzle at a location where the Mach number is 2. If the air mass flow rate through the nozzle is 0.7 kg/s, and if the pressure on the nozzle exit plane is 260 kPa, find the nozzle throat area, the nozzle exit area, the temperatures upstream and downstream of the shock wave, and the change in entropy through the nozzle.
21. Air with a stagnation pressure, and temperature of 100 kPa and 150°C is expanded through a convergent–divergent nozzle that is designed to give an exit Mach number of 2. The nozzle exit plane area is 30 cm^2. Find the mass rate of flow through the nozzle when operating at design conditions and the exit plane pressure under these design conditions. Also, find the exit plane pressure if a normal shock wave occurs in the divergent portion of the nozzle at a section where the area is half way between the throat and the exit plane areas.
22. Air flows through a convergent–divergent nozzle with an exit area to throat area ratio of 4.0. If a normal shock wave occurs in the nozzle at a location where area is 2.5 the throat area, find the Mach number on the exit plane of the nozzle.
23. Air flows through a convergent–divergent nozzle. The stagnation temperature of supply air is 200°C and the nozzle has an exit area of 2 m^2 and a throat area of 1 m^2. If the air from the nozzle is discharged to an ambient pressure of 70 kPa, find the minimum supply stagnation pressure required to produce choking in the nozzle and the mass flow rate through the nozzle when it is choked. Also, find the supply stagnation pressure that exists if a normal shock wave occurs in the divergent portion of the nozzle at a section where the area is 1.5 m^2.
24. Air flows from a large reservoir in which the pressure is 450 kPa through a convergent–divergent nozzle. A normal shock wave occurs in the diverging portion of the nozzle at a point where the nozzle area is twice the throat area. Find the Mach numbers on each side of this shock wave. If the Mach number on the exit plane of the nozzle is 0.2, find the back-pressure required to maintain the shock at this location.
25. The stagnation pressure and temperature of the inlet to a supersonic wind tunnel are 100 kPa and 30°C, respectively. The Mach number in the test section of the tunnel is 2. If the cross-sectional area of the test section is 1.2 m^2, find the throat areas of the nozzle and diffuser.
26. Air flows through a convergent nozzle. At a section within this nozzle at which this cross-sectional area is 0.01 m^2, the pressure is 300 kPa and the temperature is 30°C. If the velocity at this section of the nozzle is 150 m/s, find the Mach number at this section, the stagnation temperature and pressure, and the mass flow rate through the nozzle. If the nozzle is choked, find the area, pressure, and temperature at the exit of the nozzle.
27. Carbon dioxide flows through an 8-cm–inside diameter pipe. To determine the mass flow rate, a venturi meter with a throat diameter of 5 cm is installed in the pipe. The pressure and temperature just upstream of the venturi meter are 600 kPa

and 40°C, respectively. The difference between the pressure just upstream of the venturi meter and the pressure at the throat of the venturi meter is 15 kPa. Find the mass flow rate of carbon dioxide.

28. A large rocket engine designed to propel a satellite launcher has a thrust of 10^6 lbf when operating at sea level, the exit plane pressure being equal to the ambient pressure under these conditions. The combustion chamber pressure and temperature are 500 psia and 4500°F, respectively. If the products of combustion can be assumed to have the properties of air and if the flow through the nozzle can be assumed to be isentropic, find the throat and exit diameters of the nozzle.
29. Air, at a pressure of 700 kPa and a temperature of 80°C, flows through a convergent–divergent nozzle. The inlet area is 0.005 m^2 and the pressure on the exit plane is 40 kPa. If the mass flow rate through the nozzle is 1 kg/s, find, assuming one-dimensional isentropic flow, the Mach number, temperature, and velocity of the air at the discharge plane.
30. A small jet aircraft designed to cruise at Mach 1.5 has an intake diffuser with a fixed area ratio. Find the ideal area ratio for this diffuser and the Mach number to which the aircraft must be taken to swallow the normal shock wave if the diffuser has this ideal area ratio.
31. A fixed supersonic convergent–divergent diffuser is designed to operate at Mach 1.7. To what Mach number would the inlet have to be accelerated to swallow the shock during startup?
32. A jet aircraft is designed to fly at Mach 1.9. It is fitted with a variable-area diffuser. If the diffuser just "swallows" the shock wave at the design Mach number and the throat area is then reduced to give a "shockless" flow, find the percentage reduction in diffuser throat area that is required.
33. A small jet aircraft designed to cruise at Mach 3 has an intake diffuser with a variable area ratio. Find the ratio of the throat area under these cruise conditions to the throat area required when the aircraft is flying at Mach 1.5. Assume the diffuser intake area does not change.
34. A convergent–divergent supersonic diffuser is to be used at Mach 3.0. The diffuser is to use a variable throat area so as to swallow the starting shock. What percentage increase in throat area will be necessary?
35. A wind tunnel, designed for a test section Mach number of 4, is fitted with a variable area diffuser. Find the ratio of the diffuser throat area when operating under ideal running conditions to the diffuser throat area during starting when there is a shock wave in the working section, the Mach number ahead of this shock being 4.
36. Air flows from a tank in which the pressure is kept at 750 kPa and the temperature is kept at 30°C through a converging nozzle that discharges the air to the atmosphere. If the throat area of this nozzle is 0.6 cm^2, find the rate at which the air is discharged from the tank in kg/s.
37. In transonic wind tunnel testing, the small area decrease caused by placing the model in the test section, i.e., by the model blockage, can cause relatively large changes in the flow in the test section. To illustrate this effect, consider a tunnel that has an empty test section Mach number of 1.08. The test section has an area of 1 m^2 and the stagnation temperature of the air flowing through the test section is 25°C. If a model with a cross-sectional area of 0.005 m^2 is placed in this test section,

find the percentage change in test section velocity. Assume one-dimensional isentropic flow.

38. Consider one-dimensional isentropic flow through a convergent–divergent nozzle that has a throat area of 10 cm^2. The pressure at the throat is 310 kPa and the flow goes from subsonic to supersonic velocities in the nozzle. Find the pressures and Mach numbers at points in the nozzle upstream and downstream of the throat where the nozzle cross-sectional area is 29 cm^2.
39. A moving piston forces air from a well insulated 15-cm-diameter pipe through a convergent nozzle fitted to the end of the pipe. The nozzle has an exit diameter of 4 mm and the air is discharged to the atmosphere. If the force on the piston is 3700 N and the air temperature is 30°C, estimate the velocity on the nozzle exit plane, the piston velocity, and the mass flow rate at which the air is discharged from the nozzle.
40. A convergent–divergent nozzle with an exit area to throat area ratio of 3 is supplied with air from a reservoir in which the pressure is 350 kPa. The air from the nozzle is discharged into another large reservoir. It is found that the flow leaving the nozzle exit is directed inward at an angle of 4° to the nozzle centerline. The velocity on the nozzle exit plane is supersonic. What is the pressure in the second reservoir?
41. A convergent–divergent nozzle has an exit area to throat area ratio of 4. It is supplied with air from a large reservoir in which the pressure is kept at 500 kPa and it discharges into another large reservoir in which the pressure is kept at 10 kPa. Expansion waves form at the exit edges of the nozzle causing the discharge flow to be directed outward. Find the angle that the edge of the discharge flow makes to the axis of the nozzle.
42. A small meteorite punches a 3-cm-diameter hole in the skin of an orbiting space laboratory. The pressure and temperature in the laboratory are 80 kPa and 20°C, respectively. Estimate the initial rate at which air flows out of the laboratory. State the assumptions you make in arriving at your estimate.
43. A jet engine is running on a test bed. The stagnation pressure and stagnation temperature just upstream of the convergent nozzle fitted to the rear of this engine are found to be 700 kPa and 700°C, respectively. The exit diameter of the nozzle is 0.5 m. If the test is being run at an ambient pressure of 101 kPa, find the mass flow rate through the engine, the jet exit velocity and the thrust that is developed by the engine. Assume the gases have the properties of air.
44. Air flows through a convergent–divergent nozzle from a large reservoir in which the pressure is 300 kPa and the temperature is 100°C. The nozzle has a throat area of 1 cm^2 and an exit area of 4 cm^2. The nozzle discharges into another large reservoir in which the pressure can be varied. For what range of back-pressures will the mass flow rate through the nozzle be constant and what will the mass flow rate be under these circumstances?
45. Air flows through a convergent–divergent nozzle, the flow becoming supersonic in the divergent section of the nozzle. A normal shock wave occurs in the divergent section. If the static pressure behind this shock wave is equal to the static pressure at the throat of the nozzle, find the ratio of the nozzle area at which the shock wave occurs to the nozzle throat area.

46. A nozzle is designed to expand air from a chamber in which the pressure and temperature are 800 kPa and 40°C, respectively, to Mach 2.5. The throat area of this nozzle is to be 0.05 m^2. Find
    a. The exit area of the nozzle
    b. The mass flow rate through the nozzle when operating at design conditions
    c. The back-pressure at which there will be a normal shock wave on the exit plane of the nozzle
    d. The range of back-pressures over which there will be a normal shock wave in the nozzle
    e. The range of back-pressures over which oblique shock waves will occur outside the nozzle
    f. The range of back-pressures over which expansion waves will occur outside the nozzle
47. A convergent–divergent nozzle through which air flows is designed to generate Mach 2.5. The supply reservoir pressure is 320 kPa. What is the design back-pressure? If the nozzle operates with a back-pressure of 100 kPa the nozzle will be overexpanded and oblique shock waves will exist at the nozzle exit. Find the Mach number and flow direction just downstream of the exit plane oblique shock waves under these conditions.
48. A convergent–divergent nozzle through which air flows is designed to generate Mach 2.2. The supply reservoir pressure is 2.2 MPa and the flow is discharged into ambient air at a pressure of 101 kPa. Under these conditions, the nozzle will be underexpanded, and as a result, expansion waves will exist at the nozzle exit. Find the flow direction just downstream of the exit plane expansion waves under these conditions. What effect does the presence of the expansion waves have on the net axial thrust produced by the nozzle?

9

Adiabatic Flow in a Duct with Friction

Introduction

In the discussion of flow through ducts given up to this point, it was assumed, in almost all cases, that the effects of viscosity were negligible. This is often an adequate assumption when dealing with flow through nozzles or short ducts. For long ducts, however, the effects of viscosity, i.e., the effects of fluid friction at the walls, can in fact be dominant. This is illustrated in Figure 9.1, which shows typical Mach number and pressure variations in a constant area duct with and without friction. In incompressible flow through a duct of constant cross-sectional area, the friction only affects the pressure, which drops in the direction of flow. The velocity in such a situation remains constant along the duct. In compressible flow, however, friction effects all of the flow variables, i.e., the changes in pressure cause changes in density, which lead to changes in velocity.

In some cases, the effects of viscosity may be negligible over part of the flow but then be very important in other parts of the flow. This is illustrated in Figure 9.2.

In this chapter, consideration will be given to the effects of viscosity on steady gas flows through ducts under such conditions that compressibility effects are important. In the analyses given in this chapter it will be assumed that the flow is adiabatic, i.e., that the duct is well insulated (a discussion of the effects of friction in the presence of heat exchange is given in the next chapter). Attention will, in this chapter, mainly be restricted to flow in a constant area duct although a brief discussion of the effects of area change will be given at the end of this chapter.

Flow in a Constant Area Duct

Attention will here be given to the effects of wall friction on adiabatic flow through a duct whose cross-sectional area does not change. This type of flow, i.e., compressible adiabatic flow in a constant area duct with frictional effects, is known as "Fanno" flow.

Consider the momentum balance for the small portion of the duct shown in Figure 9.3. Since steady flow is being considered, this gives

$$\text{Net pressure force} - \text{Force due to wall shear stress}$$
$$= \text{Mass flow rate} \times (\text{Velocity out} - \text{Velocity in})$$

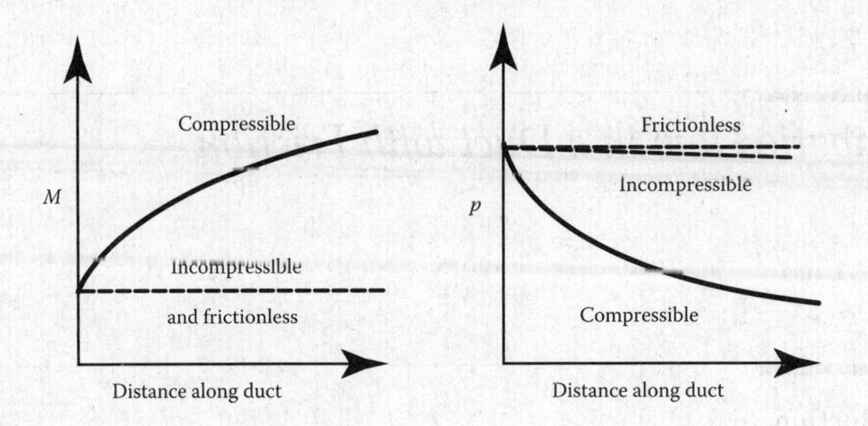

FIGURE 9.1
Effect of friction on Mach number and pressure in a duct.

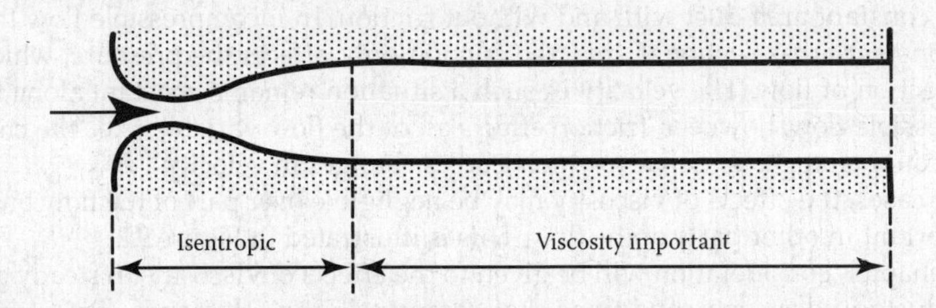

FIGURE 9.2
Effect of friction in different portions of flow.

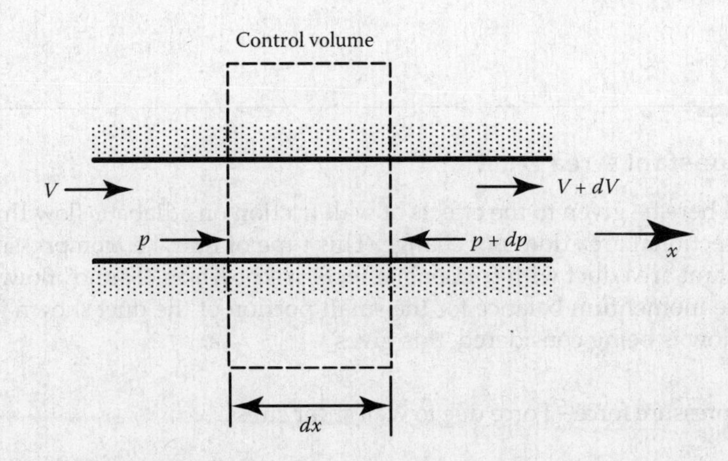

FIGURE 9.3
Control volume used in analysis of frictional flow in a duct.

i.e., since the force due to the wall shear stress is equal to the product of the shear stress and the surface area of the portion of the duct being considered,

$$pA - (p + dp)A - \tau_w(\text{perimeter})\,dx = \rho VA(V + dV - V)$$

where τ_w is the wall shear stress. Therefore,

$$-dp - \tau_w \frac{P}{A} dx = \rho V\,dV \tag{9.1}$$

where P is the perimeter of the duct and A is its cross-sectional area. In the case of a circular duct, $P = \pi D$ and $A = \pi D^2/4$ so $P/A = 4/D$. For this reason, even for noncircular ducts, the ratio P/A is usually expressed in terms of an equivalent diameter, called the "hydraulic diameter," D_H, defined by

$$D_H = \frac{4\text{ (Area)}}{\text{Perimeter}} = \frac{4A}{P} \tag{9.2}$$

Dividing Equation 9.1 by ρV^2 then gives

$$-\frac{dp}{\rho V^2} - \frac{\tau_w}{\rho V^2}\frac{P}{A} dx = \frac{dV}{V} \tag{9.3}$$

Next consider continuity of mass. This gives

$$\rho AV = \text{constant}$$

However, in the present case, since flow in a constant area duct is being considered, this reduces to

$$\rho V = \text{constant}$$

Hence, for the portion of the duct being considered,

$$\rho V = (\rho + d\rho)(V + dV)$$

From this, it follows that to first order of accuracy, i.e., by assuming that the term $d\rho \times dV$ is negligible because $d\rho$ and dV are very small.

$$\rho\,dV + V\,d\rho = 0 \tag{9.4}$$

Dividing this equation through by ρV then gives

$$\frac{dV}{V} + \frac{d\rho}{\rho} = 0 \tag{9.5}$$

Next, consider the energy equation

$$c_pT + \frac{V^2}{2} = \text{constant} = c_p(T + dT) + \frac{(V + dV)^2}{2}$$

This gives to first order of accuracy

$$c_p dT + V\,dV = 0 \tag{9.6}$$

Also, the equation of state gives

$$\frac{p}{\rho} = RT, \quad \text{i.e.,} \quad p = \rho RT$$

and

$$\frac{p + dp}{\rho + d\rho} = R(T + dT), \quad \text{i.e.,} \quad p + dp = (\rho + d\rho)R(T + dT)$$

Hence, to first order of accuracy,

$$dp = RT\,d\,\rho + \rho R\,dT \tag{9.7}$$

Further, since

$$M^2 = \frac{V^2}{a^2} = \frac{V^2}{\gamma RT}$$

and

$$(M + dM)^2 = \frac{(V + dV)^2}{(a + da)^2} = \frac{(V + dV)^2}{\gamma R(T + dT)}$$

i.e.,

$$M^2\left(1 + \frac{dM}{M}\right)^2 = \frac{V^2(1 + dV/V)^2}{\gamma RT(1 + dT/T)}$$

i.e., to first order of accuracy,

$$M^2\left(1 + 2\frac{dM}{M}\right) = \frac{V^2}{\gamma RT}\left(1 + 2\frac{dV}{V}\right)\left(1 - \frac{dT}{T}\right)$$

i.e.,

$$M^2\left(1+2\frac{dM}{M}\right)=\frac{V^2}{\gamma RT}\left(1+2\frac{dV}{V}-\frac{dT}{T}\right)$$

Hence, it follows that to first order of accuracy,

$$\frac{dM}{M}=\frac{dV}{V}-\frac{dT}{2T} \tag{9.8}$$

The above equations represent a set of five equations in five unknowns dM, dV, dp, dT, and $d\rho$, which can be solved to give expressions for dM, dV, etc. as illustrated below.

Equation 9.6 gives

$$\frac{dT}{T}+\frac{V\,dV}{c_pT}=0 \tag{9.9}$$

However, since

$$M^2=\frac{V^2}{\gamma RT}$$

and

$$(M+dM)^2=\frac{(V+dV)^2}{\gamma R(T+dT)}$$

it follows that

$$2M\,dM=\frac{2V\,dV}{\gamma RT}-\frac{V^2}{\gamma RT^2}dT$$

which can be rearranged to give

$$V\,dV=\gamma RTM\,dM+\frac{V^2}{2}\frac{dT}{T}$$

Hence, Equation 9.9 becomes

$$\frac{dT}{T}+\frac{\gamma R}{c_p}M\,dM+\frac{V^2}{2c_pT}\frac{dT}{T}=0 \tag{9.10}$$

Now, recalling that

$$R = c_p - c_v, \quad \text{i.e.,} \quad c_p = \gamma R/(\gamma - 1)$$

Equation 9.10 becomes

$$\frac{dT}{T} + (\gamma - 1)M\,dM + \frac{\gamma - 1}{2}M^2\frac{dT}{T} = 0$$

which can be rearranged to give

$$\frac{dT}{T} = -\frac{(\gamma - 1)M^2}{[1 + (\gamma - 1)M^2/2]}\frac{dM}{M} \tag{9.11}$$

Now, Equation 9.7 gives, on dividing through by $p = \rho RT$,

$$\frac{dp}{p} = \frac{d\rho}{\rho} + \frac{dT}{T} \tag{9.12}$$

Eliminating $d\rho/\rho$ between this equation and Equation 9.5 then gives

$$\frac{dp}{p} = -\frac{dV}{V} + \frac{dT}{T} \tag{9.13}$$

Hence, since Equation 9.8 gives

$$\frac{dV}{V} = \frac{dM}{M} + \frac{dT}{2T} \tag{9.14}$$

Equation 9.13 gives

$$\frac{dp}{p} = -\frac{dM}{M} + \frac{dT}{2T} \tag{9.15}$$

Using Equation 9.11 then gives

$$\frac{dp}{p} = -\left[1 + \frac{(\gamma - 1)M^2/2}{1 + (\gamma - 1)M^2/2}\right]\frac{dM}{M} \tag{9.16}$$

Next, it is noted that Equation 9.3 gives, using $a^2 = \gamma p/\rho$,

$$\frac{dV}{V} + \frac{1}{\gamma M^2}\frac{dp}{p} + \frac{\tau_w}{\rho V^2}\frac{P}{A}dx = 0 \tag{9.17}$$

Substituting for dV/V from Equation 9.8 allows this equation to be written as

$$\frac{dM}{M}+\frac{1}{2}\frac{dT}{T}+\frac{1}{\gamma M^2}\frac{dp}{p}-\frac{\tau_w}{\rho V^2}\frac{P}{A}dx=0 \tag{9.18}$$

Substituting Equations 9.11 and 9.16 into this equation then gives

$$\frac{dM}{M}-\frac{(\gamma-1)M^2/2}{[1+(\gamma-1)M^2/2]}\frac{dM}{M}-\frac{1+(\gamma-1)M^2}{[1+(\gamma-1)M^2/2]}\frac{dM}{M}+\frac{\tau_w}{\rho V^2}\frac{P}{A}dx=0$$

This is easily rearranged to give

$$\frac{dM}{M}=\frac{\gamma M^2[1+(\gamma-1)/2M^2]}{(1-M^2)}\left[\frac{\tau_w}{\rho V^2}\frac{P}{A}dx\right] \tag{9.19}$$

Substituting this equation back into Equation 9.11 then gives

$$\frac{dT}{T}=-\frac{\gamma(\gamma-1)M^4}{(1-M^2)}\left[\frac{\tau_w}{\rho V^2}\frac{P}{A}dx\right] \tag{9.20}$$

Similarly, substituting Equation 9.19 into Equation 9.16 gives

$$\frac{dp}{p}=-\frac{\gamma M^2[1+(\gamma-1)M^2]}{(1-M^2)}\left[\frac{\tau_w}{\rho V^2}\frac{P}{A}dx\right] \tag{9.21}$$

Since the wall shear stress, the velocity, and the Mach number are always positive, Equation 9.19 indicates that the sign of dM depends on the sign of $(1 - M^2)$. This equation therefore shows that if the Mach number is less than 1, friction causes the Mach number to increase, whereas if the Mach number is greater than 1, friction causes the Mach number to decrease. Viscosity therefore always causes the Mach number to tend toward 1. Since once Mach 1 is attained, changes in the downstream conditions cannot affect the upstream flow, it follows that "choking" can occur as a result of friction.

In the same way, Equations 9.20 and 9.21 show that if M is less than 1, dT and dp are negative, whereas if M is greater than 1, dT and dp are positive.

Lastly, consider the entropy change caused by friction. Since

$$ds=c_p\frac{dT}{T}-R\frac{dp}{p}$$

i.e., since

$$\frac{ds}{c_p}=\frac{dT}{T}-\frac{\gamma-1}{\gamma}\frac{dp}{p}$$

it follows, using Equations 9.20 and 9.21, that

$$\frac{ds}{c_p} = (\gamma - 1)M^2\left[\frac{\tau_w}{\rho V^2}\frac{P}{A}dx\right] \tag{9.22}$$

This shows that, as is required, the entropy always increases.

The above relations therefore together indicate that in constant area compressible duct flow with friction the flow variables change in the manner indicated in Table 9.1. These results are also summarized in Figure 9.4.

The above equations contain the wall shear stress, τ_w. This is expressed in terms of a dimensionless wall shear stress, f, which is defined by

$$f = \frac{\tau_w}{\frac{1}{2}\rho V^2} \tag{9.23}$$

The dimensionless wall shear stress, f, is termed the "Fanning friction factor." Another friction factor, f_D, termed the "Darcy friction factor," is also commonly used in the analysis of incompressible fluid flows in ducts. The two friction factors are related by

$$f = \frac{f_D}{4}$$

TABLE 9.1
Changes in Flow Variables Produced By Friction

	dM	dV	dp	dT	ds
$M < 1$	+	+	–	–	+
$M > 1$	–	–	+	+	+

Note: + quantity is increasing; – decreasing.

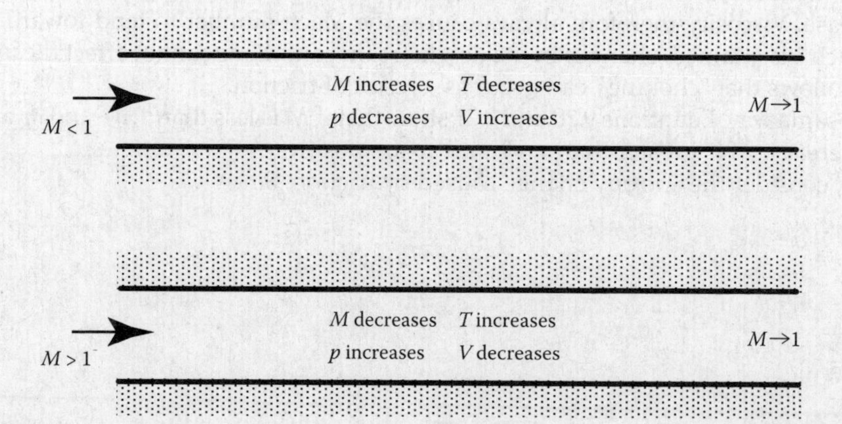

FIGURE 9.4
Changes in the flow variables produced by the effects of wall friction in adiabatic duct flow.

In general,

$$f = \text{function}\,(Re,\ \epsilon/D_H,\ M)$$

where *Re* is the Reynolds number based on the mean velocity and the hydraulic diameter and ϵ is a measure of the mean height of the wall roughness. The effect of *M* has, however, been found to be small. Therefore, for most purposes, it is adequate to assume that

$$f = \text{function}\,(Re,\ \epsilon/D_H)$$

Thus, *f* will be given by the same equations or charts that apply in low-speed duct flow. A Moody chart that can be used to give the Darcy friction factor is given in Figure 9.5. Alternatively, some easy to use expressions for the friction factor have been proposed, e.g.,

Laminar flow:

$$f = \frac{16}{Re}$$

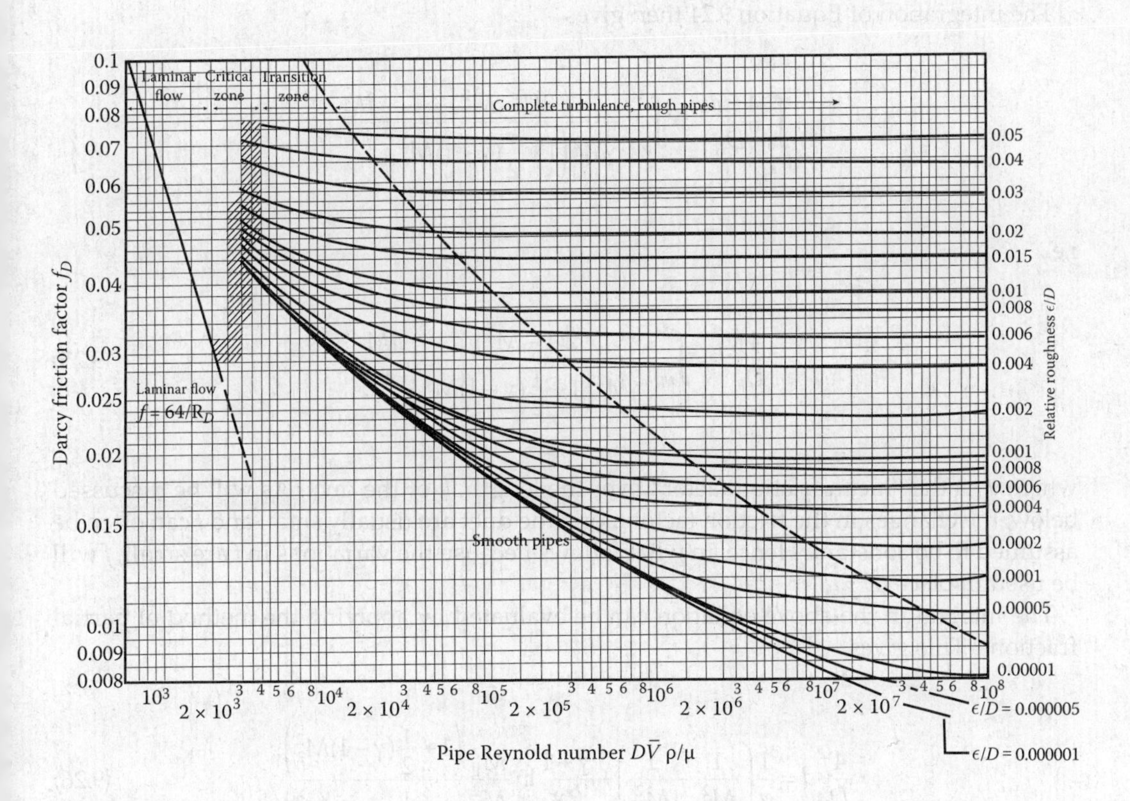

FIGURE 9.5
Moody chart. (From L.F. Moody, *Transactions of the ASME*, 671–684, Nov 1944. ASME. With permission.)

Turbulent flow:

$$f = 0.0625 \Bigg/ \left[\log\left(\frac{\epsilon}{3.7 D_H} + \frac{5.74}{Re^{0.9}} \right) \right]^2$$

Transition from laminar to turbulent flow can in most cases be assumed to occur at a Reynolds number of about 2300. Most compressible gas flows in ducts will involve turbulent flow.

The value of the wall roughness, ϵ, depends on the type of material from which the duct is made. For drawn tubing, it has a value of approximately 0.0015 mm, and for commercial steel, it has a value of approximately 0.045 mm.

In terms of f, Equation 9.19 can be written

$$\frac{4f\,dx}{D_H} = \frac{2(1-M^2)}{\gamma M^2\left(1+\frac{1}{2}(\gamma-1)M^2\right)}\frac{dM}{M} \tag{9.24}$$

where, for convenience, the "hydraulic diameter," which was discussed before and which is defined by $D_H = 4A/P$, has been introduced.

The integration of Equation 9.24 then gives

$$\int_0^l \frac{4f\,dx}{D_H} = \int_{M_1}^{M_2} \frac{2(1-M^2)}{\gamma M^2\left(1+\frac{1}{2}(\gamma-1)M^2\right)}\frac{dM}{M}$$

i.e.,

$$\frac{4\bar{f}}{D_H} l = \int_{M_1}^{M_2} \frac{2(1-M^2)}{\gamma M^2\left(1+\frac{1}{2}(\gamma-1)M^2\right)}\frac{dM}{M} \tag{9.25}$$

where $\bar{f}$ is the "mean friction factor" over the length, l, of the duct. As will be discussed below, the changes in the friction factor along the duct are usually small and f can often be assumed to be constant. Hence, in what follows because the variations in f are small, f will be used in place of $\bar{f}$.

The integral in the above equation can be evaluated by applying the method of partial fractions. This gives

$$\frac{4f}{D_H} l = \frac{1}{\gamma}\left(\frac{1}{M_1^2} - \frac{1}{M_2^2} \right) + \frac{\gamma+1}{2\gamma} \ln \frac{M_1^2}{M_2^2} \frac{\left(1+\frac{1}{2}(\gamma-1)M_2^2\right)}{\left(1+\frac{1}{2}(\gamma-1)M_1^2\right)} \tag{9.26}$$

This equation allows the change in Mach number over a given length of duct to be found. To describe the results given by this equation, it is convenient to select a reference value for M_2. Since, when friction is important, the Mach number always tends to 1, M_2 is conventionally set equal to 1 and the length of duct required to give this value of M, which is usually given the symbol l^* or l_{max}, is introduced, i.e., in presenting the values given by this equation, the duct length required to give a Mach number of 1, l^*, is introduced. With M_2 set equal to 1, the above equation gives

$$\frac{4f}{D_H} l^* = \left(\frac{1 - M^2}{\gamma M^2} \right) + \frac{\gamma + 1}{2\gamma} \ln \frac{(\gamma + 1)M^2}{2\left(1 + \frac{1}{2}(\gamma - 1)M^2\right)} \tag{9.27}$$

Similarly, integrating Equation 9.16 between a point in the duct at which the pressure is p and the real or imaginary point in the duct at which $M = 1$ and at which $p = p^*$ gives

$$\int_p^{p^*} \frac{dp}{p} = \int_M^1 \left[1 + \frac{(\gamma - 1)M^2/2}{\left(1 + \frac{1}{2}(\gamma - 1)M^2\right)} \right] \frac{dM}{M}$$

which gives

$$\frac{p}{p^*} = \frac{1}{M} \left[\frac{(\gamma + 1)/2}{1 + (\gamma + 1)M^2/2} \right]^{1/2} \tag{9.28}$$

Similarly, integrating Equation 9.11 gives

$$\int_T^{T^*} \frac{dT}{T} = -\int_M^1 \frac{(\gamma - 1)M}{1 + (\gamma - 1)M^2/2} dM$$

From this, it follows that

$$\frac{T}{T^*} = \frac{(\gamma + 1)/2}{1 + (\gamma - 1)M^2/2} \tag{9.29}$$

where, in these equations, T^* is the temperature at the real or imaginary point at which the Mach number is 1.

Using these results and the definitions of the stagnation pressure and temperature, relations for p_0/p_0^* and T_0/T_0^* can be found. For example, since

$$\frac{p_0}{p} = \left[1 + \frac{\gamma - 1}{2} M^2 \right]^{\frac{\gamma}{\gamma - 1}}$$

it follows that

$$\frac{p_0}{p_0^*} = \frac{p}{p^*}\left[\frac{1+(\gamma-1)M^2/2}{(\gamma+1)/2}\right]^{\frac{\gamma}{\gamma-1}}$$

Hence, using Equation 9.28 gives

$$\frac{p_0}{p_0^*} = \frac{1}{M}\left[\frac{1+(\gamma-1)M^2/2)}{(\gamma+1)/2}\right]^{\frac{\gamma+1}{2(\gamma-1)}} \tag{9.30}$$

The above equations have the form

$$\frac{4fl^*}{D} = \text{function } (M)$$

$$\frac{p}{p^*} = \text{function } (M)$$

$$\frac{T}{T^*} = \text{function } (M)$$

etc.

Values of these functions are available in graphs and in tables (see Appendix D) and are given by the COMPROP software. A typical chart, this being for $\gamma = 1.4$, is given in Figure 9.6.

The way in which the charts, tables, or software are used to find the changes in the values of the flow variables in ducts whose length is less than that required to give Mach 1 is illustrated in the following examples. Basically, the procedure uses the fact that if l_{12} is the actual duct length and if M_1 and M_2 are the Mach numbers at the beginning and end of the duct then $l_{12} = l_1^* - l_2^*$, similar equations applying for the changes in the other flow variables, e.g.,

$$p_2 = \frac{p_2/p^*}{p_1/p^*}\,p_1$$

Example 9.1

Air flows in a 5-cm-diameter pipe. The air enters at $M = 2.5$ and is to leave at $M = 1.5$. What length of pipe is required? What length of pipe would give $M = 1$ at the exit? Assume that $\bar{f} = 0.002$ and that the flow is adiabatic.

Solution

The flow situation being considered is shown in Figure E9.1.

For Mach numbers of 2.5 and 1.5, tables or charts or the software give $4fl^*/D = 0.432$ and 0.136, respectively. Hence, it follows that

$$\frac{4fl_{1-2}}{D} = \frac{4fl_0^*}{D} - \frac{4fl_2^*}{D} = 0.432 - 0.136$$

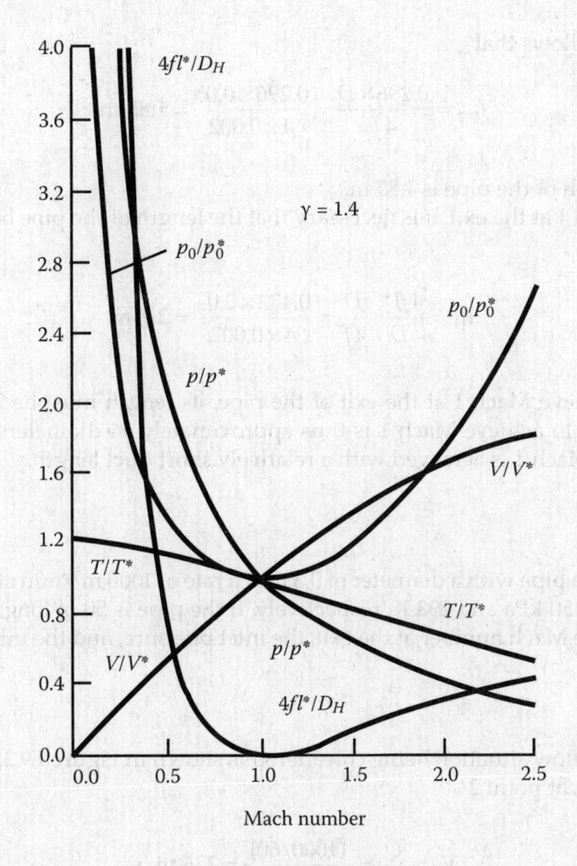

FIGURE 9.6
Variation of flow variables in adiabatic constant area duct flow with friction.

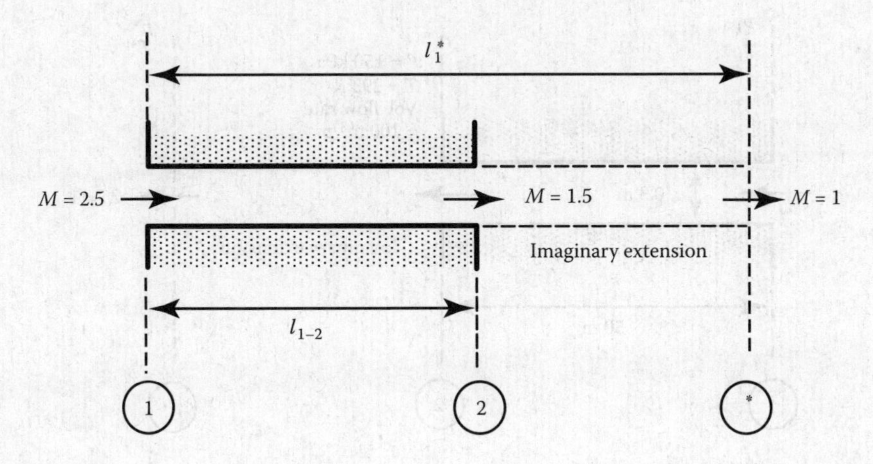

FIGURE E9.1
Flow situation considered.

from which it follows that

$$l_{1-2} = \frac{0.296 \times D}{4f} = \frac{0.296 \times 0.05}{4 \times 0.002} = 1.85 \text{ m}$$

Hence, the length of the pipe is 1.85 m.

To reach Mach 1 at the exit, it is necessary that the length of the pipe be equal to l_1^* i.e., be given by

$$l_1^* = \frac{4fl_1^*}{D}\frac{D}{4f} = \frac{0.432 \times 0.05}{4 \times 0.002} = 2.7 \text{ m}$$

Hence, to achieve Mach 1 at the exit of the pipe, its length must be 2.7 m. The pipe length required to achieve Mach 1 is thus approximately 54 diameters. In supersonic flow, therefore Mach 1 is achieved with a relatively short duct length.

Example 9.2

Air flows out of a pipe with a diameter of 0.3 m at a rate of 1000 m³/min at a pressure and temperature of 150 kPa and 293 K, respectively. If the pipe is 50 m long, find assuming that $f = 0.005$, the Mach number at the exit, the inlet pressure, and the inlet temperature.

Solution

In this case, the flow situation being considered is shown in Figure E9.2.

At the exit, i.e., at point 2

$$V_2 = \frac{Q}{A_2} = \frac{(1000/60)}{(\pi/4)(0.3)^2} = 236 \text{ m/s}$$

However,

$$a_2 = \sqrt{\gamma R T_2} = \sqrt{1.4 \times 287 \times 293} = 343 \text{ m/s}$$

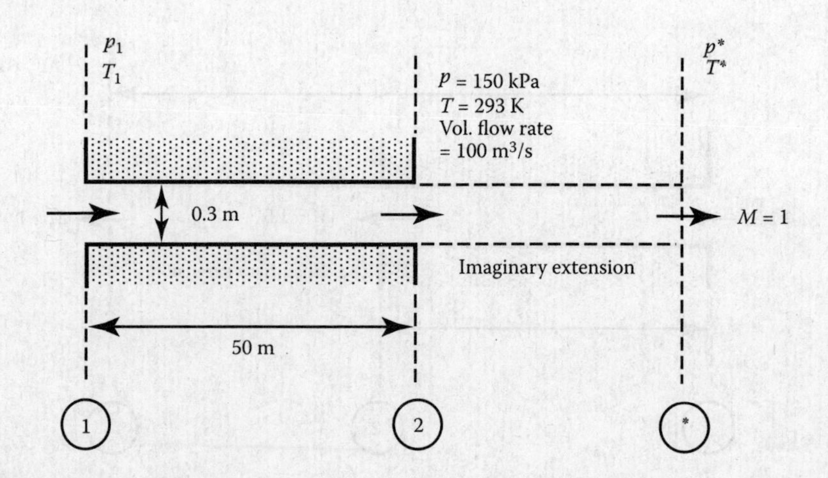

FIGURE E9.2
Flow variables in situation considered.

Hence,

$$M_2 = \frac{V_2}{a_2} = \frac{236}{343} = 0.688$$

At point 2 therefore, the relations derived above or tables or the software give

$$\frac{4fl_2^*}{D} = 0.228$$

$$\frac{p_2}{p^*} = 1.54, \quad \frac{T_2}{T^*} = 1.10$$

Now,

$$l_{1-2} = l_1^* - l_2^*$$

Hence,

$$\frac{4fl_{1-2}}{D} = \frac{4fl_1^*}{D} - \frac{4fl_2^*}{D}$$

From which it follows that

$$\frac{(4)(0.005)(50)}{0.3} = \frac{4fl_1^*}{D} - 0.228$$

which gives

$$\frac{4fl_1^*}{D} = 3.6$$

From the relations derived above or from tables or using the software for $4fl_1^*/D = 3.6$, the following is obtained:

$$M_1 = 0.345, \quad p_1/p^* = 3.14, \quad T_1/T^* = 1.17$$

Hence,

$$p_1 = \frac{p_1/p^*}{p_2/p^*} \times p_2 = \frac{3.14}{1.54} 150 = 306 \text{ kPa}$$

and

$$T_1 = \frac{T_1/T^*}{T_2/T^*} T_2 = \frac{1.17}{1.10}(293) = 312 \text{ K} = 39°\text{C}$$

Therefore, the Mach number, pressure, and temperature at the inlet are 0.345, 306 kPa, and 39°C, respectively.

Friction Factor Variations

It was assumed in the above analysis that the friction factor could be treated as constant and in the examples given above its value was assumed to be known. Now, as discussed before

$$f = \text{function}\,(Re,\ \epsilon/D_H), \quad \text{where } Re = \frac{\rho V D_H}{\mu}$$

Since the mass flow rate through the duct is given by $\dot{m} = \rho VA$, it follows that the Reynolds number is given by

$$Re = \frac{\dot{m} D_H}{A\mu} = \frac{4\dot{m}}{P\mu}$$

In a given situation, $\dot{m}$ = constant and P = constant. Therefore,

$$Re \propto \frac{1}{\mu}$$

The value of the coefficient of viscosity, μ, varies somewhat with temperature, its variation for gases being approximately described by

$$\frac{\mu_1}{\mu_2} = \left(\frac{T_1}{T_2}\right)^n$$

where T is the absolute temperature. The index n in this equation is roughly between 0.5 and 0.8 for common gases. Therefore, the changes in Re along a duct flow are usually small. Furthermore, since the flow is usually turbulent, f is only weakly dependent on Re. Hence, it is usually adequate to treat f as constant and to evaluate it using inlet conditions.

Example 9.3

Air flows through a 5-cm-diameter stainless steel pipe. The air enters the pipe at Mach 0.3 with a pressure of 150 kPa and a temperature of 40°C. Using the friction factor evaluated at the inlet conditions, determine the Mach number, temperature, and pressure at distances of 4, 8, 12, 16, 18, and 19 m from the inlet. Using these results, find the actual friction factors at these positions in the pipe and compare these values with the assumed constant value.

Solution

The flow situation here considered is shown in Figure E9.3.

Considering the inlet conditions,

$$a_1 = \sqrt{\gamma R T_1} = \sqrt{1.4 \times 287 \times 313} = 354.6 \text{ m/s}$$

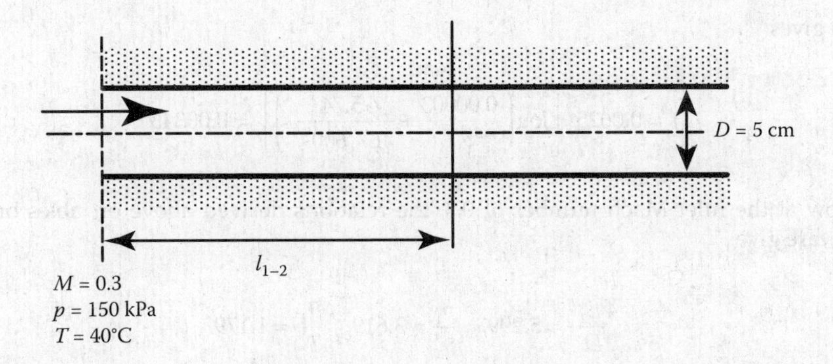

FIGURE E9.3
Flow situation considered.

Hence,

$$V_1 = M_1 \times a_1 = 0.3 \times 354.6 = 106.4 \text{ m/s}$$

The inlet density is given by

$$\rho_1 = \frac{p_1}{RT_1} = \frac{150{,}000}{287 \times 313} = 1.67 \text{ kg/m}^3$$

The kinematic viscosity μ depends only on temperature over the range of pressures here being considered and is given approximately using Sutherland's law, which, for air, gives

$$\mu = 0.00001716 \left(\frac{T}{273.1}\right)^{1.5} \frac{384.1}{T + 111} \text{N} \cdot \text{s/m}^2$$

Hence, at the inlet conditions,

$$\mu = 0.00001716 \left(\frac{313}{273.1}\right)^{1.5} \frac{384.1}{313 + 111} = 0.00001907 \text{ N} \cdot \text{s/m}^2$$

The Reynolds number based on the inlet conditions is therefore given by

$$Re = \frac{\rho VD}{\mu} = \frac{1.67 \times 106.4 \times 0.05}{0.00001907} = 467{,}600$$

The flow is thus turbulent. For stainless steel tubing, the wall roughness ϵ is 0.0015 mm. Therefore, the roughness ratio ϵ/D for the pipe being considered is 0.0015/50 = 0.00003. Using the equation given before for the friction factor, i.e.,

$$f = 0.0625 \Bigg/ \left[\log\left(\frac{\epsilon}{3.7D_H} + \frac{5.74}{Re^{0.9}}\right)\right]^2$$

then gives

$$f = 0.0625 \Bigg/ \left[\log \left(\frac{0.00003}{3.7} + \frac{5.74}{467{,}600^{0.9}} \right) \right]^2 = 0.00343$$

Now at the inlet Mach number of 0.3, the relations derived above or tables or the software give

$$\frac{4fl_1^*}{D} = 5.299, \quad \frac{p_1}{p^*} = 3.619, \quad \frac{T_1}{T^*} = 1.179$$

At any other point in the flow distance $l_{1\text{-}2}$ from the inlet,

$$l_2^* = l_1^* - l_{1\text{-}2}$$

Hence,

$$\frac{4fl_2^*}{D} = \frac{4fl_1^*}{D} - \frac{4fl_{1\text{-}2}}{D}$$

i.e., using the values derived above

$$\frac{4fl_2^*}{D} = 5.299 - \frac{4 \times 0.00343 \times l_{1\text{-}2}}{0.05} = 5.299 - 0.2736 \times l_{1\text{-}2}$$

Using the values $4fl_2^*/D$ so found, the values of M_2, p_2/p^* and T_2/T^* can be found. Then since p_1 and T_1 are known, the values of p_2 and T_2 can be found using

$$p_2 = \frac{p_2/p^*}{p_1/p^*} \times p_1$$

and

$$T_2 = \frac{T_2/T^*}{T_1/T^*} T_1$$

Using the value of T_2 so found, the speed of sound a_2 can be found and the velocity can be found by setting $V_2 = M_2 a_2$. The viscosity and density can also be found using the derived values of temperature and pressure using

$$\mu_2 = 0.00001716 \left(\frac{T_2}{273.1} \right)^{1.5} \frac{384.1}{T_2 + 111} \mathrm{N \cdot s/m^2}$$

and

$$\rho_2 = \frac{p_2}{RT_2}$$

With the density, viscosity, and velocity determined, the Reynolds number can be found and the friction factor can then be calculated using the assumed value of ϵ and the specified value of D

$$f_2 = 0.0625 \bigg/ \left[\log\left(0.0000081 + \frac{5.74}{Re_2^{0.9}} \right) \right]$$

Using this procedure in conjunction with the relations derived above or the tables or the software gives the data shown in Table E9.3 for the prescribed values of l_{1-2}.

From these results, it will be seen that the friction factor varies by little more than 1%, and therefore, the use of a constant value of f is justified.

In cases where the hydraulic diameter of the duct or the flow rate through the duct has to be found, the friction factor cannot initially be directly found and an iterative approach has to be used. The inlet Mach number is usually initially guessed and the unknown diameter or flow rate is found. This value is used to find an improved value of the Mach number and the calculation is repeated. This is continued until a converged result is obtained. This procedure is illustrated in the following example. Although sophisticated iteration procedures can be used, a straightforward procedure that involves obtaining solutions for a series of guessed values of the initial Mach number and then deducing the correct initial Mach number from the results so obtained will be used here.

Example 9.4

A worker in a protective suit in a hazardous area is supplied with oxygen at a rate of 0.06 kg/s through an "umbilical cord" that has a length of 8 m. The pressure and temperature of the oxygen at the inlet to the umbilical cord are maintained at 250 kPa and 10°C, respectively, and the pressure at the outlet end of the cord is to be 50 kPa. Find the diameter of the umbilical cord if it is made from a material that gives an effective wall roughness, ϵ, of 0.005 mm.

Solution

The situation being considered is shown schematically in Figure E9.4.

Since the inlet conditions are kept constant and since the gas involved is oxygen,

TABLE E9.3

Iterative Results Used in Obtaining Solution

l_{1-2}	$4fl_2^*/D$	M_2	p_2/p^*	T_2/T^*	p_2 (kPa)	T_2 (K)	V_2 (m/s)	Re_2	f_2
4	4.205	0.326	3.325	1.175	137.8	311.9	115.4	466,904	0.00343
8	3.110	0.361	2.996	1.170	124.2	310.6	127.5	468,380	0.00342
12	2.016	0.419	2.570	1.159	106.5	307.7	147.3	471,944	0.00342
16	0.921	0.518	2.060	1.139	85.38	302.4	180.5	478,098	0.00341
18	0.374	0.633	1.665	1.111	69.01	295.0	217.9	487,390	0.00340
19	0.101	0.771	1.343	1.073	55.76	284.9	260.8	500,597	0.00339

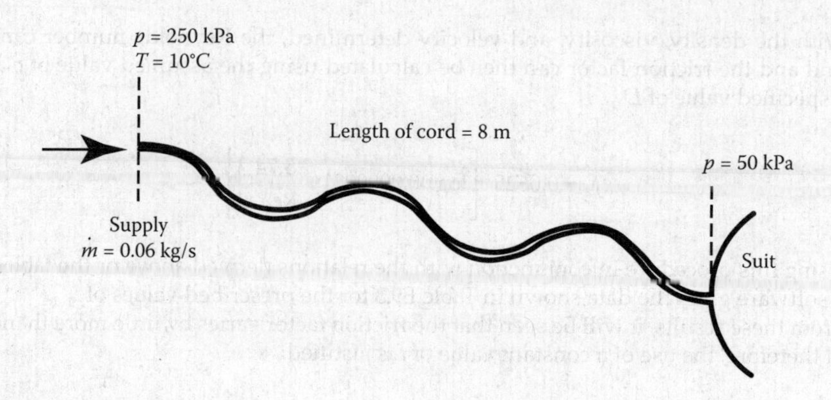

FIGURE E9.4
Flow situation considered.

$$a_1 = \sqrt{\gamma R T_1} = \sqrt{1.4 \times 260 \times 283} = 320.8 \text{ m/s}$$

$$\rho_1 = \frac{p_1}{RT_1} = \frac{250{,}000}{260 \times 283} = 3.4 \text{ kg/m}^3$$

Sutherland's equation will again be used to calculate μ, i.e.,

$$\mu_1 = 0.00001919\left(\frac{T}{273.1}\right)^{1.5} \frac{412.1}{T+139} \text{ N}\cdot\text{s/m}^2$$

$$= 0.00001919\left(\frac{283}{273.1}\right)^{1.5} \frac{412.1}{283+139} = 0.00001977 \text{ N}\cdot\text{s/m}^3$$

the constants appropriate for oxygen having been used in Sutherland's equation for μ.

The procedure is then as follows:

1. A value for the inlet Mach number, M_1, is guessed
2. The value of p_1/p^* corresponding to this value M_1 is obtained using the relations derived above or tables or the software
3. The inlet velocity is calculated using

$$V_1 = M_1 \times a_1 = M_1 \times 320.8$$

4. The diameter is calculated using

$$\rho_1\left(\frac{\pi D^2}{4}\right)V_1 = \dot{m}$$

i.e.,

$$D = \sqrt{\frac{4 \times 0.06}{\pi \times 3.4 \times V_1}}$$

5. The Reynolds number is calculated using

$$Re = \frac{\rho_1 V_1 D}{\mu_1} = \frac{3.4 V_1 D}{0.0000197}$$

6. The friction factor is found using

$$f = 0.0625 \Big/ \left[\log\left(\frac{\epsilon}{3.7D} + \frac{5.74}{Re^{0.9}}\right)\right]^2$$

ϵ being set equal to 0.000005 m

7. The value of $4fl_1^*/D$ corresponding to the chosen value of M_1 is obtained using the relations derived above or tables or the software
8. The value of $4fl_2^*/D$ is found using

$$\frac{4fl_2^*}{D} = \frac{4fl_1^*}{D} - \frac{4fl_{1-2}}{D}$$

where l_{1-2} is the length of the cord, i.e., 8 m

9. The value of p_2/p^* corresponding to this value of $4fl_2^*/D$ is found using the relations derived above or tables or the software
10. The value of p_2 is then found using

$$p_2 = \frac{p_2/p^*}{p_1/p^*} \times p_1 = \frac{p_2/p^*}{p_1/p^*} \times 250 \text{ kPa}$$

11. The value of p_2 so obtained is compared with the required value of 50 kPa.

Of course, if too large a value of M_1 is selected, a negative value of $4fl_2^*/D$ is obtained, indicating that the assumed inlet Mach number is not possible. A typical set of results is shown in Table E9.4.

From these results, it can be deduced that an umbilical cord with a diameter of 0.0169 m (16.9 mm) will give the correct discharge pressure of 50 kPa. This diameter corresponds to an inlet Mach number of 0.242.

TABLE E9.4

Iterative Results Used in Obtaining Solution

M_1	V_1 (m/s)	D	Re	f	$4fl_{1-2}/D$	l_{1-2}	$4fl_1^*/D$	$4fl_2^*/D$	p_2 (kPa)
0.1	32.08	0.0265	146,009	0.00419	5.065	66.92	61.86	10.54	240.77
0.2	64.16	0.0187	206,488	0.00393	6.717	14.53	7.816	3.58	81.78
0.3	96.24	0.0153	252,895	0.00379	7.936	5.299	−2.64	—	—
0.24	76.99	0.0171	226,196	0.00387	7.239	9.387	2.148	2.27	51.86
0.25	80.20	0.0167	230,860	0.00385	7.362	8.483	1.121	1.90	43.40
0.26	83.41	0.0164	235,432	0.00384	7.482	7.688	0.207	1.36	31.44
0.27	86.62	0.0161	239,917	0.00383	7.599	6.983	−0.62	—	—

The Fanno Line

The Fanno line has been used extensively in the past in describing the changes that occur in an adiabatic flow in a duct when wall friction is important, such flow, as noted before, being called Fanno flow. The Fanno line shows the flow process on a T–s or h–s diagram, it being noted that, since the flow of a perfect gas is being considered, $h = c_pT$ and c_p is a constant.

Now, as has been noted before:

$$\frac{ds}{c_p} = \frac{dT}{T} - \frac{(\gamma-1)}{\gamma}\frac{dp}{p}$$

Hence, since Equation 9.13 gives

$$\frac{dp}{p} = \frac{dT}{T} - \frac{dV}{V}$$

and since Equation 9.9 gives

$$\frac{dT}{T} = -\frac{V\,dV}{c_pT}$$

it follows that

$$\frac{ds}{c_p} = \frac{dT}{T} - \frac{(\gamma-1)}{\gamma}\frac{dT}{T}\left(1 + \frac{c_pT}{V^2}\right) \quad (9.31)$$

However, since adiabatic flow is being considered, the energy equation gives

$$V^2 = 2c_p(T_0 - T)$$

Combining this result with Equation 9.31 allows the variation of s with T for a given T_0 to be found since the above two equations together give

$$\frac{ds}{c_p} = \frac{dT}{T} - \frac{(\gamma-1)}{\gamma}\frac{dT}{T}\left(1 + \frac{T}{2(T_0 - T)}\right)$$

i.e.,

$$\frac{ds}{c_p} = \frac{1}{\gamma}\frac{dT}{T} - \frac{\gamma-1}{2\gamma}\frac{dT}{T_0 - T}$$

This equation can be integrated to give the variation of s with T. To do this, some arbitrary temperature T_1 at which the entropy is taken to have an arbitrary datum value of s_1 (usually taken as 0) is introduced. The integration of the above equation then gives for a given T_0

$$\frac{s-s_1}{c_p}=\frac{1}{\gamma}\int_{T_1}^{T}\frac{dT}{T}-\frac{\gamma-1}{2\gamma}\int_{T_1}^{T}\frac{dT}{T_0-T}$$

Carrying out the integration then gives

$$\frac{s-s_1}{c_p}=\ln\left[\left(\frac{T}{T_1}\right)^{\frac{1}{\gamma}}\left(\frac{T_0-T}{T_0-T_1}\right)^{\frac{\gamma-1}{2\gamma}}\right] \tag{9.32}$$

This equation allows the variation of s with T for a given T_0 and arbitrarily selected T_1 to be found. The variation resembles that shown in Figure 9.7.

The line describing the variation of s with T is called a "Fanno line" and adiabatic flow in a constant area duct with significant friction effects is, as mentioned before, termed "Fanno flow." The Fanno line shows all possible combinations of entropy and temperature that can exist in an adiabatic flow in a constant area duct at a given stagnation temperature, the stagnation temperature, of course, remaining constant in an adiabatic flow.

The maximum entropy point on the Fanno line is, for the reasons previously discussed, the point at which the Mach number is 1. The upper portion of the curve, which is associated with higher values of T, applies in subsonic flow whereas the lower portion of the curve, which is associated with lower values of T, applies in supersonic flow. Since, as discussed before, the entropy always increases, this again shows how the Mach number always moves toward 1.

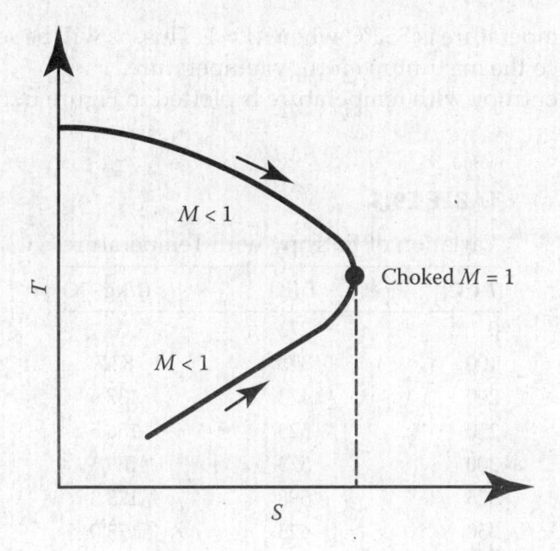

FIGURE 9.7
The Fanno line.

Example 9.5

Consider an adiabatic air flow in a constant area duct. If the stagnation temperature in the flow is 750 K and if entropy is arbitrarily taken as being equal to zero when the temperature is 0°C, plot the variation of entropy with temperature in the flow. Find the temperature at which the Mach number is 1 and show this on the plot.

Solution

Here $T_0 = 750$ K, $T_1 = 273$ K, and $s_1 = 0$, and since air flow is being considered, $c_p = 1006$ J/kg · K and $\gamma = 1.4$. Equation 9.32 therefore gives

$$\frac{s-0}{1006} = \ln\left[\left(\frac{T}{273}\right)^{\frac{1}{1.4}}\left(\frac{750-T}{750-273}\right)^{\frac{1.4-1}{2.8}}\right]$$

For any chosen value of T, this equation allows the value of s to be found. Since T must be less than 750 K, i.e., less than 477°C, results will be obtained for temperatures between 0°C and 470°C. Some values of s and T given by the above equation are shown in Table E9.5.

It has been shown several times that

$$\frac{T_0}{T} = 1 + \frac{\gamma-1}{2}M^2$$

Hence, in the present case when the Mach number is equal to 1,

$$\frac{750}{T} = 1 + \frac{1.4-1}{2}, \quad \text{i.e.,} \quad T = \frac{750}{1.2} = 625\text{ K}$$

Therefore, the temperature is 352°C when $M = 1$. This, as will be seen from the above table, corresponds to the maximum entropy temperature.

The variation of entropy with temperature is plotted in Figure E9.5.

TABLE E9.5

Variation of Entropy with Temperature

T (°C)	*T* (K)	*s* (J/kg · K)
0	273	0
100	373	82.7
200	473	137.6
250	523	156.5
300	573	169.5
325	598	173.3
350	623	175.0
400	723	167.8
450	743	124.7

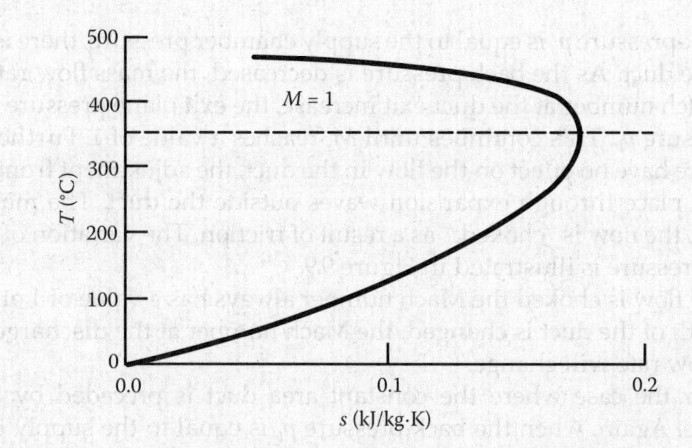

FIGURE E9.5
Variation of entropy with temperature.

Frictional Flow in a Duct Preceded by an Isentropic Nozzle

Consider the case where a constant area duct is supplied with gas that flows into the duct through a nozzle from a large chamber. The duct discharges into another large chamber. As discussed before, it is usually adequate to assume that friction effects are negligible in the nozzle. This is because the nozzle is usually relatively short and because the flow is accelerating through the nozzle. To illustrate what happens in this type of flow, it will be assumed that the conditions in the first large chamber, i.e., the stagnation conditions upstream of the nozzle, are kept constant and that the back-pressure in the large chamber into which the duct discharges is varied.

The case where the duct is preceded by a converging nozzle will first be considered. The flow situation considered in this case is shown in Figure 9.8.

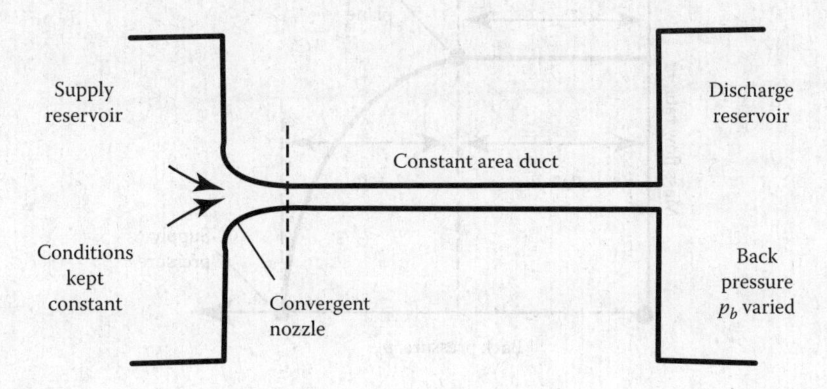

FIGURE 9.8
Convergent nozzle flow situation.

When the back-pressure p_b is equal to the supply chamber pressure, there is, of course, no flow through the duct. As the back-pressure is decreased, the mass flow rate through the duct and the Mach number at the duct exit increase, the exit plane pressure p_e being equal to the back-pressure p_b. This continues until M_e reaches a value of 1. Further decreases in the back-pressure have no effect on the flow in the duct, the adjustment from p_e to p_b in this situation taking place through expansion waves outside the duct. The mass flow rate is thus limited, i.e., the flow is "choked," as a result of friction. The variation of the mass flow rate with back-pressure is illustrated in Figure 9.9.

Since once the flow is choked the Mach number always has a value of 1 at the exit of the duct, if the length of the duct is changed, the Mach number at the discharge of the nozzle and the mass flow rate will change.

Next, consider the case where the constant area duct is preceded by a convergent–divergent nozzle. Again, when the back-pressure p_b is equal to the supply chamber pressure there is, of course, no flow through the duct. As the back-pressure is then initially decreased, the Mach number increases at the nozzle throat but then decreases again in the divergent portion of the nozzle, the flow remaining subsonic throughout the nozzle. The Mach number then increases in the constant area duct as a result of friction. As the back-pressure is further decreased, one possibility is that the Mach number will reach a value of 1.0 at the duct exit, i.e., that the flow will choke at the end of the duct as a result of friction and that the flow will remain subsonic in the nozzle no matter how low the back-pressure gets. This situation is exactly the same as that which occurs with a convergent nozzle, i.e., the same as that discussed above. A more likely situation is, however, that the system will be sized so that as the back-pressure is decreased the Mach number at the nozzle throat will reach a value of 1.0 before the Mach number at the duct exit reaches a value of 1, i.e., the nozzle will choke before choking occurs in the duct. Once the Mach number has reached a value of 1.0 at the nozzle throat, a region of supersonic flow develops downstream of the throat with further decrease in back-pressure. The region of supersonic flow is terminated by a normal shock wave. Because there is a significant region of subsonic flow near the wall, the effects of this shock wave can be spread out over a significant length of the duct in small ducts. However, the flow can usually be adequately analyzed

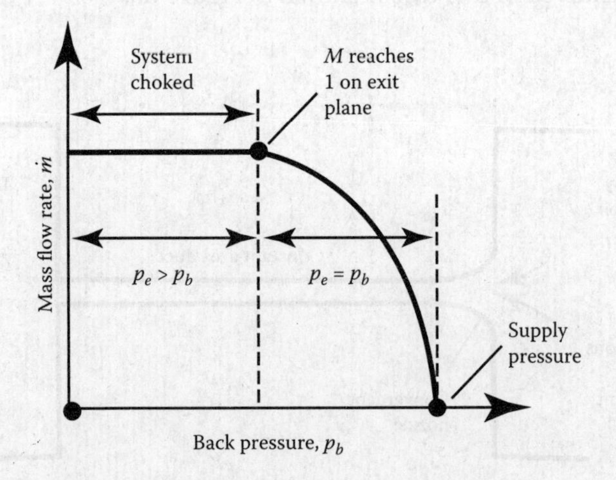

FIGURE 9.9
Variation of mass flow rate for convergent nozzle flow situation.

by assuming that a conventional normal shock wave occurs. As the back-pressure is further decreased, the shock wave moves toward the duct exit, eventually reaching the exit. Once the back-pressure has been reduced to a value at which the shock wave is at the exit plane of the duct, further reductions in the back-pressure have no effect on the flow in the duct system, the adjustment from the exit plane pressure to the back-pressure taking place through oblique shock waves or expansion waves outside the duct. The changes in the flow and the mass flow rate with back-pressure for this convergent–divergent nozzle case are illustrated in Figures 9.10 and 9.11.

The effects of back-pressure on the flow are illustrated in the following two examples.

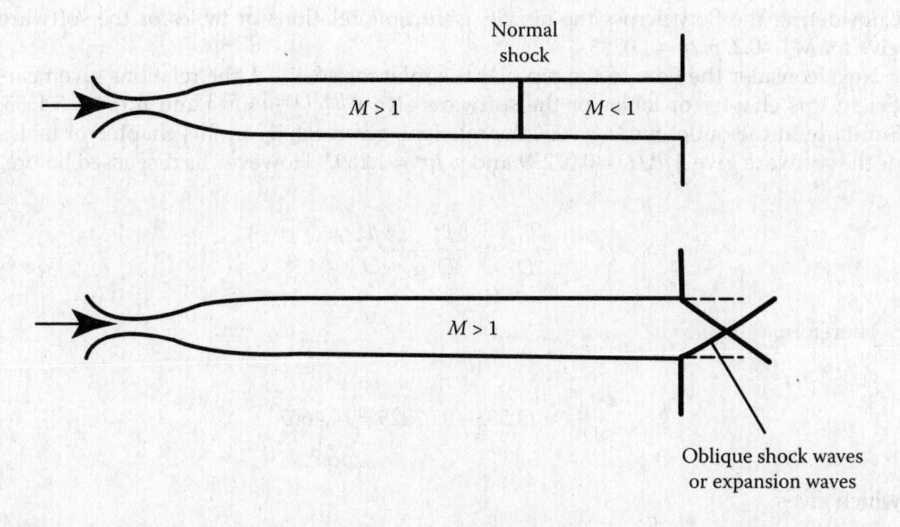

FIGURE 9.10
Variation of flow pattern with back pressure for convergent–divergent nozzle flow situation.

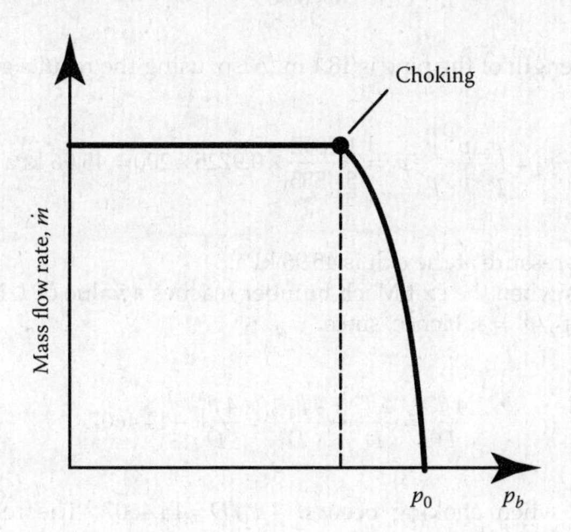

FIGURE 9.11
Variation of mass flow rate for convergent–divergent nozzle flow situation.

Example 9.6

Air, in a large reservoir, at a pressure of 200 kPa and a temperature of 30°C is expanded through a convergent nozzle. The air then flows down a pipe with a diameter of 25 mm. If the Mach number at the exit of the nozzle, i.e., at the inlet to pipe is 0.2 and the Mach number at the end of this pipe is 0.8, assuming that the flow in the nozzle is isentropic and the flow in the pipe adiabatic, find the length of the pipe and the pressure at the exit of the pipe. Also, find the pressure in the reservoir into which the pipe discharges at which choking first occurs and the inlet Mach number under these conditions. Also, plot a graph of pipe inlet and outlet Mach number against discharge reservoir pressure. It can be assumed in all calculations that $f = 0.005$.

Solution

Considering the flow across the nozzle, isentropic relations or tables or the software give for $M_1 = 0.2$, $p_0/p = 1.0283$.

Next, consider the flow in the pipe. At the inlet for $M_1 = 0.2$, the relations given earlier in this chapter or tables or the software give $4fl_1^*/D = 14.533$ and $p_1/p^* = 5.4555$. Similarly, at the outlet for $M_2 = 0.8$, the relations given earlier in this chapter, or tables or the software give $4fl_2^*/D = 0.07229$ and $p_2/p^* = 1.2892$. However, as discussed before,

$$\frac{4fl_{1-2}}{D} = \frac{4fl_1^*}{D} - \frac{4fl_2^*}{D}$$

Therefore,

$$\frac{4fl_{1-2}}{D} = 14.533 - 0.07229 = 14.4607$$

which gives

$$l_{1-2} = \frac{14.4607 \times 0.025}{4 \times 0.005} = 18.1 \text{ m}$$

Therefore, the length of the pipe is 18.1 m. Also, using the results given above

$$p_2 = \frac{p_2}{p^*}\frac{p^*}{p_1}\frac{p_1}{p_0}p_0 = \frac{1.2892}{5.4555} \times 0.9725 \times 200 = 45.96 \text{ kPa}$$

Therefore, the pressure at the exit is 45.96 kPa.

Choking occurs when the exit Mach number reaches a value of 1. Now, when $M_2 = 1$, $4fl_2^*/D = 0.0$, and $p_2/p^* = 1$. Hence, since

$$\frac{4fl_2^*}{D} = \frac{4fl_1^*}{D} - \frac{4fl_{1-2}}{D} = \frac{4fl_1^*}{D} - 14.4607$$

It follows that when choking occurs, $4fl_1^*/D = 14.4607$. The relations given earlier in this chapter, or tables or the software indicate that this value of $4fl_1^*/D$ corresponds to $M_1 = 0.2005$ and $p_1/p^* = 5.455$. Isentropic relations or tables or the software give for this value of M_1, $p_0/p_1 = 1.028$. Hence, since when the exit Mach number is 1, $p_2 = p^*$, it follows that

$$p_2 = \frac{p^*}{p_1}\frac{p_1}{p_0}p_0 = \frac{200}{1.028 \times 5.455} = 35.66 \text{ kPa}$$

It follows that an exit Mach number of 1 will be obtained when

$$p_2 = p^* = 35.66 \text{ kPa}$$

To determine the Mach number variation with back-pressure, it is recalled that, because the friction factor is assumed to remain the same, the following still applies

$$\frac{4fl_{1-2}}{D} = 14.4607$$

Hence,

$$\frac{4fl_2^*}{D} = \frac{4fl_1^*}{D} - \frac{4fl_{1-2}}{D} = \frac{4fl_1^*}{D} - 14.4607$$

Also, as before,

$$p_2 = \frac{p_2}{p^*}\frac{p^*}{p_1}\frac{p_1}{p_0}p_0 = \frac{p_2}{p^*}\frac{p^*}{p_1}\frac{p_1}{p_0} \times 200$$

For any chosen value of the nozzle exit Mach number M_1, the values of p_o/p_1, $4fl_1^*/D$, and p_1/p^* can be found using isentropic and Fanno relations or tables or the software. The value of $4fl_2^*/D$ can then be found, and then using Fanno relations or tables or the software the values of M_2 and p_2/p^* can be obtained.

A typical set of results obtained using the above procedure is shown in Table E9.6. The variations of M_1 and M_2 with p_2 is shown in Figure E9.6.

TABLE E9.6

Results for Various Values of the Nozzle Exit Mach Number

M_1	p_o/p_1	$4fl_1^*/D$	p_1/p^*	$4fl_2^*/D$	p_2/p^*	p_2 (kPa)	M_2
0.00	1.000	—	—	—	—	200.0	0.000
0.12	1.010	45.408	9.1156	30.801	7.60	165.1	0.144
0.15	1.016	27.932	7.2866	13.471	5.27	142.4	0.205
0.17	1.020	21.115	6.4253	6.6543	3.95	120.5	0.275
0.19	1.026	16.375	5.7448	1.9143	2.54	86.19	0.423
0.20	1.028	14.533	5.4555	0.0723	1.29	45.96	0.800
0.2005	1.028	14.607	5.4554	0.0000	1.00	35.66	1.000

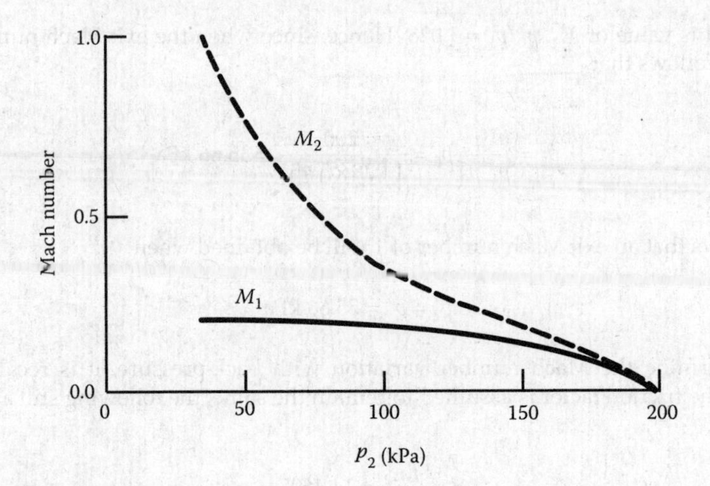

FIGURE E9.6
Variations of Mach numbers with exit pressure.

Example 9.7

Air flows steadily from a large reservoir through a convergent–divergent nozzle into a 0.3 m diameter pipe with a length of 3.5 m. The conditions in the reservoir are such that the Mach number and the pressure at the inlet to the pipe are 2 and 101.3 kPa, respectively. The average friction factor, f, for the flow in the pipe is estimated to be 0.005.

1. If no shocks occur, find M and p at the exit of the pipe
2. If there is a normal shock at the exit of the pipe, find the back-pressure in the chamber into which the pipe is discharging
3. Find the back-pressure in the chamber into which the pipe is discharging when there is a shock halfway down the pipe

Solution

1. The flow situation considered is shown in Figure E9.7a.

For $M_1 = 2$, the relations for Fanno flow derived above or tables or the software give

$$\frac{4fl_1^*}{D} = 0.3049, \quad \frac{p_1}{p^*} = 0.4083$$

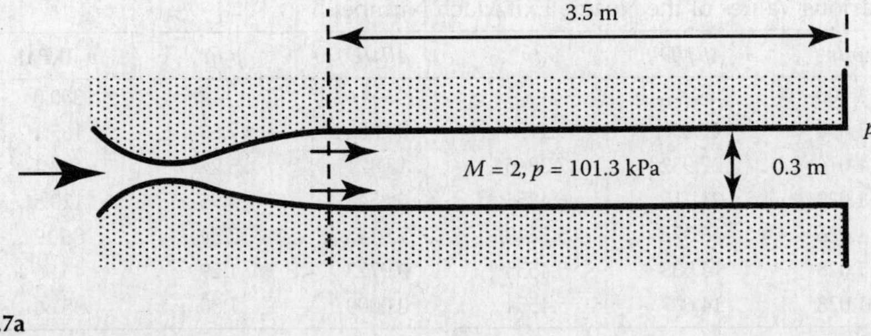

FIGURE E9.7a
Flow situation considered.

Now, at the exit of the pipe,

$$\frac{4fl_2^*}{D} = \frac{4fl_1^*}{D} - \frac{4fl_{1-2}}{D} = 0.3049 - \frac{4 \times 0.005}{0.3} \times 3.5 = 0.0717$$

For this value of $4fl_2^*/D$, the relations for Fanno flow or tables or the software give $M_2 = 1.32$ and $p_2/p^* = 0.715$. Hence,

$$p_2 = \frac{p_2}{p^*} \times \frac{p^*}{p_1} \times p_1 = \frac{0.715}{0.4083} \times 101.3 = 177.3 \text{ kPa}$$

Hence, when there are no shocks in the pipe, the pressure and the Mach number at the exit of the pipe are 177.3 kPa and 1.32, respectively.

2. When there is a shock wave on the exit plane of the pipe, the pressure and the Mach number just upstream of the shock will be 177.3 kPa and 1.32, respectively. Now for $M = 1.32$, normal shock relations or tables or software give

$$\frac{p_2}{p_1} = 1.866$$

Hence, the back-pressure is given by

$$p_{\text{back}} = 1.866 \times 177.3 = 331 \text{ kPa}$$

3. The situation being considered is shown in Figure E9.7b.

At a point halfway down the pipe,

$$\frac{4fl_2^*}{D} = \frac{4fl_1^*}{D} - \frac{4f(l_1^* - l_2^*)}{D} = 0.3049 - \frac{4 \times 0.005}{0.3}\left(\frac{3.5}{2}\right) = 0.19$$

For this value of $4fl_2^*/D$, the relations for Fanno flow or tables or the software give $M_2 = 1.65$ and $p_2/p^* = 0.534$. Hence,

$$p_2 = \frac{p_2}{p^*} \times \frac{p^*}{p_1} \times p_1 = \frac{0.534}{0.408} \times 101.3 = 132.6 \text{ kPa}$$

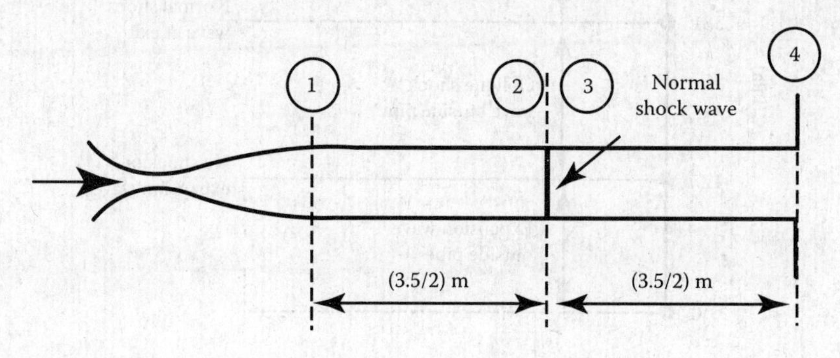

FIGURE E9.7b
Flow situation when shock wave is half way along pipe.

Now for $M = 1.65$, normal shock relations or tables or software give

$$\frac{p_3}{p_2} = 3.010, \quad M_3 = 0.654$$

Lastly, consider the subsonic flow downstream of the shock. Now for $M = 0.655$, the relations for Fanno flow derived above or tables or the software give

$$\frac{4fl_3^*}{D} = 0.31, \quad \frac{p_3}{p^*} = 1.6$$

Hence, at the exit plane,

$$\frac{4fl_4^*}{D} = \frac{4fl_3^*}{D} - \frac{4f(l_3 - l_4)}{D} = 0.31 - \left(\frac{4 \times .005}{0.3}\right)\left(\frac{3.5}{2}\right) = 0.194$$

For this value of $4fl^*/D$, the relations for Fanno flow or tables or the software give $M_4 = 0.71$ and $p_4/p^* = 1.48$. Hence,

$$p_4 = p_{\text{back}} = \frac{p_4}{p^*} \times \frac{p^*}{p_3} \times \frac{p_3}{p_2} \times p_2 = \frac{1.48}{1.6} \times 3.01 \times 132.6 = 369 \text{ kPa}$$

Hence, when the back-pressure is 369 kPa, there is a shock halfway down the pipe. The effects of the pressure on the flow are summarized in Figure E9.7c.

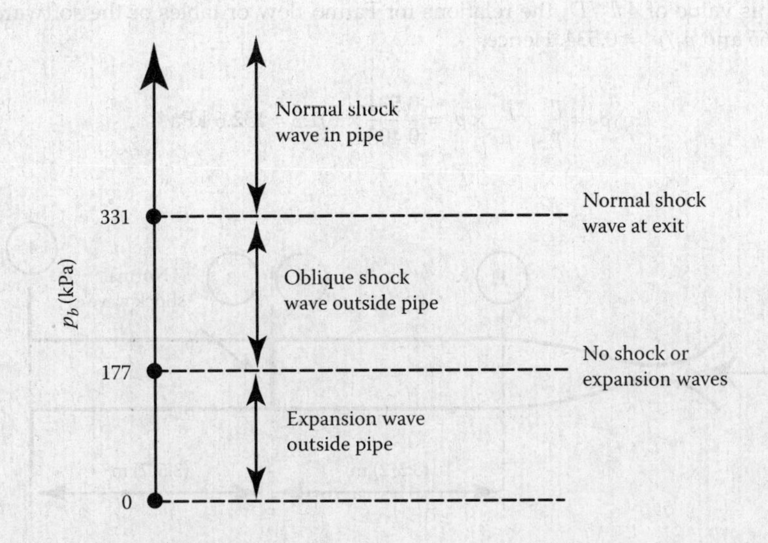

FIGURE E9.7c
Variation of flow pattern with back pressure.

The Effects of Friction on Variable Area Flow

In the discussion of the effects of friction on flow in a duct given above, it was assumed that the area of the duct remained constant. In some cases, however, flows occur in which the effects of both friction and area changes are important. This will be considered in the present section.

Consider the flow through the control volume shown in Figure 9.12. For this control volume, the continuity equation gives, as before,

$$\frac{d\rho}{\rho}+\frac{dA}{A}+\frac{dV}{V}=0 \tag{9.33}$$

whereas the momentum equation gives for this control volume,

$$pA+\left(p+\frac{dp}{2}\right)dA-(p+dp)(A+dA)-\tau_w A_s\cos\theta=\rho AV\;dV \tag{9.34}$$

The second term on the left-hand side represents the force due to the pressure on the curved wall. The fourth term on the left-hand side is the shear stress term, A_s being the wall area, i.e., the surface area, and θ the angle of inclination of the wall. Neglecting second-order terms and expressing the shear stress in terms of the friction factor then gives

$$dp+\frac{\rho V^2}{2}\frac{4f\;dx}{D_H}+\rho V\;dV=0 \tag{9.35}$$

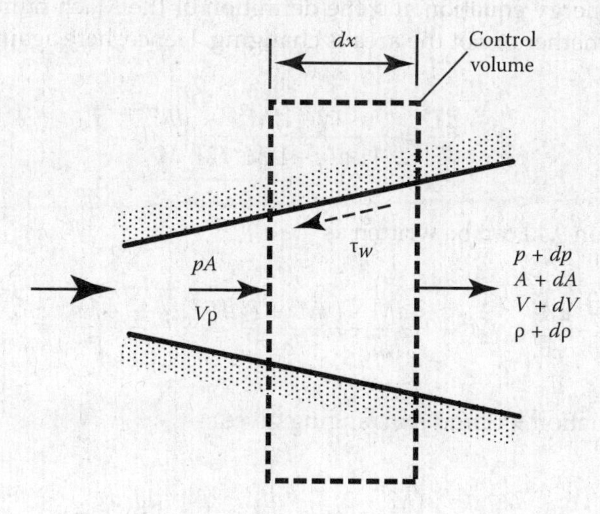

FIGURE 9.12
Control volume used in analysis of variable area flow.

which in turn can be rewritten as

$$\frac{dp}{p} + \frac{\gamma M^2}{2}\frac{4f\,dx}{D_H} + \gamma M^2 \frac{dV}{V} = 0 \tag{9.36}$$

Next, using the perfect gas law and combining it with the continuity equation result gives the following:

$$\frac{dp}{p} = \frac{d\rho}{\rho} + \frac{dT}{T} = -\frac{dA}{A} - \frac{dV}{V} + \frac{dT}{T} \tag{9.37}$$

Substituting this into the momentum equation then gives

$$-\frac{dA}{A} - \frac{dV}{V} + \frac{dT}{T} + \frac{\gamma M^2}{2}\frac{4f\,dx}{D_H} + \gamma M^2 \frac{dV}{V} = 0 \tag{9.38}$$

However, as discussed before, the definition of the Mach number gives

$$\frac{dV}{V} = \frac{1}{2}\frac{dT}{T} + \frac{dM}{M} \tag{9.39}$$

Substituting this into Equation 9.38 then gives

$$-\frac{dA}{A} - \frac{dM}{M} + \frac{1}{2}\frac{dT}{T} + \frac{\gamma M^2}{2}\frac{4f\,dx}{D_H} + \frac{\gamma M^2}{2}\frac{dT}{T} + \gamma M^2 \frac{dM}{M} = 0 \tag{9.40}$$

Now, if the derivation of Equation 9.11 is considered, it will be seen that it is based only on the use of the energy equation and the definition of the Mach number. Equation 9.11 therefore applies whether or not the area is changing. Hence, here again,

$$\frac{dT}{T} = -\frac{(\gamma - 1)M^2}{[1 + (\gamma - 1)M^2/2]}\frac{dM}{M} \tag{9.41}$$

However, Equation 9.40 can be written as

$$-\frac{dA}{A} + (\gamma M^2 - 1)\frac{dM}{M} + \frac{(\gamma M^2 + 1)}{2}\frac{dT}{T} + \frac{\gamma M^2}{2}\frac{4f\,dx}{D_H} = 0 \tag{9.42}$$

Hence, using Equation 9.41 and rearranging gives

$$-\frac{dA}{A} - \frac{(M^2 - 1)}{[1 + (\gamma - 1)M^2/2]}\frac{dM}{M} + \frac{\gamma M^2}{2}\frac{4f\,dx}{D_H} = 0$$

which can be rearranged to give

$$\frac{dM}{M} = -\left[\frac{1+(\gamma-1)M^2/2}{1-M^2}\right]\frac{dA}{A} + \left[\frac{1+(\gamma-1)M^2/2}{1-M^2}\right]\frac{\gamma M^2}{2}\frac{4f\,dx}{D_H} \tag{9.43}$$

The above equation indicates that the changes in Mach number along the duct are due to both area changes (the first term on the right-hand side) and to viscosity effects (the second term on the right-hand side). When the area is constant, the first term on the right-hand side is zero and the Fanno line equation is obtained. When friction effects are negligible, the second term on the right-hand side can be neglected and the same equation as previously used in the analysis of isentropic variable area flow is obtained. It should also be noted that it is possible for the Mach number to remain constant even when the area is changing provided the viscosity effects just balance the area change effects. In such a case,

$$\frac{dA}{dx} = \frac{2f\gamma M^2 A}{D_H}$$

Hence, in all situations, if M stays constant, the area, A, must increase.

It will be noted that both terms in Equation 9.43 contain the term $1 - M^2$ in the denominator and the equation therefore has a singularity when $M = 1$. Writing Equation 9.43 as

$$(1-M^2)\frac{dM}{M} = -\left(1+\frac{\gamma-1}{2}M^2\right)\frac{dA}{A} + \left(1+\frac{\gamma-1}{2}M^2\right)\frac{\gamma M^2}{2}\frac{4f\,dx}{D_H} \tag{9.44}$$

If the Mach number in the flow is to change from a subsonic to a supersonic value or from a supersonic to a subsonic value without a shock wave in the flow, i.e., if a sonic point occurs in the flow, Equation 9.44 shows that this sonic point must occur at the point in the flow where, since $M = 1$ at this point,

$$-\left(1+\frac{\gamma-1}{2}\right)\frac{dA}{A} + \left(1+\frac{\gamma-1}{2}\right)\frac{\gamma}{2}\frac{4f\,dx}{D_H} = 0$$

i.e.,

$$-\frac{\gamma+1}{2}\frac{dA}{A} + \frac{\gamma(\gamma+1)}{4}\frac{4f\,dx}{D_H} = 0 \tag{9.45}$$

If friction effects are negligible, i.e., if the second term in this equation is negligible, the sonic point will occur when dA is zero as discussed before. However, when friction effects are significant, because the second term in Equation 9.45 always has a positive value, the sonic point must occur at a point in the flow at which dA is positive, i.e., in a divergent portion of the flow.

In most cases, it is necessary to integrate Equation 9.43 by numerical methods simultaneously with the equations for p, T, and p. If the changes in Reynolds number due to area changes are large, it may be necessary to allow for the changes in the friction factor in this calculation.

Concluding Remarks

It has been shown that viscous effects, i.e., the effects of wall friction, in a constant area, adiabatic duct flow cause the Mach number to tend toward 1. As a result, it is possible for choking to occur due to viscous effects. An expression for the duct length required to produce choking was derived and relations for the ratio of local values of the flow variables to the values that these variables would have at the real or hypothetical point at which $M = 1$ were derived. The wall shear stress was expressed in terms of the friction factor and the way in which the friction factor is obtained was discussed. The effects of viscous friction on adiabatic flow in a variable area duct were also briefly discussed, and it was shown that in such flows the sonic point is not, in general, at a point of minimum area.

PROBLEMS

1. Air flows through a duct with a constant cross-sectional area. The pressure, temperature, and Mach number at the inlet to the duct are 180 kPa, 30°C, and 0.25, respectively. If the Mach number at the exit of the duct has risen to 0.75 as a result of friction, determine the pressure, temperature, and velocity at the exit. Assume that the flow is adiabatic.
2. Air flows through a well-insulated 4-in.-diameter pipe at the rate of 500 lbm/min. The pressure drops from 50 psia at the inlet to the pipe to a value of 40 psia at the exit. If the temperature at the inlet is 200°F, find the Mach number at the exit of the pipe.
3. Consider compressible flow through a long, well-insulated duct. At the inlet to the duct the Mach number, pressure, and temperature are 0.3, 100 kPa, and 30°C respectively. Assuming that the flow is adiabatic and that the pipe is sufficiently long to ensure that the flow is choked at the exit, find the velocity and temperature at the pipe exit.
4. Air flows through a 5-cm-diameter pipe. Measurements indicate that at the inlet to the pipe the velocity is 70 m/s, the temperature is 80°C, and the pressure 1 MPa. Find the temperature, the pressure, and the Mach number at the exit to the pipe if the pipe is 25 m long. Assume that the flow is adiabatic and that the mean friction factor is 0.005.
5. Air flows down a pipe with a diameter of 0.15 m. At the inlet to the pipe, the Mach number is 0.1, the pressure is 70 kPa, and the temperature is 35°C. If the flow can be assumed to be adiabatic and if the mean friction factor is 0.005, determine the length of the pipe if the Mach number at the exit is 0.6. Also, find the pressure and temperature at the exit to the pipe.
6. Air flows from a large tank through a well-insulated 12-mm-diameter pipe. If the air enters the pipe at Mach 0.2 and leaves at Mach 0.6, find the length of the pipe.

Assume a mean friction factor of 0.005. How much longer must the pipe be if the exit Mach number is 1? If the pipe is 75 cm longer than this latter value and if the same conditions exist in the supply chamber, what reduction in the flow rate will occur?

7. Air flows from a large tank, in which the pressure and temperature are 100 kPa and 30°C, respectively, through a 1.6-m-long pipe with a diameter of 2.5 cm. The pipe is connected to a short convergent nozzle with an exit diameter of 2.1 cm. The air from this nozzle is discharged into a large tank in which the pressure is maintained at 35 kPa. Assuming that the friction factor is equal to 0.002, find the mass flow rate through the system. The flow in the nozzle can be assumed to be isentropic and the pipe can be assumed to be heavily insulated.
8. Air at an inlet temperature of 60°C flows with a subsonic velocity through an insulated pipe having an inside diameter of 5 cm and a length of 5 m. The pressure at the exit to the pipe is 101 kPa and the flow is choked at the end of the pipe. If the average friction factor is 0.005, determine the inlet and exit Mach numbers, the mass flow rate and the change in temperature, and pressure through the pipe.
9. Hydrogen flows through a 50-mm-diameter pipe. The inlet pressure is 400 kPa, the inlet velocity is 300 m/s and the inlet temperature is 30°C. How long is the pipe if the flow is choked at the exit end? Assume a mean friction factor of 0.0058 and that the flow is adiabatic.
10. Air flows through a 0.15 m × 0.25 m rectangular duct. The Mach number, pressure, and temperature at a certain section of the duct are found to be 2, 75 kPa, and 5°C, respectively. Assuming the mean friction factor of 0.006, find the maximum length of duct that can be installed downstream of this section if no shock wave is to occur in the duct. Also, find the exit pressure and temperature that will exist with this maximum length of duct.
11. Air is stored in a tank at a pressure and temperature of 1.6 MPa and 20°C, respectively. What is the maximum possible mass rate of flow from the tank through a pipe with a diameter of 1.2 cm and a length of 30 cm? The pipe discharges to the atmosphere and the atmospheric pressure is 101 kPa. The average friction factor can be assumed to be 0.006 and the flow in the pipe can be assumed to be subsonic and adiabatic.
12. Air flows through a 12-m-long pipe which has a diameter of 25 mm. At the inlet to the pipe, the air velocity is 80 m/s, the pressure is 350 kPa and the temperature is 50°C. If the mean friction factor is 0.005, find the velocity, pressure, and temperature at the end of the pipe. Assume the flow to be adiabatic.
13. Air is expanded from a large reservoir in which the pressure and temperature are 200 kPa and 30°C, respectively, through a convergent nozzle that gives an exit Mach number of 0.2. The air then flows down a pipe with a diameter of 25 mm, the Mach number at the end of this pipe being 0.8. Assuming that the flow in the nozzle is isentropic, and the flow in the pipe adiabatic, find the length of the pipe and the pressure at the exit of the pipe. The friction factor in the pipe can be assumed to be 0.005.
14. Air flows through a 4-cm-diameter pipe. At the inlet to the pipe the stagnation pressure is 150 kPa, the stagnation temperature is 80°C, and the velocity is 120 m/s. If the mean friction factor is 0.006, and if the flow can be assumed to be adiabatic, find the maximum duct length before choking occurs.

15. Air flows through a 0.5-in.-diameter pipe at subsonic velocities. The pipe is 20 ft long and the pressure and temperature at the inlet to the pipe are 60 psia and 130°F, respectively. The pipe is discharged into a large vessel in which the pressure is kept at 20 psia. If the mean friction factor is assumed to be 0.0055 and if the flow is assumed to be adiabatic, find the mass flow rate through the pipe.
16. Air flows through a circular pipe at a rate of 8.3 kg/s. The Mach number at the inlet to the pipe is 0.15 and at the exit to the pipe is 0.5. The pressure and temperature at the inlet are 350 kPa and 38°C, respectively. Assuming the flow to be adiabatic, and the mean friction factor to be 0.005, find the length and the diameter of the duct and the pressure and temperature at the exit of the duct.
17. Air is expanded from a large reservoir, in which the pressure and temperature are 250 kPa and 30°C, respectively, through a convergent nozzle that gives an exit Mach number of 0.3. The air from the nozzle flows down a pipe having a diameter of 5 cm. The Mach number at the end of this pipe is 0.95. Find the length of the pipe and the pressure at the end of the pipe. If the actual pipe length was only 0.75 of this length, find the Mach number and the pressure that would exist at the end of the pipe. The flow in the nozzle can be assumed to be isentropic and the friction factor in the pipe can be assumed to be 0.005.
18. Air flows at a steady rate at a subsonic velocity through a pipe with an internal diameter of 26 mm and a length of 15 m. The pressure and temperature in the air at the inlet to the pipe are 140 kPa and 120°C, respectively. Assuming that the flow is adiabatic and using an average friction factor for the flow of 0.005, find the maximum possible mass flow rate through the pipe. Also, find the temperature and pressure at the exit of the pipe when the Mach number at the exit is equal to 1.
19. Air flows down a 20-mm-diameter pipe that has a length of 0.8 m. If the velocity at the inlet to the pipe is 200 m/s and its temperature is 30°C, find the average friction factor if the flow is choked at the exit to the pipe. Assume the flow to be adiabatic.
20. Air enters an insulated pipe with a diameter of 7.5 cm at Mach 3.0. As a result of friction, the Mach diameter decreases to a value of 1.5 at the exit to the pipe. If the mean friction factor is equal to 0.002, find the length of the pipe.
21. A convergent–divergent nozzle supplies air to a well-insulated constant area duct. At the inlet to the duct the Mach number is 2, the pressure is 140 kPa, and the temperature is −100°C. If the Mach number is 1 at the exit to the duct, determine the pressure and temperature at the duct exit.
22. Air flows down a constant area pipe that has a diameter of 5 cm. The Mach number at the inlet to the pipe is 2 and the inlet pressure and temperature are 80 kPa and 20°C, respectively. The flow in the pipe can be assumed to be adiabatic. If the pipe is 0.6 m long and the average friction factor is 0.005, find the Mach number, pressure, and temperature at the exit of the pipe. If, on leaving the pipe, the air flows through a convergent–divergent nozzle, which has an exit area that is three times the throat area, and if the air stream leaves the nozzle at a subsonic velocity, find the pressure and the Mach number at the exit of the nozzle if the flow in the nozzle can be assumed to be isentropic.
23. Air with a stagnation pressure of 600 kPa and a stagnation temperature of 150°C flows through a convergent–divergent nozzle, the Mach number being greater than 1 at the nozzle exit. The throat area of the nozzle is 1 cm^2. The flow from the nozzle enters a duct that has a constant area of 3 cm^2. If the flow in the nozzle can

be assumed to be isentropic and if the flow in the duct can be assumed to be adiabatic and if the mean friction factor is 0.004, find the temperature and the pressure on the exit plane of the duct.

24. Air enters a pipe having a diameter of 0.1 m and a length of 1 m with Mach 2 and a pressure of 90 kPa. Assuming the flow to be adiabatic and the mean friction factor to be 0.005, plot a graph of the pressure variation along the length of the duct.
25. An air stream enters a 2.5-cm-diameter pipe with Mach 2.5 and a pressure and temperature of 30 kPa and −15°C, respectively. The average friction factor can be assumed to be 0.005. Determine the maximum possible length of tube if there are to be no shock waves in the flow. Also, find the values of the pressure and the temperature at the tube exit for this maximum length. Assume the flow to be adiabatic.
26. Air flows from a large reservoir in which the pressure and temperature are 1 MPa and 30°C, respectively, through a convergent–divergent nozzle and into a constant area duct. The ratio of the nozzle exit area to its throat area is 3.0 and the length–diameter ratio of the duct is 15. Assuming that the flow in the nozzle is isentropic, that the flow in the duct is adiabatic, and that the average friction factor is 0.005, find the back-pressure for a normal shock to appear at the exit to the duct.
27. Air enters a pipe at Mach 2.5, a temperature of 40°C, and a pressure of 70 kPa. The pipe has a diameter of 2.0 cm, and the flow can be assumed to be adiabatic. A shock occurs in the pipe at a location where the Mach number is 2. If the Mach number at the exit from the pipe is 0.8 and if the average friction factor is 0.005, find the distance of the shock from the entrance to the pipe and the total length of the pipe. Also, find the pressure at the exit of the pipe.
28. Air with a stagnation pressure of 700 kPa flows through a convergent–divergent nozzle with an exit–throat area ratio of 3. The flow in this nozzle can be assumed to be isentropic. The air from the nozzle enters a well-insulated duct with a length–diameter ratio of 20. The mean friction factor is 0.002. The air from the duct is discharged into a large reservoir in which the pressure is 100 kPa. Find the Mach numbers at the inlet and exit of the duct.
29. Air with a stagnation pressure of 300 kPa and a stagnation temperature of 30°C enters a constant area duct at Mach number of 3. The duct has a length–diameter ratio of 60. The flow can be assumed to be adiabatic and the average friction factor is 0.0025. If the pressure at the exit to the duct is 50 kPa, determine the Mach number at the exit of the duct and the location of the shock down the duct in diameters. (Hint: Apply an iterative solution using guessed values of the shock wave position.)
30. Air flows at a steady rate through a 0.08-m-diameter × 1.5-m-long pipe. A convergent–divergent nozzle expands the air to Mach 2.25 and a pressure of 40 kPa at the inlet to the pipe. The air from the pipe is discharged into a large chamber in which the pressure can be varied. Assuming a friction factor of 0.003, find the pressure in this chamber if
    a. There are no shock or expansion waves in the flow
    b. There is a normal shock wave on the exit plane of the pipe
31. Air is expanded from a large reservoir in which the pressure and temperature are 300 kPa and 20°C, respectively, through a convergent–divergent nozzle, which

gives an exit Mach number of 1.5. The air from the nozzle flows down a 5-cm-diameter pipe to a reservoir in which the pressure is 140 kPa. A normal shock wave occurs at the end of the pipe. Find the length of the pipe. Discuss what will occur if the pressure in the reservoir at the discharge end of the pipe is decreased. The flow in the nozzle can be assumed to be isentropic and that in the pipe can be assumed to be adiabatic. The friction factor in the pipe can be assumed to be 0.002.

32. Air at an initial temperature of 45°C is to be transported through a 50-m-long, well-insulated pipe. If the mean friction factor in the pipe can be assumed to be 0.025, find the minimum pipe diameter that can be used to carry the flow without choking occurring if the inlet air velocity is 50, 100, and 400 m/s.
33. Air flows from a supersonic nozzle with a throat diameter of 6.5 cm into a 13-cm-diameter pipe. The stagnation pressure at the inlet to the pipe is 700 kPa. At distances of 5 and 33 diameters from the inlet to the pipe, the static pressures are measured and found to be 24.5 and 50 kPa, respectively. Determine the Mach numbers at these two sections and the mean friction factor between the two sections. The flow in the nozzle can be assumed to be isentropic and the flow in the pipe can be assumed to be adiabatic.
34. An apparatus that is used for determining friction factors with air flow through a pipe consists of a reservoir connected to a convergent–divergent nozzle, which in turn is connected to the pipe. The nozzle has a throat diameter of 0.6 cm, and the diameter of the pipe, which is well insulated, is 0.9 cm. In one experiment, the pressure and temperature in the reservoir are 1.7 MPa and 40°C, respectively, and the pressure at a distance of 0.15 m from the inlet to the pipe is 340 kPa. Calculate the average friction factor in the pipe. Assume that the flow in the nozzle is isentropic.
35. Air is expanded from a large reservoir in which the pressure and temperature are 600 kPa and 30°C, respectively, through a convergent–divergent nozzle, which gives an exit Mach number of 2.0. The air from the nozzle flows down a 3.5 cm diameter pipe.
    a. If the Mach number at the end of this pipe is 1.2, find the length of the pipe.
    b. If the back-pressure is changed until a normal shock wave occurs half way between the nozzle exit plane and the exit plane of the pipe, find the back-pressure. The flow in the nozzle can be assumed to be isentropic and the friction factor in the pipe can be assumed to be 0.005.
36. Air enters a linearly converging duct with a circular cross section at Mach 0.6. The inlet diameter of the duct is 10 cm and the wall makes an angle of 10° to the axis of the duct. Using numerical integration, produce a plot of Mach number along the duct up to the point where the Mach number reaches a value of 1. The mean friction factor can be assumed to be equal to 0.005 and the flow can be assumed to be adiabatic.
37. A high speed wind tunnel is supplied with air from a collection of interconnected compressed air tanks situated outside of the laboratory. The air is delivered from the tanks to the tunnel through a long 100-cm-diameter pipe. The pressure at the inlet to this supply pipe is 10 MPa and the air is to be delivered to the tunnel at a pressure of 1 MPa. How long can this supply pipe be if choking is not to occur? Assume adiabatic subsonic one-dimensional flow and a mean friction factor of 0.005.

38. Air flows down an adiabatic constant area circular pipe that has a diameter of 1.5 cm. At the inlet to the pipe, the Mach number is 0.16. Determine the pipe lengths that would give exit Mach numbers of 0.56 and 1. Assume that the friction factor, f, is 0.004. If the stagnation conditions at the inlet to the pipe are maintained at the same values, find the percentage change in the flow rate through the pipe for the two cases considered if the length of the pipe is increased by 1 m.
39. Air flows through the circular duct system shown in Figure P9.39. Find the mass flow rate through this system. Assume that the flow is adiabatic, and therefore that Fanno flow exists in the 20-m-long constant diameter section, the friction factor, f, being 0.008. Also, assume that the flows through the convergent sections at the inlet and exit of the duct are isentropic. Use an iterative solution procedure in which the duct exit pressures with various values of the discharge Mach number, M_3, are calculated and the value of M_3 that give the specified exit plane pressure is deduced.
40. Air flows through a variable area circular duct. The flow can be assumed to be adiabatic. The Mach number, temperature, and pressure at the inlet to the duct are 0.4, 450 K, and 550 kPa, respectively. The duct has an increasing cross-sectional area designed to ensure that despite the effects of friction the air temperature remains constant along the duct. If the distance between the inlet and the outlet of the duct is equal to 125 times the inlet duct diameter and if the friction factor is assumed to be 0.005, find the Mach number and pressure at the exit of the duct and the exit diameter–inlet diameter ratio.
41. Air flows from a large chamber in which the pressure and temperature are maintained at 725 kPa and 725 K, respectively, through a convergent–divergent nozzle and then into a constant diameter insulated pipe that in turn discharges into another large chamber in which the pressure is kept constant. The ratio of the nozzle exit area to the throat area of the nozzle is 2.4. The pipe has a diameter of 1.25 cm and a length of 30 cm. A friction factor of 0.005 can be assumed for the pipe flow. Find
    a. The mass flow rate through the system if the pressure in the reservoir into which the flow is discharged is kept at 0 kPa.
    b. The pressure in the reservoir into which the flow is discharged if there is a normal shock wave on the nozzle exit plane.
    c. The maximum pressure in the reservoir into which the flow is discharged at which choked flow will exist in the nozzle–pipe system.

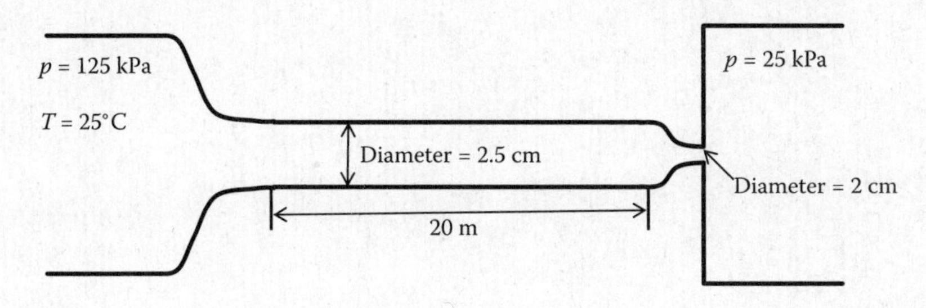

FIGURE P9.39
Flow situation considered.

42. Air flows from a large chamber in which the pressure and temperature are maintained at 100 psia and 500°R, respectively, through a convergent–divergent nozzle and then into a constant diameter insulated pipe that in turn discharges into another large chamber in which the pressure is kept constant. The ratio of the nozzle exit area to the throat area of the nozzle is 3. The pipe has a diameter of 1 ft and a length of that is equal to 15 times its diameter. A friction factor of 0.005 can be assumed for the pipe flow. The pipe discharges into another large chamber in which the pressure, the back pressure, is kept constant. A normal shock wave occurs in the pipe at a distance equal to 3 times the pipe diameter downstream of the nozzle exit. Assuming isentropic flow in the nozzle and Fanno flow in the pipe, determine the back-pressure. Sketch the process that occurs in this flow system on a Fanno line. Also, sketch the pressure variation along the length of the flow system.

10

Flow with Heat Transfer

Introduction

Most of the previous chapters in this book have been devoted to a discussion of flows in which the effects of viscosity could be neglected over most of the flow field and to flows that could be assumed to be adiabatic. However, the effects of viscosity and heat transfer can be of dominant importance in some flow situations. In this chapter, consideration will be given to some illustrative situations involving compressible flows in which heat transfer has a significant effect. This heat transfer to the flow may be the result of heat transfer from the walls over which the fluid is flowing or it may be the result of chemical reactions in the flow.

This chapter is basically broken down into three sections. The first section is predominantly concerned with external flows, i.e., flows over the surface of a body, and deals with so-called aerodynamic heating and with the factors that affect the heat transfer rate in such situations. Only a brief introduction to the topic is given. The second section is concerned with heat transfer effects in internal flows of the type dealt with in the previous chapter except that here the flows considered are not adiabatic. The third section deals very briefly with two types of shock wave in which heat generation or release plays a very important role.

Aerodynamic Heating

When a gas flows over a surface, the gas in contact with the surface is brought to rest as a result of viscosity. A rise in temperature is associated with this decrease in velocity at a surface. If the gas velocity over the surface is high, this temperature rise, associated with the slowing of the flow near a surface, can become quite large. This, basically, is what is referred to as the "aerodynamic heating" of a surface. The phenomena is illustrated in Figure 10.1.

The velocity is zero at the surface because of the action of viscosity, and the temperature rise across the boundary layer is the result of the work done on the flow by the viscous forces, i.e., the temperature rise is produced by the dissipation of kinetic energy into heat as the result of the work done by the viscous forces. The temperature rise, i.e., the aerodynamic heating, is therefore said to be the result of "viscous dissipation."

Aerodynamic heating is particularly important in hypersonic flows. However, in such flows, as will be discussed in Chapter 12, changes in the specific heats and the chemical nature of the gas can occur due to the very high temperatures arising in such flows. Further, because the temperature rises that occur at the surface are so high in hypersonic

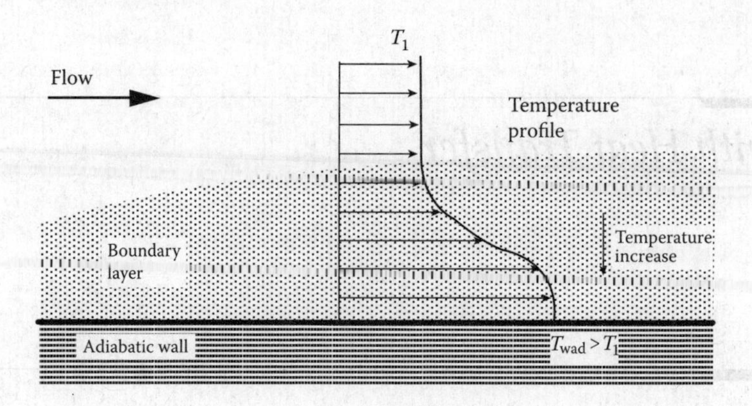

FIGURE 10.1
Temperature rise near the surface of a body.

flow, radiation heat transfer can become important. For these reasons, the analysis of aerodynamic heating in hypersonic flow can be very complex. Attention in this chapter therefore will be restricted to flows at high subsonic and at supersonic velocities. Radiation effects will be ignored in this chapter.

The Adiabatic Surface Temperature

Consider flow over a nearly flat surface at a Mach number M as shown in Figure 10.2. It is assumed that the surface is adiabatic, i.e., that there is no heat transfer to or from the surface. In this case, the surface temperature is denoted by T_{wad}. Consider two points A and B as shown in Figure 10.2.

Point A is in the freestream outside the boundary layer, whereas point B is on the surface. If the flow between points A and B is adiabatic, then, since M at point B on the surface of the plate is zero, it follows as shown before using the energy equation that

$$\frac{T_{wad}}{T_\infty} = 1 + \frac{\gamma - 1}{2} M_\infty^2 = \frac{T_0}{T_\infty}$$

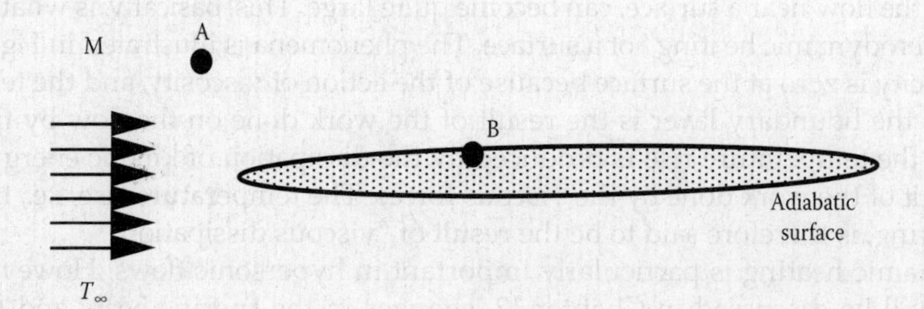

FIGURE 10.2
Flow situation considered.

where, as before, T_0 is the stagnation temperature. However, the actual process between A and B is not adiabatic. This is because as the temperature rises as the plate is approached, there is a heat transfer from the hotter gas near the surface toward the colder gas in the freestream. This is shown in Figure 10.3.

As a result of this heat transfer, the surface temperature will be lower than T_0. For this reason, it is usual to define a recovery factor, r, such that

$$\frac{T_{\text{wad}} - T_\infty}{T_0 - T_\infty} = r \tag{10.1}$$

r is thus a measure of the fraction of the adiabatic freestream to wall temperature rise that is actually "recovered" at the wall, it being defined as the ratio of the actual rise in the temperature of the gas across the boundary layer to the maximum possible rise in temperature that could occur. The value of r for a given geometrical flow situation, according to dimensionless analysis, is in general a function of the Reynolds number, the Mach number, and the Prandtl number, i.e.,

$$r = \text{function}(Re, M, Pr) \tag{10.2}$$

The Prandtl number, Pr, is a property of the fluid involved and in general varies with the fluid temperature. However, for air at moderate temperatures, the Prandtl number is approximately constant and equal to 0.7.

Experimental and analytical studies indicate that for flow over a near flat surface, the Reynolds number only affects the value of the recovery factor by determining whether the flow in the boundary layer is laminar or turbulent and that the Mach number has a negligible effect on the recovery factor. This is illustrated by the results shown in Figures 10.4 and 10.5.

Hence, these studies indicate that Equation 10.2 reduces to

$$r = \text{function}(Pr) \tag{10.3}$$

the function being different in laminar and turbulent boundary flow. Experimental and analytical studies indicate that the function is such that

$$\text{For laminar flow:} \quad r = Pr^{1/2}$$
$$\text{For turbulent flow:} \quad r = Pr^{1/3}$$

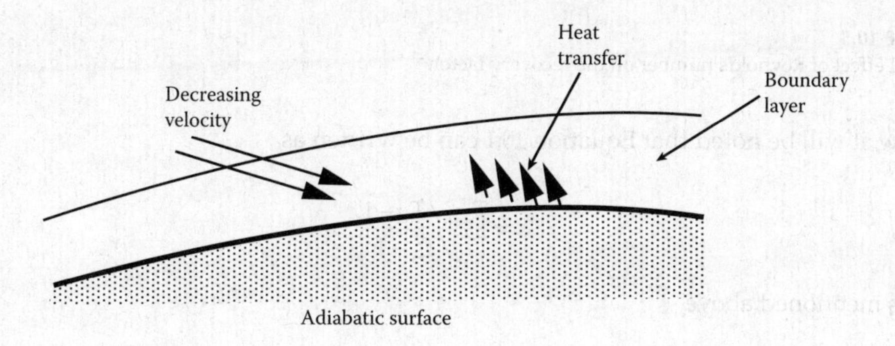

FIGURE 10.3
Heat transfer in boundary layer away from the surface.

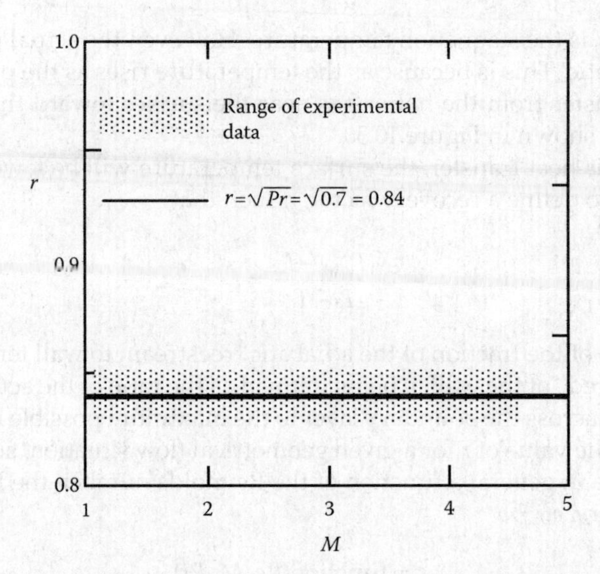

FIGURE 10.4
Typical effect of Mach number on the recovery factor for a laminar boundary layer.

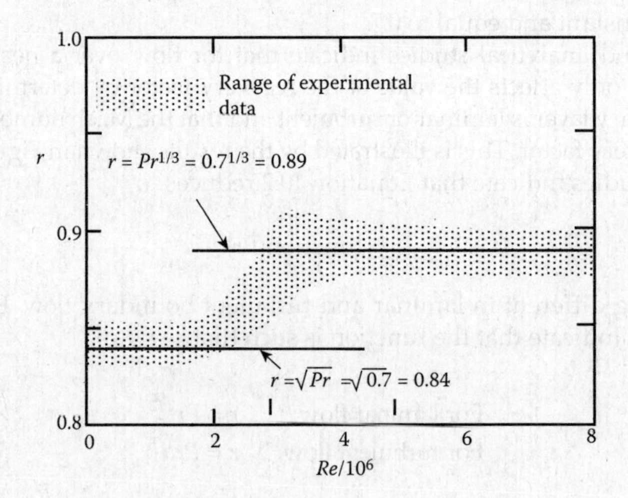

FIGURE 10.5
Typical effect of Reynolds number on the recovery factor.

Now, it will be noted that Equation 10.1 can be written as

$$r=\frac{T_{\text{wad}}/T_{\infty}-1}{T_0/T_{\infty}-1} \tag{10.4}$$

but as mentioned above

$$\frac{T_0}{T_{\infty}}=1+\frac{\gamma-1}{2}M^2 \tag{10.5}$$

Hence, Equation 10.4 gives

$$r = \frac{\dfrac{T_{wad}}{T_\infty} - 1}{\left(\dfrac{\gamma - 1}{2}\right) M^2} \tag{10.6}$$

which can be rearranged to give

$$\frac{T_{wad}}{T_\infty} = 1 + r\frac{\gamma - 1}{2} M^2 \tag{10.7}$$

where M is the Mach number in the freestream. This equation, i.e., Equation 10.7, allows the adiabatic surface temperature for a near flat surface to be determined.

Example 10.1

Consider the flow of air that has a freestream temperature of 0°C over an adiabatic flat plate. If the flow in the boundary layer can be assumed to be turbulent, determine how the temperature of the plate surface varies with Mach number.

Solution

The situation being considered is shown in Figure E10.1a.

Since the boundary layer flow is turbulent and assuming that for air $Pr = 0.7$, it follows that here

$$r = 0.7^{1/3} = 0.89$$

In the situation being considered, it then follows, assuming that for air $\gamma = 1.4$, that:

$$\frac{T_{wad}}{T_\infty} = 1 + 0.89 \times \frac{1.4 - 1}{2} M^2 = 1 + 0.178\, M^2$$

However, here, $T_\infty = 0°C = 273$ K, so the above equation gives

$$T_{wad} = 273(1 + 0.178\, M^2)$$

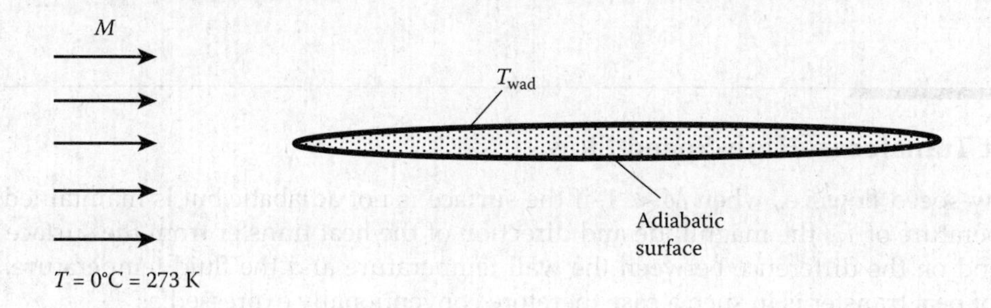

FIGURE E10.1a
Flow situation considered.

TABLE E10.1

Variation of Adiabatic Wall Temperature with Mach Number

M	T_{wad} (K)	T_{wad} (°C)
1	322	49
2	467	194
3	710	437
4	1051	778

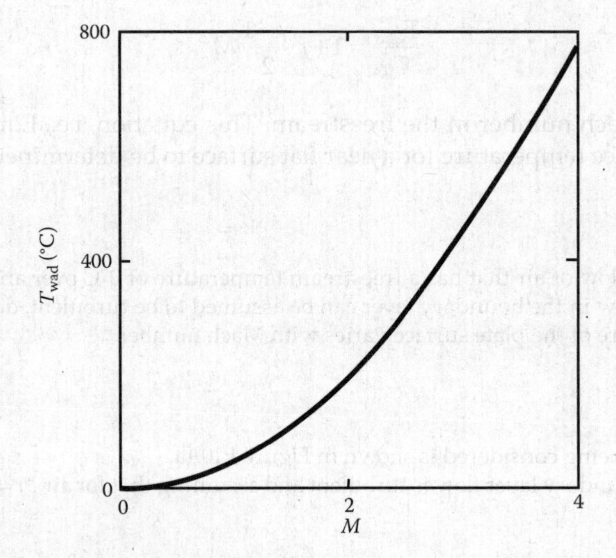

FIGURE E10.1b
Variation of adiabatic wall temperature with Mach number.

Using this equation, the values of T_{wad} listed in Table E10.1 can be obtained. These results are plotted in Figure E10.1b.

It will be seen therefore that high surface temperatures can exist at Mach numbers above ~2. These surface temperatures may, in fact, be so high that special high-temperature materials such as titanium alloys rather than conventional aluminum alloys have to be used for structural components in aircraft designed for high Mach number flight.

Heat Transfer in High-Speed Flow

In low-speed flow, i.e., when $M \ll 1$, if the surface is not adiabatic but is maintained at a temperature of T_w, the magnitude and direction of the heat transfer from the surface will depend on the difference between the wall temperature and the fluid temperature. The rate of heat transfer is in such a case therefore conventionally expressed as

$$Q = hA(T_w - T_f) \tag{10.8}$$

where T_f is the fluid temperature, A is the surface area, and h is the heat transfer coefficient. The fluid temperature T_f is usually taken as the temperature in the freestream outside the boundary layer. In high-speed flow, however, it is to be expected that the magnitude and direction of the heat transfer at the surface will depend on the difference between the wall temperature and the adiabatic wall temperature, i.e., if T_w is less than T_{wad}, there will be heat transfer from the fluid to the surface, whereas if T_w is greater than T_{wad}, there will be heat transfer from surface to the fluid. Hence, it is usual in high-speed gas flows to write

$$Q = hA(T_w - T_{wad}) \tag{10.9}$$

In low Mach number flows, T_{wad} will, according to Equation 10.7, effectively be equal to T_f, and in this case, Equation 10.9 reduces to Equation 10.8.

In high-speed flow, the heat transfer rate is given by Equation 10.9. Therefore, to find the heat transfer rate, the value of h, the heat transfer coefficient, has to be determined. Now, equations for the heat transfer coefficient are usually expressed in dimensionless form. For this purpose, the Nusselt number, Nu, is introduced, this being defined by

$$Nu = \frac{hL}{k} \tag{10.10}$$

Here, L is some appropriate measure of the size of body being considered. For example, in the case of a wide flat plate, L would be the length of the plate in the flow direction. k is the thermal conductivity of the fluid, which is, in general, temperature dependent.

Now, in low-speed flow, i.e., flow in which viscous dissipation effects are negligible, it can be shown that

$$Nu = \text{function}(Re, Pr) \tag{10.11}$$

Here, Re is the Reynolds number and Pr is, as before, the Prandtl number. Re is given by

$$Re = \frac{VL}{\nu} = \frac{\rho VL}{\mu} \tag{10.12}$$

V being the freestream velocity and ν being the kinematic viscosity, which is equal to μ/ρ, with μ being the dynamic viscosity and ρ being the density. The fluid properties in the dimensionless numbers Nu, Re, and Pr, e.g., the thermal conductivity k and the kinematic viscosity ν, are evaluated at the average of the surface and the fluid temperatures.

In high-speed gas flows, i.e., flows in which there are significant viscous dissipation effects, it would be expected that

$$Nu = \text{function}(Re, Pr, M) \tag{10.13}$$

However, it has been found using both experimental and numerical results that provided the gas properties are evaluated at a suitable mean temperature, the direct effect of the Mach number in Equation 10.13 can be neglected. The Mach number will have an indirect effect because it will, in general, influence the mean temperature used to find the gas properties. It has further been found that, provided the correct mean temperature is used to find the gas properties, the same equation that gives the Nusselt number in low-speed flow can be used in high-speed flow. For example, consider boundary layer flow over a flat plate. Experimental and numerical studies have indicated that the following approximate equations for the Nusselt number apply in low-speed flow.

For a laminar boundary layer flows,

$$Nu_L = 0.664\ Re_L^{1/2}\ Pr^{1/3} \tag{10.14}$$

For a turbulent boundary layer flows,

$$Nu_L = 0.037\ Re_L^{0.8}\ Pr^{1/3} \tag{10.15}$$

where Nu_L is the Nusselt number based on the length of the plate, L, i.e., is equal to $h\,L/k$, and Re_L is the Reynolds number based on the length of the plate L. Transition from laminar to turbulent flow occurs at approximately $Re_L = 10^6$. Theoretical and experimental results have shown that these equations can also be used in high-speed flow provided that the fluid properties are evaluated at the following temperature

$$T_{prop} = T_f + 0.5(T_w - T_f) + 0.22(T_{wad} - T_f) \tag{10.16}$$

It will be noted that in low-speed flow where, $T_{wad} = T_f$, Equation 10.16 gives

$$T_{prop} = 0.5(T_w - T_f)$$

Thus, in low-speed flow, the temperature at which the properties are evaluated is, as mentioned before, the average of the wall and fluid temperatures. This average temperature is often called the "mean film temperature" and has long been used in the computation of convective heat transfer rates in low-speed flows.

Example 10.2

Air at a temperature of 0°C flows at a velocity of 600 m/s over a wide flat plate that has a length of 1 m. The pressure in the flow is 1 atm. The flow situation is therefore as shown in Figure E10.2. Find the wall temperature if the plate is adiabatic. Also, find the heat transfer rate from surface per unit span of the plate if the plate surface is maintained at a uniform temperature, T_w, of 60°C.

Solution

In the freestream,

$$a = \sqrt{\gamma RT} = \sqrt{1.4 \times 287 \times 273} = 332\ \text{m/s}$$

Hence,

$$M = \frac{V}{a} = \frac{600}{332} = 1.81$$

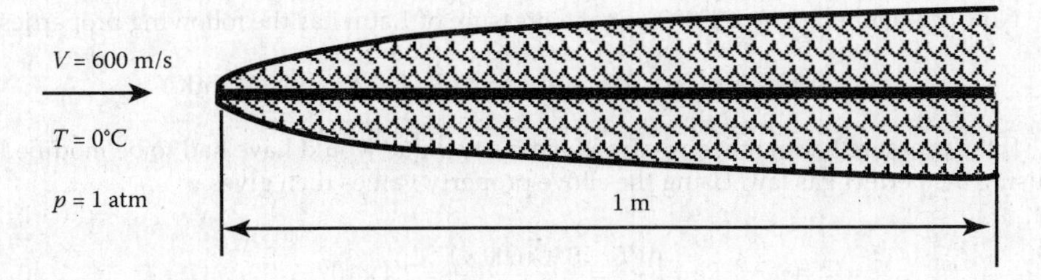

FIGURE E10.2
Flow situation considered.

Using this value then gives

$$\frac{T_{wad}}{T_\infty} = 1 + r\frac{\gamma - 1}{2}M^2 = 1 + r\frac{1.4 - 1}{2} \times 1.81^2$$

To use this equation, the value of r must be determined. To find r, it is necessary to know whether the boundary layer flow is laminar or turbulent, which requires that the value of the Reynolds number be known. To find the Reynolds number, the surface temperature has to be known. However, to find the wall temperature, the recovery factor has to be known and its value is different in laminar and turbulent flow. Therefore, an assumption as to the nature of the flow, i.e., laminar or turbulent, will be made. The wall temperature and then the gas properties will be found and then the Reynolds number, i.e.,

$$Re = \frac{\rho V L}{\mu}$$

will be evaluated and the initial assumption about the nature of the flow can then be checked.

Here, it will initially be assumed that the flow in the boundary layer is turbulent since experience suggests that this is very likely to be a correct assumption. Since the flow is assumed turbulent, it follows that, since the Prandtl number of air can be assumed to be equal to 0.7,

$$r = Pr^{1/3} = 0.7^{1/3} = 0.89$$

Using this value and assuming that $\gamma = 1.4$ then gives

$$\frac{T_{wad}}{273} = 1 + 0.89 \times \frac{1.4 - 1}{2} \times 1.81^2$$

from which it follows that

$$T_{wad} = 432 \text{ K} = 159°\text{C}$$

The air properties are found at the following temperature:

$$T_{prop} = T_f + 0.5(T_w - T_f) + 0.22(T_{wad} - T_f) = 0 + 0.5 \times (159 - 0) + 0.22 \times (159 - 0) = 114°\text{C}$$

Now at a temperature of 114°C, air at a pressure of 1 atm has the following properties

$$\rho = 0.9 \text{ kg/m}^3 \quad \mu = 225 \times 10^{-7} \text{ Ns/m}^2 \quad k = 33 \times 10^{-3} \text{ W/mK}$$

If the pressure had not been 1 atm, the density value would have had to be modified using the perfect gas law. Using the above property values then gives

$$Re = \frac{\rho VL}{\mu} = \frac{0.9 \times 600 \times 1}{225 \times 10^{-7}} = 2.4 \times 10^7$$

At this Reynolds number, the flow in the boundary layer will, as assumed, be turbulent so the assumed recovery factor is, in fact, the correct value. Hence, the adiabatic wall temperature is 159°C.

When the wall is maintained at a temperature of 60°C, the air properties are found at the following temperature

$$T_{prop} = T_f + 0.5(T_w - T_f) + 0.22(T_{wad} - T_f) = 0 + 0.5 \times (60 - 0) + 0.22 \times (159 - 0) = 65°C$$

Now at a temperature of 65°C, air at a pressure of 1 atm has the following properties:

$$\rho = 0.99 \text{ kg/m}^3 \quad \mu = 208 \times 10^{-7} \text{ Ns/m}^2 \quad k = 30 \times 10^{-3} \text{ W/mK}$$

In this case then,

$$Re = \frac{\rho VL}{\mu} = \frac{0.99 \times 600 \times 1}{208 \times 10^{-7}} = 4.3 \times 10^7$$

The Nusselt number is therefore given by

$$Nu_L = 0.037\, Re_L^{0.8}\, Pr^{1/3} = 0.037 \times (4.3 \times 10^7)^{0.8} \times 0.7^{1/3} = 41{,}980$$

From the definition of the Nusselt number, it then follows that

$$\frac{hL}{k} = 41{,}980$$

and therefore that

$$h = \frac{41{,}980 \times 30 \times 10^{-3}}{1} = 1259 \text{ W/m}^2\text{K}$$

Therefore, if there is heat transfer from both sides of the plate, the total heat transfer rate from the plate is given by

$$Q = h\,A(T_w - T_{wad}) = 1259 \times (2 \times 1 \times 1) \times (60 - 159) = -250{,}000 \text{ W}$$

The negative sign means that heat is transferred to the plate. Hence, the net rate of heat transfer to the plate is 250 kW.

The discussion of heat transfer in high-speed flow given above was concerned with flow over a flat or near-flat surface. The equations presented above can give acceptable results for preliminary design purposes in certain situations; for example, for the flow over the blades in turbomachines outside the area of the stagnation point and for the flow over the wings and fuselage of an aircraft outside the area of the stagnation point. This is illustrated in Figure 10.6.

Related approaches can be used in other flow situations but the value of the recovery factor may be different from that which applies to flat plate flows and the equation for the Nusselt number will normally be different from that for flat plate flow. For example, consider the heat transfer rate near the stagnation point of a body placed in a supersonic gas flow. This situation is shown in Figure 10.7. As shown in Figure 10.7, an effectively normal shock forms ahead of the stagnation point.

The flow through this shock is adiabatic, and the flow downstream of this shock is subsonic. The flow outside the boundary layer in the stagnation point region will therefore be near the stagnation temperature and the increase in temperature across this boundary layer due to viscous dissipation will be small with the result that the heat transfer rate across the boundary layer resulting from this temperature increase will be small. This means that if the wall in the stagnation point region is adiabatic, it will effectively be at the stagnation temperature. Hence, for flow in the stagnation point region, the recovery factor is effectively equal to 1. This means that if the wall is not adiabatic but is kept at a temperature T_w, the heat transfer rate per unit area, q, in the stagnation point region should be written as

$$q = h(T_w - T_0)$$

It has been found that the value of h can be obtained by using the equations for low-speed flow near a stagnation point and because the flow outside the boundary layer is subsonic in this stagnation point region, with the gas properties being evaluated at the average of the wall and the temperature behind the normal shock wave. If the boundary layer near the stagnation point is two-dimensional and can be assumed to be laminar, the heat transfer coefficient in this region is approximately given using the following equation:

$$Nu_D = 1.14\, Re_D^{0.5}\, Pr^{0.4} \tag{10.17}$$

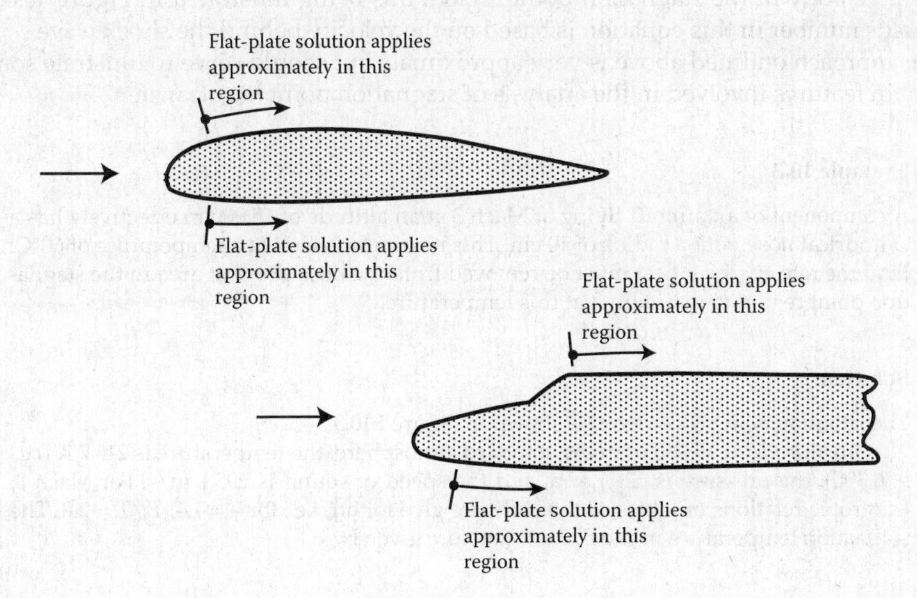

FIGURE 10.6
Applicability of flat plate equations.

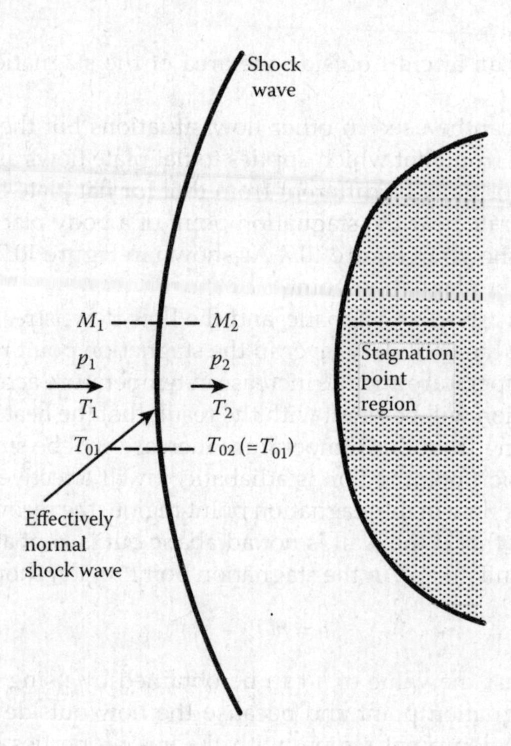

FIGURE 10.7
Stagnation point region in supersonic flow over a blunt-nosed body.

Here Nu_D and Re_D are the Nusselt and Reynolds numbers based on the effective diameter of the body in the stagnation point region, this being illustrated in Figure 10.8. The Reynolds number in this equation is based on the velocity behind the shock wave.

The approach outlined above is very approximate but should serve to illustrate some of the main features involved in the analysis of stagnation point heat transfer.

Example 10.3

A component of an aircraft flying at Mach 3 at an altitude of 15,000 m effectively has a cylindrical nose with a radius of 10 cm. This nose area is kept at a temperature of 60°C. Find the rate at which heat must be removed from the nose per unit area in the stagnation point region to maintain it at this temperature.

Solution

The situation being considered is shown in Figure E10.3.

At an altitude of 15,000 m in the standard atmosphere, the temperature is 216.7 K (i.e., −56.3°C), the pressure is 12.11 kPa, and the speed of sound is 295.1 m/s. For Mach 3, isentropic relations or tables or the software give for air, i.e., for $\gamma = 1.4$, $T_0/T = 2.8$. The stagnation temperature in the flow is therefore given by

$$T_0 = \frac{T_0}{T}T = 2.8 \times 216.7 = 606.8\ \text{K} = 333.8°\text{C}$$

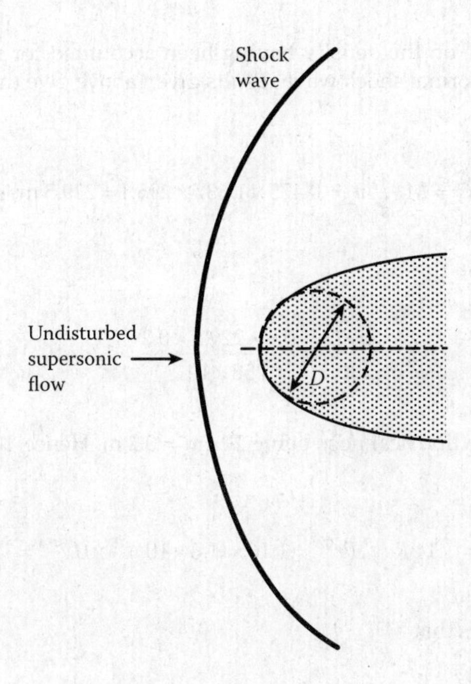

FIGURE 10.8
Stagnation point diameter.

As discussed above, the air properties for the calculation of the heat transfer rate will be obtained at a temperature of $(T_0 + T_w)/2 = (333.8 + 60)/2 = 196.9°C$.

Now normal shock relations or tables or the software give for Mach 3, $p_2/p_1 = 10.33$, $a_2/a_1 = 1.637$, and $M_2 = 0.475$. Hence, the pressure in the boundary layer will be given by

$$p_2 = \frac{p_2}{p_1} p_1 = 10.33 \times 12.11 = 125.1 \text{ kPa}$$

Therefore, the air properties will be obtained at a temperature of 196.9°C and a pressure of 125.1 kPa. At this temperature and pressure, air has the following properties:

$$\rho = 0.90 \text{ kg/m}^3 \quad \mu = 258 \times 10^{-7} \text{ Ns/m}^2 \quad k = 38 \times 10^{-3} \text{ W/mK}$$

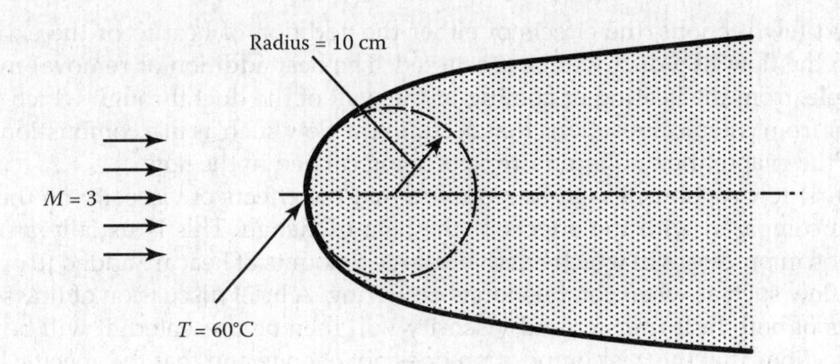

FIGURE E10.3
Flow situation considered.

the effect of pressure on the density having been accounted for using the perfect gas law. Also, since the normal shock wave results given above give the velocity behind the shock wave as

$$V_2 = M_2 \frac{a_2}{a_1} a_1 = 0.475 \times 1.637 \times 295.1 - 229.5 \text{ m/s}$$

it follows that

$$Re_D = \frac{\rho V D}{\mu} = \frac{0.9 \times 229.5 \times 0.2}{258 \times 10^{-7}} = 1.6 \times 10^6$$

the diameter of the cylindrical nose being 20 cm = 0.2 m. Hence, the Nusselt number is given by

$$Nu_D = 1.14\, Re_D^{0.5}\, Pr^{0.4} = 1.16 \times (1.6 \times 10^6)^{0.5} \times 0.7^{0.4} = 1250.3$$

From this, it follows that

$$\frac{hD}{k} = 1250.3; \quad \text{thus, } h = \frac{1250.3 \times 38 \times 10^{-3}}{0.2} = 237.6 \text{ W/m}^2\text{K}$$

Hence,

$$q = h(T_W - T_0) = 237.6 \times (60 - 333.8) = -65,043 \text{ W/m}^2$$

The negative sign means that heat is transferred to the nose. Hence, the net rate of heat transfer to the nose per unit area near the stagnation point is 65.0 kW/m².

Internal Flows with Heat Addition or Removal

In the next few sections, the effects of either the addition of heat to or the extraction of heat from the flow in a duct will be discussed. The heat addition or removal may result, for example, from the heating or cooling of the wall of the duct through which the gas is flowing or from chemical reactions that occur in the flow such as in a combustion chamber or due to the evaporation of liquid droplets being carried in the flow.

In the first few sections, it will be assumed that the effects of viscosity on the flow are negligible compared with the effects of the heat exchange. This is usually an adequate assumption in processes in which relatively large amounts of heat are added to or removed from the flow such as when combustion is occurring. A brief discussion of flows in which the effects of both heat transfer and viscosity will then be presented. It will be assumed in these sections that the gas composition does not change and that the specific heat ratio, γ therefore does not change. This assumption may not be adequate in some combustion systems.

One-Dimensional Flow in a Constant Area Duct Neglecting Viscosity

The effects of heat exchange on one-dimensional flow in which the effects of viscosity are negligible through a constant area duct will first be examined. Consider the flow through the control volume shown in Figure 10.9.

The conservation of mass gives,

$$\rho AV = \text{constant}$$

However, in the present case, since flow in a constant area duct is being considered, this reduces, as in the previous chapter, to

$$\rho V = \text{constant}$$

Hence, for the portion of the duct being considered

$$\rho V = (\rho + d\rho)(V + dV)$$

From this, it follows that to first order of accuracy, i.e., by assuming that the term $d\rho \times dV$ is negligible because $d\rho$ and dV are very small

$$\rho\, dV + V\, d\rho = 0$$

Dividing this equation through by ρV then gives

$$\frac{dV}{V} + \frac{d\rho}{\rho} = 0 \tag{10.18}$$

Similarly, the conservation of momentum applied to the control volume shown in Figure 10.9, gives

$$pA - (p + dp)A = \rho VA(V + dV - V)$$

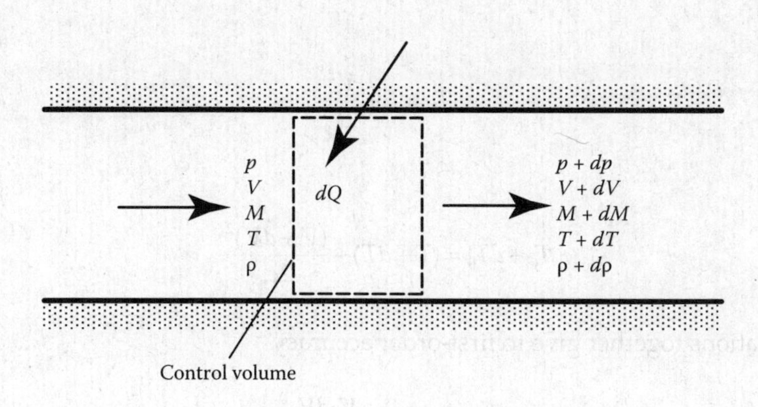

FIGURE 10.9
Control volume used in analysis of flow through a duct with heat exchange.

i.e.,

$$dp + \rho V\, dV = 0 \tag{10.19}$$

Lastly, the conservation of energy gives, since no work is being done on the system,

$$dQ = \dot{m}\left[c_p(T + dT) + \frac{(V + dV)^2}{2} - c_pT - \frac{V^2}{2}\right] \tag{10.20}$$

It should be noted that heat exchange is, thermodynamically, strictly the energy transfer that results from temperature differences. In combustion systems, it would be more correct to write the energy equation as

$$\dot{m}\Delta h_f = \dot{m}\left[c_p(T + dT) + \frac{(V + dV)^2}{2} - c_pT - \frac{V^2}{2}\right]$$

where Δh_f is the difference between the heats of formation of the reactants and the products. However, this can be quite adequately dealt with by treating $\dot{m}\Delta h_f$ as the "heat produced as the result of the combustion," and there is then no reasons in most analyses to consider the physical origin of the "heat."

It is convenient to define dq as the rate of heat addition per unit mass flow rate, i.e., to define

$$dq = \frac{dQ}{\dot{m}}$$

The conservation of energy equation, Equation 10.20, can then be written as

$$dq = c_p dT + V\, dV \tag{10.21}$$

Alternatively, consider the definition of the stagnation temperature, T_0, given earlier, i.e.,

$$T_0 = T + \frac{V^2}{2c_p}$$

or

$$T_0 + dT_0 = (T + dT) + \frac{(V + dV)^2}{2c_p}$$

These equations together give to first-order accuracy

$$dT_0 = dT + \frac{V\, dV}{c_p}$$

Hence, it will be seen that Equation 10.21 can be written as

$$dq = c_p dT_0 \tag{10.22}$$

The above equations can now be used to relate the changes in the flow variables to the rate of heat addition. Equation 10.21 gives

$$\frac{dq}{dV} = c_p \frac{dT}{dV} + V \tag{10.23}$$

However, the perfect gas law gives

$$\frac{p}{\rho} = RT, \quad \text{i.e.,} \quad p = \rho RT$$

and

$$\frac{p + dp}{\rho + d\rho} = R(T + dT), \quad \text{i.e.,} \quad p + dp = (\rho + d\rho)R(T + dT)$$

which can be used to give

$$\frac{dp}{p} = \frac{d\rho}{\rho} + \frac{dT}{T} \tag{10.24}$$

However, Equation 10.18 gives

$$\frac{d\rho}{\rho} = -\frac{dV}{V}$$

and Equation 10.19 gives

$$-\frac{dp}{p} = -\frac{\rho V\, dV}{p}$$

so Equation 10.24 can be written as

$$-\frac{\rho V\, dV}{p} = -\frac{dV}{V} + \frac{dT}{T}$$

Noting that the perfect gas law gives $p/\rho = RT$, this equation can be written as

$$-\frac{V\, dV}{RT} = -\frac{dV}{V} + \frac{dT}{T}$$

i.e.,

$$\frac{dT}{dV} = \frac{T}{V} - \frac{V}{R} \tag{10.25}$$

Substituting this into Equation 10.23 then gives

$$\frac{dq}{dV} = c_p \frac{T}{V} - V\left(\frac{c_p}{R} - 1\right)$$

Hence, since $R = c_p - c_v$ so that, as shown before, $R/c_p = (\gamma - 1)/\gamma$, it follows that

$$\frac{dq}{dV} = c_p \frac{T}{V} - \frac{V}{\gamma - 1} \tag{10.26}$$

The first term in this equation decreases with increasing V, whereas the second term increases with increasing V. At very low velocities, the first term dominates, and therefore, since T and V are positive, dq/dV is positive, i.e., dV has the same sign as dq. This means that at low velocities, adding heat increases the velocity while removing heat decreases the velocity. At high velocities, the second term in Equation 10.26 will dominate and dq/dV will be negative. Thus, at high velocities, dV has the opposite sign to dq. This means that at high velocities, adding heat decreases the velocity while removing heat increases the velocity. Transition from one form of behavior (i.e., dq/dV being positive) to the other (i.e., dq/dV being negative) will occur when dq/dV is zero. From Equation 10.26, it follows that this will occur when

$$c_p \frac{T}{V} = \frac{V}{\gamma - 1}$$

i.e., when

$$V = \sqrt{(\gamma - 1)c_p T} = \sqrt{\gamma R T} = a \tag{10.27}$$

This shows that when $M < 1$, dq/dV is positive and when $M > 1$, dq/dV is negative. When $M = 1$, dq/dV is zero. These results are summarized in Table 10.1.

Now it will be recalled that since $V = Ma$ and $V + dV = (M + dM)(a + da) = Ma + M\,da + a\,dM$, it follows that

$$\frac{dV}{V} = \frac{da}{a} + \frac{dM}{M}$$

Therefore, since as shown above,

$$-\frac{\rho V\,dV}{p} = -\frac{dV}{V} + \frac{dT}{T}$$

TABLE 10.1

Changes in Flow Variables Produced by Heat Exchange

Mach No.	Heat Addition	Heat Removal
$M < 1$	V increases	V decreases
$M > 1$	V decreases	V increases

which can be written as

$$-\frac{\gamma V\,dV}{a^2}=-\frac{dV}{V}+\frac{dT}{T}$$

i.e.,

$$(1-\gamma M^2)\frac{dV}{V}=\frac{dT}{T}$$

it follows that

$$\frac{(1+\gamma M^2)}{2}\frac{dV}{V}=\frac{dM}{M}$$

This shows that dM has the same sign as dV. This means that adding heat will tend to move M toward 1, whereas removing the heat will tend to move M away from 1.

It should also be noted that Equation 10.25 shows that dT/dV will be positive at low velocities and negative at high velocities. dT/dV will be zero when $T/V = V/R$, i.e., when $V^2 = RT$, i.e., when

$$M=\frac{1}{\sqrt{\gamma}} \tag{10.28}$$

This shows that the maximum temperature in a flow will exist where $M=1/\sqrt{\gamma}$.

Entropy changes associated with the heat addition or removal will next be considered. Since, as discussed elsewhere,

$$\frac{ds}{c_p}=\frac{dT}{T}-\frac{(\gamma-1)}{\gamma}\frac{dp}{p} \tag{10.29}$$

it follows using Equations 10.24 and 10.18 that

$$\frac{ds}{c_p}=\frac{1}{\gamma}\frac{dT}{T}+\frac{(\gamma-1)}{\gamma}\frac{dV}{V} \tag{10.30}$$

Substituting Equation 10.25 into this equation and rearranging then gives

$$\frac{ds}{c_p}=(1-M^2)\frac{dV}{V} \tag{10.31}$$

However, as shown above, dq/dV is positive if $M < 1$ and negative if $M > 1$. Equation 10.31 therefore shows that heat addition always increases the entropy, whereas heat removal always decreases the entropy. The addition of heat to a flow thus moves the entropy toward a maximum at $M = 1$. This is indicated in Table 10.2.

TABLE 10.2

Changes in Entropy Produced by Heat Exchange

Mach No.	Heat Addition	Heat Removal
$M < 1$	s increases	s decreases
$M > 1$	s increases	s decreases

Example 10.4

Air flowing through a constant area duct is being heated. The air temperature at a certain section of the duct is 200°C. If, over a short section of the duct, the stagnation temperature increases by 1%, estimate the percentage increases in V and M and the value of ds/c_p for Mach numbers, at the section considered, of 0.4, 0.8, 1.2, and 1.6.

Solution

Equation 10.22 gives

$$\frac{dq}{c_p T_0} = \frac{dT_0}{T_0}$$

However, in the present case,

$$\frac{dT_0}{T_0} = 0.01$$

so $dq/c_p T_0 = 0.01$.

Equation 10.26 gives

$$\frac{dq/(c_p T_0)}{dV/V} = \frac{T}{T_0} - \frac{V^2}{(\gamma - 1)c_p T_0}$$

i.e.,

$$\frac{0.01}{dV/V} = \frac{T}{T_0} - \frac{V^2}{0.4 \times 1007 T_0}$$

i.e.,

$$\frac{dV}{V} = 0.01 \Big/ \left(\frac{T}{T_0} - \frac{V^2}{402.8 T_0} \right)$$

However,

$$\frac{T_0}{T} = 1 + \left(\frac{\gamma - 1}{2} \right) M^2 = 1 + 0.2M^2$$

so

$$T_0 = 473(1 + 0.2M^2)$$

TABLE E10.2

Percentage Changes in Velocity and Mach Number Produced by a 1% Increase in Stagnation Temperature

M	dV/V (%)	dM/M (%)	ds/c_p
0.4	1.23	0.75	0.63
0.8	3.12	2.96	1.06
1.2	−2.95	−4.45	1.96
1.6	−1.17	−2.42	3.02

Also,

$$V = M\sqrt{\gamma RT} = M\sqrt{1.4 \times 287 \times 473} = 435.9M \text{ m/s}$$

Hence, for any value of M, the values of V, T_0/T, and T_0 can be found. The value of dV/V can then be calculated. The fractional change in the Mach number is then found using

$$\frac{dM}{M} = \frac{(1+\gamma M^2)}{2}\frac{dV}{V} = \frac{(1+1.4M^2)}{2}\frac{dV}{V}$$

whereas the entropy change is found using Equation 10.31, i.e., using

$$\frac{ds}{c_p} = (1-M^2)\frac{dV}{V}$$

Results obtained using this procedure for the specified values of Mach number are given in Table E10.2.

It is again convenient to express the changes that occur in the flow in terms of the Mach number. Consider two points in the flow as shown in Figure 10.10.

The conservation of momentum equation applied to the control volume shown gives

$$p_1 + \rho_1 V_1^2 = p_2 + \rho_2 V_2^2$$

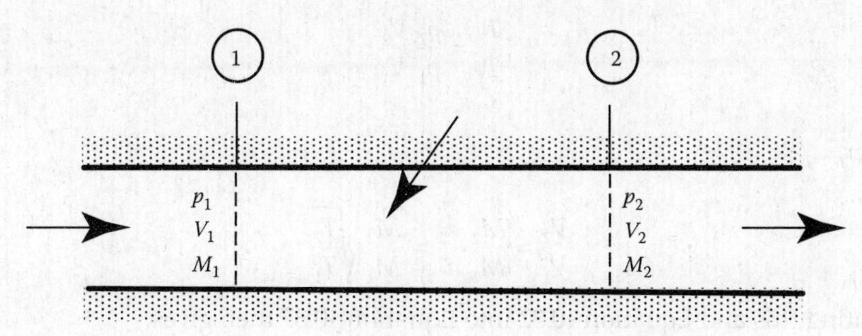

FIGURE 10.10
Points in flow considered.

Since $a^2 = \gamma p/\rho$, this equation can be written as

$$p_1(1+\gamma M_1^2) = p_2(1+\gamma M_2^2)$$

i.e.,

$$\frac{p_2}{p_1} = \frac{1+\gamma M_1^2}{1+\gamma M_2^2} \tag{10.32}$$

Now by definition, the stagnation pressure is given by

$$\frac{p_0}{p} = \left(1+\frac{\gamma-1}{2}M^2\right)^{\frac{\gamma}{\gamma-1}} \tag{10.33}$$

When there is heat addition or removal, the stagnation pressure will change. From Equations 10.32 and 10.33, it follows that

$$\frac{p_{02}}{p_{01}} = \frac{p_2}{p_1}\frac{[1+(\gamma-1)M_2^2/2]^{\gamma/\gamma-1}}{[1+(\gamma-1)M_1^2/2]^{\gamma/\gamma-1}}$$

i.e.,

$$\frac{p_{02}}{p_{01}} = \frac{[1+\gamma M_1^2]}{[1+\gamma M_2^2]}\frac{[1+(\gamma-1)M_2^2/2]^{\gamma/\gamma-1}}{[1+(\gamma-1)M_1^2/2]^{\gamma/\gamma-1}} \tag{10.34}$$

Next, it is noted that the continuity equation gives

$$\frac{\rho_2}{\rho_1} = \frac{V_1}{V_2} \tag{10.35}$$

whereas the perfect gas law gives

$$\frac{\rho_2}{\rho_1} = \frac{p_2}{p_1}\frac{T_1}{T_2} \tag{10.36}$$

These two equations together then give

$$\frac{T_2}{T_1} = \frac{p_2}{p_1}\frac{V_2}{V_1} \tag{10.37}$$

However,

$$\frac{V_2}{V_1} = \frac{M_2}{M_1}\frac{a_2}{a_1} = \frac{M_2}{M_1}\sqrt{\frac{T_2}{T_1}} \tag{10.38}$$

Substituting this and Equation 10.32 into Equation 10.37 then gives

$$\frac{T_2}{T_1} = \frac{M_2^2(1+\gamma M_1^2)^2}{M_1^2(1+\gamma M_2^2)^2} \tag{10.39}$$

Substituting this back into Equation 10.38 then gives

$$\frac{V_2}{V_1} = \frac{M_2^2(1+\gamma M_1^2)}{M_1^2(1+\gamma M_2^2)} \tag{10.40}$$

Also, substituting Equations 10.40 and 10.32 into Equation 10.36 gives

$$\frac{\rho_2}{\rho_1} = \frac{M_1^2(1+\gamma M_2^2)}{M_2^2(1+\gamma M_1^2)} \tag{10.41}$$

Lastly, it will be recalled that the stagnation temperature is given by

$$\frac{T_0}{T} = 1 + \frac{(\gamma-1)}{2}M^2$$

from which it follows that

$$\frac{T_{02}}{T_{01}} = \frac{T_2}{T_1}\frac{[1+(\gamma-1)M_2^2/2]}{[1+(\gamma-1)M_1^2/2]}$$

Substituting Equation 10.39 into this equation then gives

$$\frac{T_{02}}{T_{01}} = \frac{M_2^2}{M_1^2}\frac{(1+\gamma M_1^2)^2}{(1+\gamma M_2^2)^2}\frac{[1+(\gamma-1)M_2^2/2]}{[1+(\gamma-1)M_1^2/2]} \tag{10.42}$$

It should be noted that Equation 10.22 shows that the change in T_0 always has the same sign as q, i.e., adding heat always increases T_0, whereas removing heat always decreases T_0. Further, it was shown earlier that adding heat always moves M toward 1, whereas removing heat always moves M away from 1. Therefore, T_0 must have a maximum value when $M = 1$. The variation of T_0 with M is therefore as illustrated in Figure 10.11.

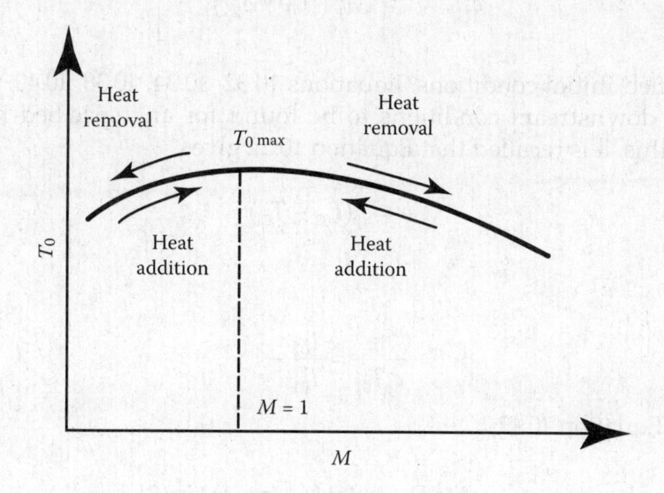

FIGURE 10.11
Variation of stagnation temperature with Mach number.

For any value of M_1, the right-hand side of Equation 10.42 has a maximum when M_2 = 1, i.e.,

$$\frac{T_{0\max}}{T_{01}} = \frac{1}{2(1+\gamma)M_1^2} \frac{(1+\gamma M_1^2)^2}{[1+(\gamma-1)M_1^2/2]} \tag{10.43}$$

Lastly, the relation between the entropy change and Mach number will be determined. Since

$$s_2 - s_1 = c_p \ln\frac{T_2}{T_1} - R\ln\frac{p_2}{p_1}$$

i.e.,

$$\frac{s_2 - s_1}{c_p} = \ln\frac{T_2}{T_1} - \frac{(\gamma-1)}{\gamma}\ln\frac{p_2}{p_1}$$

i.e.,

$$\frac{s_2 - s_1}{c_p} = \ln\left[\frac{T_2}{T_1}\left(\frac{p_1}{p_2}\right)^{\frac{(\gamma-1)}{\gamma}}\right]$$

it follows, using Equations 10.38 and 10.32, that

$$\frac{s_2 - s_1}{c_p} = \ln\left[\frac{M_2^2(1+\gamma M_1^2)^2}{M_1^2(1+\gamma M_2^2)^2}\left(\frac{1+\gamma M_2^2}{1+\gamma M_1^2}\right)^{\frac{(\gamma-1)}{\gamma}}\right]$$

i.e.,

$$\frac{s_2 - s_1}{c_p} = \ln\left[\frac{M_2^2}{M_1^2}\left(\frac{1+\gamma M_1^2}{1+\gamma M_2^2}\right)^{\frac{(\gamma+1)}{\gamma}}\right] \tag{10.44}$$

For any specified initial conditions, Equations 10.32, 10.34, 10.38, 10.40, 10.41, 10.42, and 10.44 allow the downstream conditions to be found for any specified heat addition or removal. To do this, it is recalled that Equation 10.22 gives

$$q = c_p(T_{02} - T_{01})$$

i.e.,

$$\frac{q}{c_p T_{01}} = \frac{T_{02}}{T_{01}} - 1 \tag{10.45}$$

Hence, using Equation 10.42

$$\frac{q}{c_p T_{01}} = \frac{M_2^2}{M_1^2}\frac{[1+\gamma M_1^2]}{[1+\gamma M_2^2]}\frac{[1+(\gamma-1)M_2^2/2]}{[1+(\gamma-1)M_1^2/2]} - 1 \tag{10.46}$$

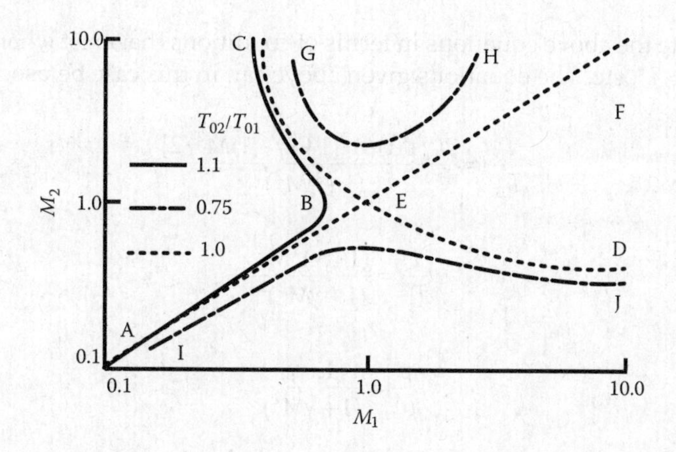

FIGURE 10.12
Variation of M_2 with M_1.

For any specified values of M_1, and q/c_pT_{01}, this equation allows M_2 to be determined. The equations given above then allow the changes in all other flow variables to be found. Since q/c_pT_{01} is by virtue of Equation 10.45, a unique function of T_{02}/T_{01}, it is usual to use the stagnation temperature ratio as the measure of the heat exchange. There is then a relationship among M_2, M_1, and T_{02}/T_{01}. The form of this relationship is illustrated in Figure 10.12.

When $T_{02}/T_{01} = 1$, the flow is adiabatic, and in subsonic flow, in this case, $M_2 = M_1$. However, when the flow is supersonic, i.e., $M_1 > 1$, there are two possible values of M_2, i.e., either $M_2 = M_1$ or $M_2 < 1$. These, of course, correspond to the possibility of no normal shock or of a normal shock occurring. Furthermore, for $M_1 < 1$ and $T_{02}/T_{01} > 1$, i.e., for heat addition, it is possible to obtain a solution that involves $M_2 > 1$. Such solutions are, for reasons discussed above, not physically possible.

The variation of entropy change with M_1 as given by Equations 10.44 and 10.46 is shown in Figure 10.13. The curves in this figure are lettered to correspond to those in Figure 10.12. As discussed above, if heat is added to the flow, the Mach number tends toward 1, whereas if heat is extracted from the flow, the Mach number moves away from 1. It is convenient

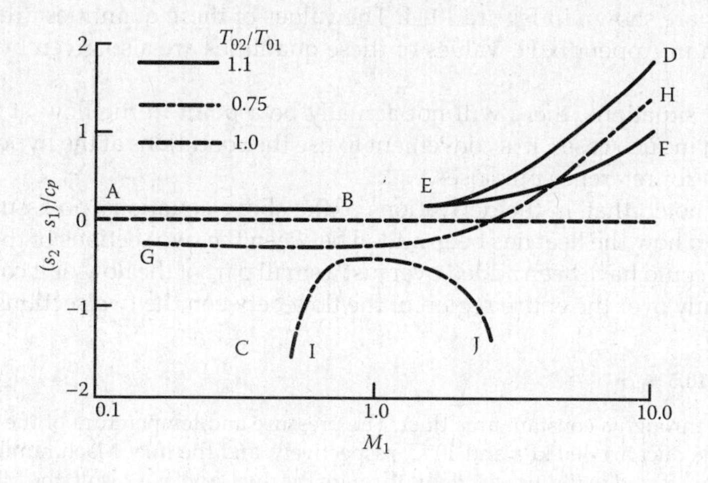

FIGURE 10.13
Variation of entropy change with M_1.

therefore to write the above equations in terms of conditions that exist when $M_2 = 1$. In this case, $p_2 = p^*$, $T_2 = T^*$, etc. The equations given above can in this case be used to give

$$\frac{T_0}{T_0^*} = \frac{2(\gamma+1)M^2[1+(\gamma-1)M^2/2]}{(1+\gamma M^2)^2} \tag{10.47}$$

$$\frac{T}{T^*} = \frac{(1+\gamma)^2 M^2}{(1+\gamma M^2)^2} \tag{10.48}$$

$$\frac{p}{p^*} = \frac{(1+\gamma)}{(1+\gamma M^2)} \tag{10.49}$$

$$\frac{p_0}{p_0^*} = \left(\frac{1+\gamma}{1+\gamma M^2}\right)\left\{\left(\frac{2}{\gamma+1}\right)\left[1+\frac{(\gamma-1)}{2}M^2\right]\right\}^{\frac{\gamma}{\gamma-1}} \tag{10.50}$$

$$\frac{V}{V^*} = \frac{(1+\gamma)M^2}{(1+\gamma M^2)} \tag{10.51}$$

$$\frac{\rho}{\rho^*} = \frac{(1+\gamma M^2)}{(1+\gamma)M^2} \tag{10.52}$$

$$\frac{s-s^*}{c_p} = \ln\left[M^2\left(\frac{1+\gamma}{1+\gamma M^2}\right)^{\frac{\gamma+1}{\gamma}}\right] \tag{10.53}$$

Thus, the values of the quantities T_0/T_0^*, T/T^*, etc. can be determined for any value of M. The variations are shown in Figure 10.14. The values of these quantities are also listed in the table given in Appendix H. Values of these quantities are also given by the software COMPROP.

In most real situations, there will not actually be a point in the flow at which $M = 1$. However, even in such cases, it is convenient to use the conditions at the hypothetical point at which $M = 1$ for reference purposes.

It should be noted that, in the derivations of the above equations, no assumptions have been made as to how the heat has been added between the two sections of the flow considered. The heat could have been added over just a small part of the flow or it could have been added uniformly over the entire region of the flow between the two sections considered.

Example 10.5

Air flows through a constant area duct. The pressure and temperature of the air at the inlet to the duct are 100 kPa and 10°C, respectively, and the inlet Mach number is 2.8. Heat is transferred to the air as it flows through the duct, and as a result, the Mach number at the exit is 1.3. Find the pressure and temperature at the exit. If no shock waves occur in the flow, find the maximum amount of heat that can be transferred to the air

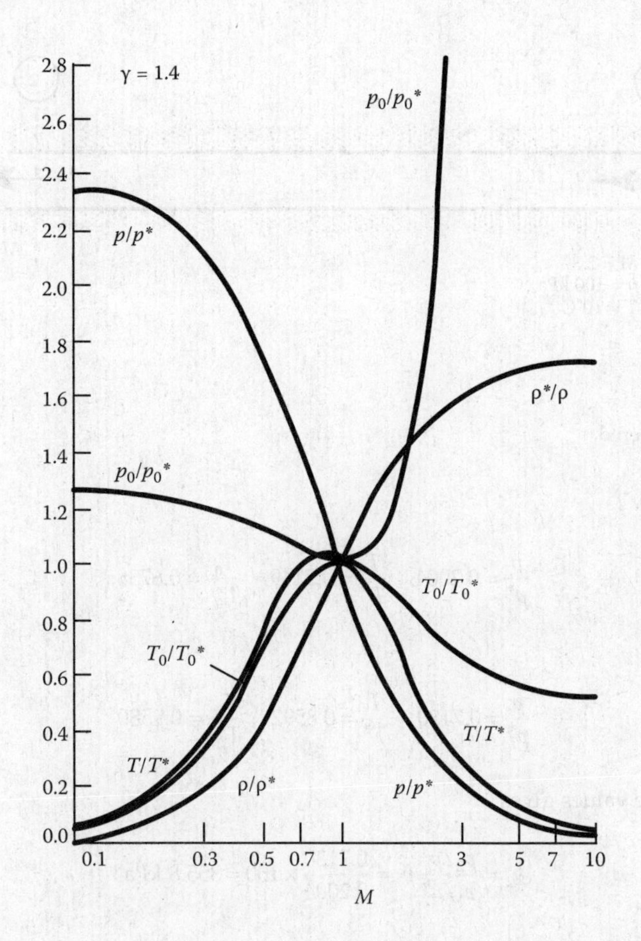

FIGURE 10.14
Variation of properties in constant area duct flow with heat exchange.

per unit mass of air. Also, find the exit pressure and temperature that would exist with this maximum heat transfer rate. Assume that the flow is steady, that the effects of wall friction can be neglected and that the air behaves as a perfect gas.

Solution

The flow situation under consideration is shown in Figure E10.5.

Now for Mach 2.8, using

$$\frac{T_0}{T} = 1 + \frac{\gamma - 1}{2}M^2 = 1 + 0.2M^2$$

or using the software or isentropic tables gives $T_{01}/T_1 = 2.568$. Hence,

$$T_{01} = 2.568 \times (273 + 10) = 726.7 \text{ K}$$

Next, using the relations given above or the tables or the software for frictionless flow in a constant area duct with heat exchange gives

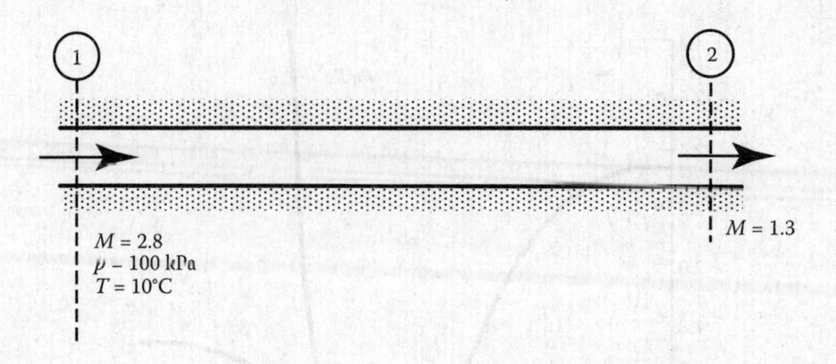

FIGURE E10.5
Flow situation considered.

For $M = 2.8$,

$$\frac{p}{p^*} = 0.2004, \quad \frac{T}{T^*} = 0.3149, \quad \frac{T_0}{T_0^*} = 0.6738$$

For $M = 1.3$,

$$\frac{p}{p^*} = 0.7130, \quad \frac{T}{T^*} = 0.8592, \quad \frac{T_0}{T_0^*} = 0.9580$$

Using these values gives

$$p_2 = \frac{p_2/p^*}{p_1/p^*} p_1 = \frac{0.7130}{0.2004} \times 100 = 355.8 \text{ kPa}$$

and

$$T_2 = \frac{T_2/T^*}{T_1/T^*} T_1 = \frac{0.8592}{0.3149} \times 283 = 772.2 \text{ K} = 499.2°\text{C}$$

Therefore, the pressure and temperature of the air at the outlet to the duct are 355.8 kPa and 499.2°C, respectively.

When the maximum amount of heat has been transferred, the Mach number at the exit is 1 and $T_{02} = T_0^*$. Hence, in this case,

$$T_{02} = \frac{T_{01}}{T_{01}/T_0^*} = \frac{726.7}{0.6738} = 1078.5 \text{ K}$$

However,

$$q = c_p(T_{02} - T_{01}) = 1.007 \times (1078.5 - 726.7) = 354.3 \text{ kJ/kg}$$

it having been assumed that $c_p = 1.007$ kJ/kg.

Also, when the maximum amount of heat has been transferred,

$$p_2 = p^* = \frac{p_1}{p_1/p^*} = \frac{100.7}{0.2004} = 499.0 \text{ kPa}$$

and

$$T_2 = T^* = \frac{T_1}{T_1/T^*} = \frac{283}{0.3149} = 898.7 \text{ K} = 625.7°\text{C}$$

Therefore, when the maximum amount of heat is transferred, the heat added is 354.3 kJ for every kilogram of air flowing through the duct, and under these circumstances, the pressure and temperature of the air at the outlet to the duct are 499.0 kPa and 625.7°C, respectively.

Example 10.6

Air flows through a constant area duct whose walls are kept at a low temperature. The air enters the pipe at Mach 0.52, a pressure of 200 kPa, and a temperature of 350°C. The rate of heat transfer from the air to the walls of pipe is estimated to be 400 kJ/kg of air. Find the Mach number, temperature, and pressure at the exit of the pipe. Assume that the flow is steady, that the effects of wall friction are negligible, and that the air behaves as a perfect gas.

Solution

For Mach 0.52, using

$$\frac{T_0}{T} = 1 + \frac{\gamma - 1}{2} M^2 = 1 + 0.2M^2$$

or using the software or isentropic tables gives $T_{01}/T_1 = 1.054$. Hence,

$$T_{01} = 1.054 \times (350 + 10) = 656.6 \text{ K}$$

Next, using the relations given above or the tables or the software for frictionless flow in a constant area duct with heat exchange gives

For $M = 0.52$,

$$\frac{p}{p^*} = 1.7414, \quad \frac{T}{T^*} = 0.8196, \quad \frac{T_0}{T_0^*} = 0.7199$$

However,

$$q = c_p(T_{02} - T_{01}) = 1.007 \times (T_{02} - 656.6)$$

it again having been assumed that $c_p = 1.007$ kJ/kg. Therefore, because $q = -400$ kJ/kg (heat is transferred from the air to the walls),

$$400 = 1.007(T_{02} - 656.6), \text{ i.e., } T_{02} = 259.4 \text{ K} = -13.6°\text{C}$$

Hence,

$$\frac{T_{02}}{T_0^*} = \frac{T_{02}}{T_{01}} \frac{T_{01}}{T_0^*} = \frac{259.4}{656.6} 0.7199 = 50.2844$$

For this value of T_{02}/T_0^*, using the relations given above or the tables or the software for frictionless flow in a constant area duct with heat exchange gives

$$M = 0.2656, \quad \frac{p}{p^*} = 2.184, \quad \frac{T}{T^*} = 0.3365$$

Hence,

$$p_2 = \frac{p_2/p^*}{p_1/p^*} p_1 = \frac{2.184}{1.741} \times 200 = 250.9 \text{ kPa}$$

and

$$T_2 = \frac{T_2/T^*}{T_1/T^*} T_1 = \frac{0.3365}{0.8196} \times 623 = 256 \text{ K} = -17°\text{C}$$

Therefore, the Mach number, pressure, and temperature of the air at the outlet to the duct are 0.2656, 250.9 kPa, and −17°C, respectively.

Example 10.7

Air enters a combustion chamber at a velocity of 80 m/s with a pressure and temperature of 180 kPa and 120°C. Find the maximum amount of heat that can be generated in the combustion chamber per unit mass of air. If the fuel has a heating value of 45 MJ/kg, find the air–fuel ratio. If the air–fuel ratio is adjusted until it is 90% of this value, find the reduction in the mass flow rate through the combustion chamber that must occur if the inlet stagnation pressure and stagnation temperature remain the same. Assume that the flow is steady, that the effects of wall friction can be neglected, that the effects of the mass of the fuel can be neglected, and that the air behaves as a perfect gas.

Solution

Using the prescribed inlet conditions gives

$$M_1 = \frac{V_1}{\sqrt{\gamma R T_1}} = \frac{80}{\sqrt{1.4 \times 287 \times 393}} = 0.1981$$

For this value of M_1, using

$$\frac{T_0}{T} = 1 + \frac{\gamma - 1}{2} M^2 = 1 + 0.2M^2$$

or using the software or isentropic tables gives $T_{01}/T_1 = 1.008$. Hence,

$$T_{01} = 1.008 \times 393 = 396.1 \text{ K}$$

Also, for this value of M_1, using the relations given above or the tables or the software for frictionless flow in a constant area duct with heat exchange gives

$$\frac{p}{p^*} = 2.2754, \quad \frac{T}{T^*} = 0.2029, \quad \frac{T_0}{T_0^*} = 0.1704$$

Now, when the maximum amount of heat is generated, the value of M_2 will be equal to 1 and $T_{02} = T_0^*$. Hence, in this case,

$$T_{02} = \frac{T_{01}}{T_{01}/T_0^*} = \frac{396.1}{0.1704} = 2324.5 \text{ K}$$

However,

$$q = c_p(T_{02} - T_{01}) = 1.007(2324.5 - 396.7) = 1942.2 \text{ kJ/kg} = 1.9422 \text{ MJ/kg}$$

it having been assumed that $c_p = 1.007$ kJ/kg · K.

Hence, the air–fuel ratio, i.e., the ratio of the mass of air to mass of fuel, when the Mach number is 1 at the exit with the specified inlet velocity is

$$AF = \frac{\text{Heat generation per unit mass of fuel}}{\text{Heat generation per unit mass of air}} = \frac{45}{1.9422} = 23.17$$

The adjusted air–fuel ratio is 0.9 × 23.17 = 20.85. The adjusted heat generation rate per unit mass of air is therefore 1942.2/0.9 = 2158 kJ/kg. Hence, since the mass flow decreases until the Mach number at exit is again 1, it follows that

$$q = c_p(T_{02} - T_{01}) = c_p(T_0^* - T_{01})$$

Hence, since the inlet stagnation temperature is the same,

$$2158 = 1.007 \times (T_0^* - 396.1)$$

This gives $T_0^* = 2539.1$ K, so $T_{01}/T_0^* = 396.1/2539.1 = 0.1560$.

This then gives, using the relations given above or the tables or the software for frictionless flow in a constant area duct with heat exchange

$$M_1 = 0.1886$$

Now the mass flow rate is given by

$$\dot{m} = \rho_1 V_1 A = \rho_0 a_0 A \frac{\rho_1}{\rho_0} \frac{a_1}{a_0} M_1$$

and isentropic relations or tables or the software give for $M = 0.198$

$$\frac{\rho}{\rho_0} = 0.980, \quad \frac{a}{a_0} = 0.996$$

and for $M = 0.1886$

$$\frac{\rho}{\rho_0} = 0.982, \quad \frac{a}{a_0} = 0.997$$

Hence, since the stagnation conditions are the same and the area is the same in the two cases

$$\frac{\text{Adjusted mass flow rate}}{\text{Original mass flow rate}} = \frac{0.982}{0.980} \times \frac{0.997}{0.996} \times \frac{0.1886}{0.198} = 0.955$$

Therefore, the mass flow rate is decreased by 4.5%.

Entropy–Temperature Relations

Traditionally, the variation of entropy with temperature has been used in examining one-dimensional flows with heat addition. To establish the relation between these quantities, it is noted that for a given flow along a duct, as discussed above

$$\frac{s - s^*}{c_p} = \ln\left[\frac{T}{T^*}\right] - \frac{(\gamma - 1)}{\gamma} \ln\left[\frac{p}{p^*}\right] \tag{10.54}$$

To establish the relation between $(s - s^*)/c_p$ and T/T^*, it is necessary to express p/p^* in terms of T/T^*. Now, from Equation 10.32, the following is obtained

$$M^2 = \frac{(1+\gamma)}{\gamma}\left[\frac{p^*}{p}\right] - \frac{1}{\gamma}$$

whereas, from Equations 10.31 and 10.32 it will be seen that

$$\frac{p}{p^*} = \frac{1}{M}\left(\frac{T}{T^*}\right)^{0.5}$$

Combining these two equations then gives

$$\left(\frac{p}{p^*}\right)^2 \left[(1+\gamma)\frac{p^*}{p} - 1\right] = \gamma \frac{T}{T^*}$$

Rearranging this equation gives

$$\left(\frac{p}{p^*}\right)^2 - (1+\gamma)\frac{p}{p^*} + \gamma \frac{T}{T^*} = 0$$

Solving this quadratic equation for p^*/p then gives

$$\frac{p}{p^*} = \frac{(1+\gamma)^2}{2} \pm \frac{\sqrt{(1+\gamma)^2 - 4\gamma(T/T^*)}}{2} \tag{10.55}$$

Substituting this into Equation 10.37 gives the required relationship as

$$\frac{s-s^*}{c_p} = \ln\left[\frac{T}{T^*}\right] - \left[\frac{\gamma-1}{\gamma}\right]\ln\left\{\frac{(1+\gamma)^2}{2} \pm \frac{\sqrt{(1+\gamma)^2 - 4\gamma(T/T^*)}}{2}\right\} \tag{10.56}$$

This allows the variation of T/T^* with $(s - s^*)/c_p$ to be found for any specified value of γ. The variation resembles that shown in Figure 10.15.

Such a curve, which basically gives the variation of T with s, is termed the "Rayleigh line," and the type of flow here being considered, i.e., one-dimensional flow with heat addition and negligible friction, is often termed "Rayleigh flow." It will be seen from Figure 10.7 that for any value of $(s - s^*)/c_p$, there are two possible values of T/T^*, one involving subsonic flow and the other involving supersonic flow, as indicated in the figure. The discussion given earlier—see the discussion of Equation 10.31—indicated that s will be a maximum when $M = 1$. The point A in Figure 10.7 is therefore the point at which $M = 1$ and at which $s = s^*$ and $T = T^*$. Since heat addition moves the Mach number toward 1, whereas heat removal moves it away from 1, the effects of q are as shown in Figure 10.15. Point B in Figure 10.15 is the point at which T is a maximum. The value of M at which this occurs can be found by differentiating Equation 10.31 to give dT/dM and then setting this equal to zero to give the value of M at which T is a maximum. The procedure gives

$$\frac{2M(1+\gamma)^2}{(1+\gamma M^2)^2} - \frac{4\gamma M^3(1+\gamma)^2}{(1+\gamma M^2)^3} = 0$$

which, on rearrangement, gives the following for the Mach number at which the temperature is a maximum

$$M = \sqrt{\frac{1}{\gamma}} \tag{10.57}$$

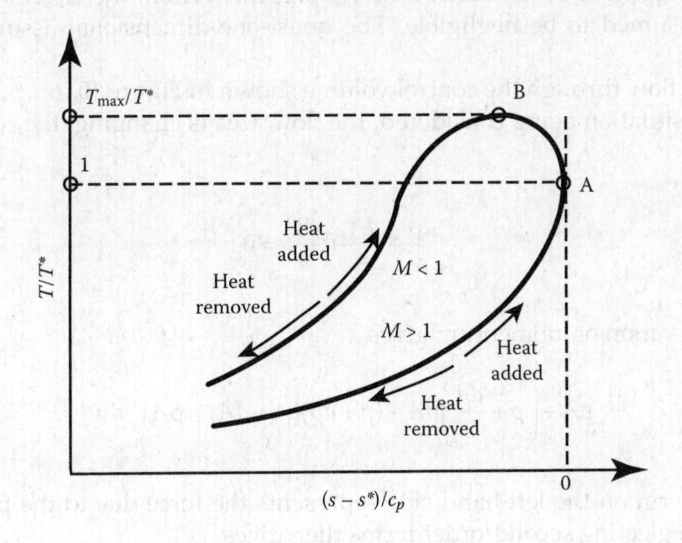

FIGURE 10.15
Rayleigh line.

TABLE 10.3

Changes in Flow Variables Produced by Heat Exchange

Mach No.	q	T_0	M	p	V	s
$M < 1$	+	+	+	–	+	+
$M > 1$	+	+	–	+	–	+
$M < 1$	–	–	–	+	–	–
$M > 1$	–	–	+	–	+	–

Note: + = quantity is increasing, – = decreasing.

This result was previously given in Equation 10.28.

Substituting Equation 10.57 back into Equation 10.48 then gives the maximum value of T as

$$\frac{T_{max}}{T^*} = \frac{(1+\gamma)^2}{4\gamma} \tag{10.58}$$

The effects of the heat addition on the flow variables in Rayleigh flow are summarized in Table 10.3.

Variable Area Flow with Heat Addition

The analysis given in the previous section assumed that the flow area, A, was constant. In some real situations, the effects of both changing flow area and heat exchange are important. This situation will be considered in the present section, the effects of wall friction again being assumed to be negligible. The quasi-one-dimensional assumption will be used.

Consider the flow through the control volume shown in Figure 10.16.

Since, in the situation being considered, the flow area is changing, the continuity equation gives,

$$\frac{d\rho}{\rho} + \frac{dA}{A} + \frac{dV}{V} = 0 \tag{10.59}$$

whereas conservation of momentum gives

$$pA + \left(p + \frac{dp}{2}\right)dA - (p+dp)(A+dA) = \rho AV\ dV$$

The second term on the left-hand side represents the force due to the pressure on the curved wall. Neglecting second-order terms then gives

$$dp + \rho V\, dV = 0 \tag{10.60}$$

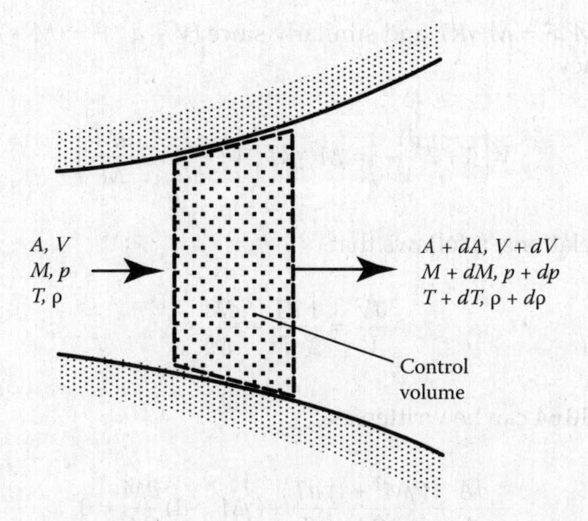

FIGURE 10.16
Control volume used in variable area analysis.

Also, as discussed before in this chapter, the equation of state gives

$$\frac{dp}{p} = \frac{d\rho}{\rho} + \frac{dT}{T} \tag{10.61}$$

whereas, as also discussed earlier in this chapter, the conservation of energy gives

$$dq = c_p dT_0 \tag{10.62}$$

where

$$dT_0 = dT + \frac{V\,dV}{c_p} \tag{10.63}$$

Equations 10.60 and 10.61 together give

$$\frac{d\rho}{\rho} + \frac{dT}{T} + \frac{\rho V\,dV}{p} = 0$$

Hence, since $a^2 = \gamma p/\rho$, it follows from this equation that

$$\frac{d\rho}{\rho} + \frac{dT}{T} + \gamma M^2 \frac{dV}{V} = 0$$

Using Equation 10.59 with this equation then gives

$$-\frac{dA}{A} + \frac{dT}{T} + (\gamma M^2 - 1)\frac{dV}{V} = 0 \tag{10.64}$$

Now since $V^2 = M^2a^2 = M^2\gamma RT$ and similarly since $(V + dV)^2 = (M + dM)^2\gamma R(T + dT)$, i.e., to first-order accuracy

$$V\left(1+2\frac{dV}{V}\right)=M^2\gamma RT\left(1+\frac{dT}{T}+2\frac{dM}{M}\right)$$

From the above relations, it follows that

$$\frac{dV}{V}=\frac{1}{2}\frac{dT}{T}+\frac{dM}{M}$$

Hence, Equation 10.64 can be written as

$$-\frac{dA}{A}+\frac{(\gamma M^2+1)}{2}\frac{dT}{T}+(\gamma M^2-1)\frac{dM}{M}=0 \quad (10.65)$$

It is next, once again, noted that

$$T_0=T\left[1+\frac{\gamma-1}{2}M^2\right]$$

and that

$$T_0+dT_0=(T+dT)+\left(\frac{\gamma-1}{2}\right)(M+dM)^2$$

so that

$$\frac{dT_0}{T}=\left[1+\frac{(\gamma-1)}{2}M^2\right]\frac{dT}{T}+(\gamma-1)M^2\frac{dM}{M}$$

which, using Equation 10.62, gives

$$\frac{dq}{c_pT}=\left[1+\frac{(\gamma-1)}{2}M^2\right]\frac{dT}{T}+(\gamma-1)M^2\frac{dM}{M} \quad (10.66)$$

This equation can then be combined with Equation 10.65 to give

$$-\frac{dA}{A}+\frac{(\gamma M^2+1)/2}{[1+(\gamma-1)M^2/2]}\frac{dq}{c_pT}+\frac{(M^2-1)}{[1+(\gamma-1)M^2/2]}\frac{dM}{M}=0 \quad (10.67)$$

Alternatively, this equation can be written as

$$-\frac{dA}{A}+\frac{(\gamma M^2-1)/2}{[1+(\gamma-1)M^2/2]}\frac{dT_0}{T}+\frac{(M^2-1)}{[1+(\gamma-1)M^2/2]}\frac{dM}{M}=0$$

The first term in these equations effectively represents the effect of the area change on the Mach number, whereas the second term in the equations effectively represents the effect of the heat exchange on the Mach number. If the heat exchange is zero, i.e., if $dq = 0$, then, as discussed in Chapter 8, when $M = 1$ and dM is nonzero, i.e., the Mach number is not a maximum or a minimum, dA must be zero, i.e., the point at which $M = 1$ in a flow in which M changes from a subsonic to a supersonic value or from a supersonic to a subsonic value corresponds to the point at which the area is a minimum. However, when there is heat addition or removal, the point at which $M = 1$ corresponds to the point at which

$$-\frac{dA}{A} + \frac{(\gamma - 1)}{[1 + (\gamma - 1)/2]} \frac{dq}{c_p T} = 0 \tag{10.68}$$

Because $\gamma > 1$, the coefficient of dq/c_pT in this equation is always positive. This means that if there is heat addition, i.e., if $dq > 0$, then dA will be positive when $M = 1$, whereas if there is heat extraction, i.e., $dq < 0$, then dA will be negative when $M = 1$. Therefore, when there is heat addition or removal, the section at which the Mach number is one does not coincide with the section of minimum flow area.

Example 10.8

Air enters a convergent duct at Mach 0.8 with a pressure of 600 kPa and a temperature of 50°C. If the exit area of the duct is 60% of that at the inlet and if the shape of the duct is such that the Mach number remains constant in the duct, find the stagnation temperature and the temperature at the exit of the duct and the amount of heat being transferred to the air per unit mass of air. Assume that the flow is steady, that the effects of wall friction are negligible, and that the air behaves as a perfect gas.

Solution

The situation being considered is shown in Figure E10.8.

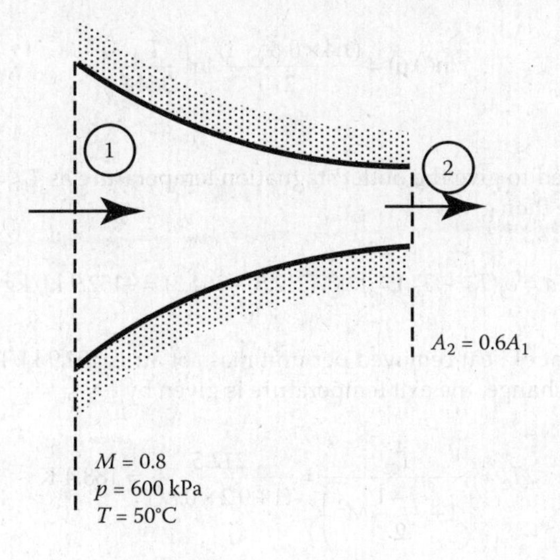

FIGURE E10.8
Flow situation considered.

Using the prescribed inlet conditions,

$$T_{01} = T_1\left(1+\frac{\gamma-1}{2}M^2\right) = 323(1+0.2\times 0.8^2) = 323\times 1.128 = 364.3 \text{ K}$$

Now since the Mach number remains constant, i.e., since $dM = 0$, Equation 10.67 gives

$$\frac{dA}{A} = \frac{(\gamma M^2+1)/2}{[1+(\gamma-1)M^2/2]}\frac{dT_0}{T}$$

but since

$$\frac{T_0}{T} = 1+\frac{(\gamma-1)}{2}M^2$$

this equation can be written as

$$\frac{dA}{A} = \frac{(\gamma M^2+1)}{2}\frac{dT_0}{T_0}$$

Because M is constant, integrating this equation between the inlet and the outlet gives

$$\ln\left(\frac{A_2}{A_1}\right) = \frac{(\gamma M^2+1)}{2}\ln\left(\frac{T_{02}}{T_{01}}\right)$$

Using the prescribed conditions then gives

$$\ln(0.6) = \frac{(1.4\times 0.8^2+1)}{2}\ln\left(\frac{T_{02}}{364.3}\right)$$

which can be solved to give the outlet stagnation temperature as $T_{02} = 212.5$ K. Using this value then gives

$$q = c_p(T_{02} - T_{01}) = 1.007(212.5 - 364.3) = -152.9 \text{ kJ/kg}$$

Thus, the amount of heat removed per unit mass of air is 152.9 kJ/kg. Since the Mach number does not change, the exit temperature is given by

$$T_2 = \frac{T_{02}}{\left(1+\frac{\gamma-1}{2}M^2\right)} = \frac{212.5}{(1+0.2\times 0.8^2)} = 188.4 \text{ K}$$

i.e., the air leaves at a temperature of 188.4 K.

The situation discussed in the above example is a very particular one and Equation 10.67 cannot, in general, be directly integrated to give the variation of M with A when dq is nonzero. However, it is quite easy to integrate this equation numerically to give this variation.

One-Dimensional Constant Area Flow with Both Heat Exchange and Friction

In some situations that are of practical importance, the effects of both heat exchange and the shear stress at the wall resulting from viscosity are important. To illustrate how such flows can be analyzed, attention will be given to one-dimensional flow in a duct of constant cross-sectional area. Consider, once again, the flow through the control volume shown in Figure 10.17.

The form of the continuity equation is, of course, the same whether heat exchange or friction is considered. In this situation, it therefore again gives

$$\frac{dV}{V} + \frac{d\rho}{\rho} = 0 \tag{10.69}$$

Similarly, since the energy balance is not affected by the friction, it again gives

$$dq = c_p dT + V\,dV = c_p dT_0 \tag{10.70}$$

Therefore, since

$$\frac{T_0}{T} = 1 + \frac{(\gamma - 1)}{2} M^2$$

it can easily be shown, as was done in the previous section, that

$$\frac{dT_0}{T} = \left[1 + \frac{\gamma - 1}{2} M^2\right] \frac{dT}{T} + (\gamma - 1) M^2 \frac{dM}{M}$$

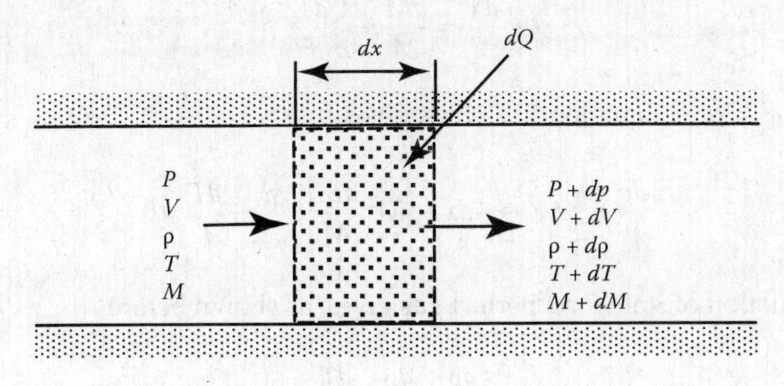

FIGURE 10.17
Control volume used in analysis of flow with heat exchange and friction.

which, when substituted into Equation 10.70, gives

$$\frac{dq}{c_p T} = \left[1 + \frac{\gamma - 1}{2} M^2\right] \frac{dT}{T} + (\gamma - 1) M^2 \frac{dM}{M} \tag{10.71}$$

Because the viscous stresses at the wall are important, the conservation of momentum equation gives for the control volume being considered

$$dp + \tau_w \frac{P}{A} dx + \rho V\, dV = 0 \tag{10.72}$$

where P is again the perimeter of the duct and A is its cross-sectional area.

The wall shear stress, τ_w, is, as in the previous chapter, expressed in terms of the dimensionless friction factor, f, as follows

$$\tau_w = f \frac{1}{2} \rho V^2 \tag{10.73}$$

and the ratio P/A is, as before, expressed in terms of the hydraulic diameter, D_H, using

$$D_H = \frac{4\ (\text{Area})}{\text{Perimeter}} = \frac{4A}{P} \tag{10.74}$$

Equation 10.72 can then be written as

$$dp + \frac{1}{2} \rho V^2 \frac{f}{D_H} dx + \rho V\, dV = 0 \tag{10.75}$$

Now as shown above—see discussion of Equation 10.65—since

$$V^2 = M^2 a^2 = M^2 \gamma R T$$

it follows that

$$\frac{dV}{V} = \frac{dM}{M} + \frac{1}{2} \frac{dT}{T} \tag{10.76}$$

Noting that $a^2 = \gamma p / \rho$, Equation 10.75 thus gives

$$\frac{dp}{p} + \frac{\gamma}{2} M^2 \frac{f}{D_H} dx + \gamma M^2 \frac{dM}{M} + \frac{\gamma M^2}{2} \frac{dT}{T} = 0 \tag{10.77}$$

Also, the equation of state for a perfect gas gives, as shown before

$$\frac{dp}{p} = \frac{d\rho}{\rho} + \frac{dT}{T}$$

This can be combined with the continuity equation, i.e., Equation 10.69, to give

$$\frac{dp}{p} = -\frac{dV}{V} + \frac{dT}{T} \tag{10.78}$$

Substituting Equation 10.76 into Equation 10.78 then gives

$$\frac{dp}{p} = -\frac{dM}{M} + \frac{1}{2}\frac{dT}{T} \tag{10.79}$$

This, in turn, can be substituted into Equation 10.77 to give

$$-\frac{dM}{M} + \frac{1}{2}\frac{dT}{T} + \frac{\gamma}{2}M^2\frac{f}{D_H}dx + \gamma M^2\frac{dM}{M} + \frac{\gamma M^2}{2}\frac{dT}{T} = 0$$

i.e.,

$$\frac{(1+\gamma M^2)}{2}\frac{dT}{T} + \frac{\gamma}{2}M^2\frac{f}{D_H}dx - (1-\gamma M^2)\frac{dM}{M} = 0 \tag{10.80}$$

Substituting the value of dT/T given by Equation 10.71 into this equation then gives

$$\left(\frac{1+\gamma M^2}{2}\right)\frac{dq}{c_pT[1+(\gamma-1)M^2/2]} + \frac{\gamma M^2}{2}\frac{f\,dx}{D_H} = \left\{(1-\gamma M^2) + \frac{(1+\gamma M^2)}{2}\frac{(\gamma-1)M^2}{[1+(\gamma-1)M^2/2]}\right\}\frac{dM}{M} \tag{10.81}$$

For specified values of dq/c_p and f/D_H, this equation together with Equations 10.80 and 10.79, can be numerically integrated to give the variations of M, T, and p along the duct.

The first term on the left-hand side of Equation 10.81 determines the relative effect of the heat exchange on the Mach number variation whereas the second term on the left-hand side determines the relative effect of the wall friction on this variation. As mentioned before, when the heat exchange is the result of combustion, the first term is usually far greater than the second term on the left-hand side, i.e., the effect of friction is negligible compared with the effect of heat exchange. It is therefore usually only in cases where the heat exchange is the result of heat transfer from the walls of the duct to the gas that both the friction term and the heat exchange term are important. Now, it is usual to express the heat transfer rate from the duct wall to the gas in terms of a heat transfer coefficient, h, defined by

$$dq_w = h\,dA_w(T_w - T) \tag{10.82}$$

where dq_w is the rate of heat transfer through the portion of the wall of surface area dA_w considered, T_w is the wall temperature at the point considered, and T is the mean gas temperature at the section considered. These quantities are defined in Figure 10.18.

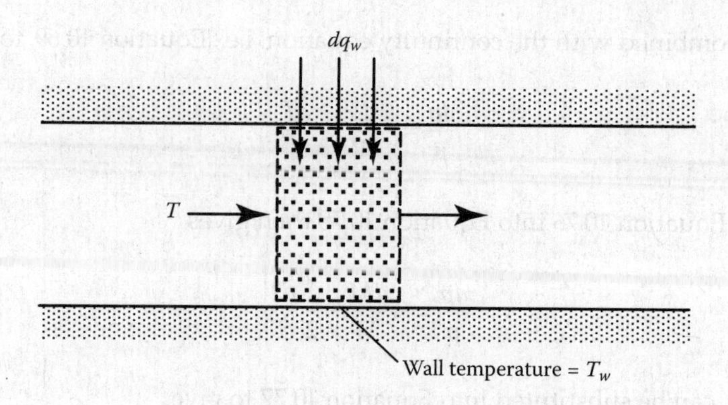

FIGURE 10.18
Wall section used in defining the heat transfer coefficient.

The gas is brought to rest at the wall so even if there is no heat transfer at the wall, i.e., if dq_w is zero, the wall temperature will be higher than T. The wall temperature in this adiabatic flow case is, as discussed earlier, termed the adiabatic wall temperature, T_{wa}. As discussed in the earlier sections of this chapter, if the process that the gas particles undergo in being brought to rest at the wall of the duct is adiabatic, the wall temperature when dq_w is zero, i.e., T_{wa}, will be the stagnation temperature T_0, which is, of course, given by

$$\frac{T_0}{T} = 1 + \frac{\gamma - 1}{2} M^2$$

However, for the reasons discussed earlier, the actual process experienced by the fluid in being brought to rest at the wall is not adiabatic and the wall temperature when dq_w is zero will be somewhat less than T_0, i.e.,

$$\frac{T_{wa} - T}{T_0 - T} = r(< 1) \tag{10.83}$$

r, as discussed before, being termed the "recovery factor." The difference between T_{wa} and T_0 is, however, comparatively small for gases that have a Prandtl number near 1, i.e., for such gases, r has a value near 1. It will therefore be assumed, here, that $T_{wa} = T_0$. In this case, the heat transfer can only be from the wall to the gas if $T_w > T_0$. If $T_w < T_0$, heat will be transferred from the gas to the wall. From this discussion it follows that the most appropriate temperature to take as the gas temperature in Equation 10.82 is T_0, i.e., to write Equation 10.82 as

$$dq_w = h\, dA_w(T_w - T_0) \tag{10.84}$$

Now, by definition

$$dq = \frac{dq_w}{\dot{m}} = \frac{dq_w}{\rho AV} \tag{10.85}$$

As shown in books on heat transfer, an analysis of the velocity and temperature fields in a flow with heat transfer indicates that there is a relationship between the heat transfer coefficient, h, and the friction factor, f. This relationship is termed the "Reynolds' analogy" and for gases that have a Prandtl number that is near 1, this analogy gives

$$h = \frac{\rho V c_p f}{8} \tag{10.86}$$

Substituting Equations 10.86 and 10.84 into Equation 10.85 then gives

$$dq = \frac{c_p f}{8} \frac{dA_w}{A} (T_w - T_0) \tag{10.87}$$

However,

$$dA_w = P\, dx$$

so using Equation 10.74, Equation 10.87 becomes

$$\frac{dq}{c_p} = \frac{f\, dx}{2D_H} (T_w - T_0) \tag{10.88}$$

Substituting this into Equation 10.81 then gives

$$\frac{(T_w - T_0)}{T_0}\left(\frac{1+\gamma M^2}{2}\right)\frac{f\, dx}{D_H} + \frac{\gamma M^2}{2}\frac{f\, dx}{D_H} = \left\{(1-\gamma M^2) + \frac{(1+\gamma M^2)}{2}\frac{(\gamma-1)M^2}{[1+(\gamma-1)M^2/2]}\right\}\frac{dM}{M} \tag{10.89}$$

The ratio of the heat transfer effect, i.e., the first term on the left-hand side of Equation 10.89, to the fluid friction effects, i.e., the second term on the left-hand side of Equation 10.89, will be seen to depend on

$$\frac{(T_w - T_0)}{T_0}\frac{(1+\gamma M^2)}{2\gamma M^2}$$

Hence, for a given situation, i.e., for a given M, this ratio will depend on $(T_w - T_0)/T_0$. If T_w is very much greater than T_0, heat transfer effects will predominate. However, if T_w is approximately equal to T_0, friction effects will dominate.

When both heat exchange and friction effects are important, Equation 10.89 must be integrated numerically to give the variation of M with x. Other equations given in this section can then be used to simultaneously determine the variations of the other variables, e.g., Equation 10.80 can be used to determine the changes in T, Equation 10.79 can be used to find the changes in p, and Equation 10.78 can be used to find the changes in V.

Isothermal Flow with Friction in a Constant Area Duct

When gases are transported through long pipelines that are not heavily insulated the gas temperature frequently remains approximately constant. Of course, this will normally require that heat is either transferred from the gas to the surroundings or that heat is transferred from the surroundings to the gas flowing through the pipe. Because the pipe is usually long in such cases and the heat transfer rates relatively low, the effects of friction cannot usually be neglected. Hence, in the present section, one-dimensional isothermal gas flow through a constant area duct with significant friction effects will be considered.

The analysis is, of course, again based on the use of the conservation of mass, momentum, and energy equations together with the perfect gas law. The same type of control volume as used in the previous section is again considered, this being shown in Figure 10.19.

For this control volume, since T is constant (i.e., $dT = 0$), the governing equations give

Conservation of mass:

$$\frac{dV}{V} + \frac{d\rho}{\rho} = 0 \tag{10.90}$$

Conservation of momentum:

$$dp + \tau_w \frac{P}{A} dx + \rho V\, dV = 0 \tag{10.91}$$

where P is again the perimeter of the duct and A is its cross-sectional area.

Conservation of energy:

$$dq = V\, dV \tag{10.92}$$

Perfect gas law:

$$\frac{dp}{p} = \frac{d\rho}{\rho} \tag{10.93}$$

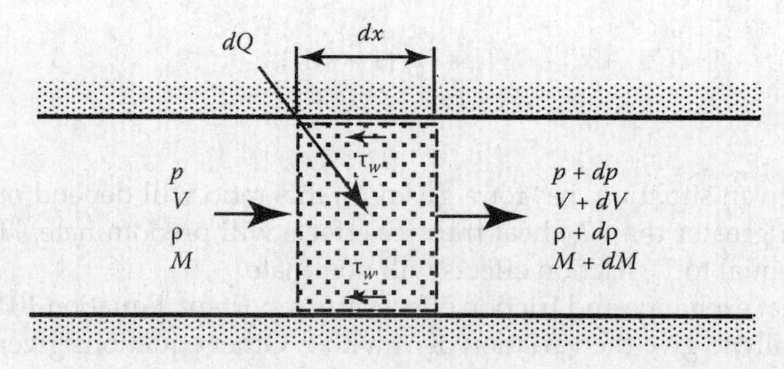

FIGURE 10.19
Control volume used in the analysis of isothermal duct flow.

Further, from the definition of the Mach number, since when T is constant, a is constant

$$\frac{dM}{M} = \frac{dV}{V} \tag{10.94}$$

The above equations represent a set of five equations involving, as variables, dV, $d\rho$, dp, dq, and dM. Before solving for these variables, the wall shear stress, τ_w is written in terms of the friction factor and the hydraulic diameter is again introduced, these being defined as discussed above by

$$\tau_w = f\frac{1}{2}\rho V^2, \quad D_H = \frac{4A}{P}$$

In terms of these, Equation 10.91 becomes

$$dp + \rho V^2 \frac{2f}{D_H} dx + \rho V\, dV = 0 \tag{10.95}$$

which can be written

$$\frac{dp}{p} + \frac{2f}{D_H} dx \frac{\rho V^2}{p} + \frac{\rho V}{p} dV = 0$$

i.e.,

$$\frac{dp}{p} + \frac{2f}{D_H} dx \gamma M^2 + \gamma M^2 \frac{dV}{V} = 0 \tag{10.96}$$

However, Equations 10.90, 10.93, and 10.94 together give

$$\frac{dV}{V} = -\frac{d\rho}{\rho} = -\frac{dp}{p} = \frac{dM}{M} \tag{10.97}$$

Therefore, Equation 10.96 gives

$$\frac{dp}{p}(1 - \gamma M^2) + \gamma M^2 \frac{2f}{D_H} dx = 0$$

i.e.,

$$\frac{dp}{p} + \left[\frac{\gamma M^2}{1 - \gamma M^2}\right] \frac{2f}{D_H} dx = 0 \tag{10.98}$$

Hence,

$$\frac{dV}{V} = -\frac{d\rho}{\rho} = -\frac{dp}{p} = \frac{dM}{M} = \left[\frac{\gamma M^2}{1 - \gamma M^2}\right] \frac{2f}{D_H} dx \tag{10.99}$$

Using this result, Equation 10.92 gives

$$\frac{dq}{V^2} = \frac{dV}{V} = \left[\frac{\gamma M^2}{1-\gamma M^2}\right]\frac{2f}{D_H}dx \tag{10.100}$$

Also, since

$$\frac{ds}{c_p T} = \frac{dT}{T} - \left[\frac{\gamma-1}{\gamma}\right]\frac{dp}{p} = -\left[\frac{\gamma-1}{\gamma}\right]\frac{dp}{p}$$

it follows from Equation 10.100 that

$$\frac{ds}{c_p T} = \left[\frac{\gamma-1}{\gamma}\right]\left[\frac{\gamma M^2}{1-\gamma M^2}\right]\frac{2f}{D_H}dx \tag{10.101}$$

Each of Equations 10.99, 10.100, and 10.101, which give the changes in the variables through the control volume, contains the term $1 - \gamma M^2$ in the denominator on the right-hand side, each of the other terms on this right side always being positive. Therefore, the changes in the variables through the control volume, i.e., *dp*, *dV*, etc., change sign when $M = 1/\sqrt{\gamma}$. The signs of these changes above and below this value are shown in Table 10.4.

The form of the variation of *dq* with *M* is illustrated in Figure 10.20.

TABLE 10.4

Changes in Flow Variables Produced by Friction in Isothermal Flow

Mach No.	*dq*	*dp*	*dρ*	*dV*	*dM*	*ds*
$M < 1/\sqrt{\gamma}$	+	–	–	+	+	+
$M > 1/\sqrt{\gamma}$	–	+	+	–	–	–

Note: + = quantity is increasing; – = decreasing.

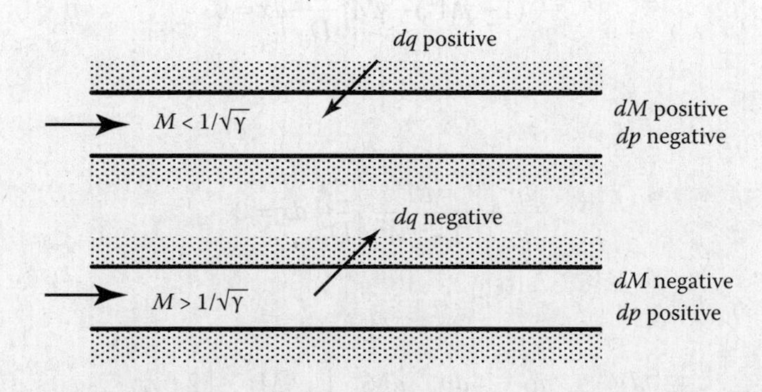

FIGURE 10.20
Variation of differential heat transfer rate with Mach number.

As M tends to $1/\sqrt{\gamma}$, dq tends to infinity, i.e., an infinite amount of heat would have to be transferred to or removed from the flow to keep the temperature of the gas constant if M were to reach $1/\sqrt{\gamma}$. Therefore, $1/\sqrt{\gamma}$ is a limiting value for M. This can further be seen from the fact that when the Mach number decreases in the flow direction whereas when $M < 1/\sqrt{\gamma}$, the Mach number increases. Hence, in all cases, M tends to $1/\sqrt{\gamma}$.

The relationships given above basically determine the rate of change of the flow variables along the duct. They must be integrated to give the change in these flow variables over a specified length of the duct l_{12} indicated in Figure 10.21.

Equation 10.97 can be directly integrated to give

$$\frac{V_2}{V_1} = \frac{\rho_1}{\rho_2} = \frac{p_1}{p_2} = \frac{M_2}{M_1} \tag{10.102}$$

To relate the changes in these variables to the length of the duct, it is noted that Equation 10.99 gives

$$\frac{dM}{\gamma M^3} - \frac{dM}{M} = \frac{2f}{D_H} dx \tag{10.103}$$

This equation can be integrated between the points 1 and 2 being considered to give

$$\left[\frac{1}{2\gamma M_1^2} - \frac{1}{2\gamma M_2^2}\right] - \ln\left[\frac{M_2}{M_1}\right] = \frac{2fl_{12}}{D_H} \tag{10.104}$$

For any value of M_1, this equation allows M_2 to be found. This determines M_2/M_1 and so allows p_2/p_1, ρ_2/ρ_1, and V_2/V_1 to be found using Equation 10.102.

Since the limiting value of M_2 is $1/\sqrt{\gamma}$, Equation 10.104 gives the following equation for the maximum duct length l^* in terms of the initial Mach number M

$$\left[\frac{1}{2\gamma M^2} - \frac{1}{2}\right] + \ln[\sqrt{\gamma} M] = \frac{2fl^*}{D_H}$$

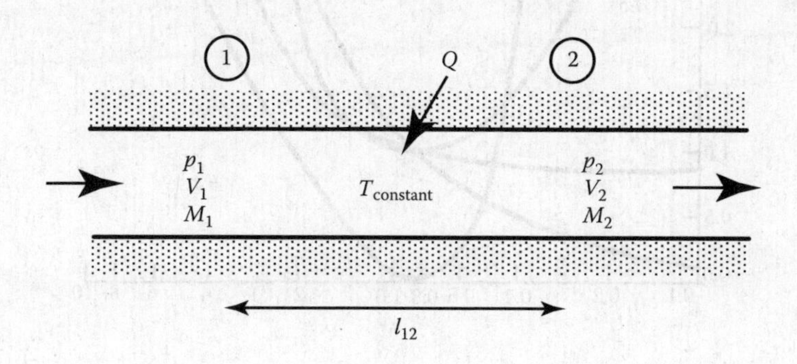

FIGURE 10.21
Changes over duct length considered.

i.e.,

$$\left[\frac{1-\gamma M^2}{\gamma M^2}\right]+\ln[\gamma M^2]=\frac{4fl^*}{D_H} \tag{10.105}$$

For any value of M, this allows the value of $4fl^*/D_H$ to be found. Hence, since $M = M^* - 1/\sqrt{\gamma}$, when $l_{12} = l^*$ so that $M^*/M = 1/\sqrt{\gamma}M$, it follows from Equation 10.102 that

$$\frac{\rho}{\rho^*}=\frac{p}{p^*}=\frac{V^*}{V}=\frac{M^*}{M}=\frac{1}{\sqrt{\gamma}M} \tag{10.106}$$

Therefore, $4fl^*/D_H$, ρ/ρ^*, p/p^*, and V/V^* are functions of M for a given value of γ. The ratios of the stagnation properties can then be obtained using the relation between the stagnation property and the Mach number, e.g., since

$$\frac{T_0}{T}=1+\frac{\gamma-1}{2}M^2$$

it follows that

$$\frac{T_0}{T_0^*}=\frac{1+(\gamma-1)M^2/2}{1+[(\gamma-1)/2](1/\gamma)}$$

i.e.,

$$\frac{T_0}{T_0^*}=\frac{2\gamma}{3\gamma-1}\left(1+\frac{\gamma-1}{2}M^2\right)$$

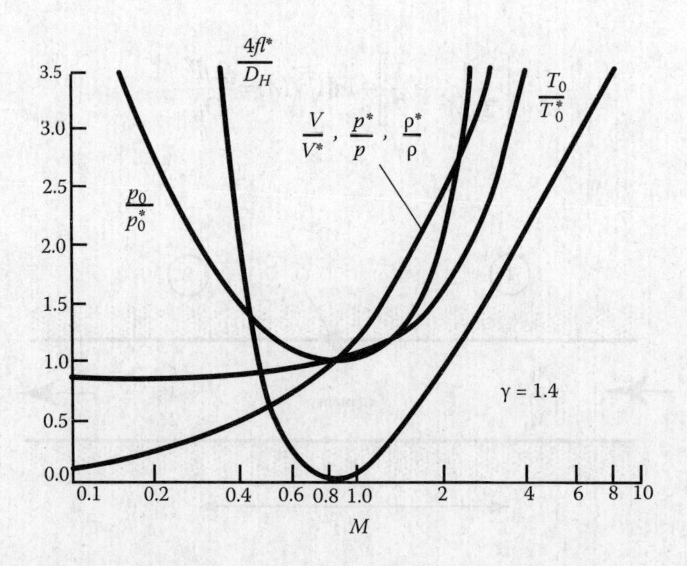

FIGURE 10.22
Variation of properties in isothermal flow for $\gamma = 1.4$.

The values given by these relations are often presented in tables (see Appendix F) and charts. A typical set of values, these being for $\gamma = 1.4$, is given in Figure 10.22. The results are also given by the COMPROP software.

The way in which these results are used is illustrated in the following example.

Combustion Waves

The discussion given thus far in this chapter has, basically, been concerned with heat transfer to or from a body to the compressible flow over the body and with the effect of heat exchange on compressible flow through a duct. There are, however, other types of compressible flows in which heat exchange is important. One such case is the flow through a combustion wave. Here, the heat addition is the result of the "release of chemical energy" associated with the combustion. It is thus dependent on the difference between the heat of formation of the initial gas mixture, i.e., the reactants, and that of the products of combustion. A combustion wave can be modeled, basically, as a composite wave consisting of a shock wave sustained by the release of chemical energy in a combustion zone immediately following the shock wave, the "combustion wave" consisting of the shock and the combustion zone. The detailed structure of the wave will not, however, be considered here. Combustion waves are often so strong that the high-temperature effects discussed in Chapter 12 become important. However, most of the main features of such flows can be adequately described by assuming that the gas is perfect. It should also be noted that the combustion wave can either be stationary with the gas flowing through it or it can be propagating through the gas. The wave will here be analyzed using a coordinator system that is fixed relative to the wave, i.e., the wave will be at rest in the analysis presented below. The results of this analysis can then be applied to a moving wave using the same procedure as was adopted in Chapter 5 to deal with moving shock waves.

Consider the flow relative to a combustion wave, i.e., consider the situation shown in Figure 10.23.

The flow through the control volume, shown in Figure 10.23, is of course analyzed by applying conservation of mass, momentum, and energy as before. The conservation of mass and conservation of momentum are unaffected by the presence of the combustion and give, as with the conventional normal shock wave flow considered in Chapter 5

$$\rho_1 V_1 = \rho_2 V_2 (= \dot{m}) \tag{10.107}$$

and

$$p_1 + \rho_1 V_1^2 = p_2 + \rho_2 V_2^2 \tag{10.108}$$

unit frontal area of the wave having again been considered.

The above two equations can be combined by noting that Equation 10.108 gives

$$p_1 + \frac{\dot{m}^2}{\rho_1} = p_2 + \frac{\dot{m}^2}{\rho_2}$$

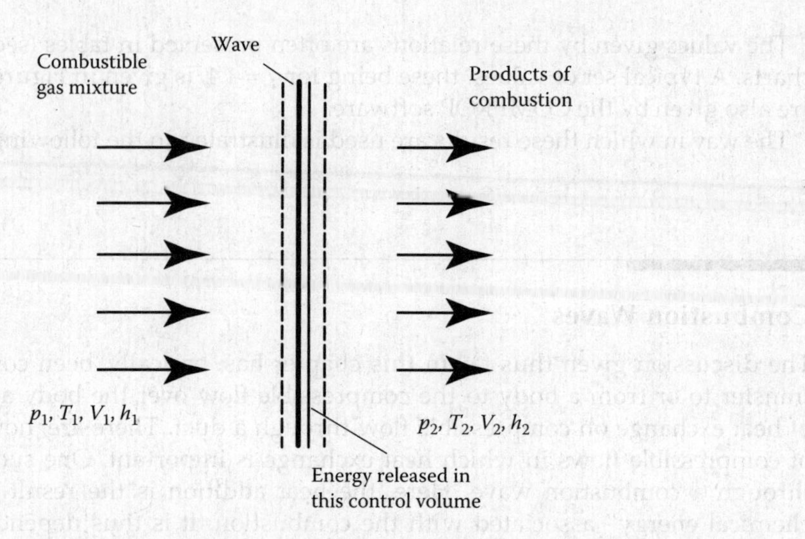

FIGURE 10.23
Control volume used in the analysis of a combustion wave.

i.e.,

$$\dot{m}^2 = \frac{p_2 - p_1}{(1/\rho_1) - (1/\rho_2)} \tag{10.109}$$

Hence, using Equation 10.107, it follows that, i.e.,

$$V_1 = \frac{1}{\rho_1}\sqrt{\frac{p_2 - p_1}{(1/\rho_1) - (1/\rho_2)}}, \quad V_2 = \frac{1}{\rho_2}\sqrt{\frac{p_2 - p_1}{(1/\rho_1) - (1/\rho_2)}} \tag{10.110}$$

The conservation of energy applied to the flow across the wave is next considered. Because of the heat generation, this gives

$$c_p T_1 + \frac{V_1^2}{2} + q = c_p T_2 + \frac{V_2^2}{2} \tag{10.111}$$

The heat generation rate per unit mass, q, depends on the chemical reaction and is assumed to be known.

Now, Equation 10.110 gives

$$V_2^2 - V_1^2 = \frac{p_2 - p_1}{(1/\rho_1) - (1/\rho_2)}\left[\frac{1}{\rho_2^2} - \frac{1}{\rho_1^2}\right]$$

i.e.,

$$V_2^2 - V_1^2 = (p_1 - p_2)\left[\frac{1}{\rho_2} + \frac{1}{\rho_1}\right] \tag{10.112}$$

Now, by using the perfect gas law, Equation 10.111 can be written as

$$V_2^2 - V_1^2 = \left[\frac{2\gamma}{\gamma - 1}\right]\left[\frac{p_1}{\rho_1} - \frac{p_2}{\rho_2}\right] + q \tag{10.113}$$

Equations 10.112 and 10.113 together then give

$$(p_1 - p_2)\left[\frac{1}{\rho_2} + \frac{1}{\rho_1}\right] = \left[\frac{2\gamma}{\gamma - 1}\right]\left[\frac{p_1}{\rho_1} - \frac{p_2}{\rho_2}\right] + q$$

i.e.,

$$\left[1 - \frac{p_2}{p_1}\right]\left[\frac{\rho_1}{\rho_2} + 1\right] = \left[\frac{2\gamma}{\gamma - 1}\right]\left[1 - \left(\frac{p_2}{p_1}\right)\left(\frac{\rho_1}{\rho_2}\right)\right] + \frac{\rho_1 q}{p_1} \tag{10.114}$$

This equation is the equivalent of Equation 5.13, which was derived for the flow through an adiabatic normal shock wave. The equivalents of the other adiabatic normal shock wave relations given in Chapter 5 can easily be derived for a detonation wave using the same basic procedure as adopted in Chapter 5.

Equation 10.114 determines the variation of p_2/p_1 with ρ_1/ρ_2 for any value of $\rho_1 q/p_1$. The form of the variation is shown in Figure 10.24.

In Figure 10.24, the curve corresponding to $q = 0$, i.e., to adiabatic flow, applies to conventional normal shock waves. The point on this $q = 0$ at which $p_2/p_1 = 1$ and $\rho_1/\rho_2 = 1$ represents conditions in the flow upstream of the combustion wave. It is not possible to determine from this graph what conditions will exist downstream of the wave for given upstream

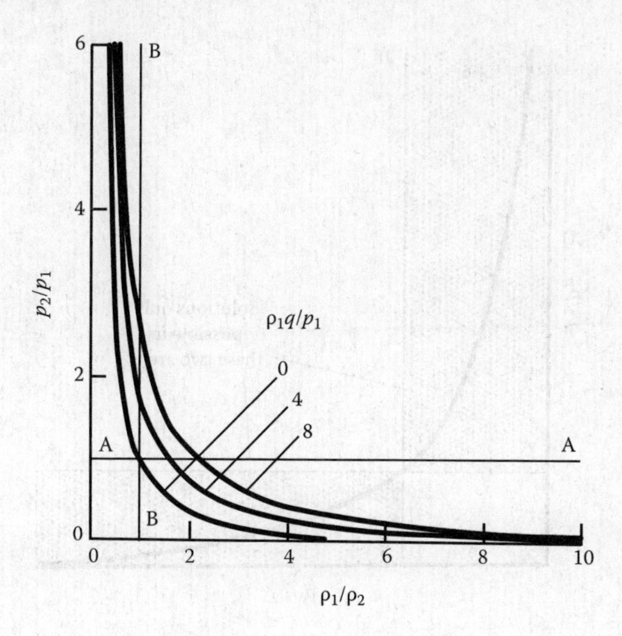

FIGURE 10.24
Relation between pressure and density ratios across a combustion wave.

conditions and given q. Therefore, the limits on possible downstream solutions have to be considered. Consider Equation 10.110. Since the velocities both before and after the wave must be positive, it follows that $p_2 - p_1$ must have the same sign as $(1/\rho_1 - 1/\rho_2)$. Therefore, a combustion wave must be such that $p_2/p_1 > 1$ and $\rho_2/\rho_1 > 1$ or $p_2/p_1 < 1$ and $\rho_2/\rho_1 < 1$.

Considering Figure 10.24, it will be seen that this requires that only solutions to the left of BB or below AA are possible, i.e., only the solutions indicated in Figure 10.25 are possible.

To decide whether there are further limits on the possible solution, the entropy change across the wave is considered. This is given by

$$\frac{s_2 - s_1}{c_p} = \ln\frac{T_2}{T_1} - \frac{(\gamma - 1)}{\gamma}\ln\frac{p_2}{p_1}$$

i.e., using the perfect gas law

$$\frac{s_2 - s_1}{c_p} = \ln\left[\left(\frac{p_2}{p_1}\right)^{1/\gamma}\left(\frac{\rho_1}{\rho_2}\right)\right] \tag{10.115}$$

This equation in conjunction with Equation 10.114 allows the variation of $s_2 - s_1/c_p$ with p_2/p_1 for any value of $\rho_1 q/p_1$ to be determined. The form of the variation is shown in Figure 10.26.

The curve corresponding to $q = 0$, i.e., to adiabatic flow, applies to conventional normal shock waves. It will be seen from Figure 10.26 that for the $q = 0$ case, the entropy increases across the wave when $p_2/p_1 > 1$ and decreases across the wave when $p_2/p_1 < 1$. When $q = 0$, therefore, only waves across which the density and, hence, the pressure increases are possible, i.e., only a compressive shock is possible. This result was previously derived in

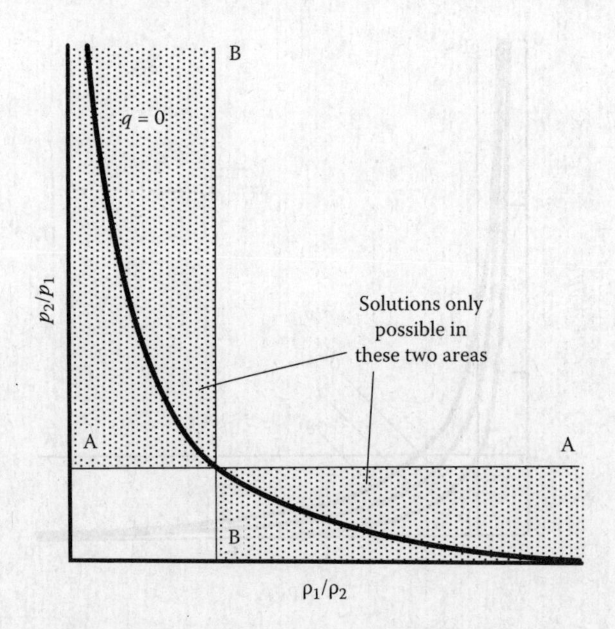

FIGURE 10.25
Possible detonation wave solutions.

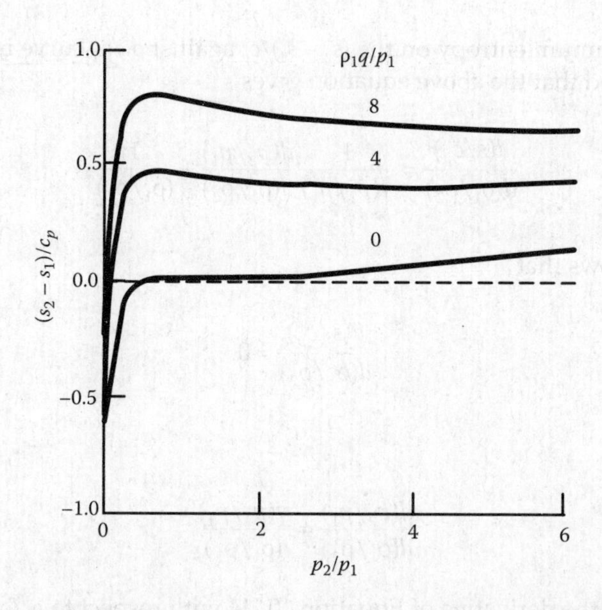

FIGURE 10.26
Entropy change across a combustion wave.

Chapter 5. However, when $q > 0$, it will be seen from Figure 10.25 that an entropy increase can occur both when $p_2/p_1 > 1$ and when $p_2/p_1 < 1$. When there is heat addition, then both compressive and expansive waves are possible. Combustion waves across which $p_2/p_1 > 1$ and therefore across which $\rho_1/\rho_2 < 1$ are termed "detonation" waves whereas combustion waves across which $p_2/p_1 < 1$, and therefore across which $\rho_1/\rho_2 > 1$ are termed "deflagration" waves. The propagation of deflagration waves is mainly governed by the rate of diffusion of heat and mass and their characteristics will not be considered further here. Therefore, attention will here be restricted to detonation waves, i.e., to waves across which $\rho_1/\rho_2 < 1$ and $p_2/p_1 > 1$, the conditions across such waves being determined by the curves given in Figure 10.24. However, the actual pressure and density ratios across the detonation wave for any value of q have still not been defined. Now it will be seen from Figure 10.26 that for a detonation wave, the entropy increase passes through a minimum as ρ_1/ρ_2 decreases from 1 toward 0. It has been proposed and experimentally verified that the conditions behind detonation waves of the type here being considered correspond to those at this minimum entropy increase point. Such detonation waves are termed "Chapman–Jouguet waves." They occur, for example, when combustion is initiated at the closed end of a long duct containing a combustible gas and when the combustion behind the wave results from the temperature rise across a normal shock propagating down a duct into a combustible mixture.

The fact that the entropy rise across such waves is a minimum allows the conditions behind a detonation wave to be determined. To do this, it is noted that Equation 10.115 can be written

$$\frac{s_2 - s_1}{c_p} = \frac{1}{\gamma} \ln\left[\frac{p_2}{p_1}\right] + \ln\left[\frac{\rho_1}{\rho_2}\right] \tag{10.116}$$

The point of minimum entropy on the $(s_2 - s_1)/c_p$ against ρ_1/ρ_2 curve is required. To find this point, it is noted that the above equation gives

$$\frac{d(s/c_p)}{d(\rho_1/\rho_2)} = \frac{1}{\gamma(p_2/p_1)}\frac{d(p_2/p_1)}{d(\rho_1/\rho_2)} + \frac{1}{(\rho_1/\rho_2)} \tag{10.117}$$

From this, it follows that

$$\frac{d(s/c_p)}{d(\rho_1/\rho_2)} = 0$$

when

$$\frac{d(p_2/p_1)}{d(\rho_1/\rho_2)} = \frac{\gamma(p_2/p_1)}{(\rho_1/\rho_2)} \tag{10.118}$$

However, taking the derivative of Equation 10.114 with respect to ρ_1/ρ_2 for a given value of $\rho_1 q/p_1$ gives on rearrangement

$$1 + \frac{\gamma+1}{\gamma-1}(p_2/p_1) = \left[1 - \frac{\gamma+1}{\gamma-1}\left(\frac{\rho_1}{\rho_2}\right)\right]\frac{d(p_2/p_1)}{d(\rho_1/\rho_2)}$$

Substituting for $d(p_2/p_1)/d(\rho_1/\rho_2)$ from Equation 10.118 then gives

$$1 + \frac{\gamma+1}{\gamma-1}(p_2/p_1) = \left[-\gamma\frac{(p_2/p_1)}{(\rho_1/\rho_2)} + \frac{\gamma(\gamma+1}{\gamma-1}\left(\frac{p_2}{p_1}\right)\right]$$

This equation gives, on rearrangement,

$$\left(\frac{\rho_1}{\rho_2}\right) = \frac{\gamma(p_2/p_1)}{(\gamma+1)(p_2/p_1) - 1} \tag{10.119}$$

In addition, since

$$M_1^2 = \frac{V_1^2}{a_1^2} = \frac{V_1^2 \rho_1}{\gamma p_1}$$

Equation 10.110 gives

$$M_1^2 = \left(\frac{1}{\gamma}\right)\frac{p_2/p_1 - 1}{1 - \rho_1/\rho_2} = \left(\frac{1}{\gamma}\right)\left[(\gamma+1)\left(\frac{p_2}{p_1}\right) - 1\right]$$

This equation, in conjunction with Equations 10.114 and 10.119, defines the values of p_2/p_1, ρ_1/ρ_2, and M_1 for a detonation wave for any specified value of $\rho_1 q/p_1$. The form of the pressure and density ratio variations with $\rho_1 q/p_1$ is illustrated in Figure 10.27, which gives results for $\gamma = 1.4$. Very high pressure ratios across a detonation wave will be seen to be possible.

Equation 10.110 gives the velocity behind the detonation wave as

$$V_2^2 = \left(\frac{p_1\rho_1}{\rho_2^2}\right)\frac{p_2/p_1 - 1}{1-\rho_1/\rho_2}$$

i.e.,

$$\frac{V_2^2}{\gamma p_2/p_2} = \frac{\rho_1/\rho_2}{\gamma(p_2/p_1)}\frac{p_2/p_1 - 1}{1-\rho_1/\rho_2}$$

i.e.,

$$M_2^2 = \frac{\rho_1/\rho_2}{\gamma(p_2/p_1)}\frac{p_2/p_1 - 1}{1-\rho_1/\rho_2}$$

Using Equation 10.119 to give the density ratio in terms of the pressure ratio then gives for a detonation wave, i.e., for a wave for which the entropy rise is a minimum

$$M_2^2 = 1, \quad \text{i.e.,} \quad M_2 = 1 \tag{10.120}$$

Hence, the Mach number behind a Chapman–Jouguet wave is always equal to 1.

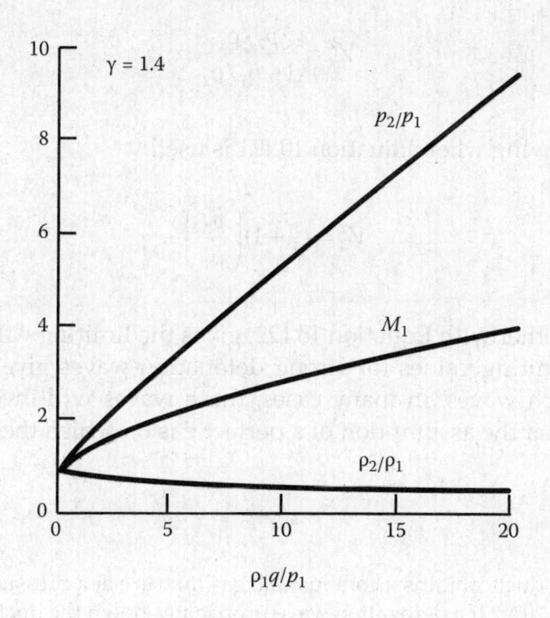

FIGURE 10.27
Pressure and density ratios across a detonation wave.

It should be noted for very strong detonation waves, i.e., for high values of p_2/p_1, the 1 in the denominator on the right-hand side of Equation 10.119 will be negligible compared with the first term in the denominator. Therefore, for a very strong detonation wave, the density ratio reaches a limit of

$$\frac{\rho_1}{\rho_2} = \frac{\gamma}{\gamma + 1} \tag{10.121}$$

Further, for strong detonation waves, Equation 10.114 can be approximated by

$$-\left(\frac{p_2}{p_2}\right)\left(\frac{\rho_1}{\rho_2} + 1\right) = -\left(\frac{2\gamma}{\gamma - 1}\right)\left(\frac{p_2}{p_1}\right)\left(\frac{\rho_1}{\rho_2}\right) + \frac{\rho_1 q}{p_1}$$

Substituting the limiting value of this into ρ_1/ρ_2 as given by Equation 10.121 into this equation and rearranging gives the limiting pressure ratio for very strong detonation waves as

$$\frac{p_2}{p_1} = (\gamma - 1)(\rho_1 q/p_1) \tag{10.122}$$

When the limiting values of the pressure and density ratios as given by Equations 10.122 and 10.121 are compared with the results given in Figure 10.27, it will be seen that the results do tend to these values at large values of $\rho_1 q/p_1$. The limiting velocity ahead of a very strong detonation wave, i.e., the limiting velocity of propagation of a strong moving detonation wave, is given by noting that when p_2/p_1 is large, Equation 10.110 gives

$$V_1^2 = \frac{p_2/\rho_1}{1 - \rho_1/\rho_2}$$

which gives the following when Equation 10.121 is used:

$$V_1^2 = (\gamma + 1)\left(\frac{p_2}{\rho_1}\right) \tag{10.123}$$

This equation, together with Equation 10.122, gives the limiting value of V_1.

Although these limiting values for strong detonation waves give an indication of the characteristics of such waves, in many cases, these waves will involve such large temperature increases that the assumption of a perfect gas on which these equations is based becomes invalid.

Example 10.9

A long insulated duct contains a combustible gas mixture at a pressure of 120 kPa and a temperature of 20°C. If a detonation wave propagates down the duct as a result of the ignition of the gas at one end of the duct, find the velocity at which the wave is moving down the duct and the pressure behind the wave. Also, find the velocity behind the wave relative to the wave. The combustion causes a heat "release" of 2 MJ/kg of gas. Assume that the gas mixture has the properties of air.

Solution

The equation of state gives

$$\rho = \frac{p}{RT} = \frac{120{,}000}{287 \times 293} = 1.427 \text{ kg/m}^3$$

Hence, here

$$\frac{\rho_1 q}{p_1} = \frac{1.427 \times 2{,}000{,}000}{120{,}000} = 23.78$$

Equation 10.97 therefore gives

$$\left[1 - \frac{p_2}{p_1}\right]\left[\frac{\rho_1}{\rho_2} + 1\right] = 7\left[1 - \left(\frac{p_2}{p_1}\right)\left(\frac{\rho_1}{\rho_2}\right)\right] + 23.78$$

whereas Equation 10.102 gives

$$\left(\frac{\rho_1}{\rho_2}\right) = \frac{1.4(p_2/p_1)}{2.4(p_2/p_1) - 1}$$

These two equations together determine the values of ρ_1/ρ_2 and p_2/p_1. Many algorithms and much software exists that allows the rapid determination of these values. The simplest approach is to guess a series of values of p_2/p_1 and then to use the second of the above equations to determine the corresponding value of ρ_1/ρ_2. Using these values, the left- and right-hand sides of the first of the above equations are found, and these values compared and the value of p_2/p_1 that makes the two sides of the first equation equal can be deduced. Whatever procedure is adopted, the following results are obtained

$$\frac{p_2}{p_1} = 7.73 \quad \text{and} \quad \frac{\rho_1}{\rho_2} = 0.617$$

so that

$$p_2 = 7.73 \times 120 = 927.6 \text{ kPa}$$

The pressure behind the wave is therefore 927.6 kPa.

Then, since

$$M_1^2 = \left(\frac{1}{\gamma}\right)\frac{p_2/p_1 - 1}{1 - \rho_1/\rho_2}$$

the following is obtained

$$M_1^2 = \left(\frac{1}{1.4}\right)\frac{7.73 - 1}{1 - 0.617}$$

From which it follows that

$$M_1 = 3.53$$

and so

$$V_1 = 3.53 \times \sqrt{1.4 \times 287 \times 293} = 1211.2 \text{ m/s}$$

This is the velocity relative to the wave upstream of the wave and is therefore equal to the velocity at which the wave is propagating, i.e., the wave is moving at a speed of 1211.2 m/s.

Now, the equation of state gives

$$\frac{T_2}{T_1} = \left(\frac{p_2}{p_1}\right)\left(\frac{\rho_1}{\rho_2}\right)$$

so

$$T_2 = 293 \times 7.73 \times 0.617 = 1397.4 \text{ K}$$

Then, since the Mach number behind the wave is always equal to 1, i.e.,

$$M_2 = 1$$

it follows that

$$V_1 = 1 \times \sqrt{1.4 \times 287 \times 1397.4} = 749.3 \text{ m/s}$$

Therefore, the velocity behind the wave relative to the wave is 749.3 m/s.

Condensation Shocks

A condensation shock is a region of relatively rapid change that occurs when there is a vapor-to-liquid phase change in a supersaturated gas flow. To understand how a condensation shock can be formed, consider the flow of moist air, i.e., air-containing water vapor, through a convergent–divergent nozzle. As the air flows down the nozzle, its velocity increases and its temperature decreases. However, the maximum amount of water vapor that air can contain per unit mass decreases with temperature. Thus, although the air is initially unsaturated, i.e., contains less than the maximum possible amount of water vapor, a point can be reached in the nozzle at which the air becomes saturated with water vapor. This is indicated schematically on a psychrometric chart in Figure 10.28.

Up until the point at which condensation starts to occur, the mass of water vapor per unit mass of air, i.e., the humidity ratio, remains constant. Therefore, the process in the nozzle before condensation starts to occur can be represented by the horizontal line **asb** shown in Figure 10.28. Once point **s** is reached, condensation would start to occur if the air–water vapor mixture remained in thermodynamic equilibrium, i.e., the process would

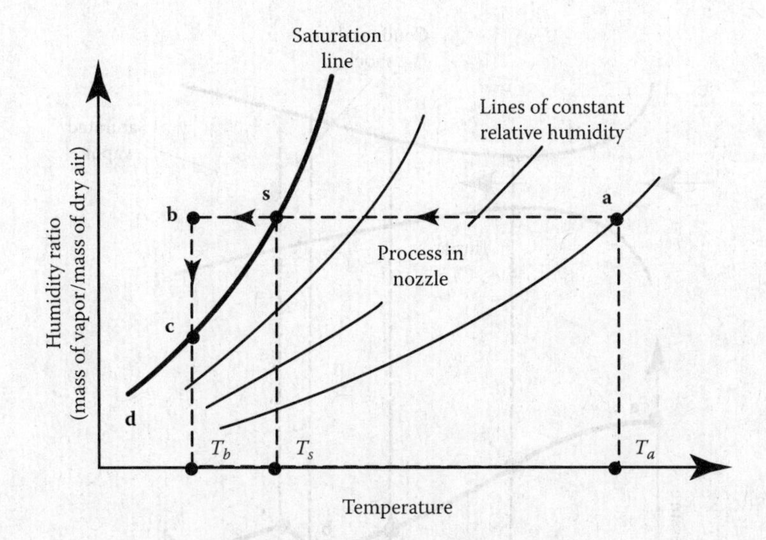

FIGURE 10.28
Variation of humidity ratio with temperature during flow through a nozzle.

be represented by line **sd** in Figure 10.28. However, the growth of the liquid water drops in the flow that should occur once point **s** is reached is a relatively slow process and does not, in fact, start until point **b** in Figure 10.28 is reached. Between points **s** and **b** in Figure 10.28, the air is supersaturated because it contains more water vapor than would be possible if the air was in thermodynamic equilibrium at the same temperature. Once condensation starts, however, the condensation process is fairly rapid and occurs over a relatively short distance, this region of condensation being termed the condensation "shock." The process through the shock is represented by the line **bc** in Figure 10.28, the air being saturated and with a very low moisture ratio after it has reached point **c**, the rest of the flow process being along the line **cd**. The flow process is shown schematically in Figure 10.29. Because the degree of supersaturation that usually occurs with nozzle flows is high, almost all of the water initially in the air is condensed out in the condensation shock.

The enthalpy change through a condensation shock is therefore approximately given by

$$h_b - h_d = c_p(T_b - T_d) + \omega L$$

where ω is the initial humidity ratio and L is the latent heat of evaporation. The specific heat c_p can be taken as that of air. The term ωL behaves in the same way as a heat generation term, i.e., a condensation shock is similar to the combustion waves discussed in the previous section. However, the conditions at which the condensation shock occurs are determined by the rate at which the condensation develops and not by entropy change considerations such as those associated with a detonation wave.

Although referred to as a condensation "shock wave," the nature of the process is really quite different from that which occurs in a conventional normal shock wave. The pressure change across a condensation shock is usually relatively small and the flow behind the "shock wave" is usually still supersonic. A condensation shock is also usually much thicker than a conventional shock wave.

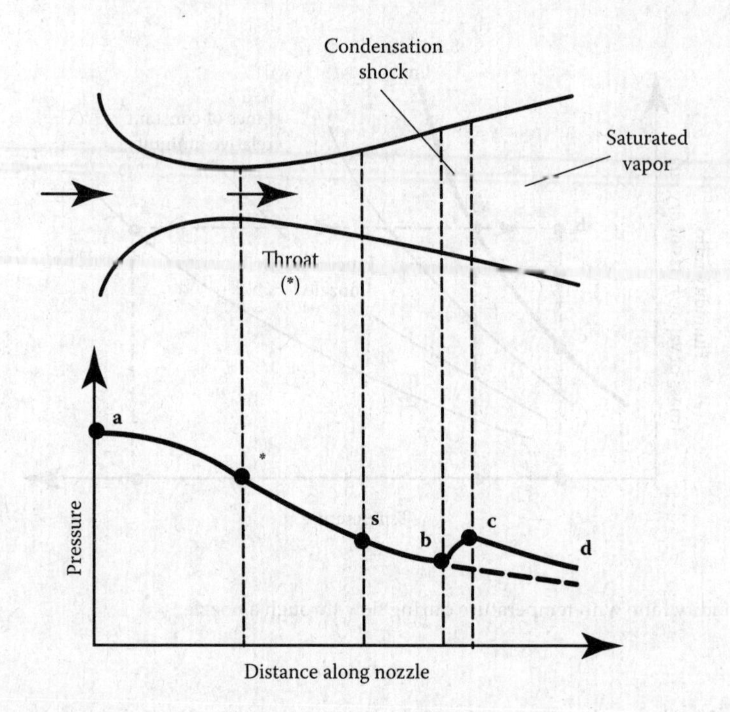

FIGURE 10.29
Formation of condensation shock in nozzle.

Condensation shocks can also occur with steam flow through a nozzle. As the steam flows through the nozzle, the temperature and pressure decrease, and the temperature eventually reaches the saturation temperature. However, condensation does not then normally begin, the flow continuing as a supersaturated vapor until a point is reached at which relative rapid condensation and a return to thermodynamic equilibrium occurs, i.e., a condensation shock occurs.

Concluding Remarks

Three forms of flow, all involving heat transfer, were considered in this chapter. The conclusions for each of these three types of flow are presented below.

External Flows

When a gas flows over a surface, the velocity at the surface over which the gas is flowing is, as a result of viscosity, zero. Therefore, the gas velocity decreases as the surface is approached. As a consequence, the gas temperature rises causing so-called aerodynamic heating of the surface. If the surface is adiabatic, i.e., if there is no heat transfer to or from the surface, the surface temperature attained is termed the adiabatic surface temperature. The so-called recovery factor has been used to relate the adiabatic surface temperature to the stagnation temperature. In high-speed flow, the convective heat transfer rate that results when the surface is either heated or cooled will depend on the difference between the surface temperature and the adiabatic surface temperature. Applications to flat plate and stagnation point flows have been considered.

Internal Flows

It was shown that in the case of constant area flows, when there is heat exchange and when the wall friction effects are negligible, the addition of heat to the flow will cause the Mach number to move toward 1, i.e., it will cause the Mach number to increase in a subsonic flow and decrease in a supersonic flow. Heat addition therefore can cause the Mach number to reach one, i.e., can cause choking to occur. It was also shown that flows in which there are changes in flow area and in which the effects of friction at the wall, i.e., of viscosity, are not negligible can be analyzed using the same basic approach as adopted when dealing with constant area flow with viscosity effects neglected but that, in these cases, simple analytical expressions for the variations for the flow properties could not be derived.

Combustion and Condensation Waves

Consideration was given to a combustion waves in which there is effectively a normal shock wave and an associated region of heat release. Two types of combustion wave were shown to be possible, but detailed attention was only given to detonation waves that are normally associated with the release of large amounts of chemical energy. Expressions for the changes across such waves were derived, and it was shown that the Mach number behind such waves is always one. Brief consideration was also given to a condensation waves that are associated with the sudden condensation of liquid in a subcooled vapor flow.

PROBLEMS

1. A rocket ascends vertically through the atmosphere with a velocity that can be assumed to increase linearly with altitude from zero at sea level to 1800 m/s at an altitude of 30,000 m. If the surface of this rocket is assumed to be adiabatic, estimate the variation of the skin temperature with altitude at a point on the surface of the rocket a distance of 3 m from the nose of the rocket. Use the flat plate equations given in this chapter and assume that at the distance from the nose considered, the Mach number, and temperature outside the boundary layer are the same as those in the freestream ahead of the rocket.
2. A flat plate, which can be assumed to be adiabatic, is placed in a wind tunnel, the plate being aligned with the flow. The test section Mach number is 3 and the static temperature is 15°C. During the startup of the tunnel, a normal shock wave occurs in the test section halfway along the plate. The boundary layer on the plate is turbulent both before and after the shock wave. What are the plate temperatures before and after the shock wave under these circumstances? Discuss the result obtained, considering the stagnation temperatures before and after the shock wave.
3. Air, at Mach 3 and a temperature of −30°C flows over a flat plate that is aligned with the flow. The plate is kept at a temperature of 25°C. Find the mean heat transfer rate from the plate surface per unit area.

4. Air is expanded through a convergent–divergent nozzle from Mach 0.2 to Mach 2. The overall nozzle length is 0.7 m, and the throat of the nozzle is at a distance of 0.25 m from the inlet. The convergent and divergent sections are of such a shape that the cross-sectional area of the nozzle varies linearly with distance in both sections. The initial temperature and pressure are 1000 kPa and 800, respectively. It can be assumed that the boundary layer on the walls of the nozzle is turbulent and that it effectively originates at the entrance section of the nozzle and it can be assumed that the properties of this boundary layer are described by the flat plate equations. If the walls of the nozzle are effectively adiabatic, estimate the variation of wall temperature with distance along the nozzle. Assume that the variation of the properties of the freestream flow outside the boundary layer can be obtained by assuming one-dimensional isentropic flow and the effect of the boundary layer on the effective flow area can be ignored. (Hint: Calculate the Mach number and temperature variation in the freestream with distance along the nozzle. Then use the definition of the recovery factor applied at each section to find the adiabatic wall temperature.)
5. If instead of being adiabatic, the wall of the nozzle described in Problem 10.4 is kept at a uniform temperature of 200°C, estimate the variation of the local wall heat transfer rate, q, per unit wall area with distance along the nozzle.
6. A flat plate with a length of 0.8 m and a width of 1.2 m is placed in the working section of a wind tunnel in which the Mach number is 4, the temperature is –70°C and the pressure is 3 kPa. If the surface temperature of the plate is kept at 30°C by an internal cooling system, find the rate at which heat must be added to or removed from the plate. Consider both the top and the bottom of the plate.
7. Air flows though a constant area duct. The air has a temperature of 20°C and Mach 0.5 at the entrance to the duct. It is desired to transfer heat from the wall of the duct to the flow such that at the exit of the duct the stagnation temperature is 1180°C. Is this possible? If not, what adjustment must be made to Mach number at the entrance to give a discharge stagnation temperature of 1180°C? Ignore the effects of friction.
8. Air with stagnation temperature of 430°C and a stagnation pressure of 1.6 MPa enters a constant area duct in which heat is transferred to the air. The Mach number at the inlet to the duct is 3 and the Mach number at the exit to the duct is 1. Determine the stagnation temperature and the stagnation pressure at the exit for the case the flow is shock-free and for the case where there is a normal shock at the duct inlet. Ignore the effects of friction.
9. Air at a temperature of 100°C with a pressure of 101 kPa enters a constant area combustion chamber at a velocity of 130 m/s. Determine the maximum amount of heat that can be transferred to the air flow per unit mass of air. Assume friction effects are negligible.
10. Heat is supplied to air flowing through a short tube causing the Mach number to increase from an initial value of 0.3 to a final value of 0.6. If the initial air temperature is 40°C and if the effects of friction are neglected, find the heat supplied per unit mass.
11. Air flows through a 10-cm-diameter pipe at a rate of 0.18 kg/s. If the air enters the pipe at a temperature of 0°C and a pressure of 60 kPa, how much heat can be

added to the air per kilogram without choking the flow? The effects of friction can be assumed to be negligible.

12. Air enters a constant area combustion chamber at a pressure of 101 kPa and a temperature of 70°C with a velocity of 130 m/s. By ignoring the effects of friction, determine the maximum amount of heat that can be transferred to the flow per unit mass of air.
13. Air flows through a 0.25-m-diameter duct. At the inlet the velocity is 300 m/s, and the stagnation temperature is 90°C. If the Mach number at the exit is 0.3, determine the direction and the rate of heat transfer. For the same conditions at the inlet, determine the amount of heat that must be transferred to the system if the flow is to be sonic at the exit of the duct.
14. Air flows through a constant area combustion chamber that has a diameter of 0.15 m and a length 5 m. The inlet stagnation temperature is 335 K, the inlet stagnation pressure is 1.4 MPa, and the inlet Mach number is 0.55. Find the maximum rate at which heat can be added to the flow. Neglect the effects of friction.
15. Air enters a constant area duct at Mach 0.15, a pressure of 200 kPa, and a temperature of 20°C. Heat is added to the air it flows through the duct at a rate of 60 kJ/kg of air. Assuming that the flow is steady and that the effects of wall friction can be ignored, find the temperature, pressure, and Mach number at which the air leaves the duct. Assume that the air behaves as a perfect gas.
16. At the inlet to a constant area combustion chamber the Mach number is 0.2 and the stagnation temperature is 120°C. What is the amount of heat transfer to the gas per unit mass if the Mach number is 0.7 at the exit of the chamber? What is the maximum possible amount of heat transfer? The gas can be assumed to have the properties of air.
17. Air flows through a constant area duct. At the inlet to the duct the stagnation pressure is 600 kPa and the stagnation temperature 200°C. If the Mach number at the inlet of the duct is 0.5 and if the flow is choked at the exit of the duct, determine the heat transfer per unit mass and the exit temperature. Assume that friction effects can be neglected.
18. Air enters a constant diameter pipe at a pressure of 200 kPa. At the exit of the pipe the pressure is 120 kPa, the Mach number is 0.75, and the stagnation temperature is 330°C. Determine the inlet Mach number and the heat transfer per unit mass of air.
19. Air flows through a 4-in.-diameter pipe. The pressure and temperature at the inlet to the pipe are 15 psia and 70°F. The velocity at the inlet is 200 ft/s. If the temperature at the exit to the pipe is 1300°F, how much heat must be added to the air and what will be the exit pressure, velocity, and Mach number? Ignore the effects of viscosity.
20. Air flows though a constant area duct. The air enters the duct at a pressure of 1 MPa, Mach 0.5 and stagnation temperature of 45°C. As a result of friction, the Mach number at the duct exit is 0.90 and the stagnation temperature at this point is 160°C. Find the amount of heat transferred per unit mass to or from the air in the duct. Also, find the pressure at the duct exit. Assume that the effects of viscosity are negligible.

21. Air enters a constant area duct at a pressure of 620 kPa and a temperature of 300°C, the velocity at the inlet being 100 m/s. If the velocity at the exit to the duct is 210 m/s, determine the pressure, temperature, stagnation pressure, and stagnation temperature at the exit to the duct. Also, find the heat transfer per unit mass in the duct. Assume that the effects of viscosity are negligible.
22. Air enters a pipe at a pressure of 200 kPa. At the exit of the pipe, the pressure is 120 kPa, the Mach number is 0.75, and the stagnation temperature is 300°C. Determine the inlet Mach number and the heat transfer rate to the air assuming that the effects of viscosity can be ignored.
23. Air with a stagnation pressure of 600 kPa and a stagnation temperature of 200°C enters a constant area pipe with a diameter of 2 cm. Heat is transferred to the air as it flows through the duct at a rate of 100 kJ/kg. Plot the mass flow rate of air as a function of back-pressure for the range 0 to 400 kPa. Assume frictionless flow.
24. Air flows through a rectangular duct with a 10 × 16-cm cross section. The air velocity, pressure, and temperature at the inlet to the duct are 90 m/s, 105 kPa, and 25°C, respectively. Heat is added to the air as it flows through the duct, and it leaves the duct with a velocity of 200 m/s. Find the pressure and temperature at the exit and the total rate at which heat is being added to the air. Ignore the effects of viscosity.
25. Air is heated as it flows through a constant area duct by an electric heating coil wrapped uniformly around the duct. The air enters the duct at a velocity of 100 m/s, a temperature of 20°C, and a pressure of 101.3 kPa, and the heat transfer rate is 40 kJ/kg per unit length of duct. Plot the variations of exit Mach number, exit temperature, and exit pressure with the length of the duct. Neglect the effects of fluid friction.
26. Air enters a combustion chamber at a velocity of 100 m/s, a pressure of 90 kPa, and a temperature of 40°C. As a result of the combustion, heat is added to the air at a rate of 500 kJ/kg. Find the exit velocity and Mach number. Also, find the heat addition that would be required to choke the flow. Neglect the effects of friction and assume that the properties of the gas in the combustion chamber are the same as those of air.
27. A fuel–air mixture that can be assumed to have the properties of air enters a constant area combustion chamber at a velocity of 10 m/s and a temperature of 100°C. What amount of heat must be added per unit mass to cause the flow at the exit to be choked? Also, find the exit Mach number and temperature if the actual heat addition due to combustion is 1000 kJ/kg.
28. Fuel and air are thoroughly mixed in the proportion of 1:40 by mass before entering a constant area combustion chamber. The pressure, temperature, and velocity at the inlet to the chamber are 50 kPa, 30°C, and 80 m/s, respectively. The heating value of the fuel is 40 MJ/kg of fuel. Assuming steady flow and that the properties of the gas mixture are the same as those of air, determine the static and the stagnation temperatures and the Mach number at the exit of the combustion chamber. Neglect the effects of friction.
29. An air–fuel mixture enters a constant area combustion chamber at a velocity of 100 m/s a pressure of 70 kPa and a temperature of 150°C. Assuming that the fuel–air ratio is 0.04, the heating value of the fuel is 30 MJ/kg, and the mixture

has the properties of air, calculate the Mach number in the gas flow after combustion is completed and the change of stagnation temperature and stagnation pressure across the combustion chamber. Neglect the effects of viscosity and assume that the properties of the gas in the combustion chamber are the same as those of air.

30. An air–fuel mixture flows through a constant area combustion chamber. The velocity, pressure, and temperature at the entrance to the chamber are 130 m/s, 170 kPa, and 120°C, respectively. If the enthalpy of reaction is 600 kJ/kg of mixture, find the Mach number and pressure at the exit of the chamber. Neglect the effects of viscosity and assume that the properties of the gas in the combustion chamber are the same as those of air.
31. An air–fuel mixture enters a constant-area combustion chamber at a Mach number of 0.2, a pressure of 70 kPa, and a temperature of 35°C. If the heat transfer to the gases in the combustion chamber is 1.2 MJ/kg of mixture, determine the Mach number at the exit of the chamber and the change in stagnation temperature through the chamber. Neglect the effects of viscosity and assume that the properties of the gas in the combustion chamber are the same as those of air.
32. In a gas turbine plant, air from the compressor enters the combustion chamber at a pressure of 420 kPa, a temperature of 110°C, and a velocity of 100 m/s. Fuel having an effective heating value of 35,000 kJ/kg is sprayed into the air stream and burnt. Two types of injection systems are available. One gives a fuel–air mass ratio of 0.015, the other a ratio of 0.021. The temperature entering the turbine should not be less than 750°C but should not exceed the temperature determined by the metallurgical limit of the blade material, which can be assumed to be 1000°C. Which of the two injection systems should be used? State the assumptions that have been made in reaching a conclusion.
33. Air enters a constant area duct with Mach 2.5, a stagnation temperature of 300°C, and a stagnation pressure of 1.2 MPa. If the flow is choked, determine the stagnation pressure and stagnation temperature at the exit to the duct and the heat transfer per unit mass for the cases where there is a normal shock at the inlet of the duct and for the case where the flow is shock free and supersonic.
34. Air at a temperature of 300 K and a Mach number of 1.5 enters a constant-area duct which feeds a convergent nozzle. At the exit of the nozzle the Mach number is 1.0 and the ratio of the nozzle exit area to the duct area is 0.98. If a normal shock wave occurs in the duct just upstream of the nozzle inlet, calculate the amount and direction of the heat exchange with the air flow through the duct. Ignore the effect of friction on the flow in the duct and assume that the flow downstream of the shock wave is isentropic.
35. A jet engine is operating at an altitude of 7000 m. The mass of air passing through the engine is 46 kg/s and the heat addition in the combustion chamber is 500 kJ/kg. The cross-sectional area of the combustion chamber is 0.5 m^2, and the air enters the chamber at a pressure of 80 kPa and a temperature of 80°C. After the combustion chamber, the products of combustion, which can be assumed to have the properties of air, are expanded through a convergent nozzle to match the atmospheric pressure at the nozzle exit. Estimate the nozzle exit diameter and the nozzle exit velocity assuming the flow in the nozzle is isentropic. State the assumptions that have been made.

36. Air enters a 7.5-cm-diameter pipe at a pressure of 1.3 kPa, a temperature of 200°C, and Mach 1.8. Heat is added to the flow as a result of a chemical reaction taking place in the duct. Find the heat transfer rate necessary to choke the flow in the pipe. Assume that the air behaves as a perfect gas with constant specific heats and neglect changes in the composition of the gas stream due to the chemical reaction.
37. Air enters a 15-cm-diameter pipe at a pressure of 1.3 Mpa, a temperature of 20°C, and a velocity of 60 m/s. Assuming that the friction factor is 0.004 and that the flow is effectively isothermal, find the Mach number at a point in the pipe where the pressure is 300 kPa.
38. Oxygen is to be pumped through a 12.5-m-long, 25-mm-diameter pipe by a compressor that delivers the oxygen to the pipe at a pressure of 1.3 MPa. The mean friction factor can be assumed to be 0.0045 and the flow in the pipe can be assumed to be isothermal, the oxygen temperature being 30°C. Find the mass flow rate.
39. Consider subsonic air flow through a pipe with a diameter of 2.5 cm which is 3 m long. At the inlet to the pipe the pressure is 200 kPa and the temperature is 30°C. If the mean value of the friction factor is assumed to be equal to 0.006, determine the mass rate of flow and the inlet and exit Mach numbers for (a) an isothermal flow in which the pressure at the exit is 130 kPa and (b) an adiabatic flow.
40. Air at a pressure of 550 kPa and a temperature of 30°C flows through a pipe with a diameter of 0.3 m and a length of 140 m. If the mass flow rate is at its maximum value, find, assuming that the flow is isothermal and that the average friction factor, is 0.0025, the Mach number at the inlet to the pipe, the mass flow rate through the pipe, the pressure at the exit of the pipe, and the rate of heat transfer to the duct.
41. In long pipelines, such as those used to convey natural gas, the temperature of the gas can usually be considered constant. In one such case, the gas leaves a pumping station at pressure of 320 kPa and a temperature of 25°C with Mach 0.10. At some other point in the flow, the pressure is measured and found to be 130 kPa. Calculate the Mach number of the flow at this section and determine how much heat has been added to or removed from the gas per unit mass between the pumping station and the point where the measurements are made. The gas can be assumed to have the properties of methane.
42. Natural gas is to be pumped through a 36-in.-diameter pipe connecting two compressor stations 40 miles apart. At the upstream station the pressure is not to exceed 100 psia and at the downstream station it is to be at least 25 psig. Calculate the maximum allowable mass rate of flow through the pipe. Assume that there is sufficient heat transfer through the pipe to maintain the gas at a temperature of 70°F. The gas can be assumed to have a molal mass of 18 and a specific heat ratio of 1.3.
43. A 3-km-long pipeline with a diameter of 0.10 m is used to transport methane at a rate of 1.0 kg/s. If the gas remains essentially at a constant temperature of 10°C and if the pressure at the exit to the pipe is 150 kPa, find the inlet pressure. Assume an average friction factor of 0.004. Methane has a molal mass of 16 and a specific heat ratio of 1.3.

44. Air enters a convergent duct at Mach 0.75, a pressure of 500 kPa, and a temperature of 35°C. The exit area of the duct is half the inlet area. If, as a result of heat transfer, the Mach number at the duct exit is also 0.75, find the heat transfer rate per kilogram of air. Neglect the effects of viscosity.
45. A tube containing a combustible gas mixture is contained in a long insulated pipe at a pressure of 150 kPa and a temperature of 30°C. The gas is ignited at one end of the tube leading to the propagation of a detonation wave down the pipe. If the combustion causes a heat "release" of 1 MJ/kg of gas, find the pressure and temperature behind the detonation wave and the velocity at which the wave is moving down the pipe. Assume that the gas mixture has the properties of air.

11

Hypersonic Flow

Introduction

Hypersonic flow was loosely defined in Chapter 1 as flow in which the Mach number is greater than about 5. No real reasons were given at that point as to why supersonic flows at high Mach numbers were different from those at lower Mach numbers. However, it is the very existence of these differences that really defines hypersonic flow. The nature of these hypersonic flow phenomena, and therefore the real definition of "hypersonic flow," will be presented in the next section.

Hypersonic flows have, up to the present, mainly been associated with the reentry of orbiting and other high altitude bodies into the atmosphere. For example, a typical Mach number with altitude variation for a reentering satellite is shown in Figure 11.1. It will be seen from this figure that because of the high velocity that the craft had to possess to keep it in orbit, very high Mach numbers—values that are well into the hypersonic range—exist during reentry.

Discussions and studies of passenger aircraft that can fly at hypersonic speeds at high altitudes have also been undertaken. A typical proposed such vehicle is shown in Figure 11.2.

This chapter, which presents a brief introduction to hypersonic flow, is the first of three interrelated chapters. One of the characteristics of hypersonic flow is the presence of so-called high-temperature gas effects, and these effects will be discussed more fully in the next chapter. Hypersonic flow is also conventionally associated with high altitudes where the air density is very low, and "low-density flows" will be discussed in Chapter 13.

Characteristics of Hypersonic Flow

As mentioned above, hypersonic flows are usually loosely described as flows at very high Mach numbers, say greater than about 5. However, the real definition of hypersonic flows is that they are flows at such high Mach numbers that phenomena occur that do not exist at low supersonic Mach numbers. These phenomena are discussed in this section.

One of the characteristics of hypersonic flow is the presence of an interaction between the oblique shock wave generated at the leading edge of the body and the boundary layer on the surface of the body. Consider the oblique shock wave formed at the leading edge of a wedge in a supersonic flow as shown in Figure 11.3.

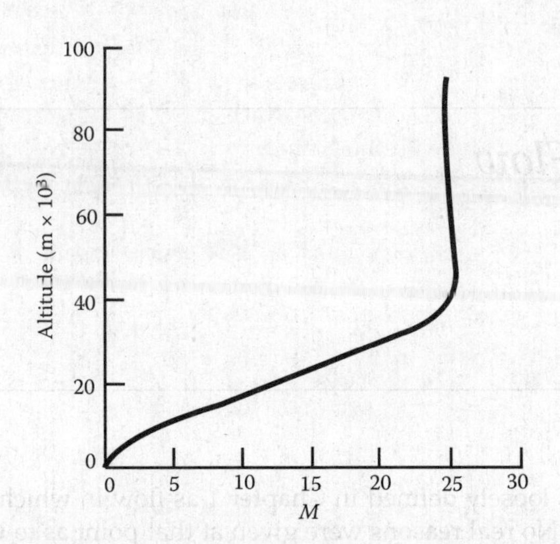

FIGURE 11.1
Typical variation of Mach number with altitude during reentry.

FIGURE 11.2
Proposed hypersonic passenger aircraft. (Copyright © Boeing.)

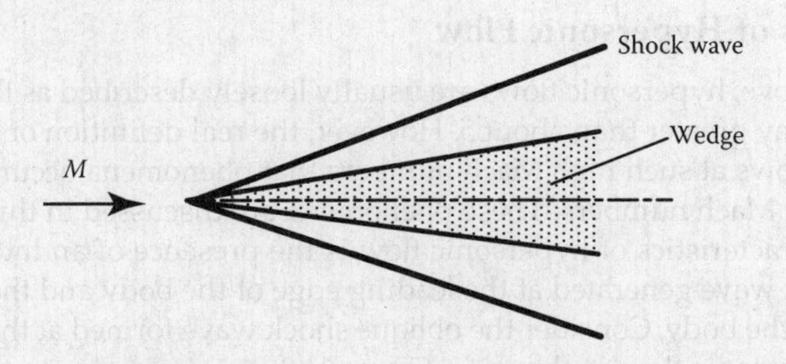

FIGURE 11.3
Flow over a wedge.

As the Mach number increases, the shock angle decreases and the shock therefore lies very close to the surface at high Mach numbers. This is illustrated in Figure 11.4.

Because the shock wave lies close to the surface at high Mach numbers, there is an interaction between the shock wave and the boundary layer on the wedge surface. To illustrate this shock wave–boundary layer interaction, consider the flow of air over a wedge having a half-angle of 5° at various Mach numbers. The shock angle for any selected value of M can be obtained from the oblique shock relations or charts or using the software (see Chapter 6). The angle between the shock wave and the wedge surface is then given by the difference between the shock angle and the wedge half-angle. The variation of this angle with Mach number is shown in Figure 11.5.

It will be seen from Figure 11.5 that as the Mach number increases, the shock wave lies closer and closer to the surface. Further, hypersonic flow normally only exists at relatively

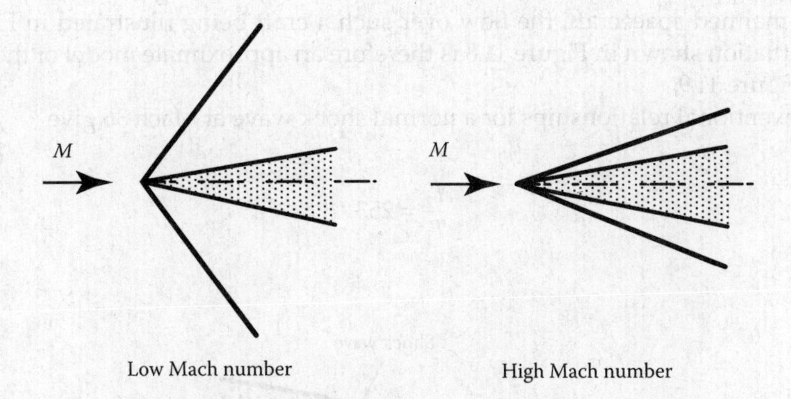

FIGURE 11.4
Shock angle at low and high supersonic Mach number flow over a wedge.

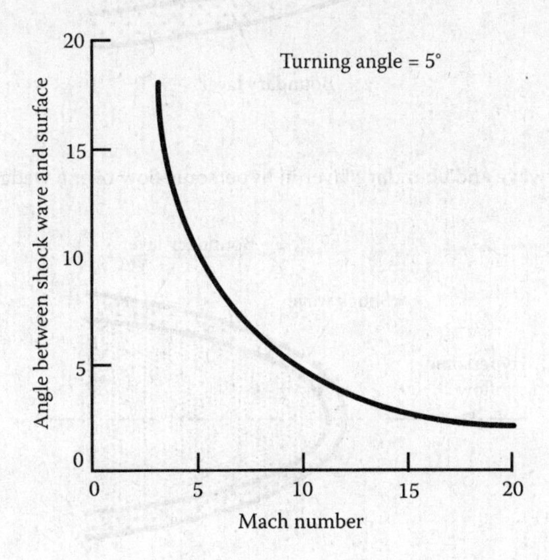

FIGURE 11.5
Variation of angle between shock wave and wedge surface with Mach number for flow over a wedge.

low ambient pressures, which means that the Reynolds numbers tend to be low and the boundary layer thickness therefore tends to be relatively large. In hypersonic flow, then the shock wave tends to lie close to the surface and the boundary layer tends to be thick. As a consequence, interaction between the shock wave and the boundary layer flow usually occurs, the shock being curved as a result and the flow resembling that shown in Figure 11.6.

The above discussion used the flow over a wedge to illustrate interaction between the shock wave and the boundary layer flow in hypersonic flow. This interaction occurs, in general, in hypersonic flow for all body shapes as illustrated in Figure 11.7.

Another characteristic of hypersonic flows is the high temperatures that are generated behind the shock waves in such flows. To illustrate this, consider the flow through a normal shock wave occurring ahead of a blunt body at Mach 36 at an altitude of 59 km in the atmosphere. The flow situation is shown in Figure 11.8.

These were approximately the conditions that occurred during the reentry of some of the earlier manned spacecraft, the flow over such a craft being illustrated in Figure 11.9. The flow situation shown in Figure 11.8 is therefore an approximate model of the situation shown in Figure 11.9.

Now, conventional relationships for a normal shock wave at Mach 36 give

$$\frac{T_2}{T_1} = 253$$

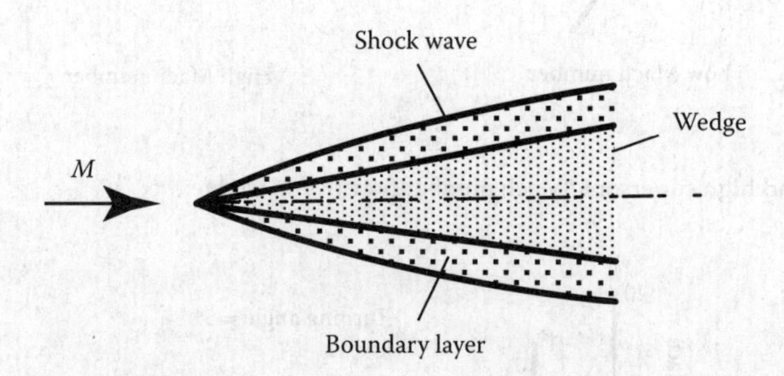

FIGURE 11.6
Interaction between shock wave and boundary layer in hypersonic flow over a wedge.

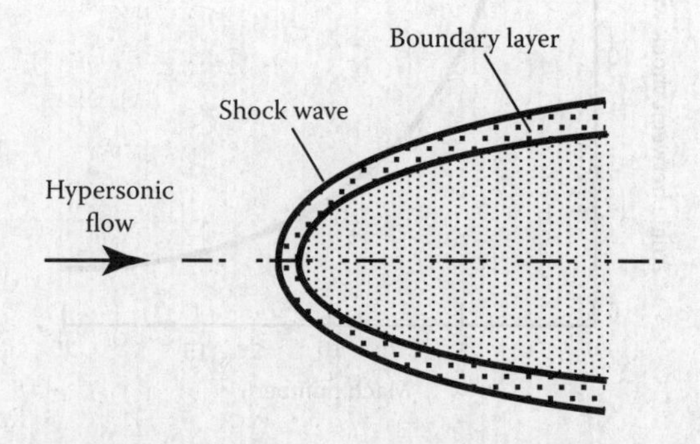

FIGURE 11.7
Interaction between shock wave and boundary layer in hypersonic flow over a curved body.

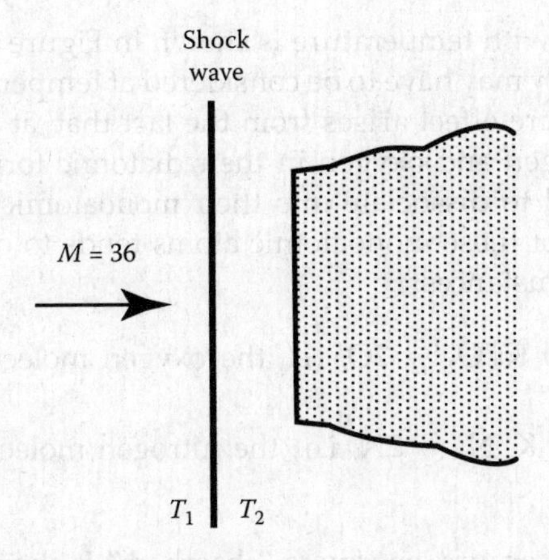

FIGURE 11.8
Normal shock wave in situation considered.

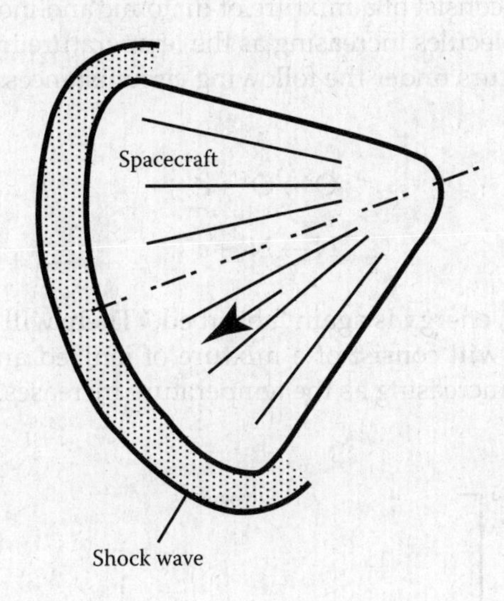

FIGURE 11.9
Flow over early spacecraft during reentry.

However, at 59 km in atmosphere

$$T = 258 \text{ K (i.e., } -15°\text{C)}$$

Hence, the conventional normal shock wave relations give the temperature behind the shock wave as

$$T_2 = 258 \times 253 = 65{,}200 \text{ K}$$

At temperatures as high as these a number of so-called high temperature gas effects will become important. For example, the values of the specific heats, c_p and c_v, and their ratio, γ, change at higher temperatures, their values depending on temperature. For example, the variation of γ of nitrogen with temperature is shown in Figure 11.10. It will be seen from this figure that changes in γ may have to be considered at temperatures above about 500°C.

Another high-temperature effect arises from the fact that, at ambient conditions, air is made up mainly of nitrogen and oxygen in their diatomic form. At high temperatures, these diatomic gases tend to dissociate into their monoatomic form, and at still higher temperatures, ionization of these monoatomic atoms tends to occur. Dissociation occurs under the following circumstances:

For 2000 K < T < 4000 K: $O_2 \rightarrow 2O$, i.e., the oxygen molecules break down to O molecules

For 4000 K < T < 9000 K: $N_2 \rightarrow 2N$, i.e., the nitrogen molecules break down to N molecules

When such dissociation occurs, energy is "absorbed." It should also be clearly understood that the range of temperatures given indicates that not all of the air is immediately dissociated once a certain temperature is reached. Over the temperature ranges indicated above the air will, in fact, consist of a mixture of diatomic and monoatomic molecules, the fraction of monatomic molecules increasing as the temperature increases.

Similarly, ionization occurs under the following circumstances:

For $T > 9000$ K

$$O \rightarrow O^{+} + e^{-}$$

$$N \rightarrow N^{+} + e^{-}$$

When ionization occurs, energy is again "absorbed." There will again be a range of temperatures over which air will consist of a mixture of ionized and unionized atoms, the fraction of ionized atoms increasing as the temperature increases.

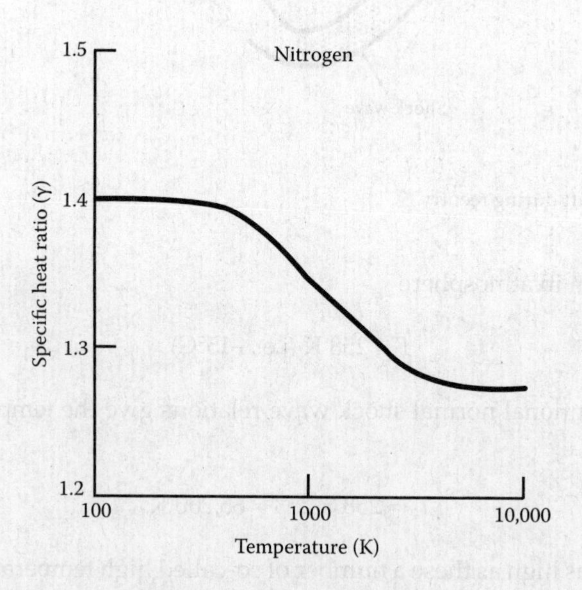

FIGURE 11.10
Variation of specific heat ratio of nitrogen with temperature.

Other chemical changes can also occur at high temperatures, e.g., there can be a reaction between the nitrogen and the oxygen to form nitrous oxides at high temperatures. This and the other effects mentioned above are illustrated by the results given in Figure 11.11.

Figure 11.11 shows the variation of the composition of air with temperature. It will be seen therefore that at high Mach numbers, the temperature rise across a normal shock may be high enough to cause specific heat changes and dissociation, and at very high Mach numbers, ionization. As a result of these processes, conventional shock relations do not apply. For example, for the case discussed above, of a normal shock wave at Mach 36 at an altitude of 59 km in the atmosphere, as a result of these high temperature gas effects the actual temperature behind the shock wave is approximately

$$T_2 \approx 11{,}000 \text{ K}$$

rather than the value of 65,200 K indicated by the normal shock relations for a perfect gas.

There are several other phenomena that are often associated with high Mach number flow and whose existence helps define what is meant by a hypersonic flow. For example, as mentioned above, since most hypersonic flows occur at high altitudes, the presence of low-density effects such as the existence of "slip" at the surface, i.e., of a velocity jump at the surface (see Figure 11.12) is often taken as an indication that hypersonic flow exists.

The details of these high-temperature and low-density effects will be discussed in Chapters 12 and 13. The remainder of this chapter will be devoted to a discussion of an

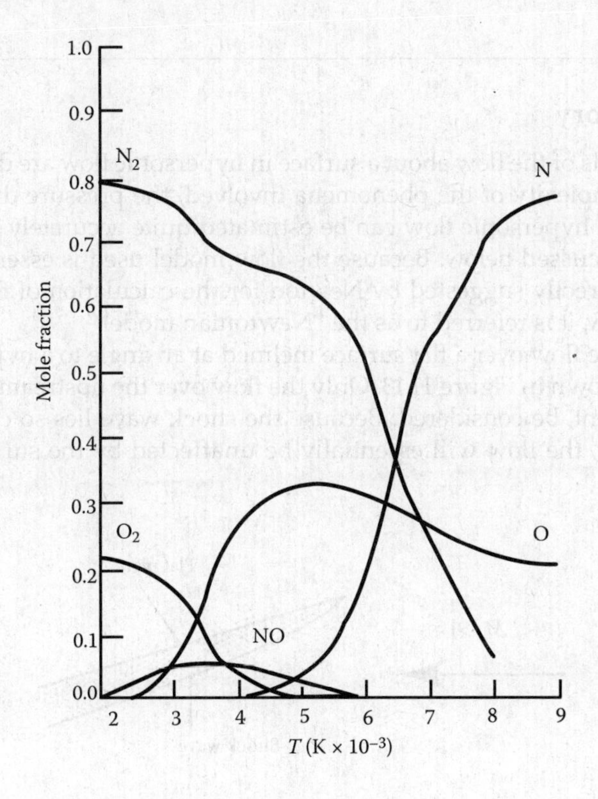

FIGURE 11.11
Variation of equilibrium composition of air with temperature.

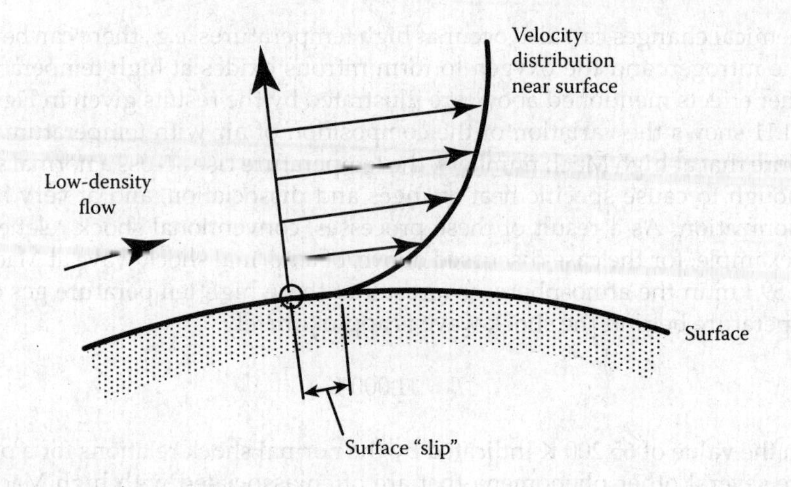

FIGURE 11.12
Surface slip in low-density flow.

approximate method of calculating the pressures and forces acting on a surface placed in a hypersonic flow.

Newtonian Theory

Although the details of the flow about a surface in hypersonic flow are difficult to calculate because of the complexity of the phenomena involved, the pressure distribution about a surface placed in a hypersonic flow can be estimated quite accurately using the approximate approach discussed below. Because the flow model used is essentially the same as one that was incorrectly suggested by Newton for the calculation of forces on bodies in incompressible flow, it is referred to as the "Newtonian model."

First, consider the flow over a flat surface inclined at an angle to a hypersonic flow. This flow situation is shown in Figure 11.13. Only the flow over the upstream face of the surface will, for the moment, be considered. Because the shock wave lies so close to the surface in hypersonic flow, the flow will essentially be unaffected by the surface until the flow

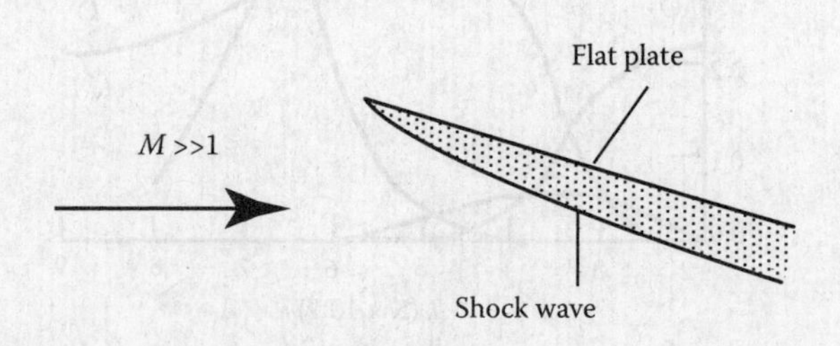

FIGURE 11.13
Hypersonic flow over a plane surface.

reaches the surface, i.e., until it "strikes" the surface, at which point it will immediately become parallel to the surface. Hence, the flow over the upstream face of a plane surface at hypersonic speeds resembles that shown in Figure 11.14.

To find the pressure on the surface, consider the momentum equation applied to the control volume shown in Figure 11.15.

Because the flow at the surface is all assumed to be turned parallel to the surface, no momentum leaves the control volume in the n direction, so the force on the control volume in this direction is equal to the product of the rate mass enters the control volume and the initial velocity component in the n direction, i.e., is given by

$$\text{Mass flow rate} \times \text{Velocity in } n \text{ direction} = (\rho_\infty V_\infty A \sin\theta)(V_\infty \sin\theta)$$
$$= \rho_\infty V_\infty^2 \sin^2\theta A$$

where A is the area of the surface.

Now if p is the pressure acting on the upstream face of the surface, the net force acting on the control volume in the n direction is given by

$$pA - p_\infty A$$

In deriving this result, it has been noted that since the flow is not affected by the presence of the surface until it effectively reaches the surface, the pressure on ABCDE (see Figure 11.15) is everywhere equal to p_∞ and that the forces on BC and DE are therefore equal and opposite and cancel. The force on AB is $p_\infty A \cos\theta$, which has a component in the

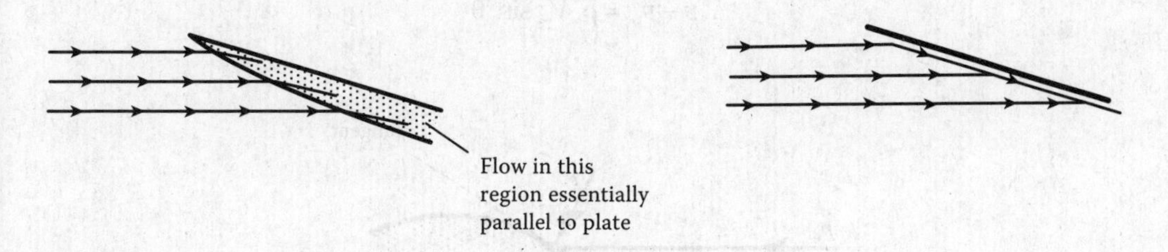

FIGURE 11.14
Newtonian model of hypersonic flow over a plane surface. The actual flow is shown on the left and the Newtonian model flow is shown on the right.

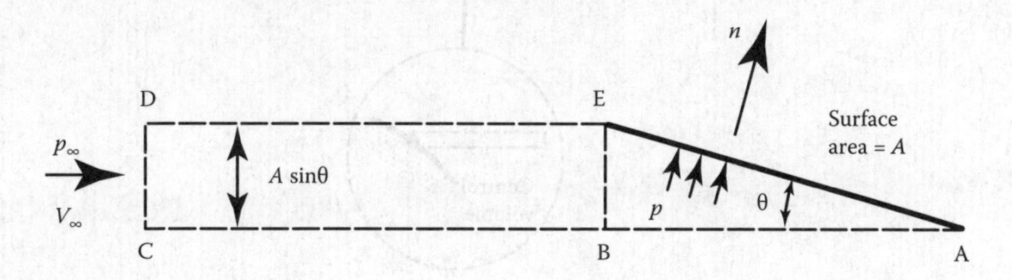

FIGURE 11.15
Control volume considered.

direction normal to the surface of $p_\infty A \cos^2\theta$. Similarly, the force on CD is $p_\infty A \sin\theta$, which has a component in the direction normal to the surface of $p_\infty A \sin^2\theta$. Hence, the sum of the forces on AB and CD in the direction normal to the surface is $p_\infty A \cos^2\theta + p_\infty A \sin^2\theta$, i.e., equal to $p_\infty A$.

Combining the above two results then gives

$$p - p_\infty = \rho_\infty V_\infty^2 \sin^2\theta \tag{11.1}$$

This result can be expressed in terms of a dimensionless pressure coefficient, defined as before by

$$C_p = \frac{p - p_\infty}{\frac{1}{2}\rho_\infty V_\infty^2} \tag{11.2}$$

to give

$$C_p = 2\sin^2\theta \tag{11.3}$$

From the above analysis, it therefore follows that the pressure coefficient is determined only by the angle of the surface to the flow.

The above analysis was for flow over a flat surface. However, it will also apply to a small portion of a curved surface such as that shown in Figure 11.16.

Therefore, the local pressure acting at any point on the surface will be given as before by

$$p - p_\infty = \rho_\infty V_\infty^2 \sin^2\theta$$

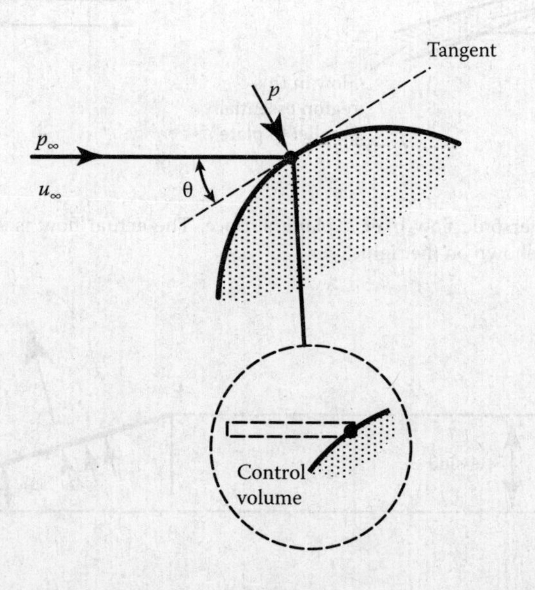

FIGURE 11.16
Control volume considered in dealing with flow over a curved surface.

from which it can be deduced that

$$C_p = 2\sin^2\theta \tag{11.4}$$

where θ is the local angle of inclination of the surface and C_p is the local pressure coefficient.

Equation 11.4 can be written as

$$\frac{p - p_\infty}{p_\infty} = \frac{\rho_\infty}{p_\infty} V_\infty^2 \sin^2\theta$$

Since $a_\infty^2 = \gamma p_\infty / \rho_\infty$, this equation gives

$$\frac{p - p_\infty}{p_\infty} = \gamma M_\infty^2 \sin^2\theta$$

i.e.,

$$\frac{p}{p_\infty} = 1 + \gamma M_\infty^2 \sin^2\theta \tag{11.5}$$

Example 11.1

Using the Newtonian model, find the pressure distribution on the body shown in Figure E11.1. The flow is two-dimensional and the Mach number and the pressure in the freestream ahead of the body are 10 and 7 Pa, respectively.

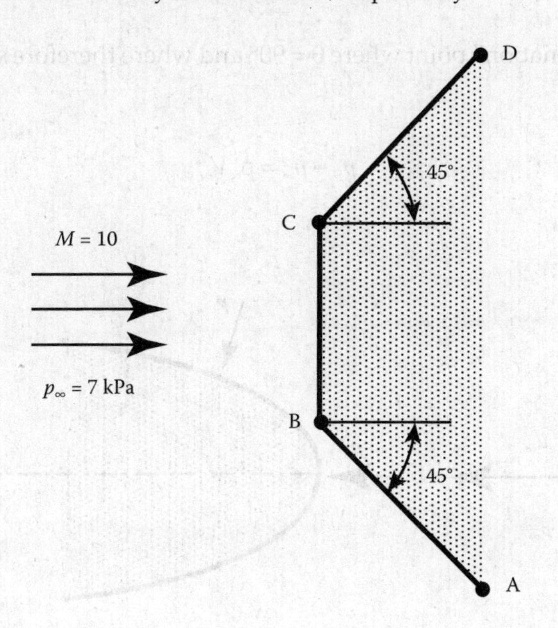

FIGURE E11.1
Flow situation considered.

Solution

Because each surface of the body is flat, the pressure is constant along each surface. Now, as shown above, for any surface

$$p = p_\infty[1 + \gamma M_\infty^2 \sin^2\theta]$$

For surface BC, θ = 90°, whereas for surfaces AB and CD, θ = 45°. Hence,

$$p_{AB} = p_{CD} = 7[1 + 1.4 \times 10^2 \times \sin^2 45°] = 497 \text{ Pa}$$

and

$$p_{BC} = 7[1 + 1.4 \times 10^2 \times \sin^2 90°] = 987 \text{ Pa}$$

Hence, the pressures on surfaces AB and CD are 497 Pa and the pressure on surface BC is 987 Pa.

Modified Newtonian Theory

Consider hypersonic flow over a symmetrical body of arbitrary shape such as that shown in Figure 11.17.

At any point on the surface, as shown above, the pressure is given by

$$p - p_\infty = \rho_\infty V_\infty^2 \sin^2\theta$$

Hence, at the "stagnation" point where θ = 90° and where therefore sin θ = 1 the pressure, p_S, is given by

$$p_S - p_\infty = \rho_\infty V_\infty^2$$

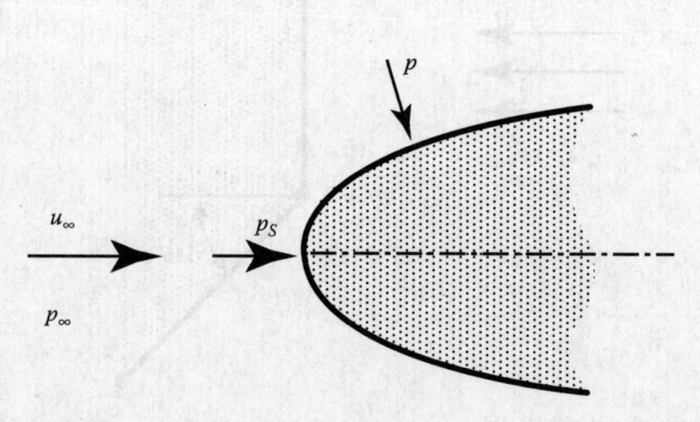

FIGURE 11.17
Form of body being considered.

Hence,

$$\left(\frac{p_S - p_\infty}{\frac{1}{2}\rho_\infty V_\infty^2}\right) = 2$$

i.e., the pressure coefficient at the stagnation point is given by

$$C_{p_S} = 2 \tag{11.6}$$

From these relations, it follows that the pressure distribution about the surface can be written as

$$\frac{C_p}{C_{p_S}} = \sin^2\theta \tag{11.7}$$

or as

$$\frac{p - p_\infty}{p_S - p_\infty} = \sin^2\theta \tag{11.8}$$

Now the Newtonian theory does not really apply near the stagnation point. However, the shock wave in this region is, as previously discussed, effectively a normal shock wave and therefore the pressure on the surface at the stagnation point can be found using normal shock relations and then the Newtonian relation can be used to determine the pressure distribution around the rest of the body. This means, for example, that Equation 11.7 can be written as

$$\frac{C_p}{C_{p_{S_N}}} = \sin^2\theta \tag{11.9}$$

where $C_{p_{S_N}}$ is the pressure coefficient at the stagnation point as given by the normal shock relations. This is, basically, the modified Newtonian equation.

Now it will be recalled from Chapter 5 that the normal shock relations give

$$\frac{p_S}{p_\infty} = \frac{[(\gamma+1)M_\infty^2/2]^{\gamma/(\gamma-1)}}{[2\gamma M_\infty^2/(\gamma+1) - (\gamma - 1/\gamma+1)]^{1/(\gamma-1)}} \tag{11.10}$$

i.e., since

$$C_p = \frac{p - p_\infty}{\frac{1}{2}\rho_\infty V_\infty^2} = \frac{(p/p_\infty) - 1}{\gamma M_\infty^2/2} \tag{11.11}$$

Therefore, the normal shock relation, i.e., Equation 11.10 gives

$$C_{p_{S_N}} = \frac{[(\gamma+1)M_\infty^2/2]^{\gamma/(\gamma-1)}}{[2\gamma M_\infty^2/(\gamma+1)-(\gamma-1)/(\gamma+1)]^{1/(\gamma-1)}[\gamma M_\infty^2/2]} \tag{11.12}$$

If M_∞ is very large, the above equation tends to

$$C_{p_{S_N}} = \frac{[(\gamma+1)/2]^{\gamma/(\gamma-1)}}{[2\gamma/(\gamma+1)]^{1/(\gamma-1)}[\gamma/2]} \tag{11.13}$$

For $\gamma = 1.4$, this gives the limiting value of $C_{p_{S_N}}$ for large values of M_∞ as 1.839. Hence, assuming a perfect gas and a large freestream Mach number, the modified Newtonian theory gives

$$C_p = 1.839 \sin^2\theta \tag{11.14}$$

As discussed in the first section of this chapter, when the Mach number is very large, the temperature behind the normal shock wave in the stagnation point region becomes so large that high-temperature gas effects become important and these affect the value of $C_{p_{S_N}}$. The way in which the flow behind a normal shock wave is calculated when these high temperature gas effects become important will be discussed in the next chapter. The relation between the perfect gas normal shock results, the normal shock results with high temperature effects accounted for, and the Newtonian result is illustrated by the typical results shown in Figure 11.18.

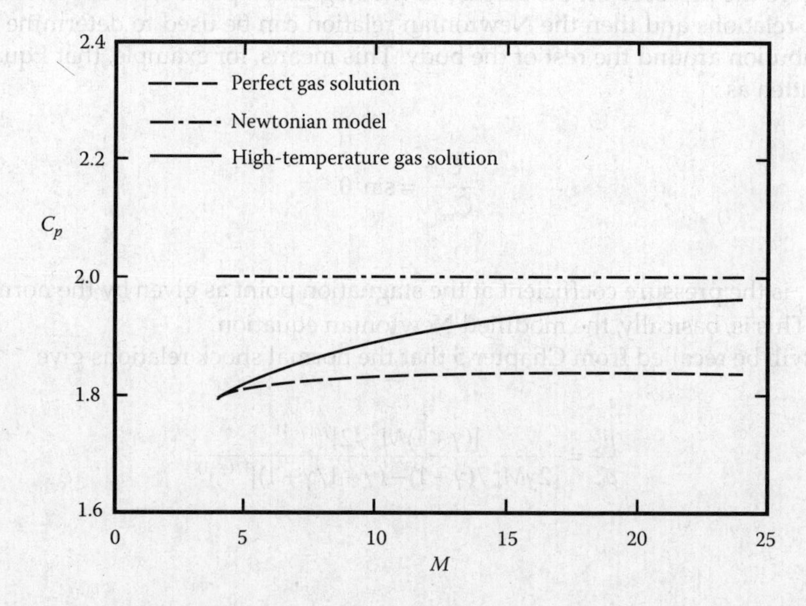

FIGURE 11.18
Typical variation of stagnation point pressure ratio with Mach number.

The results shown in Figure 11.18 and similar results for other situations indicate that the stagnation pressure coefficient given by the high Mach number form of the normal shock relations for a perfect gas, i.e., Equation 11.13, applies for Mach numbers above about 5 and that it gives results that are within 5% of the actual values for Mach numbers up to more than 10. Therefore, the modified Newtonian equation, as given in Equation 11.9, using the high Mach number limit of the perfect gas normal shock to give the stagnation point pressure coefficient, will give results that are of adequate accuracy for values of M_∞ up to more than 10. At higher values of M_∞, the unmodified Newtonian equation gives more accurate results than this form of the modified equation. Of course, the modified Newtonian equation with the stagnation pressure coefficient determined using high temperature normal shock results will apply at all hypersonic Mach numbers.

It should be noted that Equation 11.7 gives

$$\frac{p - p_\infty}{p_\infty} = C_{ps} \frac{1}{2} \frac{\rho_\infty}{p_\infty} V_\infty^2 \sin^2\theta$$

i.e., again using $a_\infty^2 = \gamma p_\infty / \rho_\infty$, the above equation gives

$$\frac{p - p_\infty}{p_\infty} = C_{ps} \frac{\gamma}{2} M_\infty^2 \sin^2\theta$$

i.e.,

$$\frac{p}{p_\infty} = 1 + C_{ps} \frac{\gamma}{2} M_\infty^2 \sin^2\theta \tag{11.15}$$

Example 11.2

Consider two-dimensional flow over a 30-cm-thick body with a circular leading edge that is moving through air at Mach 14, the ambient air pressure being 2 Pa. Using the modified Newtonian model in conjunction with the strong shock limit of the perfect gas relation for a normal shock, find the pressure variation around the leading edge in terms of the distance from the stagnation point.

Solution

The flow situation here being considered is shown in Figure E11.2a.

It was shown above that the strong shock solution for air gives $C_{ps} = 1.839$. Hence, the modified Newtonian model gives:

$$\frac{p}{p_\infty} = 1 + \frac{1.839 \times 1.4}{2} M_\infty^2 \sin^2\theta = 1 + 1.287 M_\infty^2 \sin^2\theta$$

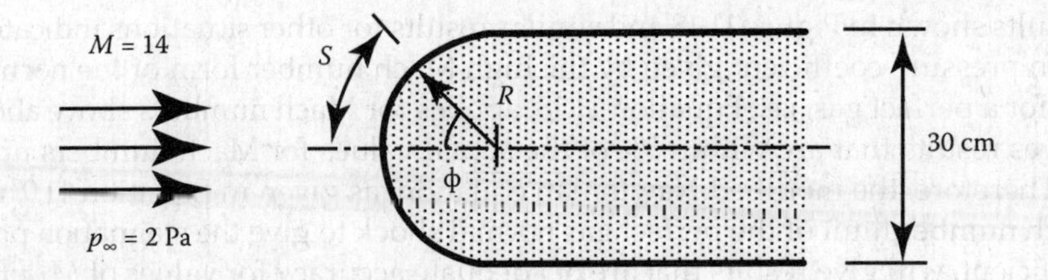

FIGURE E11.2a
Flow situation considered.

In the present situation, it will be seen from Figure E11.2a that if S is the distance around the leading edge from the stagnation point, then $\theta = \pi/2 - \phi$, so $\sin\theta = \cos\phi$. Hence, since $\phi = S/R$, the pressure variation is given by

$$p = p_\infty\left[1 + 1.287 M_\infty^2 \cos^2\left(\frac{S}{0.15}\right)\right]$$

$$= 2\left[1 + 1.287 \times 14 \times 14 \cos^2\left(\frac{S}{0.15}\right)\right]$$

$$= 2 + 504.6\cos^2\left(\frac{S}{0.15}\right)$$

It having been noted that the radius of the leading edge, R, is equal to half the thickness of the body, i.e., 30/2 cm.

Now S varies from 0 to $\pi R/2$, i.e., from 0 to 0.2356 m around the leading edge. The pressure variation is therefore as shown in Table E11.2 and plotted in Figure E11.2b.

TABLE E11.2
Pressure Variation Around Nose of Airfoil

S (m)	*p* (Pa)
0.00	506.6
0.02	497.7
0.04	471.6
0.06	430.1
0.08	376.2
0.10	313.7
0.12	246.9
0.14	180.7
0.16	119.7
0.18	68.3
0.20	29.9
0.22	7.5
0.2356	2.0

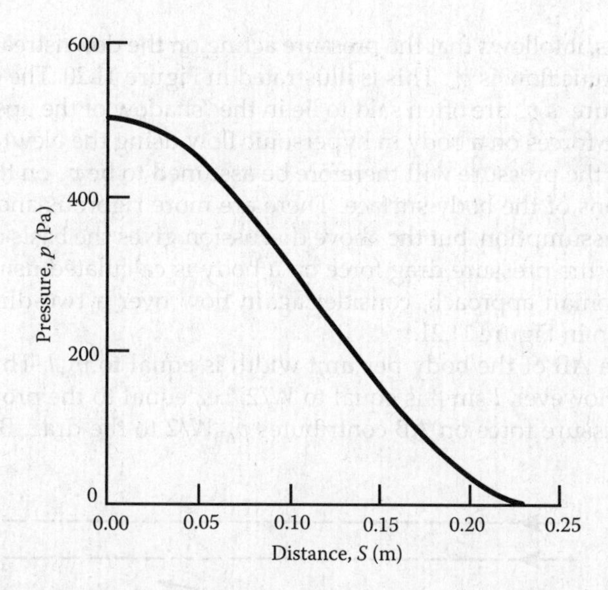

FIGURE E11.2b
Pressure variation around nose of body considered.

Forces on a Body

The Newtonian or modified Newtonian model gives the pressure distribution on the upstream faces (e.g., faces AB and BC of the two-dimensional wedge-shaped body shown in Figure 11.19) of a body in a hypersonic flow to an accuracy that is acceptable for many purposes. To find the net force acting on a body it is also necessary to know the pressures acting on the downstream faces of the body (e.g., face AC of the body shown in Figure 11.19).

Now, as discussed above, in hypersonic flow, it is effectively only when the flow reaches the surface that it is influenced by the presence of the surface. The flow that does not reach the surface is therefore unaffected by the body. The flow leaving the upstream faces of the body therefore turns parallel to the original flow direction. Since the flow is then all parallel to the original flow direction and since the pressure in the outer part of the flow that was not affected by the presence of the body is p_∞, the pressure throughout this downstream flow

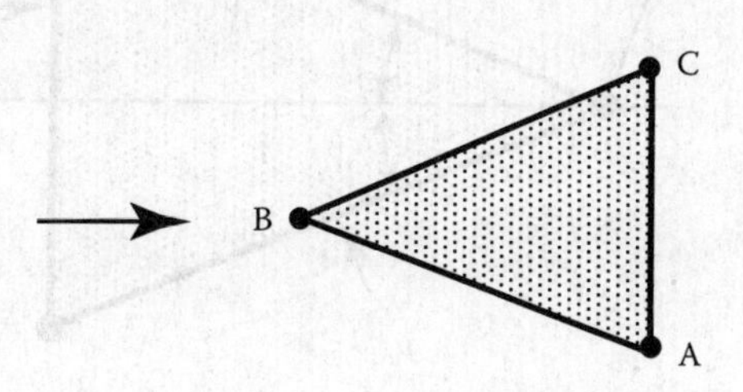

FIGURE 11.19
Two-dimensional flow over a wedge-shaped body in hypersonic flow.

will be p_∞. From this, it follows that the pressure acting on the downstream faces of body in Newtonian hypersonic flow is p_∞. This is illustrated in Figure 11.20. The downstream faces on which the pressure is p_∞ are often said to lie in the "shadow of the upstream flow."

In calculating the forces on a body in hypersonic flow using the Newtonian or modified Newtonian model, the pressure will therefore be assumed to be p_∞ on the downstream or "shadowed" portions of the body surface. There are more rigorous and elegant methods of arriving at this assumption, but the above discussion gives the basis of the argument.

To illustrate how the pressure drag force on a body is calculated using the Newtonian or modified Newtonian approach, consider again flow over a two-dimensional wedge-shaped body shown in Figure 11.21.

The force on face AB of the body per unit width is equal to $p_{AB}l$. This contributes $p_{AB}l \sin\beta$ to the drag. However, $l \sin\beta$ is equal to $W/2$, i.e., equal to the projected area of face AB. Hence, the pressure force on AB contributes $p_{AB}W/2$ to the drag. Because the wedge

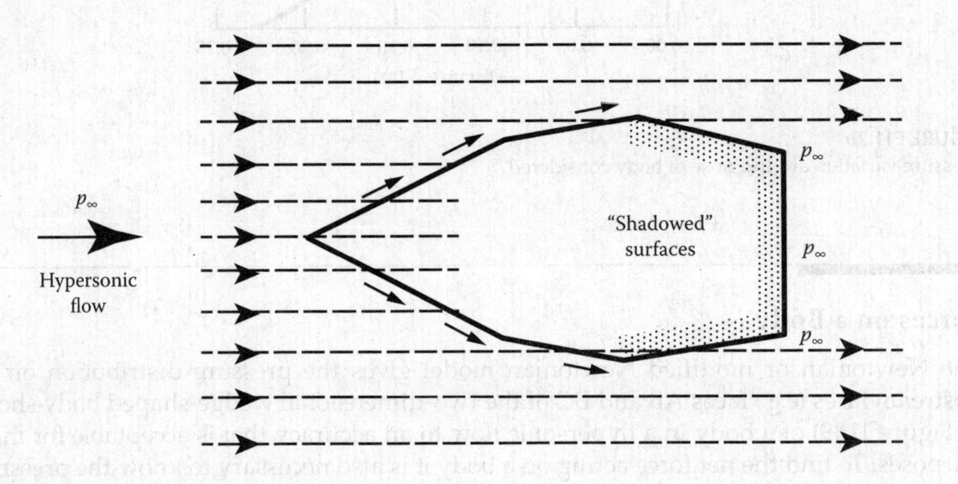

FIGURE 11.20
"Shadowed" areas of a body in hypersonic flow.

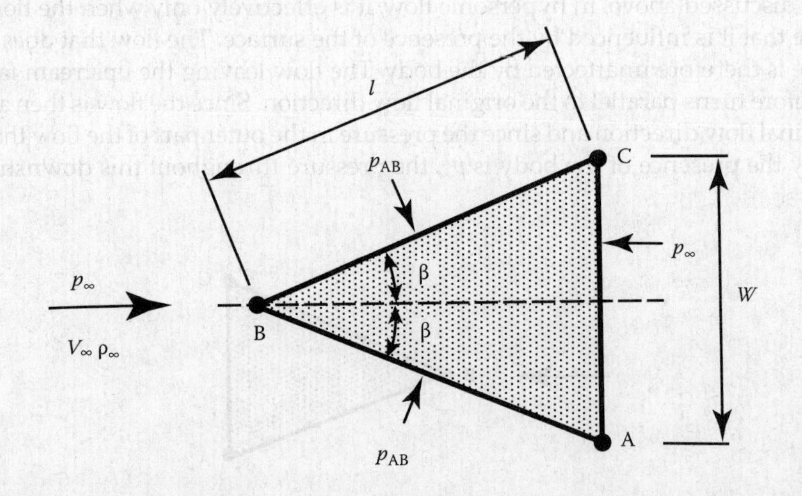

FIGURE 11.21
Pressures acting on faces of wedged-shaped body.

is symmetrically placed with respect to the freestream flow, the pressure on BC will be equal to that on AB so the pressure force on BC will also contribute $p_{AB}W/2$ to the drag. Therefore, since AC is a shadowed surface on which the pressure is assumed to be p_∞, the drag on the wedge per unit width is given by

$$D = 2\frac{p_{AB}W}{2} - p_\infty W = (p_{AB} - p_\infty)W \tag{11.16}$$

Now the drag coefficient for the type of body being considered is defined by

$$C_D = \frac{D}{\frac{1}{2}\rho_\infty V_\infty^2 \times \text{projected area}}$$

but since unit width is being considered, the projected area normal to the freestream flow direction is equal to W, hence

$$C_D = \frac{D}{\frac{1}{2}\rho_\infty V_\infty^2 W} \tag{11.17}$$

Combining Equations 11.16 and 11.17 then gives

$$C_D = \frac{p_{AB} - p_\infty}{\frac{1}{2}\rho_\infty V_\infty^2}$$

Using Equation 11.8, this equation then gives

$$C_D = \frac{(p_S - p_\infty)\sin^2\beta}{\frac{1}{2}\rho_\infty V_\infty^2} = C_{p_S}\sin^2\beta \tag{11.18}$$

where β is the half-angle of the wedge.

The Newtonian model gives $C_{p_S} = 2$ so the Newtonian model gives for flow over a wedge

$$C_D = 2\sin^2\beta \tag{11.19}$$

It must be stressed that the above analysis only gives the pressure drag on the surface. In general, there will also be a viscous drag on the body. However, if the body is relatively blunt, i.e., if the wedge angle is not very small, the pressure drag will be much greater than the viscous drag.

To illustrate how the drag on an axisymmetric body is calculated, consider flow over a conical body symmetrically placed with respect to the freestream. Consider a small portion of the surface of this body on the forward face of this body and the equivalent small portion of the surface on the shadowed face as shown in Figure 11.22.

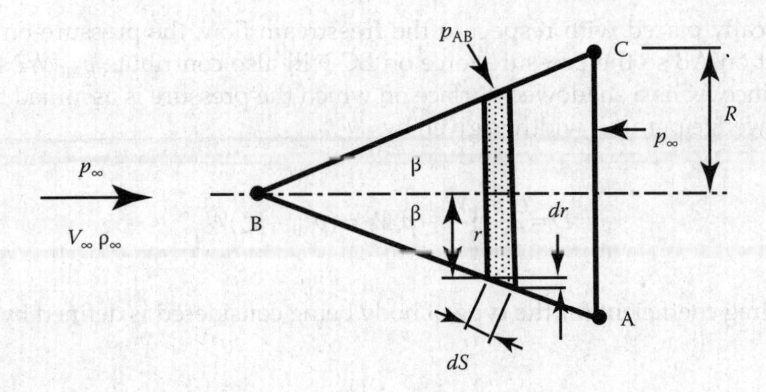

FIGURE 11.22
Surface sections used in determining drag force on a conical body.

The force on the forward-facing section is equal to $p_{AB}2\pi r\, dS$. This contributes an amount equal to $p_{AB}2\pi r\, dS \sin\theta$ to the drag. However, $2\pi r\, dS \sin\beta$ is equal to $2\pi r\, dr$, i.e., equal to the projected area of the section. Hence, since the pressure on the shadowed wall section is assumed to be p_∞, the net contribution to the drag force by the two wall sections is given by

$$dD = (p_{AB} - p_\infty)2\pi r\, dr$$

This can be integrated to get the total drag force acting on the cone. This gives

$$D = \int_0^R (p_{AB} - p_\infty)2\pi r\ dr$$

i.e., since p_{AB} is a constant

$$D = (p_{AB} - p_\infty)\pi R^2 \tag{11.20}$$

Using Equation 11.8, this gives

$$D = (p_S - p_\infty)\sin^2 \beta\pi R^2 \tag{11.21}$$

The drag coefficient based on the projected frontal area πR^2 is therefore given by

$$C_D = \frac{(p_S - p_\infty)\sin^2\beta}{\frac{1}{2}\rho_\infty V_\infty^2} = C_{p_S} \sin^2\beta \tag{11.22}$$

Again, since the Newtonian model gives $C_{p_S} = 2$, the Newtonian model gives for flow over a cone

$$C_D = 2\sin^2\beta \tag{11.23}$$

Example 11.3

Using the modified Newtonian model and using the limiting perfect gas normal shock result to get the stagnation point pressure, derive an expression for the drag on a cylinder whose axis is normal to a hypersonic air flow. Use this result to find the drag force per m span on a 0.1-m-diameter cylinder moving at Mach 8 through air at an ambient pressure of 10 Pa.

Solution

The flow situation being considered here is shown in Figure E11.3.

Considering the surface sections shown in Figure E11.3, the net contribution to the drag force on the cylinder resulting from the pressures acting on the two surface sections is equal to $(p - p_\infty) \times$ projected area of surface sections, i.e., equal to $(p - p_\infty)\,dy$. The total drag on the cylinder is therefore given by

$$D = 2\int_0^R (p - p_\infty)\,dy$$

Now the modified Newtonian model gives using the normal shock relations to obtain the stagnation point pressure

$$\frac{p - p_\infty}{p_{SN} - p_\infty} = \sin^2\beta$$

i.e., since $\sin\beta = \sqrt{R^2 - y^2/R}$, it follows that

$$\frac{p - p_\infty}{p_{SN} - p_\infty} = \frac{R^2 - y^2}{R^2}$$

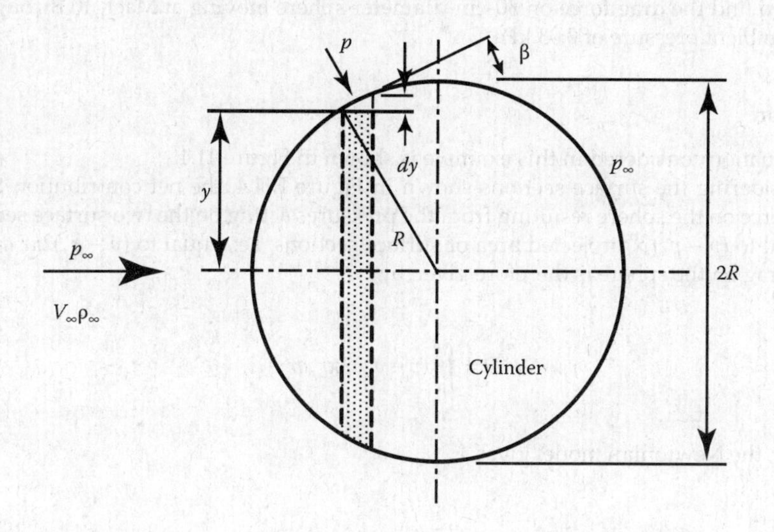

FIGURE E11.3
Flow situation considered and surface section considered.

Hence,

$$D = 2\int_0^R (p_{SN} - p_\infty)\left(\frac{R^2 - y^2}{R^2}\right)dy$$

$$= 2(p_{SN} - p_\infty)\frac{2R}{3}$$

Since the frontal area of the cylinder per unit span is 2*R*, the above result can be written in terms of the drag coefficient as

$$C_D = \frac{2}{3}C_{p_{SN}} = \frac{2 \times 1.839}{3} = 1.226$$

it having been noted that the strong shock solution for air gives $C_{p_{SN}} = 1.839$.
Now,

$$D = C_D \frac{1}{2}\rho_\infty V_\infty^2 2R = C_D \frac{p_\infty}{2}\frac{\rho_\infty V_\infty^2}{p_\infty}2R = C_D \frac{p_\infty}{2}\gamma M_\infty^2 2R$$

Hence, since C_D was shown to be 1.226 according to the form of the modified Newtonian model adopted, using the specified conditions, the drag force on the cylinder is given by

$$D = 1.266 \times 10 \times 1.4 \times 8^2 \times 0.1 = 109.9 \text{ N}$$

Therefore, the drag force on the cylinder is 109.9 N.

Example 11.4

Using the Newtonian model, derive an expression for the drag on a sphere. Use this result to find the drag force on 30-cm-diameter sphere moving at Mach 10 through air at an ambient pressure of 0.03 kPa.

Solution

The situation considered in this example is shown in Figure 11.4.

Considering the surface sections shown in Figure E11.4, the net contribution to the drag force on the sphere resulting from the pressures acting on the two surface sections is equal to $(p - p_\infty) \times$ projected area of surface sections, i.e., equal to $(p - p_\infty)2\pi r\, dr$. The total drag on the sphere is therefore given by

$$D = \int_0^R (p - p_\infty)2\pi r \; dr$$

Now, the Newtonian model gives

$$\frac{p - p_\infty}{p_S - p_\infty} = \sin^2\beta$$

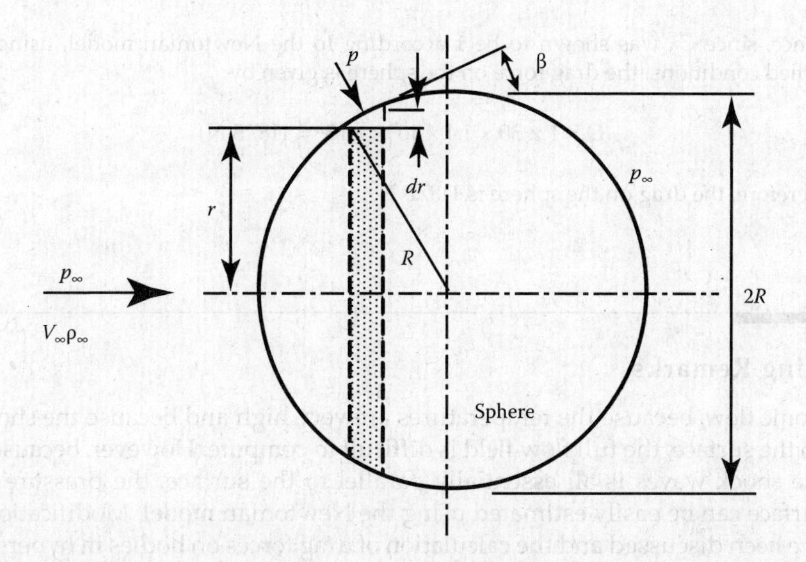

FIGURE E11.4
Flow situation considered and surface section considered.

i.e., since $\sin\beta = \sqrt{R^2 - r}/R$, it follows that

$$\frac{p - p_\infty}{p_S - p_\infty} = \frac{R^2 - r^2}{R^2}$$

Hence,

$$D = \int_0^R (p_S - p_\infty) 2\pi r \left(\frac{R^2 - r^2}{R^2} \right) dr$$

$$= (p_S - p_\infty)\pi \frac{R^2}{2}$$

Since the frontal area of the sphere is πR^2, the above result can be written in terms of the drag coefficient as

$$C_D = \frac{C_{p_S}}{2} = 1$$

it having again been noted that the Newtonian model gives $C_{p_S} = 2$.

Now,

$$D = C_D \frac{1}{2} \rho_\infty V_\infty^2 \pi R^2 = C_D \frac{p_\infty}{2} \frac{\rho_\infty V_\infty^2}{p_\infty} \pi R^2 = C_D \frac{p_\infty}{2} \gamma M_\infty^2 \pi R^2$$

Hence, since C_D was shown to be 1 according to the Newtonian model, using the specified conditions, the drag force on the sphere is given by

$$D = 1 \times 30 \times 1.4 \times 10^2 \times \pi 0.3^2 = 1187.5\ \text{N}$$

Therefore, the drag on the sphere is 1187.5 N.

Concluding Remarks

In hypersonic flow, because the temperatures are very high and because the shock waves lie close to the surface, the full flow field is difficult to compute. However, because the flow behind the shock waves is all essentially parallel to the surface, the pressure variation along a surface can be easily estimated using the Newtonian model. Modifications to this model have been discussed and the calculation of drag forces on bodies in hypersonic flow using this method has been considered.

PROBLEMS

1. A flat plate is set at an angle of 3° to an air flow at Mach 8 in which the pressure is 1 kPa. Estimate the pressure acting on the lower surface of this plate.
2. Air, at a pressure of 10 Pa, flowing at Mach 8 passes over a body that has a semi-circular leading edge with a radius of 0.15 m. Assuming the flow to be two-dimensional, find the pressure acting on this nose portion of the body at a distance of 0.1 m around the surface measured from the leading edge of the body. The flow situation is shown in Figure P11.2.
3. Using the Newtonian model, estimate the pressures acting on surfaces 1, 2, and 3 of the body shown in Figure P11.3.
4. Consider hypersonic flow over the body shape indicated in Figure P11.4. Using Newtonian theory, derive an expression for the pressure distribution around the surface of this body in terms of the distance from the stagnation point, S.
5. Consider two-dimensional air flow at Mach 7 over the body shown in Figure P11.5. The pressure in the flow ahead of the body is 12 Pa. Using the Newtonian method, find the pressures acting on the surfaces 1, 2, and 3 indicated in Figure P11.5.
6. A wedge-shaped body has an included angle of 40° and a base width of 1.5 m. Using the Newtonian model and assuming two-dimensional flow, find the drag

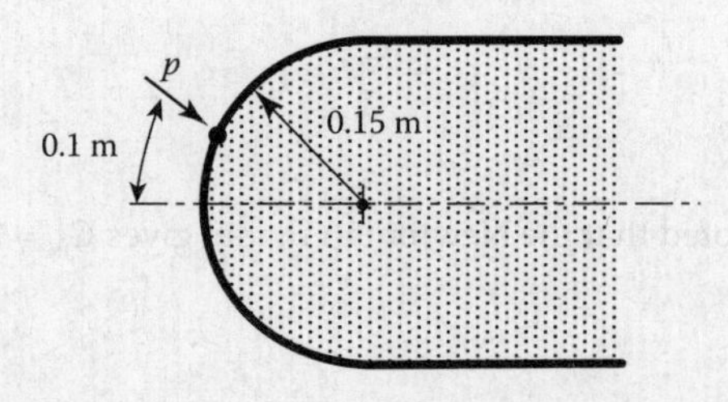

FIGURE P11.2
Flow geometry considered.

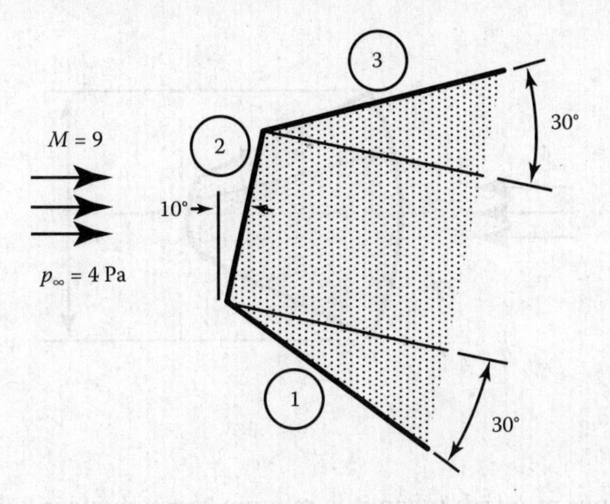

FIGURE P11.3
Flow situation considered.

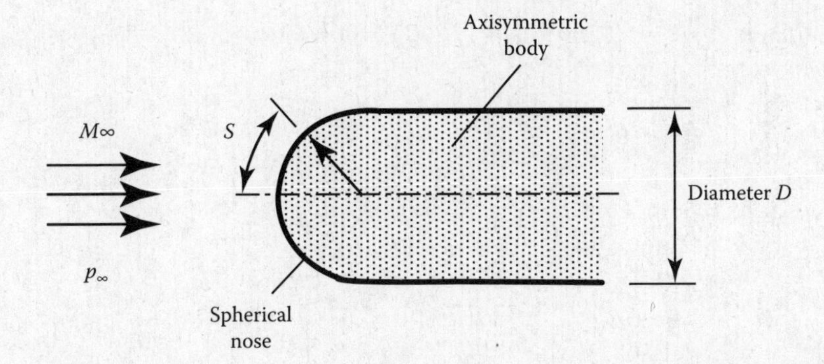

FIGURE P11.4
Flow situation considered.

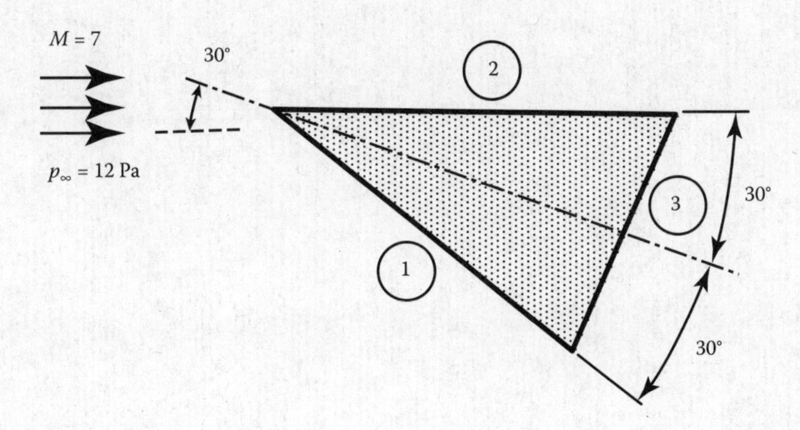

FIGURE P11.5
Flow situation considered.

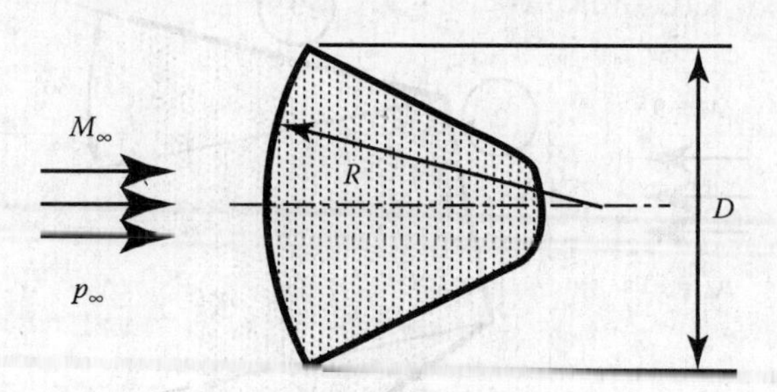

FIGURE P11.7
Flow situation considered.

on the wedge per m width when it is moving through air in which the ambient pressure is 10 Pa at Mach 7.

7. The axisymmetric body shown in Figure P11.7 is an approximate model of some earlier spacecraft. Using the Newtonian model, derive an expression for the drag coefficient for this body in hypersonic flow.

12

High-Temperature Flows

Introduction

Chapter 1 contains a discussion of some of the assumptions that are commonly adopted in the analysis of compressible gas flows, i.e., of the assumptions that are commonly used in modeling such flows. It was explained in Chapter 1 that most such analyses are based on the assumptions that

- The specific heats of the gas are constant
- The perfect gas law, $p/\rho = RT$, applies
- There are no changes in the physical nature of the gas in the flow
- The gas is in thermodynamic equilibrium

However, as discussed in the previous chapter, if the temperature in the flow becomes very high, it is possible that some of these assumptions will cease to be valid. To investigate again (a discussion of this was also given in the previous chapter) whether it is possible to get such high temperatures in a flow, consider the flow of air through a normal shock wave. The air will be assumed to have a temperature of 216.7 K (i.e., –56.3°C) ahead of the shock. This is the temperature in the so-called standard atmosphere between an altitude of ~11,000 and ~25,000 m. The situation considered is therefore as shown in Figure 12.1a.

If the assumptions discussed above apply, it was shown in Chapter 5 that

$$\frac{T_2}{216.7} = \frac{[2\gamma M_1^2 - (\gamma - 1)][2 + (\gamma - 1)M_1^2]}{(\gamma + 1)^2 M_1^2}$$

Hence, since $\gamma = 1.4$ for air, the temperature downstream of the shock will be given by

$$\frac{T_2}{216.7} = \frac{(2.8M_1^2 - 0.4)(2 + 0.4M_1^2)}{5.76M_1^2} \tag{12.1}$$

The variation of T_2 with M_1 given by this equation is shown in Figure 12.2.

If instead of passing through a normal shock, the flow is brought to rest isentropically (see Figure 12.1b), the temperature attained, i.e., the stagnation temperature, is given by

$$\frac{T_2}{216.7} = 1 + \frac{\gamma - 1}{2} M_1^2$$

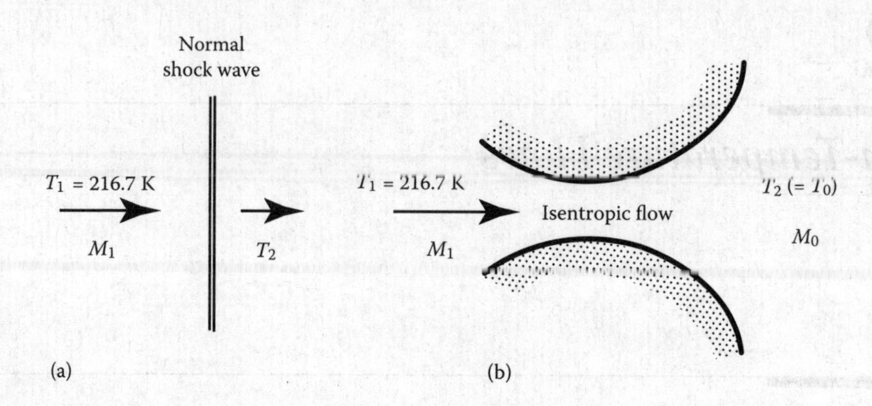

FIGURE 12.1
Flow situations considered: (a) normal shock; (b) isentropic deceleration.

i.e., since the case of $\gamma = 1.4$ is being considered

$$\frac{T_2}{216.7} = 1 + 0.2M_1^2 \tag{12.2}$$

The variation of T_2 with M_1 given by the equation is also shown in Figure 12.2. It will be seen from Figure 12.2 that at Mach numbers of roughly 5 or greater, T_2 exceeds 1000°C in both types of flow considered. When temperatures as high as this exist in a flow, it seems prudent to investigate the applicability of the assumptions on which the analysis of the flow is based, i.e., in particular, to consider whether at such temperatures the specific heats can still be assumed to be independent of temperature, whether the perfect gas law is still applicable, and whether dissociation of the gas molecules and, perhaps, even ionization of the atoms is likely to occur. These effects will be examined in this chapter. However, only a brief introduction to the very important topic of high temperature gas flows can be given

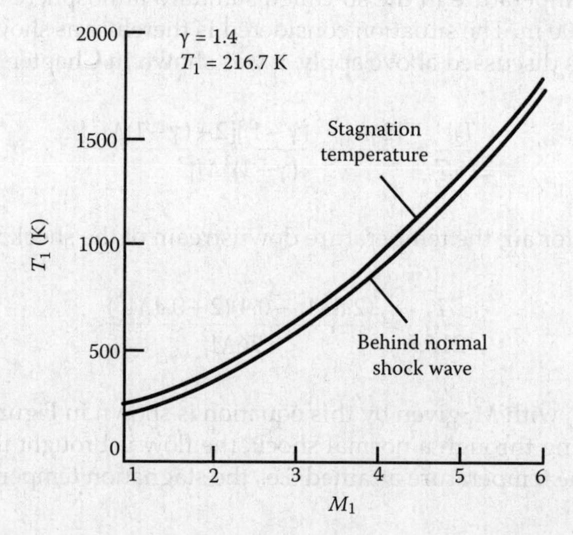

FIGURE 12.2
Mach number dependence of temperature behind normal shock wave and in isentropic flow considered.

in this chapter despite the fact that such flows occur in a number of situations of great practical importance.

Effect of Temperature on Specific Heats

The first high temperature gas effect considered here is the possibility of changes in the specific heats of the gas at high temperatures. The increase of internal energy that results when the temperature of a gas is increased is associated with an increase in the energy possessed by the gas molecule. Now the increase in the energy of a molecule can be associated with an increase in the translational kinetic energy or with an increase in the rotational kinetic energy or with an increase in the vibrational kinetic energy of the molecule. This is illustrated in Figure 12.3.

In addition, at high temperatures, changes in the energy associated with the electron motion can occur. Therefore,

$$\Delta e = \Delta e_{trans} + \Delta e_{rot} + \Delta e_{vib} + \Delta e_{el} \tag{12.3}$$

where Δe is the change in internal energy, Δe_{trans} is the change in translational energy, Δe_{rot} is the change in rotational energy, Δe_{vib} the change in vibrational energy, and Δe_{el} is the change in electron energy. Measuring e from 0 at absolute zero temperature this gives

$$e = e_{trans} + e_{rot} + e_{vib} + e_{el} \tag{12.4}$$

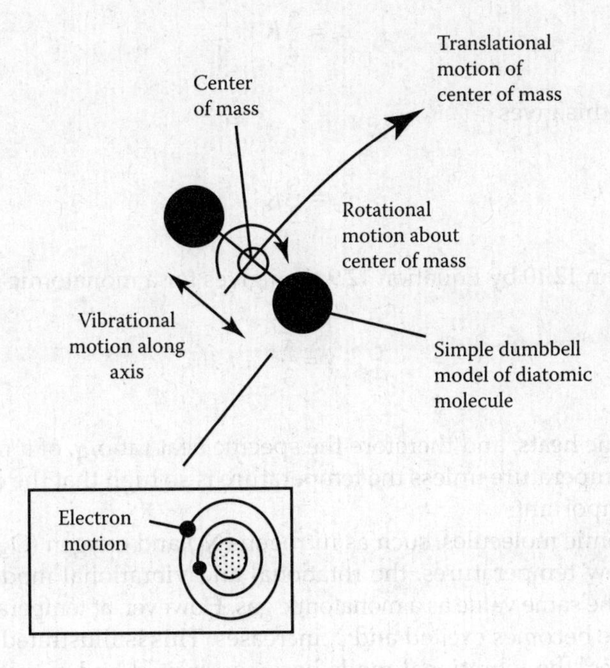

FIGURE 12.3
Excitation modes of a diatomic molecule.

The specific heat at constant volume, c_v, is given by

$$c_v = \frac{\partial e}{\partial T} \tag{12.5}$$

it follows that

$$c_v = \frac{\partial e_{\text{trans}}}{\partial T} + \frac{\partial e_{\text{rot}}}{\partial T} + \frac{\partial e_{\text{vib}}}{\partial T} + \frac{\partial e_{\text{el}}}{\partial T} \tag{12.6}$$

The effect of e_{el} can be neglected for most high temperature gas flow applications. This assumption will be adopted in the following analysis.

First, consider a monatomic gas such as helium (He), argon (Ar), or neon (Ne). The molecules (atoms) of a monatomic gas have no rotational or vibrational energy so for such a gas

$$c_v = \frac{\partial e_{\text{trans}}}{\partial T} \tag{12.7}$$

Statistical thermodynamics gives

$$e_{\text{trans}} = \frac{3}{2}RT \tag{12.8}$$

Thus, for a monatomic gas, Equation 12.7 gives

$$c_v = \frac{3}{2}R \tag{12.9}$$

Since $c_p - c_v = R$, this gives

$$c_p = \frac{5}{2}R \tag{12.10}$$

Dividing Equation 12.10 by Equation 12.9 then gives for a monatomic gas

$$\gamma = \frac{5}{3} \tag{12.11}$$

Hence, the specific heats, and therefore the specific heat ratio, γ, of a monatomic gas, do not change with temperature unless the temperature is so high that the change in electron energy becomes important.

Gases with diatomic molecules, such as nitrogen (N_2) and oxygen (O_2) will next be considered. At very low temperatures, the rotational and vibrational modes are not excited and c_v and c_p have the same value as a monatomic gas. However, at temperatures above 100 K, the rotational mode becomes excited and c_v increases. This is illustrated in Figure 12.4. At higher temperatures, the vibrational mode becomes excited and c_v then again increases. This is also shown in Figure 12.4.

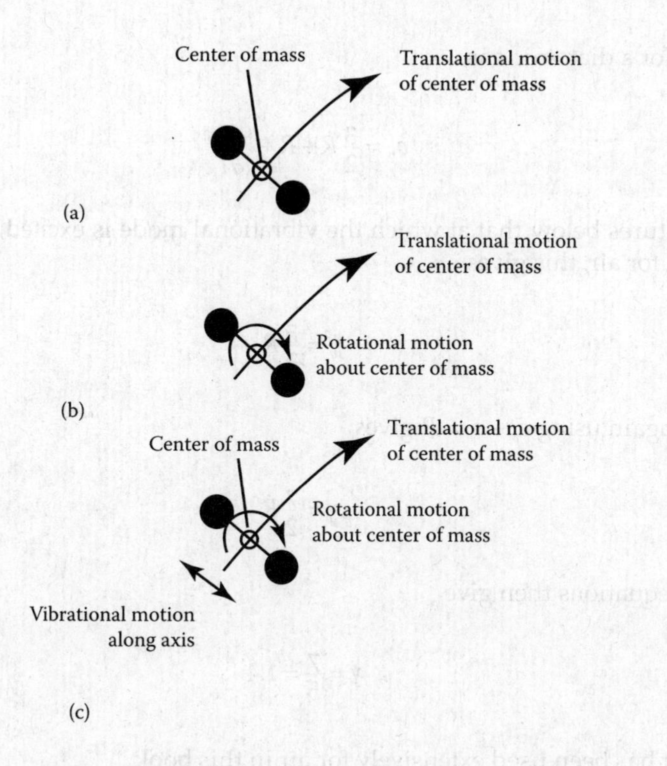

FIGURE 12.4
Effect of temperature on excitation modes of a diatomic molecule: (a) "low" temperatures, (b) "intermediate" temperatures, and (c) "high" temperatures.

Approximate temperatures at which the rotational energy mode becomes fully excited for various diatomic gases are listed in Table 12.1. These results indicate that under all conditions conventionally encountered in gas flows, the rotational mode will be fully excited.

Now, statistical thermodynamics indicates that for a diatomic gas

$$e_{\text{rot}} = RT \tag{12.12}$$

Thus, for a diatomic gas,

$$e = \frac{3}{2}RT + RT + e_{\text{vib}} \tag{12.13}$$

TABLE 12.1

Rotational Excitational Temperatures for Various Gases

Gas	Rotational Excitation Temperature (K)
Hydrogen (H_2)	86
Nitrogen (N_2)	3
Oxygen (O_2)	2
Carbon monoxide (CO)	3

Therefore, for a diatomic gas,

$$c_v = \frac{3}{2}R + R + \frac{\partial e_{vib}}{\partial T} \tag{12.14}$$

At temperatures below that at which the vibrational mode is excited, e.g., very roughly below 2000 K for air, this gives

$$c_v = \frac{5}{2}R \tag{12.15}$$

Therefore, again using $c_p - c_v = R$ gives

$$c_p = \frac{7}{2}R \tag{12.16}$$

These two equations then give

$$\gamma = \frac{7}{5} = 1.4 \tag{12.17}$$

the value that has been used extensively for air in this book.

At higher gas temperatures, the vibrational mode becomes excited. A simple model of a diatomic molecule gives

$$\frac{\partial e_{vib}}{\partial T} = R\left[\frac{\theta_{vib}}{2}\right]^2 \frac{e^{\theta_{vib}/T}}{[e^{\theta_{vib}/T} - 1]^2} \tag{12.18}$$

This is an approximate equation and may not be adequate in some situations. It only applies to a diatomic gas. More complex equations are available for gases with more complex molecules. θ_{vib} depends on the type of gas involved and some approximate values are given in Table 12.2.

Substituting Equation 12.18 into Equation 12.14 then gives, if electronic energy is neglected

$$c_v = \frac{5}{2}R + R\left[\frac{\theta_{vib}}{T}\right]^2 \frac{e^{\theta_{vib}/T}}{[e^{\theta_{vib}/T} - 1]^2} \tag{12.19}$$

TABLE 12.2

θ_{vib} Values for Various Gases

Gas	**θ_{vib} (K)**
Hydrogen (H_2)	6140
Oxygen (O_2)	2260
Nitrogen (N_2)	3340
Carbon monoxide (CO)	3120

Again using $c_p - c_v = R$, this equation indicates that if the vibration mode is excited

$$c_p = \frac{7}{2}R + R\left[\frac{\theta_{vib}}{T}\right]^2 \frac{e^{\theta_{vib}/T}}{[e^{\theta_{vib}/T} - 1]^2} \tag{12.20}$$

Hence, using Equations 12.19 and 12.20

$$\gamma = 1.4\left\{\frac{1 + \dfrac{2}{7}\left[\dfrac{\theta_{vib}}{T}\right]^2 e^{\theta_{vib}/T}/[e^{\theta_{vib}/T} - 1]^2}{1 + \dfrac{2}{5}\left[\dfrac{\theta_{vib}}{T}\right]^2 e^{\theta_{vib}/T}/[e^{\theta_{vib}/T} - 1]^2}\right\} \tag{12.21}$$

It will be noted that if the temperature, T, is high, i.e., if (θ_{vib}/T) is small, then since e^x tends to $1 + x$ for small x

$$\left[\frac{\theta_{vib}}{T}\right]^2 \frac{e^{\theta_{vib}/T}}{[e^{\theta_{vib}/T} - 1]^2} \approx \frac{[e^{\theta_{vib}/T} - 1]^2}{e^{\theta_{vib}/T}} \approx 1 \tag{12.22}$$

Hence, the above equations indicate that the following limiting values will be reached by a diatomic gas at high temperatures

$$c_v = \frac{7}{2}R, \quad c_p = \frac{9}{2}R, \quad \gamma = \frac{9}{7} \tag{12.23}$$

Thus, for a diatomic gas the specific heats are as follows

$$\begin{aligned}
&T \text{ very low:} && c_v = \frac{3}{2}R, && c_p = \frac{5}{2}R, && \gamma = \frac{5}{3}\\
&T \text{ intermediate values:} && c_v = \frac{5}{2}R, && c_p = \frac{7}{2}R, && \gamma = \frac{7}{5}\\
&T \text{ high:} && c_v = \frac{7}{2}R, && c_p = \frac{9}{2}R, && \gamma = \frac{9}{7}
\end{aligned} \tag{12.24}$$

This is illustrated by the results for nitrogen shown in Figure 12.5.

To illustrate the effect of vibrational energy on high-temperature flows, consider the variation of the stagnation temperature with M. Since in generating the stagnation conditions the flow is brought to rest adiabatically, the energy equation gives

$$h + \frac{V^2}{2} = h_0 \tag{12.25}$$

which gives

$$\begin{aligned}
\frac{V^2}{2} = h_0 - h = \int_T^{T_0} c_p\,dT &= R\int_T^{T_0}\left[\frac{7}{2} + \left(\frac{\theta_{vib}}{T}\right)^2 \frac{e^{\theta_{vib}/T}}{(e^{\theta_{vib}/T} - 1)^2}\right]dT\\
&= R\left[\frac{7}{2} + \frac{\theta_{vib}}{e^{\theta_{vib}/T} - 1}\right]_T^{T_0}
\end{aligned} \tag{12.26}$$

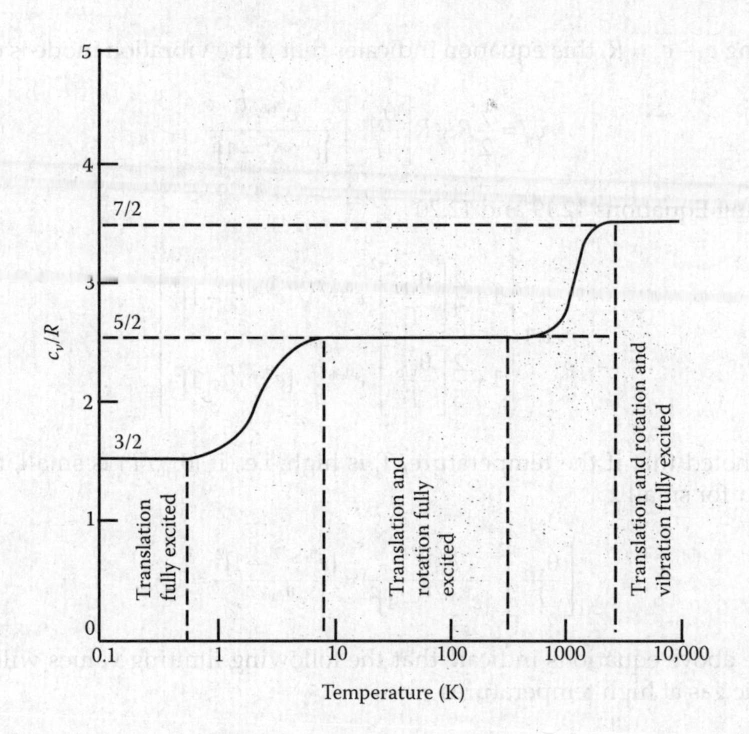

FIGURE 12.5
Specific heat variation with temperature for nitrogen.

i.e.,

$$\frac{V^2}{2} = R\left\{\frac{7}{2}(T_0 - T) + \theta_{\text{vib}}\left[\frac{1}{e^{\theta_{\text{vib}}/T} - 1} - \frac{1}{e^{\theta_{\text{vib}}/T} - 1}\right]\right\} \tag{12.27}$$

The speed of sound is, as before, given by

$$a^2 = \gamma RT$$

Thus, the above equation can be written as

$$M^2 = \left(\frac{2}{\gamma}\right)\left\{\frac{7}{2}\left[\frac{T_0}{T} - 1\right] + \left(\frac{\theta_{\text{vib}}}{T}\right)\left[\frac{1}{e^{\theta_{\text{vib}}/T_0} - 1} - \frac{1}{e^{\theta_{\text{vib}}/T} - 1}\right]\right\} \tag{12.28}$$

The specific heat ratio, γ, is given by Equation 12.21.

Equation 12.28 gives the relation between T_0, M, and T. For the situation discussed above, i.e., for flow with $T = 216.7$ K, it gives the variation of T_0 with T that is shown in Figure 12.6. Also, shown in this figure is the variation given by assuming that c_p is constant and that $\gamma = 1.4$. The vibrational excitation will be seen to have a significant effect on T_0 when M is high.

The stagnation pressure can be found by noting that the entropy change between any two points in the flow is given by

$$s_2 - s_1 = \int_{T_1}^{T_2} c_p \frac{dT}{T} - \int_{p_1}^{p_2} R\frac{dp}{p} \tag{12.29}$$

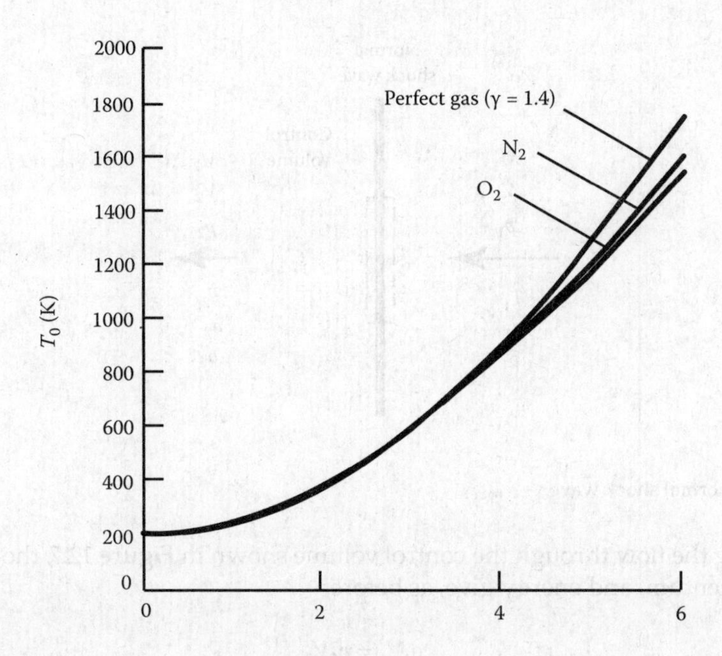

FIGURE 12.6
Typical variation of stagnation temperature with Mach number.

Since the flow being considered is isentropic, i.e., $s_2 - s_1 = 0$, and because R is constant, this equation gives

$$R\ln\left(\frac{p_0}{p}\right) = \int_T^{T_0} c_p \frac{dT}{T} \tag{12.30}$$

Using Equation 12.20 then gives

$$\ln\left(\frac{p_0}{p}\right) = \int_T^{T_0}\left[\frac{7}{2} + \left(\frac{\theta_{vib}}{T}\right)^2 \frac{e^{\theta_{vib}/T}}{(e^{\theta_{vib}/T}-1)^2}\right]\frac{dT}{T} = \frac{7}{2}\ln\left(\frac{T_0}{T}\right) + \left[\left(\frac{\theta_{vib}}{T}\right)\frac{e^{\theta_{vib}/T}}{(e^{\theta_{vib}/T}-1)} - \ln(e^{\theta_{vib}/T}-1)\right]_T^{T_0}$$

i.e.,

$$\frac{p_0}{p} = \left[\frac{T_0}{T}\right]^{7/2}\left[\frac{e^{\theta_{vib}/T_0}-1}{e^{\theta_{vib}/T}-1}\right]\exp\left\{\left(\frac{\theta_{vib}}{T_0}\right)\left[\frac{e^{\theta_{vib}/T_0}}{e^{\theta_{vib}/T_0}-1}\right] - \left(\frac{\theta_{vib}}{T}\right)\left[\frac{e^{\theta_{vib}/T}}{e^{\theta_{vib}/T}-1}\right]\right\} \tag{12.31}$$

Since (T_0/T) is given by Equation 12.28, this equation allows p_0 to be found for any values of M, T, and p. If the density is required, it can then be found, using $p/\rho = RT$, i.e., using $\rho_0/\rho = (p_0/p)(T/T_0)$.

As another example of high-temperature effects on gas flows consider the flow through a normal shock wave under such conditions that the effects of vibrational excitation are significant. The variables are defined in Figure 12.7. Explicit expressions for the changes across the shock wave will not be derived. Instead, a procedure that allows the downstream conditions to be found for any prescribed values of the upstream conditions will be discussed.

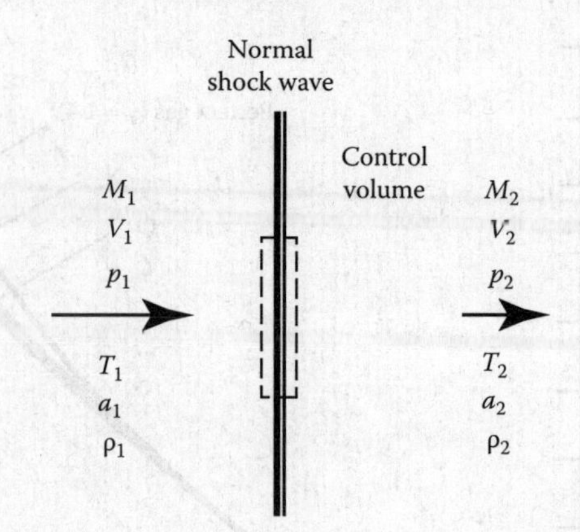

FIGURE 12.7
Flow through a normal shock wave.

Considering the flow through the control volume shown in Figure 12.7, the conservation of mass, momentum, and energy give, as before,

$$\rho_1 V_1 = \rho_2 V_2 \tag{12.32}$$

$$p_1 + \rho_1 V_1^2 = p_2 + \rho_2 V_2^2 \tag{12.33}$$

$$h_1 + \frac{V_1^2}{2} = h_2 + \frac{V_2^2}{2} \tag{12.34}$$

The continuity equation, (12.32) gives

$$V_2 = \rho_1 V_1 / \rho_2 \tag{12.35}$$

Substituting this into the momentum equation, (12.33) then gives

$$p_1 + \rho_1 V_1^2 = p_2 + \rho_1^2 V_1^2 / \rho_2$$

i.e.,

$$p_2 = p_1 + \rho_1 V_1^2 \left[1 - \frac{\rho_1}{\rho_2} \right] \tag{12.36}$$

Substituting Equation 12.35 into the energy equation, (12.34) gives

$$\frac{V_1^2}{2} \left[1 - \left(\frac{\rho_1}{\rho_2} \right)^2 \right] = h_2 - h_1 \tag{12.37}$$

However,

$$h_2 - h_1 = \int_{T_1}^{T_2} c_p \, dT \tag{12.38}$$

Thus, using Equation 12.20 and the integral discussed in the derivation of Equation 12.27, the following is obtained

$$h_2 - h_1 = R\left\{\frac{7}{2}(T_2 - T_1) + \theta_{vib}\left[\left(\frac{1}{e^{\theta_{vib}/T_2} - 1}\right) - \left(\frac{1}{e^{\theta_{vib}/T_1} - 1}\right)\right]\right\} \quad (12.39)$$

Lastly, it is noted that the perfect gas equation gives $p/\rho = RT$, i.e., gives

$$\left(\frac{p_2}{p_1}\right)\left(\frac{\rho_1}{\rho_2}\right) = \left(\frac{T_2}{T_1}\right) \quad (12.40)$$

Equations 12.36, 12.37, 12.39, and 12.40 constitute a set of four simultaneous equations in the four unknowns p_2, ρ_2, $(h_2 - h_1)$, and T_2. These can be solved to give the values of the variables using, for example, an iterative technique. For example, T_2 can be guessed, e.g., it could be set equal to $2T_1$, and Equation 12.39 can then be used to solve for $h_2 - h_1$. Substituting this value into Equation 12.37 then allows the corresponding value of ρ_2 to be found. Substituting this into Equation 12.36 then gives p_2. Substituting these values of p_2 and ρ_2 into Equation 12.40 then gives a new value of T_2. The whole procedure can then be repeated until the values of T_2 ceases to change. Much more efficient and elegant solution procedures are available but that given here should indicate the basic ideas involved. This procedure has been used to determine the variation of T_2 with M_1 for the case of $T_1 = 216.7$ K discussed before. The results are compared in Figure 12.8 with those obtained before assuming constant specific heats.

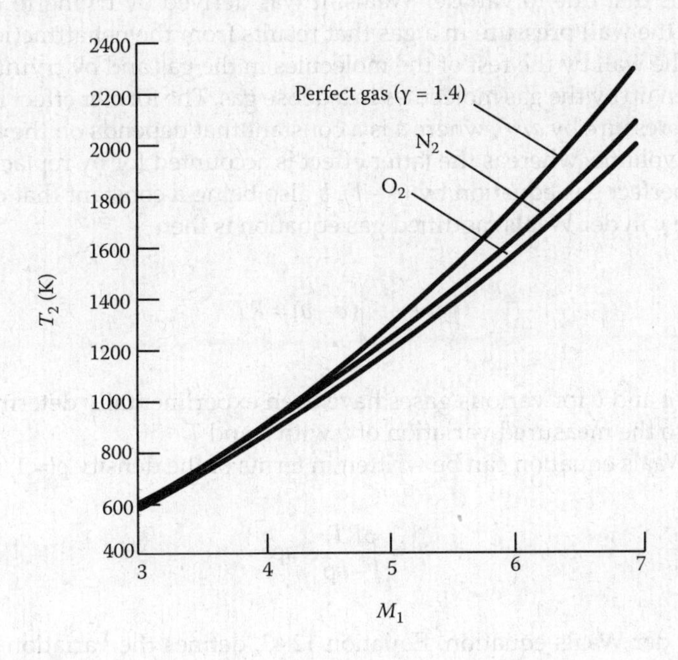

FIGURE 12.8
Results for a normal shock wave in air with $T_1 = 216.7$ K.

All the other flows discussed earlier, such as isentropic flow through a nozzle, can be analyzed for the case where the vibrational excitation of a diatomic gas becomes important using the approach outlined above. These analyses, and those discussed above, indicate that the effects of specific heat changes, caused by the excitation of the vibrational modes in diatomic gases and in gases with more complex molecules, are likely to be important in high Mach number flows.

Perfect Gas Law

The perfect gas law gives

$$\frac{p}{\rho RT} = 1 \tag{12.41}$$

Therefore, the quantity,

$$Z = \frac{p}{\rho RT} \tag{12.42}$$

which is termed the "compressibility factor," can be used as a measure of the deviation of the behavior of the gas from that of a perfect gas. For a gas that obeys the perfect gas law $Z = 1$.

One of the most widely discussed gas equation that accounts for the deviations from Equation 12.41 is that due to van der Waals. It was derived by trying to account for the modification of the wall pressure in a gas that results from the net attraction exerted on a molecule near the wall by the rest of the molecules in the gas and by trying to account for the volume taken up by the gas molecules in a dense gas. The former effect is accounted for by altering the pressure by a/v^2, where a is a constant that depends on the type of gas and v is the specific volume, whereas the latter effect is accounted for by replacing the specific volume in the perfect gas equation by $(v - b)$, b also being a constant that depends on the type of gas. The van der Waals modified gas equation is then

$$\left(p + \frac{a}{v^2}\right)(v - b) = RT \tag{12.43}$$

The values of a and b for various gases have been experimentally determined by fitting Equation 12.43 to the measured variation of v with p and T.

The van der Waals equation can be written in terms of the density $\rho(= 1/v)$ as

$$p = \frac{\rho RT}{1 - b\rho} - a\rho^2$$

Now, the van der Waals equation, Equation 12.43, defines the variation of p with v for a given value of T. The form of the variation for various values of T is as shown in Figure 12.9. For temperatures below a certain value, the curves display a minimum whereas

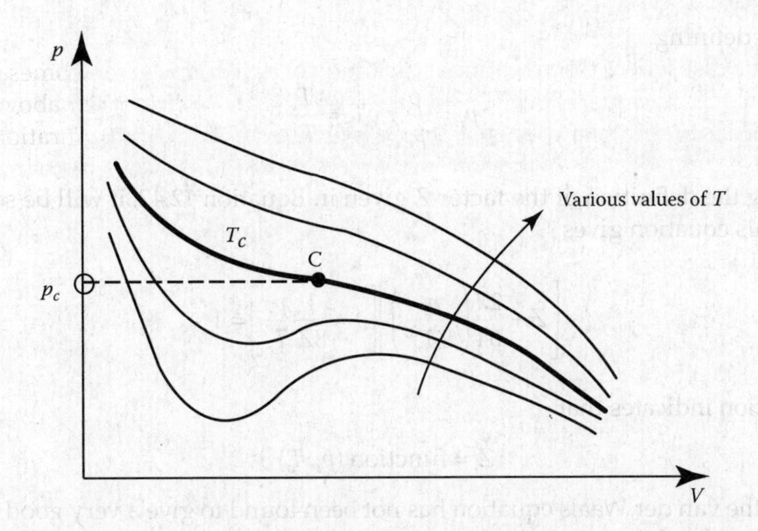

FIGURE 12.9
Isotherms given by the van der Waals equation.

for temperatures above this value, the curves have no minimum. The temperature that divides the two types of behavior is indicated in Figure 12.9 as T_c. Now, on the T_c curve, there will be a point, marked C in Figure 12.9, at which the slope is zero, i.e., at which $dp/dv = 0$. Because the slope of the curve, dp/dv, is never positive on the T_c curve, the slope has a maximum at point C, i.e., d^2p/dv^2 is 0 at point C.

Thus, point C is that at which $dp/dv = 0$ and $d^2p/dv^2 = 0$. Using the van der Waals equation, Equation 12.43, thus shows that p_c and T_c are such that

$$0 = \frac{RT_c}{(v_c - b)^2} + \frac{2a}{v_c^3}$$
$$0 = \frac{RT_c}{(v_c - b)^3} - \frac{6a}{v_c^4} \quad (12.44)$$

Solving between these then gives

$$a = \frac{27}{64}\frac{R^2T_c^2}{p_c}, \quad b = \frac{RT_c}{8p_c} \quad (12.45)$$

Substituting these values back into Equation 12.43 then gives

$$\left(p + \frac{27}{64}\frac{R^2T_c^2}{p_c v^2}\right)\left(v - \frac{RT_c}{8p_c}\right) = RT$$

i.e.,

$$\left[\frac{pv}{RT} + \frac{27}{64}\left(\frac{RT}{pv}\right)\left(\frac{T_c}{T}\right)^2\left(\frac{p}{p_c}\right)\right]\left[1 - \frac{1}{8}\left(\frac{RT}{pv}\right)\left(\frac{T_c}{T}\right)\left(\frac{p}{p_c}\right)\right] = 1 \quad (12.46)$$

Therefore, defining

$$p_r = \frac{p}{p_c}, \quad T_r = \frac{T}{T_c} \tag{12.47}$$

and recalling the definition of the factor Z given in Equation 12.42, it will be seen that the van der Waals equation gives

$$\left[Z + \frac{27}{64}\left(\frac{p_r}{ZT_r^2}\right)\right]\left[1 - \frac{1}{8Z}\frac{p_r}{T_r}\right] = 1 \tag{12.48}$$

This equation indicates that

$$Z = \text{function}\ (p_r, T_r) \tag{12.49}$$

Although the van der Waals equation has not been found to give a very good description of the behavior of real gases over a wide range of temperatures, the form of behavior derived from this equation and given in Equation 12.49 does describe the form of the behavior of real gases very well. The values of p_c and T_c for various common gases are listed in Table 12.3.

The form of the variation of Z with p_r and T_r for all gases obtained using such values of p_c and T_c is then as shown in Figure 12.10. This graph, it must be stressed, applies quite accurately for almost all gases provided the correct values of p_c and T_c are used. It will be noted from Table 12.3 that for air, p_c and T_c are approximately 3910 kPa and 132 K, respectively.

For a perfect gas, $Z = 1$. It will be seen from Figure 12.10 that the greatest deviations from perfect gas behavior occur when T_r is near 1 and p_r is near 2. In most compressible flows encountered in practice if p_r is low T_r also tends to be low and when p_r is high, say of the order 2 to 3, T_r tends also to be high. As a result, it will be seen that in such flows, Z tends to be always near 1. This means that the use of the perfect gas equation will give good results in such flows.

If the circumstances are such that Z is likely to deviate significantly from 1, a more complex equation has to be used to describe the gas behavior. The van der Waals equation discussed above is quite adequate for this purpose in many cases. However, other equations that provide a better description of the gas behavior over wider ranges of pressure and temperature have been developed. The Beattie–Bridgeman equation is typical of these equations. It gives

$$p = \frac{RT(1-\epsilon)}{v^2}(v + B) - \frac{A}{v^2} \tag{12.50}$$

TABLE 12.3
Critical Temperature and Pressure Values

Gas	T_c (K)	p_c (atm)
Air	132.41	37.25
Helium (He)	5.19	2.26
Hydrogen (H_2)	33.24	12.80
Nitrogen (N_2)	126.2	33.54
Oxygen (O_2)	154.78	50.14
Carbon monoxide (CO)	132.91	34.26
Carbon dioxide (CO_2)	304.20	72.90

Note: 1 atm = 101.3 kPa.

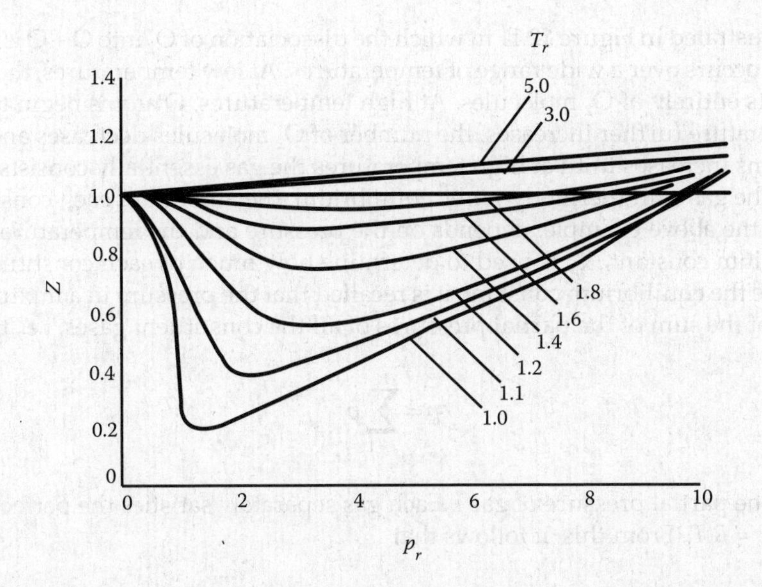

FIGURE 12.10
Variation of compressibility factor with reduced pressure, p_r.

where

$$A = A_0(1-a/v), \quad B = B_0(1-b/v), \quad \epsilon = c/vT^3 \tag{12.51}$$

A_0, B_0, a, b, and c are constants that depend on the type of gas involved.

The general conclusion that can be reached from the discussion given in this section is that the use of $p/\rho = RT$ will provide a quite adequate description of most compressible flows encountered in practice. As a check on the adequacy of the results obtained using this equation, the variations of pressure and temperature in such flows can be calculated using the perfect gas equation and then the results can be used to deduce the variations of p_r and T_r in the flow. Figure 12.10 can then be used to decide whether a more complex gas behavior equation should have been utilized in the analysis of the flow.

Dissociation and Ionization

At high temperatures, the molecules of gases that consist of two or more atoms can start to break down into simpler molecules and into the atoms of which these molecules consist, i.e., dissociation of the molecules can occur. Examples of dissociation are

$$\begin{gathered} N_2 \rightarrow N + N \\ O_2 \rightarrow O + O \\ 2H_2O \rightarrow 2H_2 + O_2 \end{gathered} \tag{12.52}$$

As a very rough guide, if the temperature of a gas goes above 2000 K, the possibility that dissociation is occurring has to be considered.

Now, as illustrated in Figure 12.11 in which the dissociation of O_2 into O + O is illustrated, dissociation occurs over a wide range of temperatures. At low temperatures, the gas essentially consists entirely of O_2 molecules. At high temperatures, O atoms begin to exist and as the temperature further increases, the number of O_2 molecules decreases and the number of O atoms increases until at high temperatures the gas essentially consists entirely of O atoms. If the gas is in thermodynamic equilibrium, the amount of each constituent, i.e., O_2 and O in the above example, depends on the pressure and the temperature of the gas. The equilibrium constant, K_p, is used to determine how much of each constituent is present. To define the equilibrium constant, it is recalled that the pressure in a mixture of gases is made up of the sum of the partial pressures of all the constituent gases, i.e., that

$$p = \sum_{i=1}^{N} p_i \tag{12.53}$$

where p_i is the partial pressure of gas i. Each gas separately satisfies the perfect gas equation, i.e., $p_i/\rho_i = R_i T_i$. From this, it follows that

$$\frac{p_i}{p} = \frac{n_i}{n_T} \tag{12.54}$$

where n_i is the number of moles of constituent i and n_T is the total number of moles.

If the dissociation equation is of the form

$$i_A A = i_B B + i_C C \tag{12.55}$$

then the equilibrium constant, K_p, is defined as

$$K_p = \frac{p_B^{i_B} p_C^{i_C}}{p_A^{i_A}} \tag{12.56}$$

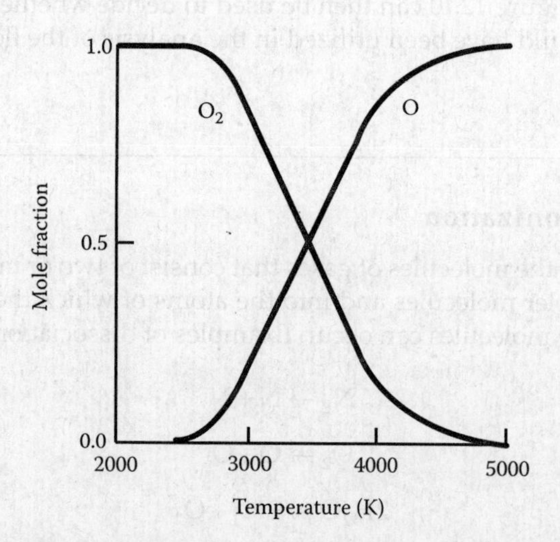

FIGURE 12.11
Effect of temperature on composition of oxygen at 1 atm.

TABLE 12.4

Variation of K_p for Dissociation with Temperature

T (K)	$\log_{10} K_p$ for $O_2 \rightarrow 2O$	$\log_{10} K_p$ for $N_2 \rightarrow 2N$
1000	–44.71	–99.08
2000	–14.41	–41.61
3000	–4.19	–22.32
4000	0.95	–12.62
5000	4.05	–6.75
10,000	10.36	5.37

where p_A, p_B, and p_C are the partial pressures of constituents A, B, and C, respectively. For example, for the reaction

$$O_2 \rightarrow O + O$$

i.e.,

$$1O_2 \rightarrow 1O + 1O$$

it will be seen that $i_A = 1$, $i_B = 1$, and $i_C = 1$. It follows therefore that

$$K_p = p_O^1 p_O^1 / p_{O_2}^1 = p_O^2 / p_{O_2}$$

Returning to the discussion of Equation 12.56, in view of Equation 12.54, Equation 12.56 gives

$$K_p = \left[\frac{n_B}{n_T}\right]^{i_B} p^{i_B} \left[\frac{n_C}{n_T}\right]^{i_C} p^{i_C} \Bigg/ \left[\frac{n_A}{n_T}\right]^{i_A} p^{i_A} = \left[\frac{n_B^{i_B} n_C^{i_C}}{n_A^{i_A}}\right] \frac{p^{i_B + i_C - i_A}}{n_T^{i_B + i_C - i_A}} \tag{12.57}$$

The equilibrium constant, K_p, is a function of temperature. Values for $O_2 = O + O$ and for $N_2 = N + N$ are listed in Table 12.4. The pressure used in expressing these values is in atmospheres. It should be noted that Table 12.4 gives values of $\log_{10} K_p$. It will be seen that K_p is therefore very small for both O_2 and N_2 below 2000 K and that it remains small for N_2 up to 4000 K.

The way in which these values of K_p are used to find the amount of dissociated gas present is illustrated in the following example.

Example 12.1

Oxygen kept at a pressure of 25 kPa is heated to temperatures of 3000, 4000, and 5000 K. Find the amounts of O_2 and O present at each temperature.

Solution

The values of K_p at each temperature are as follows, the values given in Table 12.4 being used:

$$T = 3000\text{ K}: \quad K_p = 10^{-4.19} = 6.46 \times 10^{-5}$$

$$T = 4000\text{ K}: \quad K_p = 10^{0.95} = 8.91$$

$$T = 5000\text{ K}: \quad K_p = 10^{4.05} = 1.12 \times 10^4$$

TABLE E12.1

Values of a and b with Temperature

T (K)	a	b
3000	1	0
4000	0.053	1.895
5000	0	2

The reaction being considered is

$$O_2 \rightarrow aO_2 + bO$$

but mass balance requires

$$2 = 2a + b \quad \text{(a)}$$

whereas the definition of the equilibrium constant K_p gives, as discussed above,

$$K_p = p_O^2/p_{O2} = \frac{(b/a+b)^2 p^2}{(a/a+b)p} = \frac{b^2}{a(a+b)}p$$

However, here $p = 25/101 = 0.248$ atm; thus,

$$K_p = \frac{0.248b^2}{a(a+b)} \quad \text{(b)}$$

Since Equation (a) gives $b = 2(1 - a)$, Equation (b) gives

$$K_p = \frac{0.992(1-a)^2}{2a-a^2}$$

For each value of K_p, this allows a to be found. b can then be found using Equation (a). The results so obtained are given in Table E12.1.

From these results, it will be seen that at this pressure there is essentially no dissociation below a temperature of 3000 K and the oxygen is fully dissociated above a temperature at 5000 K.

In many situations, the flow involves a mixture of gases, the most common example of this being the flow of air. To illustrate how such situations are dealt with, consider the case of the flow of air. It will be assumed that air consists of a mixture of N_2 and O_2 and therefore that the reactions of interest are

$$O_2 \rightarrow O + O$$

$$N_2 \rightarrow N + N$$

In fact, another reaction of great practical consequence can occur in high temperature air flows. This is the reaction $N + O \rightarrow NO$, a reaction that has extremely important environmental consequences. This reaction will, however, for simplicity, not be considered here.

The air pressure is made up of the partial pressures of the constituent gases, i.e.,

$$p = p_{O_2} + p_O + p_{N_2} + p_N \quad (12.58)$$

The partial pressures of the O and N are given by the definition of the equilibrium constant by

$$\frac{p_{O}^{2}}{p_{O_2}} = K_{pO_2} \tag{12.59}$$

$$\frac{p_{N}^{2}}{p_{N_2}} = K_{pN_2} \tag{12.60}$$

the equilibrium constant K_p being known functions of temperature.

In addition, an overall mass balance requires, since

$$aN_2 + bO_2 = cN_2 + dN + eO_2 + fO$$

that

$$2a = 2c + d \text{ and } 2b = 2e + f \tag{12.61}$$

The composition of the air in the undissociated state is assumed to be known, i.e., a and b are known. For example, under standard conditions, air consists of approximately 80% N_2 and approximately 20% O_2.

Equation 12.61 gives

$$\frac{2c+d}{2e+f} = \frac{a}{b} \tag{12.62}$$

the right-hand side of this equation being the ratio of the number of moles of nitrogen involved to the number of moles of oxygen involved. As mentioned above, it is a known quantity, equal to approximately 4.

Now, as discussed before (see Equation 12.54),

$$\frac{p_i}{p} = \frac{n_i}{n_T}$$

Hence,

$$\frac{p_{N_2}}{p} = \frac{c}{c+d+e+f}$$

i.e.,

$$c = \left[\frac{c+d+e+f}{p}\right] p_{N_2} \tag{12.63}$$

Similarly,

$$d = \left[\frac{c+d+e+f}{p}\right] p_{N} \tag{12.64}$$

$$e = \left[\frac{c+d+e+f}{p} \right] p_{O_2} \tag{12.65}$$

$$f = \left[\frac{c+d+e+f}{p} \right] p_{O} \tag{12.66}$$

Substituting these equations into Equation 12.62 then gives

$$\frac{2p_{N_2} + p_N}{2p_{O_2} + p_O} = \frac{a}{b} (\approx 4) \tag{12.67}$$

Equations 12.58, 12.59, 12.60, and 12.67 constitute a set of four equations in the four unknowns p_{O_2}, p_O, p_{N_2}, and p_N. The values of these partial pressures are therefore easily solved for using the known values of the equilibrium constants. Once the values of p_{O_2}, p_O, p_{N_2}, and p_N are found, the mole fraction of each constituent can be calculated. Typical results for air were given in the previous chapter. The results given there included the effect of NO production, which was not considered in the above discussion.

Example 12.2

Find the composition of air at a temperature of 5000 K and a pressure of 1 atm.

Solution

The equilibrium constants for O_2 and N_2 at 5000 K are

$$K_{p_{O_2}} = 10^{4.05} = 11{,}200$$

$$K_{p_{N_2}} = 10^{-6.75} = 1.778 \times 10^{-7}$$

Therefore, Equations 12.59 and 12.60 give

$$p_{O_2} = p_O^2 / 11{,}200.0 \tag{a}$$

$$p_{N_2} = p_N^2 / 0.0000001778 \tag{b}$$

Substituting for p_{O_2} and p_{N_2} in Equations 12.67 and 12.58 then gives

$$\frac{p_N^2}{0.0000000889} + p_N = 4 \left[\frac{p_O^2}{5600.0} + p_O \right]$$

and

$$\frac{p_O^2}{11{,}200.0} + p_O + \frac{p_N^2}{0.0000001778} + p_N = p$$

Solving between these two equations for p_O/p and p_N/p then gives

$$\frac{p_O}{p} = 0.333, \quad \frac{p_N}{p} = 3.44 \times 10^{-4}$$

Substituting these into Equations (i) and (ii) then gives

$$\frac{p_{O_2}}{p} \approx 0, \quad \frac{p_{N_2}}{p} = 0.666$$

From these results, it will be seen that, at this temperature, the oxygen is essentially all dissociated, while hardly any of the nitrogen is dissociated, the air then consisting essentially of $0.8N_2 + 0.2O + 0.2O$.

If very high temperatures are involved in the flow, the ionization of the atoms can occur. For example, in the case of air flow, the following ionization reactions become important:

$$\begin{aligned} N &\rightarrow N^+ + e^- \\ O &\rightarrow O^+ + e^- \end{aligned} \quad (12.68)$$

The degree of ionization that has occurred at a given temperature is found using an equilibrium constant defined in the same way as in the above discussion of dissociation. Values of K_p that apply to both of the reactions given in Equation 12.68 are listed in Table 12.5.

A consideration of the values given in Table 12.5 shows that ionization of N and O is not likely to be important at temperatures below roughly 10,000 K.

Example 12.3

Find the amounts of O^+ in oxygen that has been heated to a temperature of 15,000 K.

Solution

The reaction being considered is

$$O \rightarrow O^+ + e^-$$

TABLE 12.5
Variation of K_p for Ionization with Temperature

T (K)	$\log_{10} K_p$ for Ionization of N and O
1000	−163.16
2000	−78.09
3000	−49.30
4000	−34.69
5000	−25.80
10,000	−7.48
15,000	−1.02

there being a negligible amount of O_2 present at this temperature. Hence, writing

$$O \rightarrow aO + bO^+ + be^{-1}$$

it follows that

$$K_p = \frac{p_O^+}{p_O} = \frac{\left[\frac{a}{a+2b}\right]p}{\left[\frac{a}{a+2b}\right]p} = \frac{b}{a}$$

However, a mass balance requires

$$b + a = 1$$

Therefore,

$$K_p = \frac{b}{1-b}$$

Since, as will be seen from Table 12.5, $K_p = 0.0955$ at a temperature of 15,000 K, it follows that $b = 0.087$ and hence $a = 0.913$. Therefore, approximately 9% of the oxygen atoms have ionized at this temperature.

The discussion to this point has been concerned with the composition of a gas undergoing dissociation and ionization. In calculating compressible gas flows at high temperatures, it is the changes in the thermodynamic properties of the flow that are actually required. However, if the composition of the gas is known under any conditions, the enthalpy of the gas under these conditions can be obtained. To do this, the enthalpy of the gas is written in terms of a fixed energy, i.e., the energy of formation at a chosen temperature, plus the change in sensible enthalpy between the temperature at which the energy of formation is specified and the actual gas temperature, i.e., h is written as

$$h = h_f + \Delta h_{\text{sen}}$$

Δh_{sen} being the change in sensible enthalpy and h_f being the heat of formation. Δh_{sen} is evaluated using the procedures discussed in the previous section. Values of h_f for various gases are available.

The enthalpy of a mixture of gases is then found by adding the enthalpy of all the constituent gases. Computer programs for doing this for air are available. The results are also available in the form of a "Mollier chart." This gives the variation of h with s for various temperatures and pressures. The form of such a chart is illustrated in Figure 12.12, which applies to air.

Example 12.4

Air flowing at a velocity of 4000 m/s has a temperature of 1750°C and a pressure of 1 kPa. If it is isentropically brought to rest, find the temperature that then exists (i.e., find the stagnation temperature).

Solution

On the Mollier chart, the process is as shown in Figure E12.4.

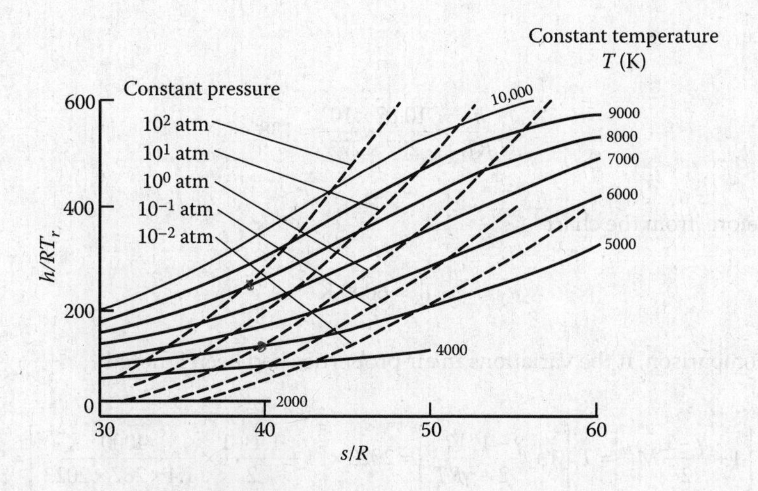

FIGURE 12.12
Form of Mollier chart for air ($T_f = 273$ K).

The energy equation gives

$$h_2 = h_1 + \frac{V^2}{2}$$

From the diagram for $T = 2023$ K and $p = 0.0099$ atm,

$$\frac{h_1}{RT_r} = 36$$

Thus,

$$h_2 = 36 \times 287 \times 273 + \frac{4000^2}{2} = 10.82 \times 10^6 \text{J/kg}$$

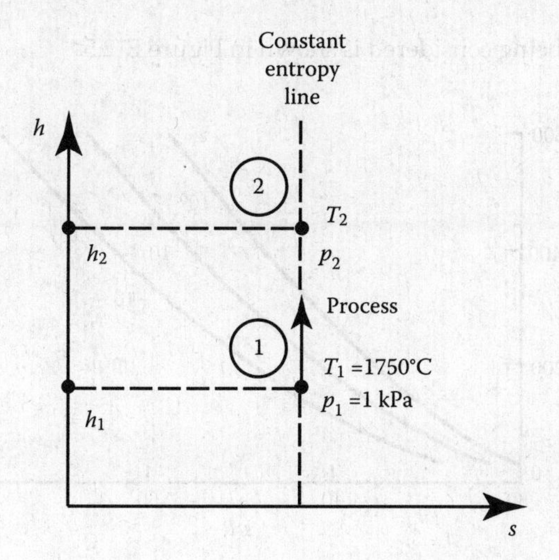

FIGURE E12.4
Process on Mollier chart.

Hence,

$$\frac{h_2}{RT_r} = \frac{10.82 \times 10^6}{287 \times 273} = 138$$

Therefore, from the chart

$$T_2 = 6000 \text{ K}$$

For comparison, if the variations in air properties had been ignored

$$T_0 = T_1\left[1 + \frac{\gamma - 1}{2} M_1^2\right] = T_1\left[1 + \frac{\gamma - 1}{2} \frac{V_1^2}{\gamma RT}\right] = 2023 \times \left[1 + \frac{1.4 - 1}{2} \times \frac{4000^2}{1.4 \times 287 \times 2023}\right] = 9987 \text{ K}$$

From these results, it will be seen that the dissociation of the air has a significant influence on the value of the stagnation temperature.

The flow through a normal shock wave can also be calculated using the Mollier chart. To do this, the density also has to be determined. A Mollier chart for air that shows some constant density lines is therefore given in Figure 12.13. The use of the Mollier charts to calculate the changes across a normal shock wave is illustrated in the following example.

Example 12.5

A normal shock wave forms ahead of a body moving through the air at a velocity 7500 m/s. If the ambient temperature, density, and pressure are 230 K, 0.018 kg/m³, and 1.2 kPa, respectively, find the pressure and temperature behind the shock wave.

Solution

The flow situation being considered is shown in Figure E12.5.

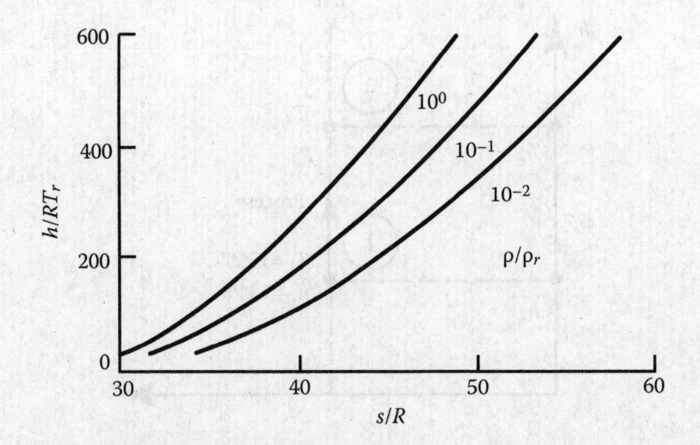

FIGURE 12.13
Constant density lines on Mollier chart for air (T_r = 273 K).

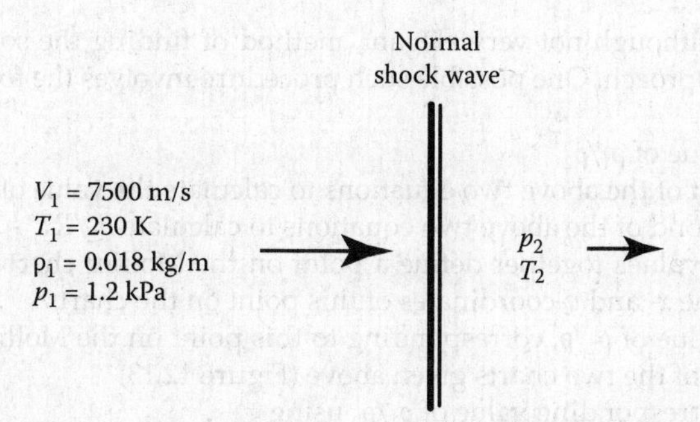

FIGURE E12.5
Flow situation considered.

As shown earlier in this chapter—see Equations 12.36 and 12.37—the following equations apply across a normal shock wave:

$$p_2 = p_1 + \rho_1 V_1^2 \left[1 - \frac{\rho_1}{\rho_2}\right]$$

and

$$\frac{V_1^2}{2}\left[1 - \left(\frac{\rho_1}{\rho_2}\right)^2\right] = h_2 - h_1$$

Here V_1 = 7500 m/s, p_1 = 1.2 kPa, ρ_1 = 0.018 kg/m³, and T_1 = 230 K. Hence, the first of the above equations gives

$$p_2 = 1200 + 0.018 \times 7500^2 \times \left[1 - \frac{\rho_1}{\rho_2}\right]$$

i.e.,

$$p_2 = 1200 + 1{,}013{,}000\left[1 - \frac{\rho_1}{\rho_2}\right]$$

i.e.,

$$\frac{p_2}{p_r} = 0.012 + 10.0\left[1 - \frac{\rho_1}{\rho_2}\right]$$

The second of the above equations gives

$$\frac{(h_2 - h_1)}{RT_r} = 359\left[1 - \left(\frac{\rho_1}{\rho_2}\right)^2\right]$$

The simplest, although not very elegant, method of finding the solution is to use a trial-and-error approach. One possible such procedure involves the following steps:

1. Guess a value of ρ_1/ρ_2
2. Use the first of the above two equations to calculate the value of p_2/p_r
3. Use the second of the above two equations to calculate h_2/RT
4. These two values together define a point on the Mollier chart (Figure 12.13). Establish the x- and y-coordinates of this point on the chart
5. Find the value of ρ_2/ρ_r corresponding to this point on the Mollier chart using the second of the two charts given above (Figure 12.13)
6. Find the corresponding value of ρ_1/ρ_2 using

$$\frac{\rho_1}{\rho_2} = \frac{\rho_1}{\rho_r}\frac{\rho_r}{\rho_2} = \frac{0.014}{\rho_2/\rho_r}$$

7. Compare the value of the density ratio so obtained with the initial guessed value
8. Repeat the procedure with different initial guessed values until the two values agree

Using this procedure gives p_2 = 950 kPa and T_2 = 8000 K. Due to the coarse scales used, it is not possible to get the result very accurately using the Mollier charts given above.

Nonequilibrium Effects

The discussion up to this point in this chapter has assumed that the gas is always in thermodynamic equilibrium, i.e., that the composition and properties depend only on the temperature and the pressure at the point in the flow being considered. In low-temperature flows, this assumption is essentially always valid unless a gas–liquid or gas–solid phase change occurs. In high-temperature flows, however, when the effects of vibrational excitation and dissociation and ionization are significant, this may not be the case. This is because the excitation of the vibrational modes of the gas molecules and dissociation and ionization proceed at a relatively slow rate, the rates being too slow to ensure that thermodynamic equilibrium always exists at all points in the flow. This situation is particularly likely to occur with flow through a shock wave. As discussed before, a shock wave is very thin. The time taken for the gas to pass through the shock wave may therefore be so short that there is no time for vibrational excitation or dissociation or ionization to occur to any significant extent, i.e., the nature of the gas remains essentially frozen in its upstream state during its passage through the shock. As a result of this, the flow immediately downstream of the shock can be calculated using the equations given in Chapter 5. In the gas flow downstream of the shock, the gas is therefore not in thermodynamic equilibrium and the vibrational excitation, dissociation, and ionization proceed at a finite rate and thermodynamic equilibrium is therefore only again attained at some distance downstream of the shock. The flow near the shock is therefore as illustrated in Figure 12.14. In such a case, it is more accurate to speak of a shock front and a shock layer rather than a shock wave. This is illustrated in Figure 12.15.

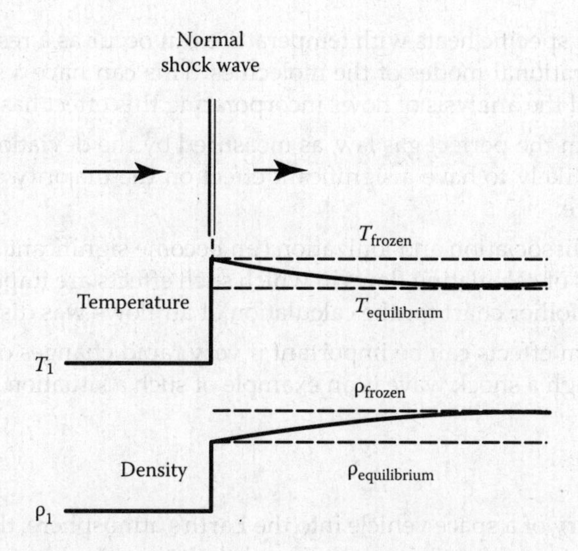

FIGURE 12.14
Nonequilibrium effects downstream of a shock wave.

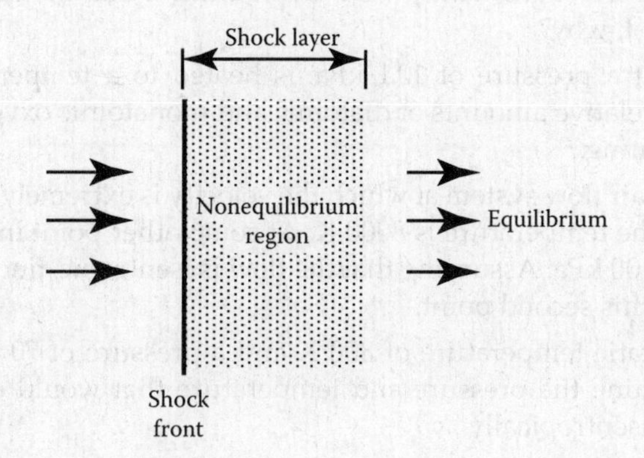

FIGURE 12.15
Shock layer in nonequilibrium flow.

The overall change through the shock layer will be given by the methods of analyses discussed earlier in this chapter. If the thickness of the shock layer and the nature of the flow in this layer are required, the rate equations for the nonequilibrium reactions have to be solved simultaneously with the flow equations.

Concluding Remarks

In high-temperature flows, which are usually associated with high Mach numbers, it has been shown that

- Changes in the specific heats with temperature can occur as a result of the excitation of the vibrational modes of the molecules. This can have a significant effect on the flow and the analysis of flows incorporating this effect has been discussed.
- Deviations from the perfect gas law, as measured by the deviation of the factor Z from 1, are unlikely to have a significant effect on the majority of flows encountered in practice.
- The effects of dissociation and ionization can become significant at high temperatures. Methods of calculating flows in which such effects are important have been discussed. A Mollier chart for the calculation of air flows was discussed.
- Nonequilibrium effects can be important if very rapid changes occur in the flow. The flow through a shock wave is an example of such a situation.

PROBLEMS

1. During the entry of a space vehicle into the Earth's atmosphere, the Mach number at a given point on the trajectory is 38 and the atmospheric temperature is 0°C. Calculate the temperature at the stagnation point of the vehicle assuming that a normal shock wave occurs ahead of the vehicle and assuming that the air behaves as a calorically perfect gas with $\gamma = 1.4$. Do you think that the value so calculated is accurate? If not, why?
2. Oxygen, kept at a pressure of 10.1 kPa, is heated to a temperature of 4000 K. Determine the relative amounts of diatomic and monatomic oxygen that are present after the heating.
3. At a point in an air flow system at which the velocity is extremely low, the pressure is 10 MPa and the temperature is 8000 K. At some other point in the flow system, the pressure is 100 kPa. Assuming that the flow is isentropic, find the temperature and velocity at this second point.
4. Nitrogen at a static temperature of 800 K and a pressure of 70 kPa is flowing at Mach 3. Determine the pressure and temperature that would exist if the gas is brought to rest isentropically.
5. Air at a pressure of 101 kPa and a temperature of 20°C has its temperature raised to 4000 K in a constant-pressure process. Determine the composition of the air at this elevated temperature. Assume the air to initially consist of 3.76 mol N/mol O.
6. As a result of an explosion, a normal shock wave moves at a velocity of 6000 m/s through still air at a pressure and temperature of 101 kPa and 25°C, respectively. Find the pressure, temperature, and air velocity behind the wave.
7. A blunt-nosed body is moving through air at a velocity of 5000 m/s. The pressure and the temperature of the air are 22 kPa and 43°C, respectively. The shock wave that exists ahead of the body can be assumed to be normal in the vicinity of the stagnation point. Find the pressure behind the shock wave.

13

Low-Density Flows

Introduction

It has been assumed in all of the preceding discussion in this book that the gas behaves as a continuum, i.e., that the molecular nature of the gas does not have to be considered in analyzing the flow of the gas. However, it may not be possible to use this continuum assumption in the analysis of the flow when the density of the gas is very low. Flows in which the density is so low that noncontinuum effects become important are often termed "rarefied gas flows."

The conditions under which noncontinuum effects become important and the nature of the changes in the flow produced by these effects is the subject of this chapter. Noncontinuum effects can have an important influence on the flow over craft operating at high altitudes at high Mach numbers. They can also have an important influence on the flow in high vacuum systems. However, because this book is intended to give a broad introduction to compressible fluid flows, no more than a very brief introduction to the topic of noncontinuum effects will be given here despite their significant practical importance in a number of situations.

Knudsen Number

A gas can be assumed to behave as a continuum if the mean free path, i.e., the average distance that a molecule moves before colliding with another molecule, λ, is small compared with the significant characteristic length, L, of the flow system. The ratio of λ/L is, of course, dimensionless and is called the Knudsen number, Kn, i.e.,

$$Kn = \frac{\lambda}{L} \tag{13.1}$$

To relate the Knudsen number to the dimensionless parameters used elsewhere in the study of compressible flows, it is convenient to be able to relate the coefficient of viscosity to the mean free path. To do this, consider three layers distance λ apart in the flow as shown in Figure 13.1.

Because molecules arriving at plane A shown in Figure 13.1 from plane B have not collided with any other molecules over the distance λ, they arrive with an excess mean velocity of $\lambda\, \partial u/\partial y$. Similarly, molecules arriving at plane A from plane C arrive with a mean velocity deficit of $\lambda\, \partial u/\partial y$. When the molecules from planes B and C arrive at plane A, they

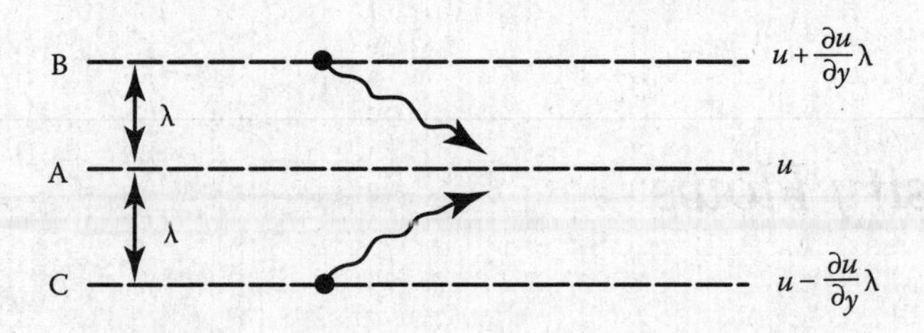

FIGURE 13.1
Layers considered in the analysis of viscosity.

collide with the molecules on this plane and attain the mean velocity on this plane. Because of the change in mean momentum that therefore occurs at plane A, there is effectively a shear force acting on plane A that is proportional to the excess or deficit of momentum with which the molecules arrive, i.e., proportional to $\lambda\, \partial u/\partial y$. The net force per unit area will then be proportional to the number of molecules arriving per unit area per unit time multiplied by $\lambda\, \partial u/\partial y$. Now the number of molecules arriving per unit area per unit time on plane A will depend on the number of molecules per unit volume, i.e., on the density, and on the mean speed of the molecules, c_m, i.e., the force per unit area, the shear stress, τ, will be given by an equation that has the form

$$\tau \propto \rho c_m \lambda \frac{\partial u}{\partial y} \tag{13.2}$$

However, by definition, the coefficient of viscosity, μ, is given by

$$\tau = \mu \frac{\partial u}{\partial y} \tag{13.3}$$

Comparing these two equations indicates that

$$\mu \propto \rho c_m \lambda \tag{13.4}$$

However, the mean molecular speed is proportional to the speed of sound, a, since a sound wave is propagated as a result of molecular collisions. Hence, Equation 13.4 gives

$$\mu \propto \rho a \lambda \tag{13.5}$$

from which it follows that

$$\lambda \propto \frac{\mu}{\rho a} \tag{13.6}$$

A more complete analysis gives $\lambda = 1.26\sqrt{\gamma}\mu/\rho a$.

Using Equation 13.6, it will be seen from the definition of the Knudsen number given in Equation 13.1 that

$$Kn \propto \frac{\mu}{\rho a L} = \frac{\mu}{\rho V L}\frac{V}{a}$$

i.e.,

$$Kn \propto \frac{M}{Re} \tag{13.7}$$

From this equation, it follows that large Knudsen numbers will be associated with high-Mach-number, low-Reynolds-number flows. These are exactly the conditions that normally exist when a body, such as a re-entering orbital craft, is passing through the upper atmosphere.

For many purposes, the size of the body L is suitable for use in defining the Knudsen number. However, if the Reynolds number is significantly above 1, a distinct boundary layer will exist adjacent to the surface of the body (see Figure 13.2), and the thickness of this boundary layer, δ, may be a more suitable length scale to compare with the near free path in defining the Knudsen number.

Now, because noncontinuum effects are associated with low density flows and because in such flows the Reynolds numbers will usually be relatively small, noncontinuum effects are usually likely to be important in flows in which the boundary layer is laminar, and in this case

$$\delta \propto \frac{L}{Re^{0.5}} \tag{13.8}$$

Therefore, in situations in which a distinct boundary layer exists, a more suitable Knudsen number to use is

$$Kn = \frac{\lambda}{\delta} \propto \frac{\mu}{\rho a} \frac{Re^{0.5}}{L} = \frac{M}{Re^{0.5}} \tag{13.9}$$

To conclude this section, it will be noted that since for many gases μ is approximately proportional to $T^{0.5}$ and since a is also proportional to $T^{0.5}$, Equation 13.6 shows that λ is dominantly dependent on 1/ρ. The interrelationship between ρ and λ will be seen by comparing the results given in Figures 13.3 and 13.4.

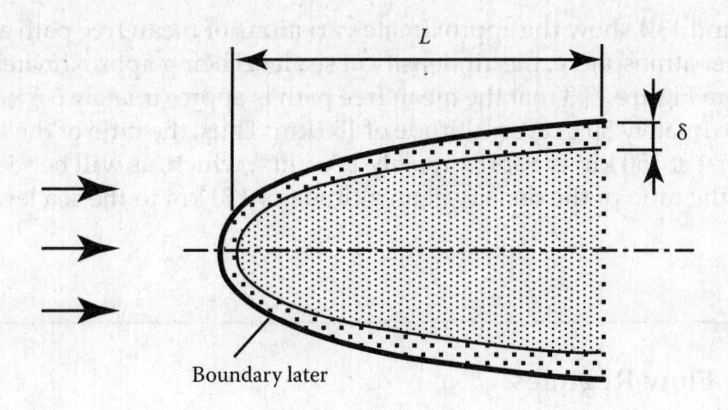

FIGURE 13.2
Boundary layer around body.

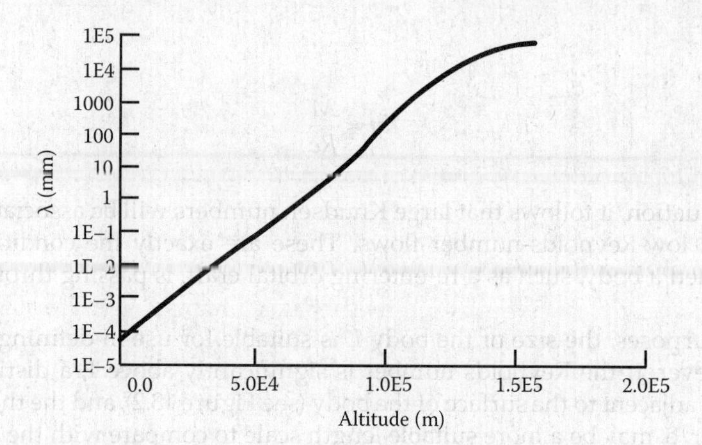

FIGURE 13.3
Variation of mean free path in upper atmosphere.

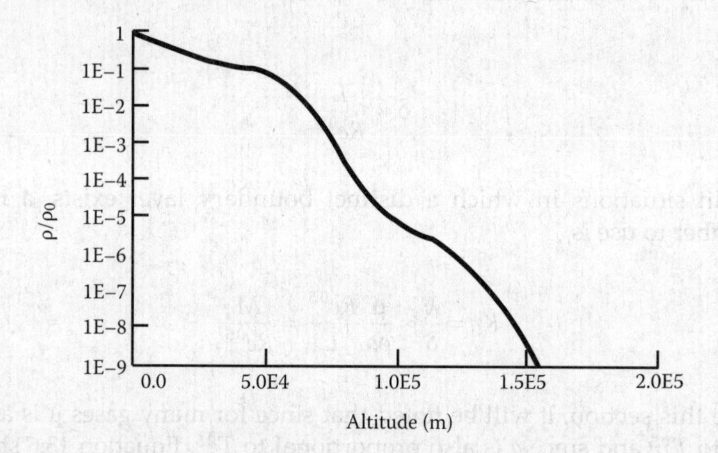

FIGURE 13.4
Variation of density ratio in upper atmosphere.

Figures 13.3 and 13.4 show the approximate variations of mean free path and air density ratio in the upper atmosphere, the air density at sea level being approximately 1.2 kg/m^3. It will be seen from Figure 13.3 that the mean free path is approximately 6.6×10^{-5} mm at sea level and approximately 50 m at an altitude of 150 km. Thus, the ratio of the mean free path at sea level to that at 150 km is approximately 1.3×10^{-9}, which, as will be seen from Figure 13.4, is close to the ratio of the density at an altitude of 150 km to the sea level density.

Low-Density Flow Regimes

As the Knudsen number in a flow over a body increases, the first observable noncontinuum flow effect is that there is an apparent jump in the velocity at the surface of the body, i.e., the gas velocity at the body surface can no longer be assumed to be equal to the surface velocity, i.e., there is a "slip" at the surface as indicated in Figure 13.5. Flows in which this effect becomes important are termed "slip flows."

Slip flow occurs roughly in the following ranges

$$\begin{aligned} \text{If } Re > 1: \quad & 0.001 < \frac{M}{Re^{0.5}} < 0.1 \\ \text{If } Re < 1: \quad & 0.001 < \frac{M}{Re} < 0.1 \end{aligned} \tag{13.10}$$

Two criteria are necessary because, as explained in the previous section, if Re is large (taken to imply $Re > 1$), the boundary layer thickness is the important length scale, whereas if Re is small (taken to imply $Re < 1$), the size of the body is the important length scale.

If the Knudsen number for the flow is very large, i.e., the mean free path is much greater than the size of the body, very infrequent intermolecular collisions occur. Molecules arriving at the surface of the body over which the gas is flowing will, essentially, therefore not have collided with any molecules reflected from the surface. The molecules therefore arrive at the surface with the full freestream velocity. In such flows, there is then essentially no velocity gradient adjacent to the surface as shown in Figure 13.6. Such flows in which there is essentially no interaction between the molecules leaving the surface and those approaching the surface are termed "free molecular flows." They can be assumed to exist if

$$\frac{M}{Re} > 3 \tag{13.11}$$

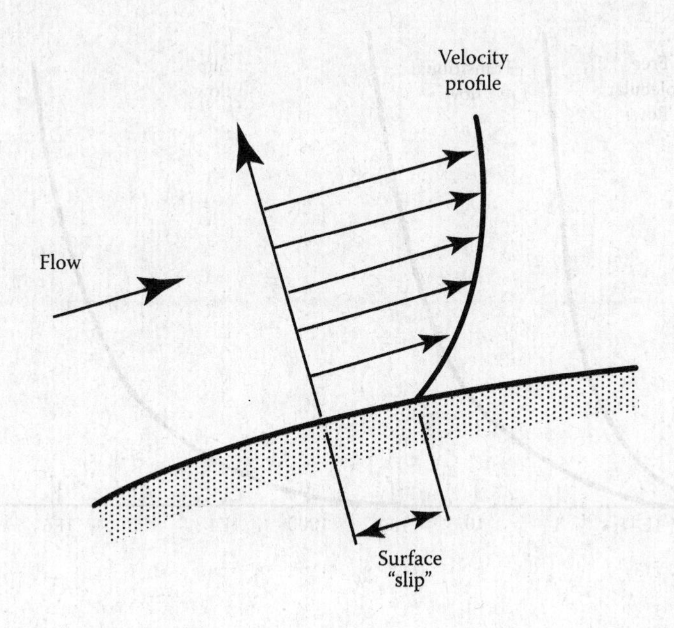

FIGURE 13.5
Velocity variation near a surface in slip flow.

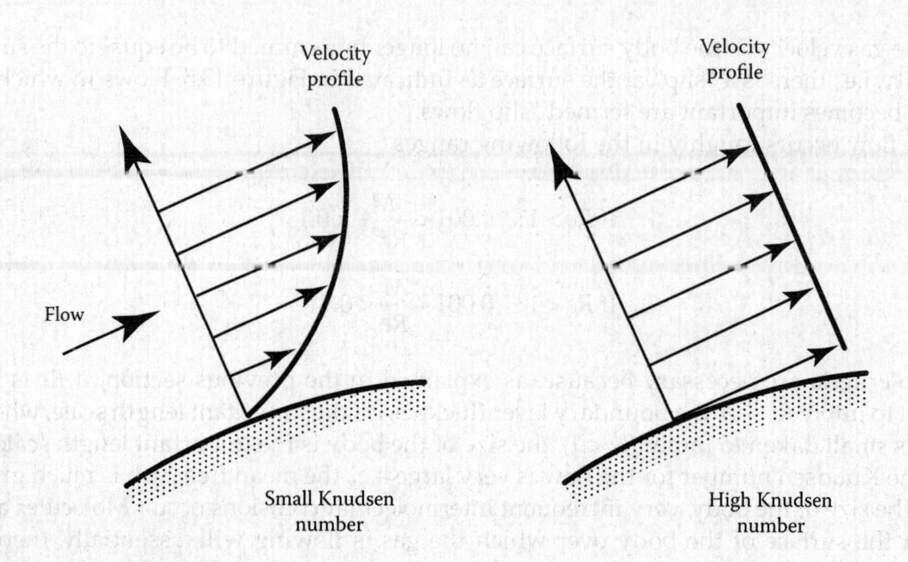

FIGURE 13.6
Effect of Knudsen number on velocity gradient in flow near a surface.

A Knudsen number based on the boundary layer thickness has no meaning in such flows because the boundary layer concept is not applicable in such flows.

Between the slip flow region and the free molecular flow region, the flow is said to be in the transition region. When no low-density effects are present, the flow is in the continuum region. Using Equations 13.10 and 13.11, the Reynolds and Mach number regions for the various flow regions can be defined and are shown in Figure 13.7.

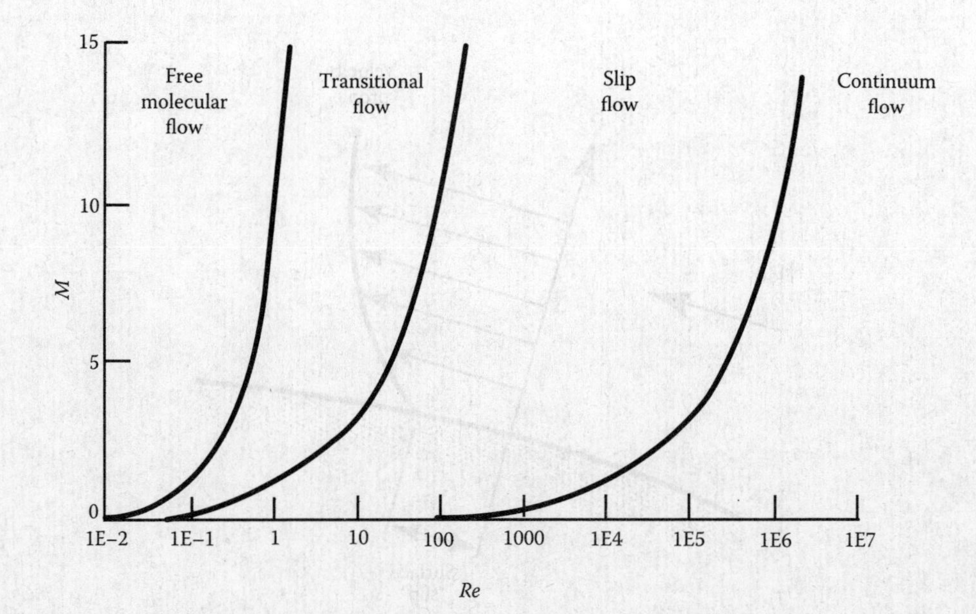

FIGURE 13.7
Effect of Mach and Reynolds numbers on low-density flow regimes.

A very brief introduction to the analysis of slip flows and free molecular flows is presented in the next two sections.

Example 13.1

The variation of velocity with altitude for an orbiting body during re-entry into the atmosphere is shown in Figure E13.1. The body has a length of 4 m. Find the altitudes at which the flow over the body passes from the free molecular regime to the transition regime, from the transition to the slip flow regime, and from the slip flow to the continuum regime.

Solution

The velocity at various altitudes can be obtained from Figure E13.1 and some typical values are shown in Table E13.1. The values of the density, the speed of sound, and the viscosity at these altitudes are also shown in this table. Using these values, the values of the Mach number, M, of the Reynolds number, Re, of $M/Re^{0.5}$, and of M/Re have been derived and are also shown in Table E13.1.

From these results, it can be seen that $M/Re^{0.5}$ becomes equal to 0.01 at an altitude, H, of about 65,000 m and becomes equal to 0.1 at an altitude of about 110,000 m. This defines the slip flow regime. $M/Re^{0.5}$ can be used to define this regime because, as will be seen from the tabulated results, the Reynolds number is greater than 1 at all altitudes considered.

It will also be seen from the tabulated results that M/Re becomes equal to 3 at an altitude of about 140,000 m. Hence,

$H < 65{,}000$ m: continuum flow
65,000 m $< H <$ 110,000 m: slip flow
110,000 m $< H <$ 140,000 m: transitional flow
$H >$ 140,000 m: free molecular flow

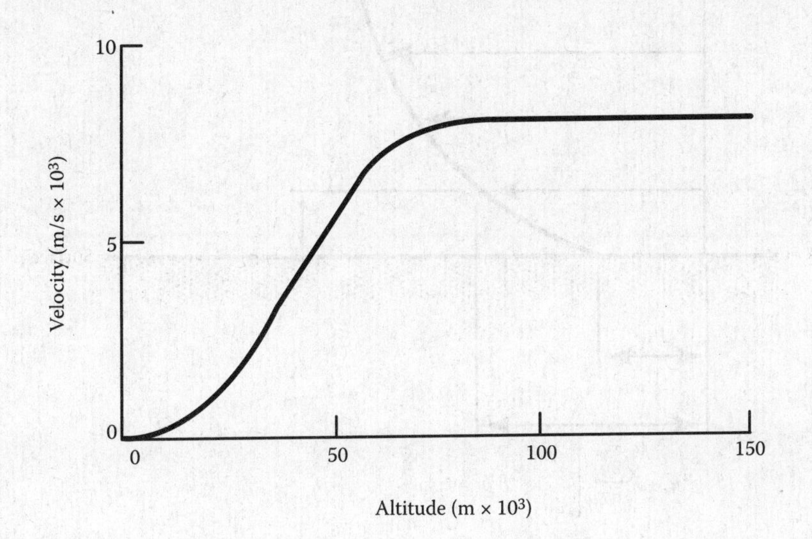

FIGURE E13.1
Variation of velocity with altitude.

TABLE E13.1

Variation of Flow Variables Used in Analysis with Altitude

H (m)	*V* (m/s)	ρ (kg/m³)	*a* (m/s)	μ (kg/m · s)	*M*	*Re*	*M*/*Re*$^{0.5}$	*M*/*Re*
30,000	2200	1.7×10^{-1}	290	1.6×10^{-5}	7.6	9.3×10^{7}	0.008	—
50,000	5500	2.6×10^{-2}	280	1.5×10^{-5}	19.6	3.8×10^{7}	0.003	—
70,000	7800	2.6×10^{-3}	260	1.3×10^{-5}	30.0	6.2×10^{6}	0.012	—
100,000	8000	3.9×10^{-6}	280	1.5×10^{-5}	28.2	8.3×10^{3}	0.310	0.003
120,000	8000	2.3×10^{-7}	400	2.5×10^{-5}	20.0	1.7×10^{2}	1.530	0.118
150,000	8000	2.0×10^{-9}	630	5.3×10^{-5}	12.6	1.2	11.50	10.50

Slip Flow

In this type of flow, the mean velocity of the molecules at the surface is significantly different from the velocity of the surface. Consider flow very close to a surface as shown in Figure 13.8.

Here, u_s is the mean velocity of the molecules at the surface. Since the molecules move a mean distance λ between collisions, the molecules arriving at the surface will do so with a mean gas velocity parallel to the surface of

$$u_s + \lambda \left.\frac{\partial u}{\partial y}\right|_{y=0} \tag{13.12}$$

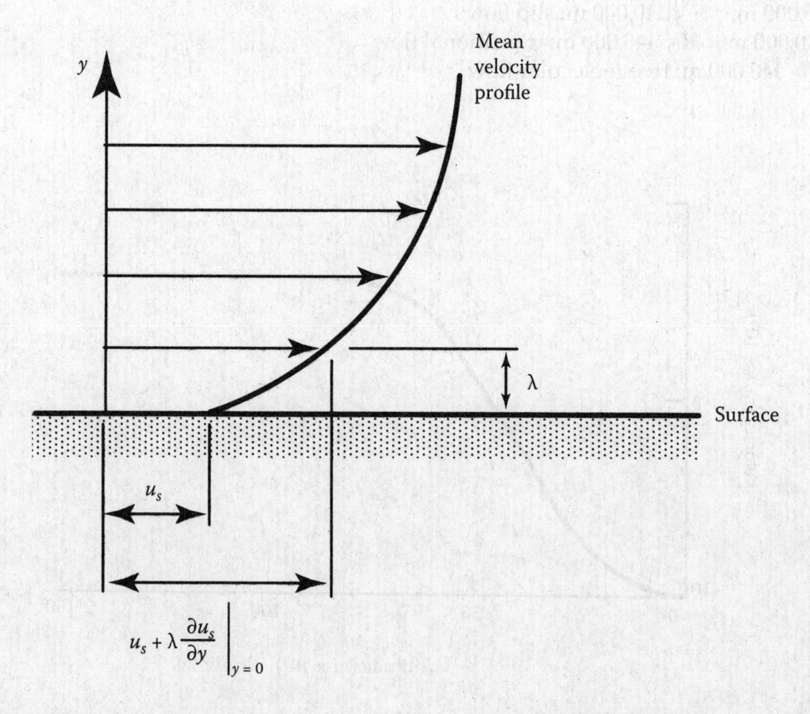

FIGURE 13.8
Slip flow near surface.

When the molecules strike the surface, they can be reflected from the surface either in a spectral manner or in a diffuse manner. In the former case, they are reflected with the same mean velocity parallel to the surface as they had before reaching the surface, i.e., $\lambda\, \partial u/\partial y$, the derivatives being evaluated at the surface. In the case of diffuse reflection, the molecules, on the average, lose their mean velocity parallel to the surface. The two types of reflection are illustrated in Figure 13.9.

With an actual surface, some of the molecules are reflected in a diffuse manner and some are reflected in a spectral manner, the relative fraction of molecules reflected in the two ways being dependent on the nature of the surface. Now, all of the molecules arrive at the surface with a mean velocity parallel to the surface that is given by Equation 13.12. Let a fraction d of these molecules leave the surface with, on the average, no mean velocity parallel to the surface, i.e., d is the fraction of the molecules reflected diffusely. A fraction $(1 - d)$ of the incident molecules will therefore leave the surface with a mean velocity parallel to the surface given by Equation 13.12, i.e., with the same velocity parallel to the wall as they had when they impinged on the wall. If all the molecules adjacent to the surface are considered, on the average, half will be about to strike the surface and half will be leaving the surface as a result of reflection. The mean velocity parallel to the surface that the molecules adjacent to the surface have, u_s, is therefore given by

$$u_s = \frac{1}{2}\left[u_s + \lambda \left.\frac{\partial u}{\partial y}\right|_{y=0}\right] + \frac{1}{2}(1-d)\left[u_s + \lambda \left.\frac{\partial u}{\partial y}\right|_{y=0}\right]$$

i.e.,

$$u_s = \left[1 - \frac{d}{2}\right]\left[u_s + \lambda \left.\frac{\partial u}{\partial y}\right|_{y=0}\right] \tag{13.13}$$

This equation can then be rearranged to give the surface velocity u_s as

$$u_s = \left[\frac{2}{d} - 1\right]\lambda \left.\frac{\partial u}{\partial y}\right|_{y=0} \tag{13.14}$$

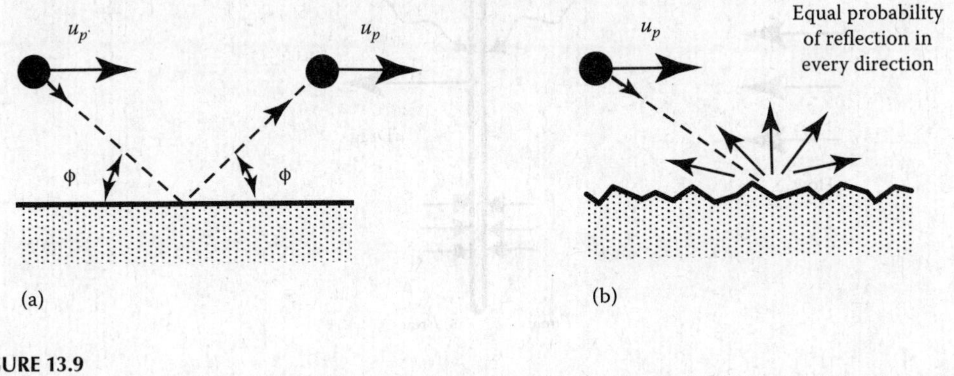

u_p u_p u_p Equal probability of reflection in every direction

φ φ

(a) (b)

FIGURE 13.9
Spectral and diffuse reflection of molecules at a surface. (a) Spectral reflection (velocity parallel to the surface is unchanged by the reflection). (b) Diffuse reflection (no mean velocity parallel to the surface after the reflection).

TABLE 13.1

Diffuse Reflection Fractions

Gas and Surface Material	d
Air on machined brass	1.0
Air on glass	0.89
Air on oil	0.90
Hydrogen on oil	0.93
Helium on oil	0.87

Thus, even in continuum flows, a slip velocity will exist. However, in such flows, λ is so small that u_s is effectively zero.

As mentioned before, d will depend on the nature of the surface. It will also depend on the type of gas involved. Typical values for air flow are listed in Table 13.1.

Free Molecular Flow

As discussed above, the free molecular flow regime is entered when the Knudsen number is large. To give a very simple introduction to the analysis of free molecular flows, the drag force on a flat plate placed at right angles to a flow will be considered, i.e., the flow situation shown in Figure 13.10 will be considered.

Since free molecular flow is being considered, the molecules reach the plate with the velocity they have well away from the plate. If there was no mean gas velocity, the molecules would arrive at the front and the rear sides of the plate with the same velocity and

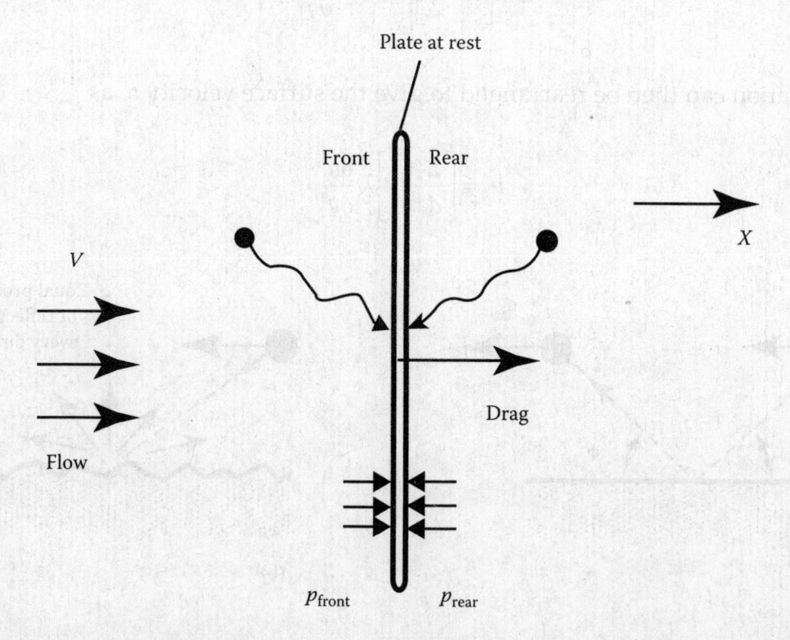

FIGURE 13.10
Free molecular flow situation considered.

the same pressure would be exerted on the two sides of the plate and there would then, of course, be no drag force acting on the plate. However, when there is a mean gas velocity, the molecules reaching the front of the plate have a higher mean velocity than those reaching the back of the plate. There is therefore an increase in the pressure on the front of the plate and a decrease in the pressure on the back, which leads the drag force.

Now the pressure force on the front of the plate will be equal to the loss of x-wise momentum that results from the fact that the plate is at rest, i.e.,

$$P_{\text{front}} = \dot{m}A(c_m + V) \tag{13.15}$$

where $\dot{m}$ is the total mass of the molecules that strike the plate per unit time per unit area and A is the frontal area of the plate. However, $\dot{m}$ will be given by

$$\dot{m} = Nm \tag{13.16}$$

where N is the number of molecules striking the plate per unit time per unit area and m is the mass of one molecule. Hence, Equation 13.33 gives

$$P_{\text{front}} = NmA(c_m + V) \tag{13.17}$$

Similarly, the pressure force on the back surface will be given by

$$P_{\text{back}} = NmA(c_m - V) \tag{13.18}$$

The net force on the plate is therefore given by

$$P = P_{\text{front}} - P_{\text{back}} = 2NmAV \tag{13.19}$$

However, N will be proportional to the number of molecules by unit volume, n, and the mean molecular velocity c_m, i.e.,

$$N \propto nc_m \tag{13.20}$$

Hence,

$$P \propto 2mnc_m AV \tag{13.21}$$

However,

$$\rho = mn \tag{13.22}$$

Thus, Equation 13.39 gives

$$P \propto 2\rho c_m AV \tag{13.23}$$

Now, the drag coefficient, C_D, is defined by

$$C_D = \frac{P}{\rho V^2 A/2} \tag{13.24}$$

Thus, Equation 13.41 gives

$$C_D \propto 4\left(\frac{c_m}{V}\right) \tag{13.25}$$

i.e., writing

$$S = V/c_m \tag{13.26}$$

Equation 13.43 gives

$$C_D \propto \frac{4}{S} \tag{13.27}$$

Since c_m is proportional to the speed of sound, S is dependent on the effective Mach number in the flow.

Equation 13.45 indicates that C_DS will be a constant. If bodies other than a flat plate are considered, e.g., if flow over a cylinder is considered, the same form of result is obtained. The measured variation of C_DS with S for a cylinder is shown in Figure 13.11. It will be seen from Figure 13.11 that the experimental results confirm that the product C_DS is approximately constant in free molecular flow.

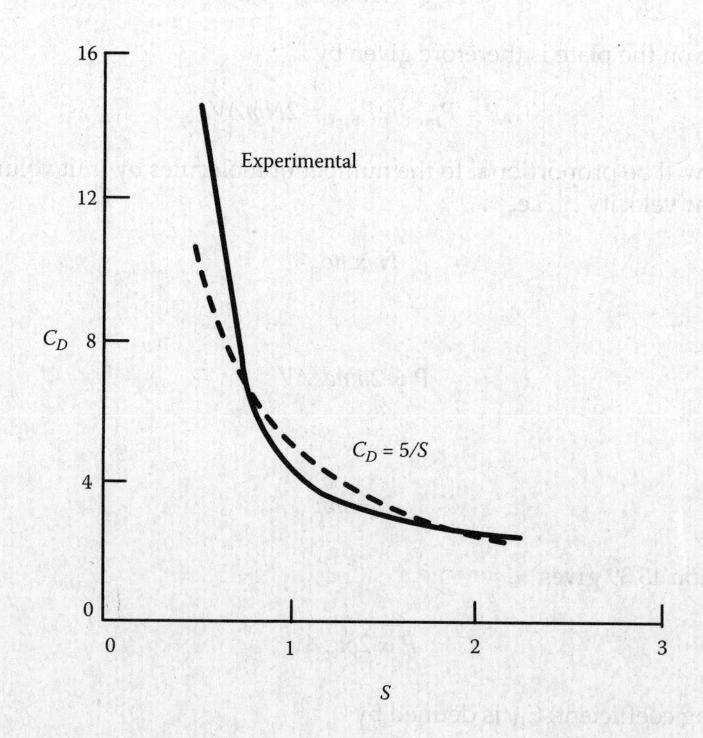

FIGURE 13.11
Drag coefficient variation for a cylinder in free molecular flow.

Example 13.2

If the mean molecular speed c_m is assumed to be equal to $\sqrt{2RT}$, find the order of magnitude of the drag force per unit frontal area on a flat plate over which air is flowing at a velocity of 8000 m/s, the plate being normal to the flow. The density of the air is 10^{-10} times the standard sea level density and the temperature of the air is 1000 K.

Solution

At a temperature of 1000 K, the speed of sound is equal to $\sqrt{\gamma R1000}$, i.e., 634 m/s, so the Mach number in the flow is given by

$$M = \frac{8000}{634} = 12.6$$

The viscosity of the air is approximately 420×10^{-7} N · s/m²; thus, since the air density at sea level is 1.16 kg/m³, the Reynolds number per meter, $\rho V/\mu$, is given by

$$Re = \frac{1.16 \times (10 \times 10^{-10}) \times 8000}{420 \times 10^{-7}} = 0.22.$$

Hence, the ratio of the Mach number to the Reynolds number per meter is equal to 12.6/0.22 = 57.2. Therefore, unless the plate is extremely large, free molecular flow exists. However, in such flow

$$C_D \propto 4\left(\frac{c_m}{V}\right)$$

and since

$$c_m = \sqrt{2RT} = \sqrt{2 \times 287 \times 1000} = 758 \text{ m/s}$$

in the present case,

$$C_D \propto 4\left(\frac{758}{8000}\right) = 0.379$$

The drag force on the plate per unit plate area is

$$C_D \rho V^2 A/2$$

Hence, again noting that the air density at sea level is 1.16 kg/m³, the order of magnitude of the drag force per unit plate area is

$$\text{Drag force per unit area} = C_D \rho V^2/2 = 0.379 \times 1.16 \times 10^{-10} \times 8000^2/2 = 0.0014 \text{ N}$$

Therefore, the order of magnitude of the drag force on the plate per unit area is 0.0014 N.

Concluding Remarks

In low-density flows, it may not be possible to assume that the gas is a continuum. The Knudsen number is the parameter that is used to determine when noncontinuum effects start to become important. The relation between the Knudsen number and the Mach and Reynolds numbers has been discussed.

The first noncontinuum effect is that "surface slip" starts to be important at the surface, i.e., that the effective gas velocity at the surface of a body in a flow is nonzero. Flows in which this is the only noncontinuum effect are called slip flows. The range of flow conditions over which the slip flow region extends has been discussed.

When the gas density is very low, there are so few molecules present that the molecular motion is essentially unaffected by the presence of a body until the molecules reach the surface. This type of flow is called free molecular flow and the conditions under which it occurs have also been discussed.

Between the slip flow region and the free molecular flow region, there is a region termed the transition region. The analysis of flow in this region is relatively complex and has not been discussed here.

PROBLEMS

1. A small rocket probing the atmosphere has a length of 3 m. It is fired vertically upward through the atmosphere, its average velocity being 1000 m/s. Consider the flow over the rocket at this average velocity at altitudes of 30,000 and 80,000 m. Can the air flow over the rocket be assumed to be continuous at these two altitudes? At an altitude of 30,000 m, the air has a temperature, pressure, and viscosity of –55°C, 120 Pa, and 1.5×10^{-5} kg/m · s, respectively, whereas at an altitude of 80,000 m, the air has a temperature, pressure, and viscosity of –34°C, 0.013 Pa, and 1.7×10^{-5} kg/m · s, respectively.
2. A small research vehicle with a length of 4 ft travels at Mach 15 at altitudes of 100,000 and 250,000 ft. Determine whether, at these two altitudes, the missile is in the continuum, slip, transition, or free molecular flow regimes. It can be assumed that at an altitude of 100,000 ft, the pressure, temperature, and viscosity are 22 psf, 340 R, and 96×10^{-7} lbm/ft-s, respectively, whereas at an altitude of 250,000 ft, they are 0.11 psf, 450 R, and 100×10^{-7} lbm/ft-s, respectively.
3. Find the drag force per unit length on a 0.5-cm-diameter cylinder placed in an air flow in which the temperature is 800 K, the density is 8×10^{-9} kg/m^3 and the velocity is 10,000 m/s.

14

An Introduction to Two-Dimensional Compressible Flow

Introduction

The discussion given in the preceding chapters has mainly been concerned with one-dimensional flows. This is because a large number of flows that occur in engineering practice can be adequately modeled, at least for preliminary design purposes, by assuming that the flow is one-dimensional and because an understanding of many features of compressible flows can be gained by considering one-dimensional flow. However, this assumption is not always adequate. There are a number of flows in which it is necessary to account for the two- or three-dimensional nature of the flow. For example, the flows shown in Figure 14.1 cannot be treated as one-dimensional.

In this chapter, attention will mainly be given to the analysis of steady two-dimensional isentropic flows, i.e., the effects of viscosity and heat transfer will be neglected in most of this chapter. The extension of the methods discussed in this chapter to three-dimensional flows is relatively straightforward and will not be discussed here. Also, although the equations for two-dimensional flow can be written in compact form using vector notation, this notation will not be used here in an effort to concentrate on developing an understanding of the meaning and implications of the equations. The governing equations will therefore be expressed in terms of Cartesian coordinates.

The governing equations will first be developed and methods of solving these equations will then be discussed.

Governing Equations

The equations governing steady, isentropic, two-dimensional compressible flow will be derived in this section.

Continuity Equation

Consider flow through the control volume shown in Figure 14.2. The conservation of mass requires that since steady flow is being considered

$$\begin{array}{c}\text{Rate mass}\\\text{enters AB}\end{array} - \begin{array}{c}\text{Rate mass}\\\text{leaves CD}\end{array} + \begin{array}{c}\text{Rate mass}\\\text{enters AC}\end{array} - \begin{array}{c}\text{Rate mass}\\\text{leaves BC}\end{array} = 0 \qquad (14.1)$$

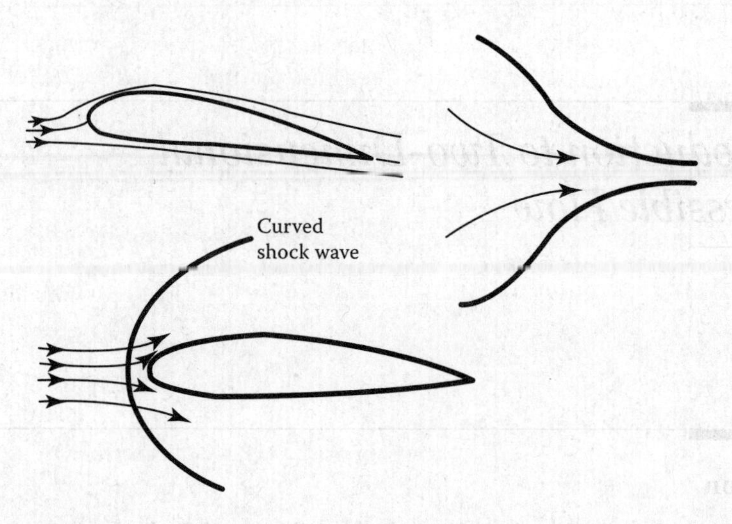

FIGURE 14.1
Two-dimensional flows.

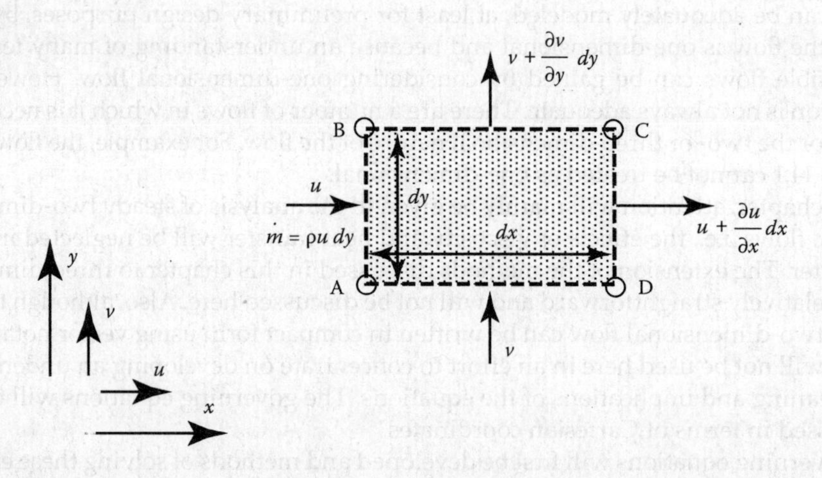

FIGURE 14.2
Control volume considered. (Photo #: l8-00478.tif, Title: McDonnell Douglas HSCT Concept in Flight, www.Boeingimages.com. Copyright © Boeing Intellectual Property Management Copyright and Trademark Licensing.)

Because two-dimensional flow is being considered, attention can be restricted to unit width of the control volume. Since the mass flow rate through AB is $\rho u \times$ area of AB $= \rho u\, dy$, and since, similarly, the mass flow rate through AD is $\rho v\, dx$, Equation 14.1 gives, since dx and dy are by assumption small,

$$\rho u\, dy - \left[(\rho u\, dy) + \frac{\partial}{\partial x}(\rho u\, dy)\, dx \right] + \rho v\, dx - \left[(\rho v\, dx) + \frac{\partial}{\partial y}(\rho v\, dx)\, dy \right] = 0$$

i.e.,

$$\frac{\partial}{\partial x}(\rho u) + \frac{\partial}{\partial y}(\rho v) = 0 \tag{14.2}$$

This is the continuity or conservation of mass equation for two-dimensional compressible flow. It basically states that the rate of change of ρu in the x direction is associated with an equal but negative rate of change of ρv in the y direction.

Momentum Equation

Since viscous effects are being neglected, the conservation of momentum equation gives in the x direction for the control volume shown in Figure 14.2.

$$\begin{matrix}\text{Rate } x \text{ momentum} \\ \text{leaves CD}\end{matrix} - \begin{matrix}\text{Rate } x \text{ momentum} \\ \text{enters AB}\end{matrix} + \begin{matrix}\text{Rate } x \text{ momentum} \\ \text{leaves BC}\end{matrix} + \begin{matrix}\text{Rate } x \text{ momentum} \\ \text{enters AD}\end{matrix}$$

$$= \begin{matrix}\text{Pressure force} \\ \text{on AB}\end{matrix} - \begin{matrix}\text{Pressure force} \\ \text{on CD}\end{matrix} \tag{14.3}$$

In writing this equation, it has been recalled that steady flow is being considered and it has been noted that the pressure acting on faces AD and BC of the control volume does not contribute any force in the x direction. All body forces such as those that could arise due to gravity or electrical and magnetic fields have also been neglected.

Since a control volume of unit width is being considered and since the rate x momentum enters through a face of the control volume is (mass flow rate through face × u), Equation 14.3 becomes

$$\left[(\rho u^2\, dy) + \frac{\partial}{\partial x}(\rho u^2\, dy)\, dx - (\rho u^2\, dy)\right] + \left[(\rho v u\, dx) + \frac{\partial}{\partial y}(\rho v u\, dx)\, dy - (\rho v u\, dx)\right]$$

$$= p\, dy - \left[p\, dy + \frac{\partial}{\partial x}(p\, dy)\, dx\right]$$

i.e., dividing through by $dx\, dy$,

$$\left[\frac{\partial}{\partial x}(\rho u^2) + \frac{\partial}{\partial x}(\rho v u)\right] = -\frac{\partial p}{\partial x} \tag{14.4}$$

The left-hand side of this equation can be written as

$$u\frac{\partial(\rho u)}{\partial x} + \rho u\frac{\partial u}{\partial x} + u\frac{\partial}{\partial y}(\rho v) + \rho v\frac{\partial u}{\partial y}$$

i.e., as

$$u\left[\frac{\partial}{\partial x}(\rho u) + \frac{\partial}{\partial y}(\rho v)\right] + \rho u\frac{\partial u}{\partial x} + \rho v\frac{\partial u}{\partial y}$$

The bracketed term is zero by virtue of the continuity equation 14.2. Equation 14.4 can therefore be written as

$$\rho u \frac{\partial u}{\partial x} + \rho v \frac{\partial u}{\partial y} = -\frac{\partial p}{\partial x} \tag{14.5}$$

Similarly, conservation of momentum in the y direction gives

Rate y momentum leaves BC $-$ Rate y momentum enters AD $+$ Rate y momentum leaves CD $-$ Rate y momentum enters AB

$=$ Pressure force on AD $-$ Pressure force on BC

i.e.,

$$\rho v^2\, dx + \frac{\partial}{\partial y}(\rho v^2\, dx)\, dy - \rho v^2\, dx + \rho uv\, dy + \frac{\partial}{\partial x}(\rho uv\, dy)\, dx - \rho uv\, dy$$
$$= p\, dx - \left[p\, dx + \frac{\partial}{\partial y}(p\, dx)\, dy \right]$$

i.e.,

$$\frac{\partial}{\partial x}(\rho uv) + \frac{\partial}{\partial y}(\rho v^2) = -\frac{\partial p}{\partial y}$$

i.e.,

$$v\left[\frac{\partial}{\partial x}(\rho u) + \frac{\partial}{\partial y}(\rho v) \right] + \rho u \frac{\partial v}{\partial x} + \rho v \frac{\partial v}{\partial y} = -\frac{\partial p}{\partial y}$$

By again using the continuity equation, this equation gives

$$\rho u \frac{\partial v}{\partial x} + \rho v \frac{\partial v}{\partial y} = -\frac{\partial p}{\partial y} \tag{14.6}$$

Energy Equation

Because adiabatic flow is being considered, the conservation of energy gives for flow through the control volume shown in Figure 14.2

$$\begin{aligned} &\underset{\text{leave CD}}{\text{Rate (enthalpy + kinetic energy)}} - \underset{\text{enter AB}}{\text{Rate (enthalpy + kinetic energy)}} \\ &+ \underset{\text{leave BC}}{\text{Rate (enthalpy + kinetic energy)}} - \underset{\text{enter AD}}{\text{Rate (enthalpy + kinetic energy)}} = 0 \end{aligned} \tag{14.7}$$

However, the rate at which (enthalpy + kinetic energy) enters AB is equal to the mass flow rate through AB × $[c_pT + (u^2 + v^2)/2]$, i.e., is given by

$$\rho u\, dy[c_pT + (u^2 + v^2)/2]$$

Similarly the rate at which (enthalpy + kinetic energy) enters AD is given by

$$\rho v\, dx[c_pT + (u^2 + v^2)/2]$$

Hence, Equation 14.7 gives

$$\begin{aligned} &\rho u\, dy\left[c_pT + \frac{(u^2+v^2)}{2}\right] + \frac{\partial}{\partial x}\left\{\rho u\, dy\left[c_pT + \frac{(u^2+v^2)}{2}\right]\right\} dx \\ &\qquad - \rho u\, dy\left[c_pT + \frac{(u^2+v^2)}{2}\right] + \rho v\, dx\left[c_pT + \frac{(u^2+v^2)}{2}\right] \\ &\qquad + \frac{\partial}{\partial y}\left\{\rho v\, dx\left[c_pT + \frac{(u^2+v^2)}{2}\right]\right\} dy - \rho v\, dx\left[c_pT + \frac{(u^2+v^2)}{2}\right] = 0 \end{aligned}$$

i.e.,

$$\frac{\partial}{\partial x}\left\{\rho u\left[c_pT+\frac{(u^2+v^2)}{2}\right]\right\}+\frac{\partial}{\partial y}\left\{\rho v\left[c_pT+\frac{(u^2+v^2)}{2}\right]\right\}=0 \tag{14.8}$$

This equation can be written as

$$\left[c_pT+\frac{(u^2+v^2)}{2}\right]\left[\frac{\partial}{\partial x}(\rho u)+\frac{\partial}{\partial y}(\rho v)\right]+u\frac{\partial}{\partial x}\left[c_pT+\frac{(u^2+v^2)}{2}\right]+v\frac{\partial}{\partial y}\left[c_pT+\frac{(u^2+v^2)}{2}\right]=0$$

However, by virtue of the continuity equation 14.2, the first term in this equation is zero. Hence, the conservation of energy equation can be written as

$$u\frac{\partial}{\partial x}\left[c_pT+\frac{(u^2+v^2)}{2}\right]+v\frac{\partial}{\partial y}\left[c_pT+\frac{(u^2+v^2)}{2}\right]=0 \tag{14.9}$$

Now consider the change in any quantity Z over a short length ds of a streamline as shown in Figure 14.3.

The change will be given by

$$dZ=\frac{\partial Z}{\partial x}dx+\frac{\partial Z}{\partial y}dy \tag{14.10}$$

The rate of change in Z as it moves along the streamline is dZ/dt where dt is the time taken for the gas particles to move from A to B. Hence, the rate of change in Z is given by

$$\frac{\partial Z}{\partial t}=\frac{\partial x}{\partial t}\frac{\partial Z}{\partial x}+\frac{\partial y}{\partial t}\frac{\partial Z}{\partial y} \tag{14.11}$$

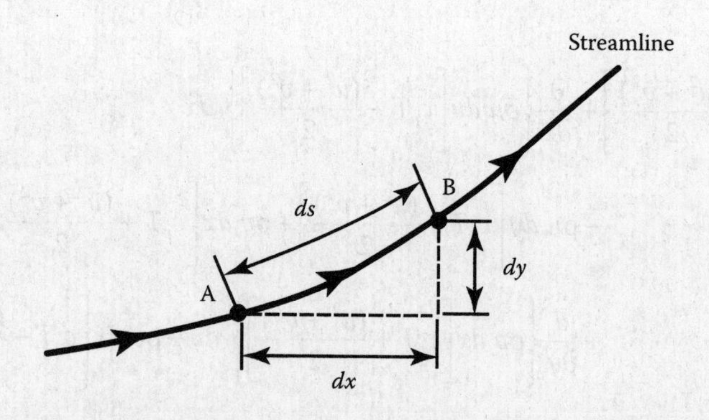

FIGURE 14.3
Changes over length of streamline considered.

i.e., since $dx/dt = u$ and $dy/dt = v$

$$\frac{\partial Z}{\partial t} = u\frac{\partial Z}{\partial x} + v\frac{\partial Z}{\partial y} \tag{14.12}$$

If the quantity Z is not changing along the streamline, i.e., if $dZ/dt = 0$, this equation shows that

$$u\frac{\partial Z}{\partial x} + v\frac{\partial Z}{\partial y} = 0 \tag{14.13}$$

Comparing this with Equation 14.9 then shows that the conservation of energy equation indicates that the quantity

$$c_pT + \frac{(u^2 + v^2)}{2}$$

is constant along a streamline. Now, in most flows, the flow can be assumed to originate from a region of uniform flow as indicated in Figure 14.4.

In such cases, the value of $[c_pT + (u^2 + v^2)/2]$ will initially be the same on all streamlines. The conservation of energy equation then shows that this quantity will remain the same everywhere in the flow. Hence, if the velocity components are determined from the continuity and momentum equations, the energy gives the temperature at any point in the flow as

$$c_pT + \left(\frac{u^2 + v^2}{2}\right) = c_pT_1 + \left(\frac{u_1^2 + v_1^2}{2}\right) = c_pT_0 \tag{14.14}$$

the subscript 1 referring to conditions in the initial uniform flow and T_0 being the stagnation temperature in this initial uniform flow.

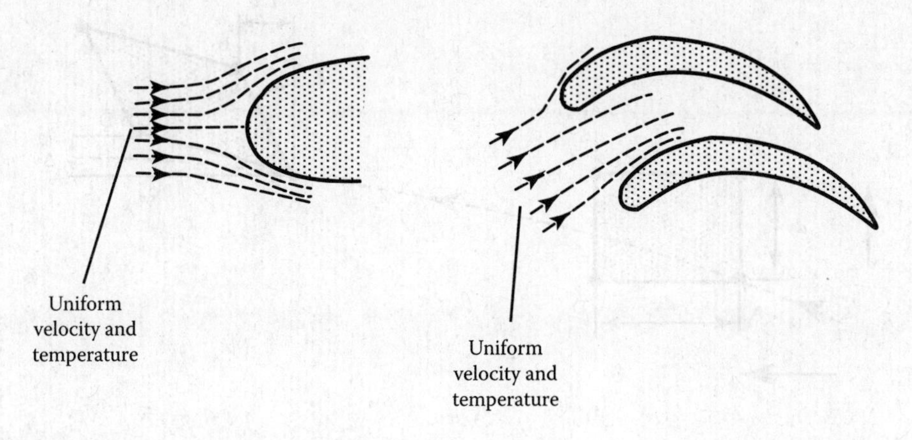

FIGURE 14.4
Uniform upstream velocity.

Vorticity Considerations

As previously discussed, attention is here being directed to flows in which the effects of viscosity are negligible. With this in mind, consider the fluid particles in the flow. Only if there are tangential forces acting on the surface of these particles can there be a change in the net rate at which the particles rotate. This is illustrated in Figure 14.5.

However, the only source of a tangential force is viscosity. Hence, if viscous effects are neglected, there can be no change in the rate at which the fluid particles rotate. If the flow originates in a uniform freestream (see Figure 14.4), the fluid particles will have no initial rotation, and so they will have no rotation anywhere in the flow.

Consider a fluid particle that is initially rectangular in shape with side lengths dx and dy. As it moves through the flow, this particle will distort as indicated in Figure 14.6.

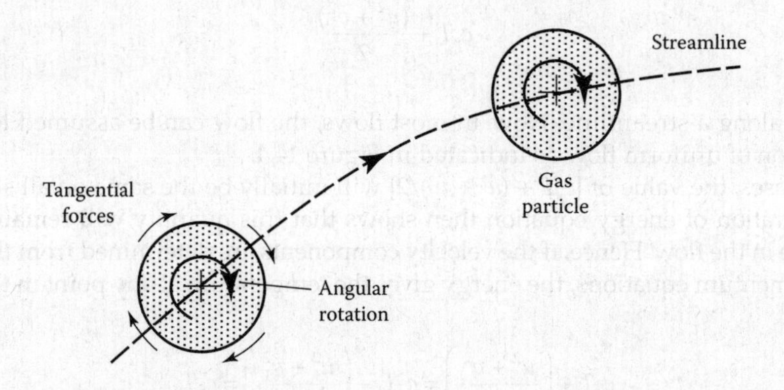

FIGURE 14.5
Changes in rotational motion of particles produced by the viscous stresses acting on them. Tangential stresses are required if the rotational motion is changing as the particle moves along the streamline.

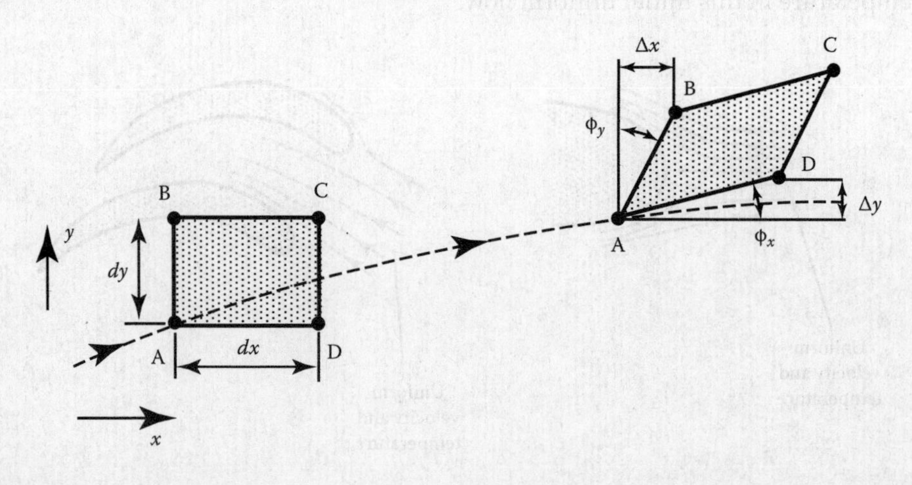

FIGURE 14.6
Distortion of fluid particles.

The net amount by which the particle has rotated in the counter clockwise direction is $(\phi_x - \phi_y)/2$. The net rate at which the fluid particles are rotating (i.e., the vorticity) is then given by

$$\omega = \frac{1}{2}\left(\frac{\partial\phi_x}{\partial t} - \frac{\partial\phi_y}{dt}\right) \tag{14.15}$$

Now, consider the velocities of corner points A, B, and D as indicated in Figure 14.7. From this, it follows that

$$\frac{\partial\phi_x}{\partial t} = \frac{\partial(\Delta y)/\partial t}{\partial x} = \left(\frac{v + (\partial v/\partial x) - v}{dx}\right) = \frac{\partial v}{\partial x} \tag{14.16}$$

and

$$\frac{\partial\phi_y}{\partial t} = \frac{\partial(\Delta x)/\partial t}{\partial x} = \left(\frac{v + (\partial v/\partial y) - u}{dy}\right) = \frac{\partial u}{\partial y} \tag{14.17}$$

Substituting these into Equation 14.15 then gives the rate of fluid particle rotation as

$$\omega = \frac{1}{2}\left(\frac{\partial v}{\partial x} - \frac{\partial u}{\partial y}\right) \tag{14.18}$$

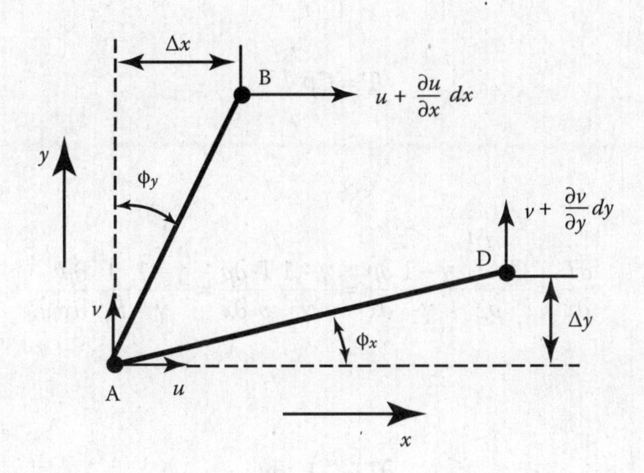

FIGURE 14.7
Velocities of corner points of fluid particle considered.

If viscous forces are negligible and if the flow initially has no rotation, it follows that everywhere in the flow

$$\frac{\partial v}{\partial x} - \frac{\partial u}{\partial y} = 0 \tag{14.19}$$

Using this, the x momentum equation, Equation 14.5 can be written as

$$\rho u \frac{\partial u}{\partial x} + \rho v \frac{\partial v}{\partial x} = -\frac{\partial p}{\partial x}$$

i.e.,

$$\frac{\partial}{\partial x}\left(\frac{u^2 + v^2}{2}\right) = -\frac{1}{\rho}\frac{\partial p}{\partial x} \tag{14.20}$$

Similarly, the y momentum equation, Equation 14.6, can be written using Equation 14.19 as

$$\rho u \frac{\partial u}{\partial y} + \rho v \frac{\partial v}{\partial y} = -\frac{\partial p}{\partial y}$$

i.e.,

$$\frac{\partial}{\partial y}\left(\frac{u^2 + v^2}{2}\right) = -\frac{1}{\rho}\frac{\partial p}{\partial y} \tag{14.21}$$

However, since isentropic flow is being considered,

$$T = Cp^{\frac{\gamma-1}{\gamma}}$$

Hence,

$$\frac{\partial T}{\partial x} = \frac{Cp^{\frac{\gamma-1}{\gamma}}}{p}\frac{\gamma-1}{\gamma}\frac{\partial p}{\partial x} = \frac{\gamma-1}{\gamma}\frac{T}{p}\frac{\partial p}{\partial x} = \frac{\gamma-1}{\gamma}\frac{1}{R\rho}\frac{\partial p}{\partial x}$$

i.e.,

$$\frac{\partial T}{\partial x} = \frac{1}{c_p \rho}\frac{\partial p}{\partial x} \tag{14.22}$$

Similarly,

$$\frac{\partial T}{\partial y} = \frac{1}{c_p \rho} \frac{\partial p}{\partial y} \tag{14.23}$$

Substituting Equations 14.22 and 14.23 into Equations 14.20 and 14.21, respectively, gives

$$\frac{\partial}{\partial x}\left(\frac{u^2 + v^2}{2}\right) + C_p \frac{\partial T}{\partial x} = 0$$

i.e., since c_p is assumed constant,

$$\frac{\partial}{\partial x}\left[c_p T + \left(\frac{u^2 + v^2}{2}\right)\right] = 0 \tag{14.24}$$

and

$$\frac{\partial}{\partial y}\left[c_p T + \left(\frac{u^2 + v^2}{2}\right)\right] = 0 \tag{14.25}$$

These two equations together indicate that the quantity

$$c_p T + \left(\frac{u^2 + v^2}{2}\right)$$

remains constant in an irrotational isentropic flow. This is the same as the result deduced from energy considerations. Thus, in irrotational isentropic flow, momentum, and energy conservation considerations give the same result. This was discussed in Chapter 4.

The Velocity Potential

The equations governing the velocity field in two-dimensional irrotational flow are the continuity equation (14.2) and the irrotationality equation (14.19), i.e.,

$$\text{Equation 14.2:} \quad \frac{\partial}{\partial x}(\rho u) + \frac{\partial}{\partial y}(\rho v) = 0$$

$$\text{Equation 14.19:} \quad \frac{\partial v}{\partial x} - \frac{\partial u}{\partial y} = 0$$

The boundary conditions on these equations are that u and v are prescribed in the initial flow, e.g., $u = u_\infty$, $v = 0$ well upstream of the body considered, and that the velocity component normal to any solid surface is zero.

If a quantity, Φ, termed the velocity potential, is introduced, Φ being such that

$$u = \frac{\partial \Phi}{\partial x}, \quad v = \frac{\partial \Phi}{\partial y} \tag{14.26}$$

then it will be seen that the left-hand side of Equation 14.19 becomes

$$\frac{\partial^2 \Phi}{\partial x\, \partial y} - \frac{\partial^2 \Phi}{\partial y\, \partial x}$$

This will always be zero so it follows that the velocity potential function, as defined by Equation 14.26, satisfies the irrotationality equation (14.19). The continuity equation (14.2) must then be used to solve for Φ. This equation gives

$$\frac{\partial}{\partial x}\left(\rho \frac{\partial \Phi}{\partial x}\right) + \frac{\partial}{\partial y}\left(\rho \frac{\partial \Phi}{\partial y}\right) = 0$$

i.e.,

$$\rho\left(\frac{\partial^2 \Phi}{\partial x^2} + \frac{\partial^2 \Phi}{\partial y^2}\right) + \left(\frac{\partial \Phi}{\partial x}\frac{\partial \rho}{\partial x} + \frac{\partial \Phi}{\partial y}\frac{\partial \rho}{\partial y}\right) = 0 \tag{14.27}$$

However, since isentropic flow is being considered,

$$\rho = C p^{\frac{1}{\gamma}} \tag{14.28}$$

Hence,

$$\frac{\partial \rho}{\partial x} = \frac{1}{\gamma}\frac{C p^{\frac{1}{\gamma}}}{p}\frac{\partial p}{\partial x} = \frac{\rho}{\gamma p}\frac{\partial p}{\partial x} = \frac{1}{a^2}\frac{\partial p}{\partial x}$$

Similarly, it can be shown that

$$\frac{\partial \rho}{\partial y} = \frac{1}{a^2}\frac{\partial p}{\partial y} \tag{14.29}$$

Using these in the momentum equations (14.5) and (14.6) and using Equation 14.26, then gives

$$\frac{\partial \rho}{\partial x} = -\frac{\rho}{a^2}\left(\frac{\partial \Phi}{\partial x}\frac{\partial^2 \Phi}{\partial x^2} + \frac{\partial \Phi}{\partial y}\frac{\partial^2 \Phi}{\partial x\,\partial y}\right) \tag{14.30}$$

and

$$\frac{\partial \rho}{\partial y} = -\frac{\rho}{a^2}\left(\frac{\partial \Phi}{\partial x}\frac{\partial^2 \Phi}{\partial x\,\partial y} + \frac{\partial \Phi}{\partial y}\frac{\partial^2 \Phi}{\partial y^2}\right) \tag{14.31}$$

Substituting these two equations into Equation 14.27 then gives

$$\frac{\partial^2 \Phi}{\partial x^2} + \frac{\partial^2 \Phi}{\partial y^2} - \frac{1}{a^2}\left[\left(\frac{\partial \Phi}{\partial x}\right)^2 \frac{\partial^2 \Phi}{\partial x^2} + \frac{\partial \Phi}{\partial x}\frac{\partial \Phi}{\partial y}\frac{\partial^2 \Phi}{\partial x\,\partial y}\right] - \frac{1}{a^2}\left[\frac{\partial \Phi}{\partial y}\frac{\partial \Phi}{\partial x}\frac{\partial^2 \Phi}{\partial x\,\partial y} + \left(\frac{\partial \Phi}{\partial y}\right)^2 \frac{\partial^2 \Phi}{\partial y^2}\right] = 0$$

i.e.,

$$\frac{\partial^2 \Phi}{\partial x^2} + \frac{\partial^2 \Phi}{\partial y^2} - \frac{1}{a^2}\left[\left(\frac{\partial \Phi}{\partial x}\right)^2 \frac{\partial^2 \Phi}{\partial x^2} + 2\frac{\partial \Phi}{\partial x}\frac{\partial \Phi}{\partial y}\frac{\partial^2 \Phi}{\partial x\,\partial y} + \left(\frac{\partial \Phi}{\partial y}\right)^2 \frac{\partial^2 \Phi}{\partial y^2}\right] = 0 \tag{14.32}$$

This can be written as

$$\frac{\partial^2 \Phi}{\partial x^2} + \frac{\partial^2 \Phi}{\partial y^2} - \frac{1}{a^2}\left[u^2 \frac{\partial^2 \Phi}{\partial x^2} + 2uv\frac{\partial^2 \Phi}{\partial x\,\partial y} + v^2 \frac{\partial^2 \Phi}{\partial y^2}\right] = 0 \tag{14.33}$$

Beside Φ, this equation contains the speed of sound a. An expression relating a to Φ is therefore required. This is supplied by the energy equation, which, as discussed above, gives (see Equation 14.14)

$$c_p T + \left(\frac{u^2 + v^2}{2}\right) = c_p T_0 \tag{14.34}$$

where T_0 is the stagnation temperature, which is a constant throughout the flow. Hence, since

$$a^2 = \gamma RT \text{ and } C_p = \gamma R/(\gamma - 1)$$

Equation 14.34 gives

$$a^2 + \frac{\gamma - 1}{20}(u^2 + v^2) = a_0^2$$

i.e., using Equation 14.26,

$$a^2 = a_0^2 - \left(\frac{\gamma - 1}{2}\right)\left[\left(\frac{\partial \Phi}{\partial x}\right)^2 + \left(\frac{\partial \Phi}{\partial y}\right)^2\right] \tag{14.35}$$

Equations 14.32 and 14.35 together describe the variation of Φ in irrotational, isentropic, flow. In low-speed flow (i.e., $M \ll 1$), the density variation is negligible and Equation 14.27 gives

$$\frac{\partial^2 \Phi}{\partial x^2} + \frac{\partial^2 \Phi}{\partial y^2} = 0 \tag{14.36}$$

Hence, in low-speed flow, the variation of Φ is governed by Laplace's equations. In incompressible flow, then, it is relatively easy to determine Φ, Equation 14.36 being a linear equation. In compressible flow, however, the compressibility effects give rise to the nonlinear terms in Equation 14.32, i.e., terms such as $(\partial\Phi/\partial x)^2(\partial^2\Phi/\partial x^2)$, which involve the product of functions of Φ. This makes the determination of Φ in compressible flows significantly more difficult than in incompressible flows. To solve for Φ in compressible flows, the following methods can be used:

- Full numerical solutions
- Transformation of variables to give a linear governing equation
- Linearized solutions

The second method is only applicable in a few situations and will not be discussed here. Linearized solutions will be discussed in the next section and a very brief discussion of numerical methods is given in a later section.

Linearized Solutions

To keep the drag low on objects in high-speed flows, the objects are usually kept relatively slender to minimize the disturbance they produce in the flow. With such slender objects, the differences between the values of the flow variables near the object and the values of these variables in the freestream flow ahead of the object are small, e.g., consider the situation shown in Figure 14.8.

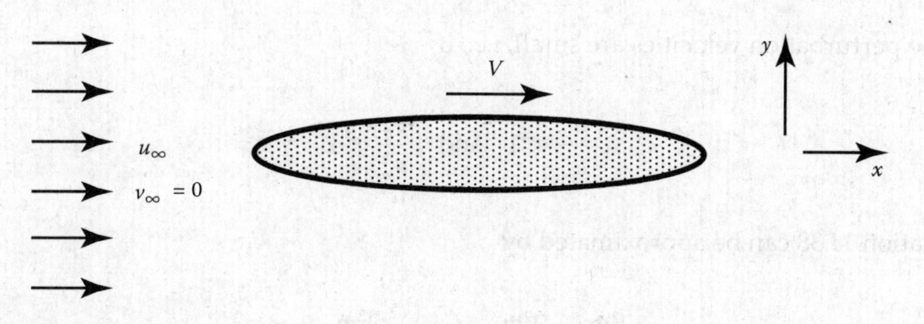

FIGURE 14.8
Flow situation considered.

If the components in the x- and y-directions of the velocity V are $u_\infty + u_p$ and v_p then, for a slender object, the perturbation velocities u_p and v_p will be very small compared with u_∞. This assumption is the basis for the analysis given in the present section.

Now, in the undisturbed flow ahead of the object, $u = u_\infty$ and $v = 0$, so the velocity potential is, by virtue of Equation 14.26, given by

$$\frac{\partial \Phi}{\partial x} = u_\infty \quad \text{and} \quad \frac{\partial \Phi}{\partial y} = 0$$

i.e.,

$$\Phi = u_\infty x$$

The velocity potential in the flow will therefore be written as

$$\Phi = u_\infty x + \Phi_p$$

where Φ_p is the perturbation velocity potential, which must be such that

$$u_p = \frac{\partial \Phi_p}{\partial x}, \quad v_p = \frac{\partial \Phi_p}{\partial y} \tag{14.37}$$

Substituting the above relations into the potential function equation in the form given in Equation 14.33 leads to

$$\frac{\partial^2 \Phi_p}{\partial x^2} + \frac{\partial^2 \Phi_p}{\partial y^2} - \frac{1}{a^2}\left[(u_\infty + u_p)^2 \frac{\partial^2 \Phi_p}{\partial x^2} + 2(u_\infty + u_p)v_p \frac{\partial^2 \Phi_p}{\partial x\, \partial y} + v_p^2 \frac{\partial^2 \Phi_p}{\partial y^2}\right] = 0$$

i.e.,

$$\frac{\partial^2 \Phi_p}{\partial x^2} + \frac{\partial^2 \Phi_p}{\partial y^2} - \left(\frac{u_\infty}{a}\right)^2 \left[\left(1 + \frac{u_p}{u_\infty}\right)^2 \frac{\partial^2 \Phi_p}{\partial x^2} + 2\left(1 + \frac{u_p}{u_\infty}\right)\left(\frac{v_p}{u_\infty}\right) \frac{\partial^2 \Phi_p}{\partial x\, \partial y} + \left(\frac{v_p}{u_\infty}\right)^2 \frac{\partial^2 \Phi_p}{\partial y^2}\right] = 0 \tag{14.38}$$

If the perturbation velocities are small, i.e., if

$$\frac{u_p}{u_\infty} \ll 1 \quad \text{and} \quad \frac{v_p}{u_\infty} \ll 1 \tag{14.39}$$

Equation 14.38 can be approximated by

$$\frac{\partial^2 \Phi_p}{\partial x^2} + \frac{\partial^2 \Phi_p}{\partial y^2} - \left(\frac{u_\infty}{a}\right)^2 \frac{\partial^2 \Phi_p}{\partial x^2} = 0$$

i.e.,

$$\left[1 - M_\infty^2 \left(\frac{a_\infty}{a}\right)^2\right] \frac{\partial^2 \Phi_p}{\partial x^2} + \frac{\partial^2 \Phi_p}{\partial y^2} = 0 \tag{14.40}$$

However, it was shown above that $c_pT + V^2/2$ is the same everywhere in the flow, so

$$c_pT_\infty + \frac{u_\infty^2}{2} = c_pT + \frac{(u_\infty + u_p)^2 + v_p^2}{2}$$

i.e.,

$$c_p(T_\infty - T) = \frac{2u_pu_\infty + u_p^2 + v_p^2}{2} \tag{14.41}$$

However, since

$$a^2 = \gamma RT = \gamma c_p\left(1 - \frac{1}{\gamma}\right)T = (\gamma - 1)c_pT$$

Equation 14.41 gives

$$\left(\frac{a}{a_\infty}\right)^2 = 1 - (\gamma - 1)\left(\frac{u_pu_\infty}{a_\infty^2} + \frac{1}{2}\frac{u_p^2}{a_\infty^2} + \frac{1}{2}\frac{v_p^2}{a_\infty^2}\right) = 1 - (\gamma - 1)M_\infty^2\left[\frac{u_p}{u_\infty} + \frac{1}{2}\left(\frac{u_p}{u_\infty}\right)^2 + \frac{1}{2}\left(\frac{v_p}{u_\infty}\right)^2\right]$$

For small u_p/u_∞ and v_p/u_∞, this gives

$$\left(\frac{a}{a_\infty}\right)^2 = 1 - (\gamma - 1)M_\infty^2 \frac{u_p}{u_\infty} \tag{14.42}$$

Substituting this into Equation 14.40 then gives

$$\left[1-M_\infty^2+(\gamma-1)M_\infty^4\frac{u_p}{u_\infty}\right]\frac{\partial^2\phi_p}{\partial x^2}+\frac{\partial^2\phi_p}{\partial y^2}=0 \tag{14.43}$$

Provided that M_∞ is not very large and provided that M_∞ is not near 1

$$(\gamma-1)M_\infty^4\frac{u_p}{a_\infty}\ll 1-M_\infty^2$$

and Equation 14.43 then gives

$$(1-M_\infty^2)\frac{\partial^2\Phi_p}{\partial x^2}+\frac{\partial^2\Phi_p}{\partial y^2}=0 \tag{14.44}$$

This is the linearized velocity potential equation. It applies provided M_∞ is not very near 1 or very large. Equation 14.44 is a linear equation and is therefore much easier to solve than the full, nonlinear velocity potential equation.

To solve Equation 14.44, it is necessary to specify the boundary conditions on the solution. Well upstream of the body, as indicated in Figure 14.9, the flow is undisturbed and Φ_p is zero. On the surface of the body, there is no velocity component normal to the surface, i.e., the velocity vector is tangent to the surface. Hence, if θ is the angle the surface makes to the x direction as indicated in Figure 14.9,

$$\tan\theta=\left.\frac{dy}{dx}\right|_s=\left.\frac{v_p}{u_\infty+u_p}\right|_s \tag{14.45}$$

The subscript s indicates conditions on the surface.

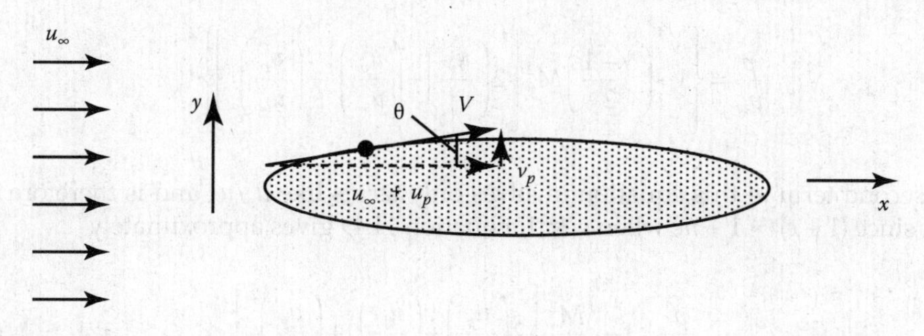

FIGURE 14.9
Boundary conditions.

Now,

$$\frac{v_p}{u_\infty + u_p} = \frac{v_p/u_\infty}{1+u_p/u_\infty} = \frac{v_p}{u_\infty} - \left(\frac{v_p}{u_\infty}\right)\left(\frac{u_p}{u_\infty}\right) \tag{14.46}$$

Hence, since u_p/u_∞ and v_p/u_∞ are $\ll 1$, these two equations give

$$\frac{v_p\big|_s}{u_\infty} = \left.\frac{dy}{dx}\right|_s \tag{14.47}$$

v_p, of course, being equal to $\partial\Phi_\infty/\partial y$.

The solution of the velocity potential equation allows the velocity components at all points in the flow to be determined. The pressure variation through the flow can then be deduced from the velocity distribution. To do this, it is noted that the energy equation (14.41) gives

$$T_\infty - T = \frac{2u_p u_\infty + u_p^2 + v_p^2}{2c_p}$$

i.e.,

$$\frac{T}{T_\infty} = 1 - \left(\frac{\gamma-1}{2}\right)M_\infty^2\left[2\left(\frac{u_p}{u_\infty}\right) + \left(\frac{u_p}{u_\infty}\right)^2 + \left(\frac{v_p}{u_\infty}\right)^2\right] \tag{14.48}$$

However, since isentropic flow is being considered,

$$\frac{p}{p_\infty} = \left(\frac{T}{T_\infty}\right)^{\frac{\gamma}{\gamma-1}}$$

Thus, Equation 14.48 gives

$$\frac{p}{p_\infty} = \left\{1 - \left(\frac{\gamma-1}{2}\right)M_\infty^2\left[2\left(\frac{u_p}{u_\infty}\right) + \left(\frac{u_p}{u_\infty}\right)^2 + \left(\frac{v_p}{u_\infty}\right)^2\right]\right\}^{\frac{\gamma}{\gamma-1}} \tag{14.49}$$

The second term in this equation involves only terms like u_p/u_∞ and is therefore small. Hence, since $(1+\epsilon)^n \approx 1 + n\epsilon$ when $\epsilon \ll 1$, Equation 14.49 gives approximately

$$\frac{p}{p_\infty} = 1 - \frac{\gamma M_\infty^2}{2}\left[2\left(\frac{u_p}{u_\infty}\right) + \left(\frac{u_p}{u_\infty}\right)^2 + \left(\frac{v_p}{u_\infty}\right)^2\right] \tag{14.50}$$

Further, since v_p/u_∞ and u_p/u_∞ are small, the second two terms in the bracket are much smaller than the first, so approximately

$$\frac{p}{p_\infty} = 1 - \gamma \frac{M_\infty^2}{2}\left(2\frac{u_p}{u_\infty}\right) \tag{14.51}$$

It is usual to express the pressure distribution in terms of a pressure coefficient defined by

$$C_p = \frac{p - p_\infty}{\frac{1}{2}\rho_\infty u_\infty^2} = \frac{(p/p_\infty) - 1}{\frac{1}{2}(\rho_\infty/p_\infty)u_\infty^2}$$

i.e.,

$$C_p = \frac{(p/p_\infty) - 1}{\gamma M_\infty^2/2} \tag{14.52}$$

because $a_\infty^2 = \gamma p_\infty/\rho_\infty$.

Combining this with Equation 14.51 then gives for linearized flows

$$C_p = -\frac{2u_p}{u_\infty} \tag{14.53}$$

Using Equation 14.37, this can be written as

$$C_p = -\frac{2}{u_\infty}\frac{\partial \Phi_p}{\partial x} \tag{14.54}$$

Thus, once the distribution of Φ_p has been determined, the variation of C_p through the flow field can be found, i.e., the variation of C_p about the surface of a body in the flow can be determined.

Linearized Subsonic Flow

It is to be expected that in subsonic flow over a body the solution for Φ_p in the actual flow can be related to the solution for the flow that would exist over the same body if the flow was incompressible. To show that this is indeed the case, Equation 14.44, which governs the actual flow, is written as

$$\beta^2 \frac{\partial^2 \Phi_p}{\partial x^2} + \frac{\partial^2 \Phi_p}{\partial y^2} = 0 \tag{14.55}$$

where

$$\beta^2 = 1 - M_\infty^2 \tag{14.56}$$

The coordinates in the real space, x and y, are now transformed into coordinates ζ and η in a transformed space using

$$\zeta = x, \quad \eta = \beta y \tag{14.57}$$

Moreover, let the linearized velocity potential in the transformed space, $\overline{\Phi_p}$, be related to the linearized velocity potential in the real space, Φ_p by

$$\overline{\Phi_p} = \beta \Phi_p \tag{14.58}$$

This is illustrated in Figure 14.10.

Substituting Equations 14.57 and 14.58 into Equation 14.55 and noting that ζ depends only on x and that η depends only on y gives

$$\beta^2 \frac{\partial^2(\overline{\Phi_p}/\beta)}{\partial \zeta^2}\left(\frac{d\zeta}{dx}\right)^2 + \frac{\partial^2(\overline{\Phi_p}/\beta)}{\partial \eta^2}\left(\frac{d\eta}{dy}\right)^2 = 0$$

i.e., since $d\eta/dy = \beta$,

$$\beta \frac{\partial^2 \overline{\Phi_p}}{\partial \zeta^2} + \beta \frac{\partial^2 \overline{\Phi_p}}{\partial \eta^2} = 0$$

i.e.,

$$\frac{\partial^2 \overline{\Phi_p}}{\partial \zeta^2} + \frac{\partial^2 \overline{\Phi_p}}{\partial \eta^2} = 0 \tag{14.59}$$

Thus, the variation of $\overline{\Phi_p}$ in the transformed plane is governed by Laplace's equation.

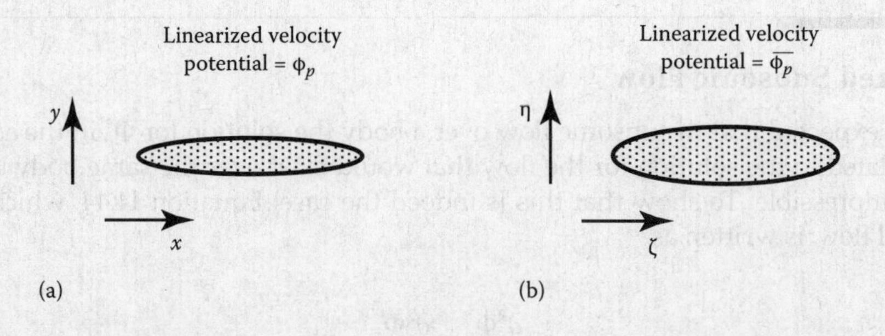

FIGURE 14.10
(a) Actual flow and (b) transformed flow.

Now consider the boundary conditions at the surface as discussed in the derivation of Equation 14.47. Let

$$y = f(x)$$

describe the shape of the actual body and let

$$\eta = F(\zeta)$$

describe the shape of the body in the transformed plane.

The boundary condition in the actual flow is

$$\left.\frac{\partial \Phi_p}{\partial y}\right|_s = u_\infty \frac{df}{dx} \tag{14.60}$$

whereas in the transformed plane the boundary condition is

$$\left.\frac{\partial \overline{\Phi_p}}{\partial \eta}\right|_s = u_\infty \frac{dF}{d\zeta} \tag{14.61}$$

However,

$$\left.\frac{\partial \overline{\Phi_p}}{\partial \eta}\right|_s = \left[\frac{\partial(\Phi_p\beta)}{\partial y}\frac{\partial y}{\partial \eta}\right]_s = \left[\frac{\partial(\Phi_p\beta)}{\partial y}\frac{1}{\beta}\right]_s$$

i.e.,

$$\left.\frac{\partial \overline{\Phi_p}}{\partial \eta}\right|_s = \left.\frac{\partial \Phi_p}{\partial y}\right|_s \tag{14.62}$$

Equation 14.62 therefore shows that the lef-hand side of Equations 14.60 and 14.61 are equal, so

$$\frac{df}{dx} = \frac{dF}{d\zeta} \tag{14.63}$$

This means that the shape of the body is the same in real and the transformed planes, i.e., the function that relates y to x on the surface of the body in the real plane is identical to the function that relates η to ζ on the surface of the body in the transformed plane. Since $\overline{\Phi_p}$ is determined by Equation 14.59, which is identical to the equation that applies in incompressible flow, and since the shape of the body is the same in the real and transformed planes, it follows that $\overline{\Phi_p}$ is the same as the linearized velocity potential function that would exist in incompressible flow over the body being considered. Hence, if Φ_p is determined from the solution for incompressible flow over the body, the actual linearized velocity potential and x and y being given by

$$\Phi_p = \overline{\Phi_p}/\beta = \overline{\Phi_p}/\sqrt{1-M^2}, \quad x = \zeta, \quad y = \eta/\beta = \eta/\sqrt{1-M^2}$$

Now, Equation 14.54 gives

$$C_p = -\frac{2}{u_\infty}\frac{\partial \Phi_p}{dx} = -\frac{2}{u_\infty}\frac{\partial(\overline{\Phi_p}/\beta)}{d\zeta}$$

i.e.,

$$C_p = \frac{1}{\beta}\left[-\frac{2}{u_\infty}\frac{\partial \overline{\Phi_p}}{d\zeta}\right] \tag{14.64}$$

However, the variation of $\overline{\Phi_p}$ with ζ is the same as in incompressible flow over the body shape being considered, i.e., by virtue of Equation 14.54, Equation 14.64 gives

$$C_p = \frac{1}{\beta}C_{p0} = \frac{C_{p0}}{\sqrt{1-M_\infty^2}} \tag{14.65}$$

where C_{p0} is the value of the pressure coefficient that would exist in incompressible flow over the body being considered. This means that if the pressure coefficient distribution is determined for incompressible irrotational flow, the pressure coefficient distribution is compressible flow at Mach number M_∞ can be found by applying a "compressibility correction factor" equal to $1/\sqrt{1-M_\infty^2}$. This only applies in subsonic flows, i.e., to flows in which $M_\infty < 1$.

Consider the lift force on a body, i.e., the force normal to the upstream flow, resulting from the variation in the pressure over the surface of the body. If L is the lift per unit span, it will be seen from Figure 14.11 that

$$L = \int (p - p_\infty)\cos\theta \; ds$$

the integral being carried out over the surface of the body.

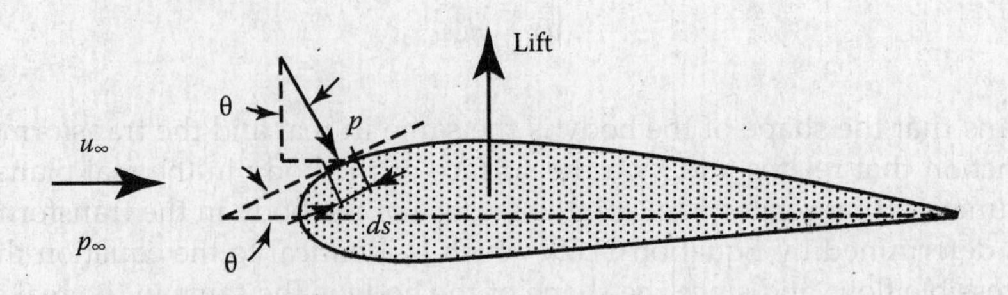

FIGURE 14.11
Calculation of lift from pressure distribution.

This equation can be written as

$$C_L\left(=\frac{L}{\frac{1}{2}\rho_\infty u_\infty^2 c}\right)=\int C_p\cos\theta\, d\left(\frac{s}{c}\right) \quad (14.66)$$

where C_L is the lift coefficient and c is the wing chord. From Equation 14.65, it follows that

$$C_L=\frac{C_{L0}}{\sqrt{1-M_\infty^2}} \quad (14.67)$$

C_{L0} being the lift coefficient that would exist in incompressible flow. Since $1-M_\infty^2<1$, the above equations indicate that compressibility increases the coefficient of lift.

Now for smaller angles of attack, α, this being defined in Figure 14.12, $C_L = a\alpha$ in incompressible flow. In compressible flow, then $C_L = a\alpha/\sqrt{1-M_\infty^2}$. This is shown in Figure 14.13.

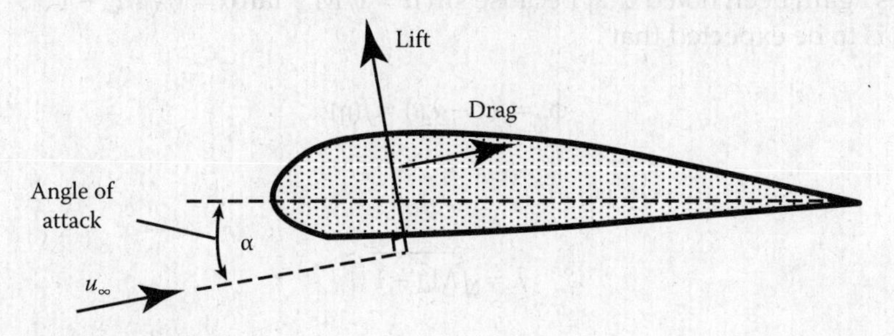

FIGURE 14.12
Definition of airfoil angle of attack.

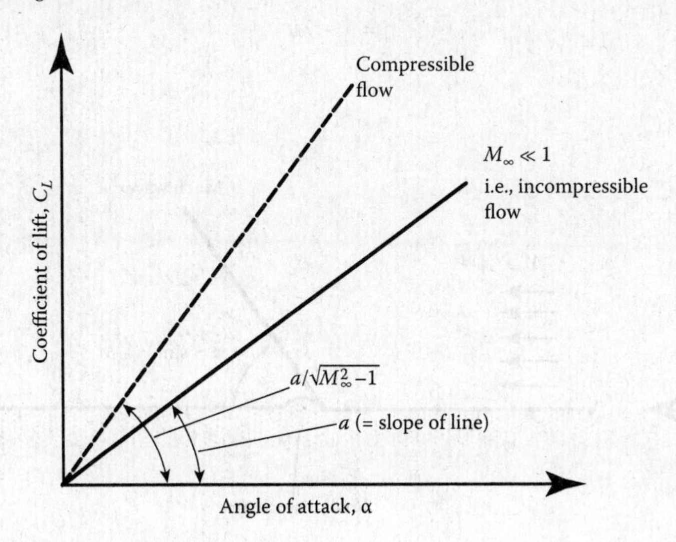

FIGURE 14.13
Effect of compressibility on lift coefficient variation.

Linearized Supersonic Flow

In supersonic flow disturbances are, as discussed in Chapter 2, propagated along Mach lines as indicated in Figure 14.14.

The flow upstream of the Mach line is undisturbed by the presence of the wave. It is to be expected therefore that the perturbation velocity potential, Φ_p, will, in supersonic flow, be constant along a Mach line. Now, along a Mach line

$$\frac{y}{x} = \frac{1}{\sqrt{M_\infty^2 - 1}}$$

i.e.,

$$x = \sqrt{M_\infty^2 - 1}\, y$$

where it has again been noted that because $\sin\alpha = 1/M_\infty$, $\tan\alpha = 1/\sqrt{M_\infty^2 - 1}$.

Hence, it is to be expected that

$$\phi_p = f(x - \lambda y) = f(\eta) \tag{14.68}$$

where

$$\lambda = \sqrt{M_\infty^2 - 1} \tag{14.69}$$

and

$$\eta = x - \lambda y \tag{14.70}$$

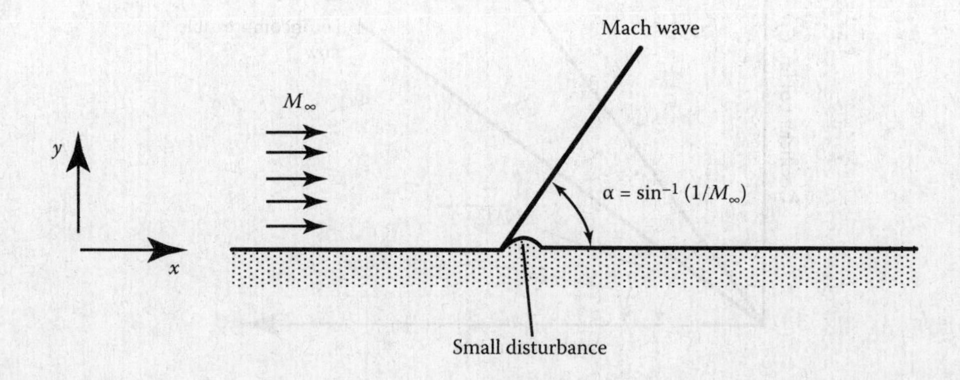

FIGURE 14.14
Mach wave.

If this is, indeed, a solution it must satisfy Equation 14.44, which can be written as

$$\lambda^2 \frac{\partial^2 \Phi_p}{dx^2} - \frac{\partial^2 \Phi_p}{dy^2} = 0 \tag{14.71}$$

Now, Equation 14.68 gives

$$\frac{\partial \Phi_p}{dx} = \frac{df}{d\eta}\frac{\partial \eta}{dx} = \frac{df}{d\eta}$$

from which it follows that

$$\frac{\partial^2 \Phi_p}{dx^2} = \left(\frac{d^2 f}{d\eta^2}\right)\left(\frac{\partial \eta}{dx}\right)^2 = \frac{d^2 f}{d\eta^2} \tag{14.72}$$

Similarly, Equation 14.68 gives

$$\frac{\partial^2 \Phi_p}{dy^2} = \left(\frac{d^2 f}{d\eta^2}\right)\left(\frac{\partial \eta}{dy}\right)^2 = \lambda^2 \frac{d^2 f}{d\eta^2} \tag{14.73}$$

Substituting these last two equations into the left-hand side of Equation 14.71 gives this left-hand side as

$$\lambda^2 \frac{d^2 f}{d\eta^2} - \lambda^2 \frac{d^2 f}{d\eta^2}$$

which is always zero. This proves that Equation 14.68 is, indeed, a solution to the linearized velocity potential equation for supersonic flow.

Next consider the pressure coefficient at the surface of a body in a supersonic flow. It will be recalled that the linearized boundary condition at the surface as given in Equation 14.47 requires that

$$\frac{v_p\big|_s}{u_\infty} = \tan\theta \tag{14.74}$$

However, since linearized flow is being considered, θ must remain small and $\tan\theta$ can be approximated by θ. Equation 14.74 can therefore, in linearized flow, be written as

$$\frac{v_p\big|_s}{u_\infty} = \theta \tag{14.75}$$

It is then noted that using Equation 14.68 gives

$$u_p = \frac{\partial \Phi_p}{\partial x} = \frac{df}{d\eta}\frac{\partial \eta}{\partial x} = \frac{df}{\partial \eta}, \quad v_p = \frac{\partial \Phi_p}{\partial y} = \frac{df}{d\eta}\frac{\partial \eta}{\partial y} = -\lambda\frac{df}{\partial \eta}$$

Dividing these two equations gives

$$\frac{u_p}{v_p} = -\frac{1}{\lambda}$$

i.e.,

$$u_p = -\frac{v_p}{\lambda} \tag{14.76}$$

Combining this with Equation 14.75 then gives

$$u_p\big|_s = -\frac{u_\infty}{\lambda}\theta \tag{14.77}$$

However, Equation 14.53 gives $C_p = -2u_p/u_\infty$ so at the surface of a body in supersonic flow

$$C_p = \frac{2\theta}{\lambda} = \frac{2\theta}{\sqrt{M_\infty^2 - 1}} \tag{14.78}$$

This equation shows that, provided the assumption of linearized flow is applicable, the pressure coefficient on the surface of a body in supersonic flow is proportional to the angle the surface makes to the oncoming flow.

It should be noted that the above analysis only applies to flow in which the Mach waves run upward as shown in Figure 14.14. Consider the waves generated on the underside of a surface as shown schematically in Figure 14.15.

On the lower surface of a body, it is to be expected then that

$$\Phi_p = h(x + \lambda y) = h\zeta \tag{14.79}$$

where

$$\zeta = x + \lambda y \tag{14.80}$$

Applying the same type of analysis as that presented above for the upper surface then gives for the lower surface

$$C_p = \frac{-2\theta}{\sqrt{M_\infty^2 - 1}} \tag{14.81}$$

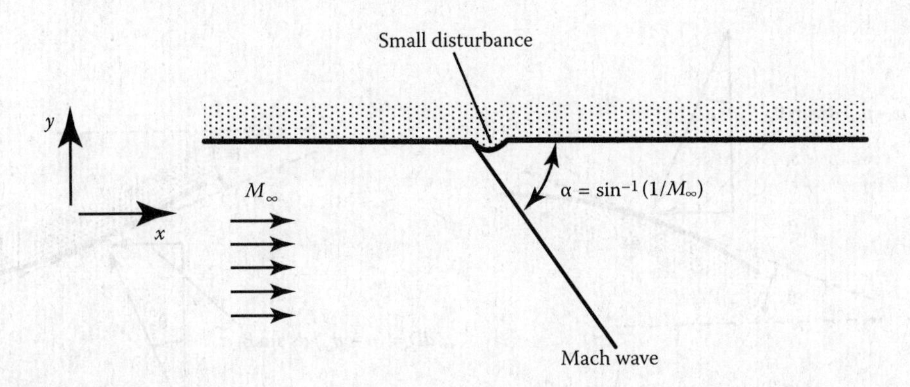

FIGURE 14.15
Mach waves on lower surface.

Equations 14.78 and 14.81 define the distribution of the pressure coefficient over a slender body in supersonic flow. Consider a body of the type shown in Figure 14.16.

On the forward part of the body between A and B, θ is positive, and C_p is, by virtue of Equation 14.78, positive. Also, on the forward part of the body between A and D, θ is negative, and C_p is thus also, by virtue of Equation 14.81, positive. Hence, C_p is positive everywhere along BAD. On the rear portion of the body, θ is negative between B and C and positive between D and C. Hence, by virtue of Equations 14.78 and 14.81, C_p is negative everywhere along BCD.

Next, consider the drag force, D, on a body such as that shown in Figure 14.16. The drag force arises because of the pressure variation about the surface and, considering the forces acting on a small portion of the surface indicated in Figure 14.17, it will be seen that

$$D = \int (p - p_\infty) \sin\theta \; ds \tag{14.82}$$

the integration being carried out over the surface of the body, the upper and lower surfaces being separately considered.

Since linearized theory is being used,which applies only to slender bodies, $\sin\theta \approx \theta$ and $ds \approx dx$. Hence, Equation 14.82 can be written as

$$D = \int (p - p_\infty)\theta \; dx \tag{14.83}$$

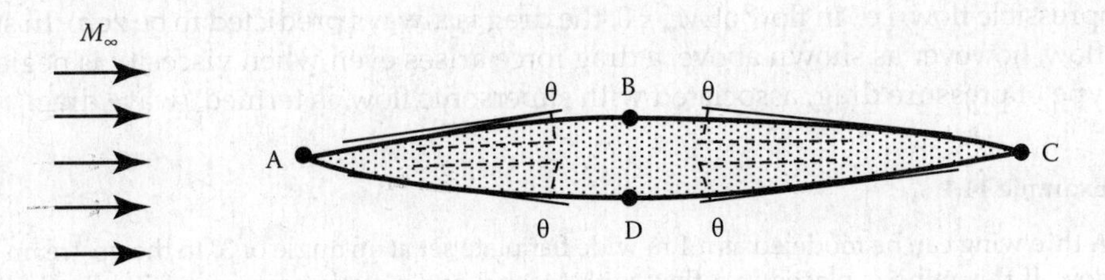

FIGURE 14.16
Type of body considered.

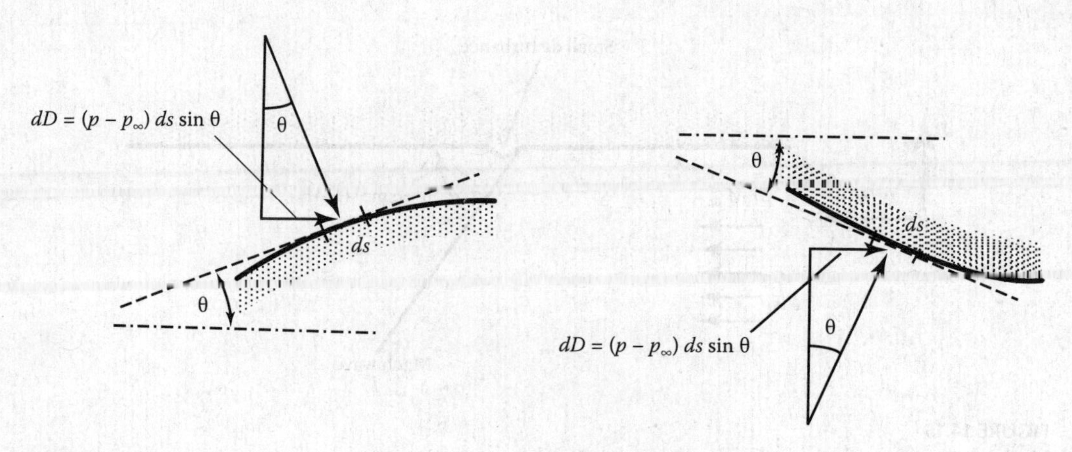

FIGURE 14.17
Generation of drag force. $p - p_\infty$ is the pressure relative to freestream pressure.

The drag coefficient C_D is defined as usual in two-dimensional flow by

$$C_D = \frac{D}{\frac{1}{2}\rho u_\infty^2 c} \tag{14.84}$$

where c is the chord of the body. The unit span, i.e., unit distance at right angles to the flow, has been considered.

Equation 14.83 thus gives

$$C_D = \int_{\text{upper surface}} C_p \theta \, d(x/c) + \int_{\text{lower surface}} C_p \theta \, d(x/c) \tag{14.85}$$

Consider the body shown in Figure 14.16. Since C_p is positive between A and B and between A and C, this portion BAC of the surface will give a positive contribution to the drag. However, C_p is negative between B and D and between C and D, and therefore because of the way the surface slopes, this also gives a positive contribution to the drag. Thus, there will be the net drag force on the body. Now, viscosity is being neglected in the present analysis. When this assumption, i.e., that viscous effects are negligible, is made in incompressible flow, i.e., in flow at $M_\infty \approx 0$, the drag is always predicted to be zero. In supersonic flow, however as shown above, a drag force arises even when viscosity is neglected. This type of pressure drag, associated with supersonic flow, is termed "wave drag."

Example 14.1

A thin wing can be modeled as a 1 m wide flat plate set at an angle of 3° to the upstream flow. If this wing is placed in a flow with Mach 3 and a static pressure of 50 kPa, find using linearized theory the pressure on the upper and lower surfaces of the airfoil and the lift and drag per meter span.

Solution

The flow situation here being considered is shown in Figure E14.1.

The angle that the upper surface makes to the flow is –3° = –0.0524 radians. Hence, Equation 14.78 gives

$$C_{p\ \mathrm{upper}} = \frac{-2 \times 0.0524}{\sqrt{9-1}} = -0.0371$$

i.e.,

$$\frac{p_{\mathrm{upper}} - p_\infty}{\frac{1}{2}\rho_\infty u_\infty^2} = -0.0371$$

However,

$$\frac{1}{2}\rho_\infty u_\infty^2 = \frac{p_\infty}{2}\frac{\rho_\infty}{p_\infty}u_\infty^2 = \frac{\gamma p_\infty M_\infty^2}{2}$$

Thus,

$$\frac{p_{\mathrm{upper}} - 50}{1.4 \times 50 \times 9/2} = -0.0371$$

Therefore,

$$p_{\mathrm{upper}} = 38.31 \text{ kPa}$$

Similarly, since θ for the lower surface is also –3°, Equation 14.81 gives

$$\frac{p_{\mathrm{lower}} - 50}{1.4 \times 50 \times 9/2} = +0.0371$$

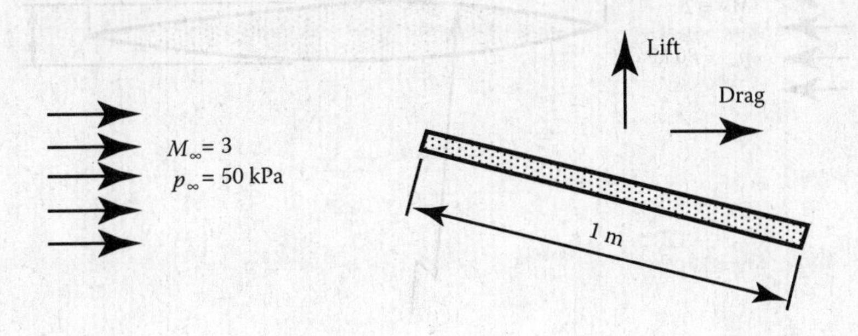

FIGURE E14.1
Flow situation considered.

Therefore,

$$p_{\text{lower}} = 61.69 \text{ kPa}$$

The lift will be given by

$$L = (p_{\text{lower}}A - p_{\text{upper}}A)\cos\theta$$

where A is the planform area of the wing, i.e., 1 × 1 m. Since θ is small, it is consistent with the previous assumptions to set cos θ = 1. Hence,

$$L = (61.69 - 38.31) \times 1 \times 1 = 23.38 \text{ kN}$$

Similarly, if sin θ is approximated by θ, the drag in the airfoil is given by

$$D = (p_{\text{lower}}A - p_{\text{upper}}A)\sin\theta = (p_{\text{lower}} - p_{\text{upper}})A\theta = (61.69 - 38.31) \times 0.0524 = 1.23 \text{ kN}$$

Hence, the pressures on the upper and lower surfaces are 38.31 and 61.69 kPa, respectively, and the lift and drag are 23.38 and 1.23 kN, respectively.

If the oblique shock and expansion wave results given in Chapters 6 and 7 are used, the values of the pressures and the lift and drag obtained are very close to the values given by the approximate linearized theory.

Example 14.2

An airfoil of the shape shown in Figure E14.2a is placed in a flow at Mach 2 and a pressure of 80 kPa. Derive expressions for the pressure variations along the upper and lower surfaces.

Solution

Consider the geometrical parameters shown in Figure E14.2b.

Considering the triangle ABC shown in Figure E14.2b, it will be seen that

$$R^2 = \left(\frac{c}{2}\right)^2 + \left(R - \frac{t}{2}\right)^2 = \frac{c^2}{4} + R^2\left(1 - \frac{t}{2R}\right)^2$$

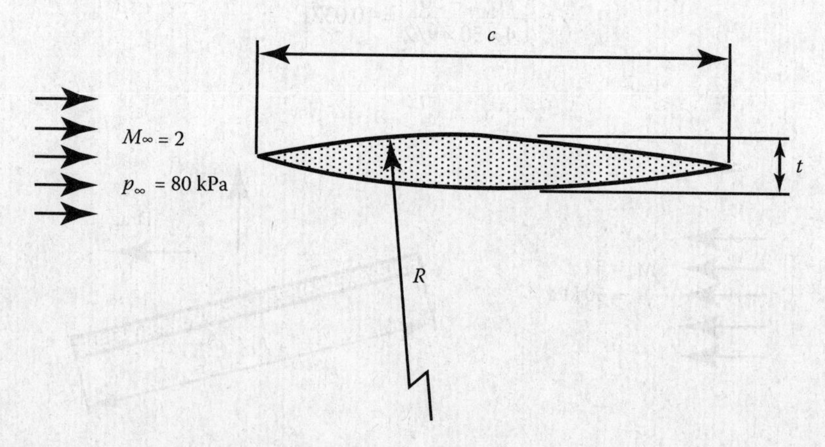

FIGURE E14.2a
Flow situation considered and airfoil shape.

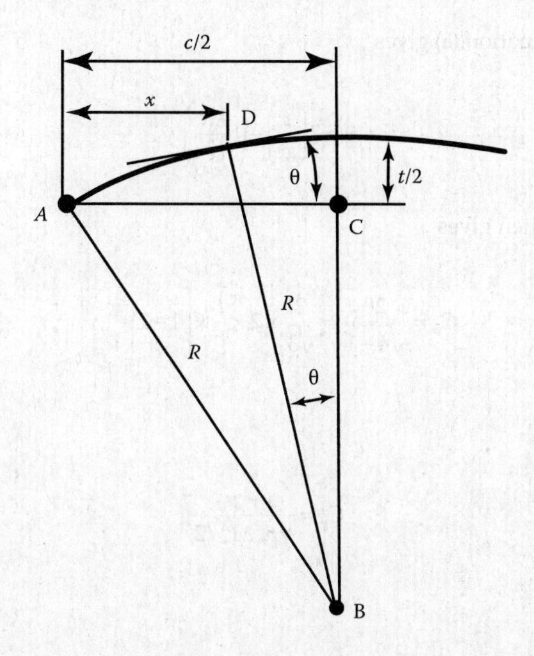

FIGURE E14.2b
Geometrical factors defining shape of airfoil.

However, by assumption, $t/R \ll 1$, so this equation gives approximately

$$R^2 = \frac{c^2}{4} + R^2 - \frac{2tR}{2}$$

i.e.,

$$R = \frac{c^2}{4t} \tag{a}$$

Next consider point D, which lies distance x from the leading edge. It will be seen that

$$x = \frac{c}{2} - R\sin\theta$$

which, since $\sin\theta \approx \theta$ gives

$$x = \frac{c}{2} - R\theta$$

i.e.,

$$\theta = \frac{c/2 - x}{R}$$

Hence, using Equation (a) gives

$$\theta = 2\left(\frac{t}{c}\right) - \left(\frac{4xt}{c^2}\right) \tag{b}$$

Equation 14.78 then gives

$$C_p = \frac{20}{\sqrt{4-1}} = \frac{2}{\sqrt{3}} \times 2 \times \frac{t}{c} \times \left[1 - 2\left(\frac{x}{c}\right)\right]$$

i.e., since

$$C_p = \frac{p - p_\infty}{\gamma p_\infty M_\infty^2/2}$$

it follows that

$$p = p_\infty + \frac{\gamma p_\infty M_\infty^2}{2} \times \frac{4}{\sqrt{3}} \times \left(\frac{t}{c}\right) \times \left[1 - 2\left(\frac{x}{c}\right)\right]$$

$$= 80 + \frac{1.4 \times 80 \times 4}{2} \times \frac{4}{\sqrt{3}} \times \left(\frac{t}{c}\right) \times \left[1 - 2\left(\frac{x}{c}\right)\right]$$

$$= 80 + 517\left(\frac{t}{c}\right)\left[1 - 2\left(\frac{x}{c}\right)\right]$$

The pressure on the surface therefore varies linearly with x/c. The variation is shown in Figure E14.2c. The pressure variations along the upper and lower surfaces are of course the same.

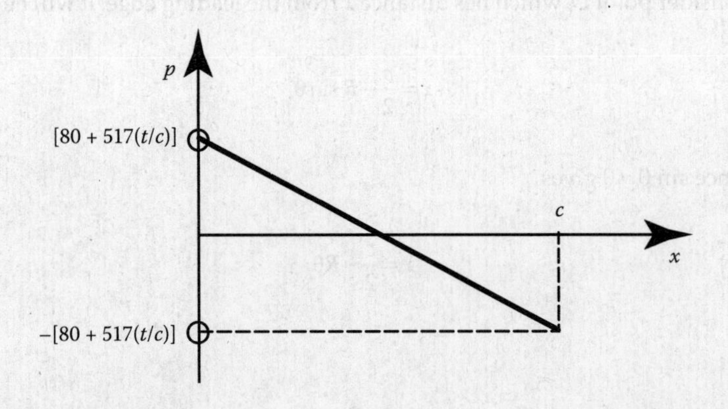

FIGURE E14.2c
Pressure variations on surfaces of airfoil.

Numerical Solutions

The calculation of two- and three-dimensional compressible flows is today usually undertaken using numerical methods, i.e., computational fluid dynamics (CFD) methods. Some general illustrations of this approach are given in Appendix B.

Concluding Remarks

The equations for two-dimensional compressible flow have been derived in this chapter. Some methods of solving these equations based on the use of approximate linearized methods that assume that the disturbances produced in the flow are small, have been discussed.

PROBLEMS

1. Air flows with Mach 2.8 over a flat plate that is set at an angle of 7° to the upstream flow. The pressure in the upstream flow is 100 kPa. Find the lift and drag coefficients using linearized theory.
2. A thin symmetrical supersonic airfoil has parabolic upper and lower surfaces with a maximum thickness occurring at midchord. Using linearized theory, compute the drag coefficient on this airfoil when it is set at an angle of attack of 0°.
3. The pressure coefficient at a certain point on a two-dimensional airfoil in a very low Mach number air flow is found to be –0.5. Using linearized theory, estimate the pressure coefficients that would exist at the same point on this airfoil in flows at Mach numbers of 0.5 and 0.8.
4. A thin airfoil can be approximated as a flat plate. The airfoil is set at an angle of 10° to an air flow with Mach 2, a temperature of –50°C, and a pressure of 50 kPa. Using linearized theory, find the pressures on the upper and lower surfaces of this wing.
5. An airfoil has a triangular cross-sectional shape. The lower surface of the airfoil is flat and the ratio of the maximum thickness to the chord is 0.1. The maximum thickness occurs at a distance of 0.3 times the chord downstream of the leading edge. If this airfoil is placed with its lower surface at an angle of attack of 2° to an airflow in which the Mach number is 3, use linearized theory to determine the distribution of the pressure coefficient over the surface of the airfoil.
6. A symmetrical double-wedge airfoil has a maximum thickness equal to 0.05 times the chord. This airfoil is placed at an angle of attack of 5° to an airstream with Mach 2, a pressure of 50 kPa and a temperature of –50°C. Find the lift and drag acting on the airfoil using linearized theory and using shock wave and expansion wave results.

Appendix A: COMPROP2 Code for Compressible Flow Properties

Introduction

COMPROP2 is an interactive Window-based software program for the calculation of compressible flow properties. This program was developed by Dr. A. J. Ghajar, regents professor of mechanical and aerospace engineering at Oklahoma State University, Dr. L. M. Tam, professor of electromechanical engineering at University of Macau, and C. W. Pau, former research associate of electromechanical engineering at University of Macau.

The program consists of six modules, isentropic flow, normal shock wave, oblique shock wave, Fanno flow (adiabatic and isothermal flow with friction), and Rayleigh flow (flow with heat transfer), and airfoil (supersonic airfoil analysis). By default, a specific heat ratio of 1.4 (the value for air at standard conditions) is assumed. However, the value of the specific heat ratio can be changed in all modules in the specific heat ratio (gamma) EditBox.

The use of each of the modules will now be briefly discussed. The COMPROP2 program and a detailed User's Manual providing a more detailed explanation of each of the six modules and 17 worked out example problems are available through the book website at www.crcpress.com.

Isentropic Flow Module

Click the icon on the left panel and the Isentropic Flow window will pop up. This module allows the user to input one of the following six parameters:

1. Mach number (M)
2. Stagnation-to-static pressure ratio (p_0/p)
3. Stagnation-to-static temperature ratio (T_0/T)
4. Stagnation-to-static density ratio (ρ_0/ρ)
5. Area ratio (A/A^*)
6. Prandtl–Meyer angle

To enter a value, click on the corresponding EditBox, enter the value, and press ENTER. The corresponding values of the other five parameters are then displayed. All results will also print on the datasheet at the right, and a log file for later analysis or printout is generated.

Instead of inputting the value in the EditBox, the user can alternatively pick the input data from a graph. The procedure in this case is as follows:

Click the button, a graph will pop up. The graph properties are as follows:

X-axis: Mach number
Y-axis: numerical value of required property
Aqua line: stagnation-to-static pressure ratio (p_0/p)
Blue line: stagnation-to-static density ratio (ρ_0/ρ)
Red line: stagnation-to-static temperature ratio (T_0/T)
Green line: area ratio (A/A^*)

Move the mouse cursor over the graph; slide the black line to the desired Mach number. As the line is sliding, corresponding properties for that Mach number are interactively displayed on right panel. When the desired Mach number or property is selected, the user can click the mouse and the result will be recorded in the datasheet and in the log file.

Normal Shock Wave Module

Click the icon on the left panel and the Normal Shock Wave window will pop up. This module allows the user to input one of the following six parameters:

1. Mach number ahead of the shock wave (M_1)
2. Mach number behind the shock wave (M_2)
3. Stagnation pressure ratio across the shock wave (p_{02}/p_{01})
4. Static pressure ratio across the shock wave (p_2/p_1)
5. Static temperature ratio across the shock wave (T_2/T_1)
6. Stagnation-to-static pressure ratio across the shock wave (p_{02}/p_1)

To enter a value, click on the corresponding EditBox, enter the value, and press ENTER. The corresponding values of the other five parameters and of the static density ratio across the shock wave (ρ_2/ρ_1) are then displayed. All results will also print on the datasheet at the right, and a log file for later analysis or printout is generated.

Instead of inputting the value in the EditBox, the user can again alternatively pick the input data from a graph. The procedure in this case is as follows:

Click the button, a graph will pop up. The graph properties are as follows:

X-axis: Mach number ahead of the shock wave
Y-axis: log scale of numerical value
Aqua line: static pressure ratio across the shock wave (p_2/p_1)
Blue line: static density ratio across the shock wave (ρ_2/ρ_1)
Red line: static temperature ratio across the shock wave (T_2/T_1)
Fuchsia line: stagnation pressure ratio across the shock wave (p_{02}/p_{01})
Green line: stagnation-to-static pressure ratio across the shock wave (p_{02}/p_1)
Lime line: Mach number behind the shock wave (M_2)

Move the mouse cursor over the graph; slide the black line to the desired Mach number. As the line is sliding, corresponding properties for that Mach number is displayed on right panel interactively. When the desired Mach number is selected, the user can do a mouse click and result will be recorded in the datasheet and in the log file.

Oblique Shock Wave Module

Click the icon on the left panel and the Oblique Shock Wave window will pop up. This module allows the user to input two of the following three combinations of parameters:

1. Mach number ahead of the shock wave (M_1) and the shock angle
2. Shock and turning angles
3. Turning angle and the Mach number ahead of the shock wave (M_1)

To enter a pair of values, click on the chosen EditBox on the left, enter the first value, and press ENTER. The cursor will then be switched to the corresponding second EditBox, and the user can enter the second value and press ENTER. Tabulated weak and/or strong shock solutions for shock angle, turning angle, M_1, M_{1n}, M_{1t}, M_{2n}, M_{2t}, M_2, p_2/p_1, T_2/T_1, and p_{02}/p_{01} are then given. In addition, for the specified conditions, the program also gives the maximum turning angle and the shock angle corresponding to the maximum turning angle. The user can switch between the strong shock solution and the weak shock solution by pressing the button if available.

Instead of inputting the value in the EditBox, the user can again alternatively pick the input data from a graph. The procedure in this case is as follows:

Click the button, a graph will pop up. The graph properties are as follows:

X-axis: turning angle
Y-axis: shock angle
Red line: constant Mach number line
Blue line: below blue line, weak shock wave occurs. Above blue line, strong shock wave occurs

Move the mouse cursor over the graph to the position where desired Mach number, shock angle, and turning angle is located. As the cursor is moving, corresponding properties are displayed on right panel interactively. When the desired point is selected, clicking the mouse will cause the result to be recorded in the datasheet and the log file.

Fanno Flow Module

Click the icon on the left panel and the Fanno Flow window will pop up. This module allows results to be obtained for ADIABATIC and/or ISOTHERMAL friction flow in a constant area duct. The user can switch between adiabatic and isothermal flow by clicking on the corresponding tab on the top of the window.

Adiabatic Flow

This option allows the user to input one of the following four parameters:

1. Mach number (M)
2. Friction factor term ($4fl^*/D$)
3. Static temperature ratio (T/T^*)
4. Static pressure ratio (p/p^*)

To enter a value, click on the corresponding EditBox, enter the value, and press ENTER. The corresponding values of the other parameters and of ρ/ρ^* and p_0/p_0^* are then displayed. All results will also print on the datasheet at the right, and a log file for later analysis or printout is generated.

Instead of inputting the value in the EditBox, the user can alternatively again pick the input data from a graph. The procedure in this case is as follows:

Click the button, a graph will pop up. The graph properties are as follows:

X-axis: Mach number

Y-axis: numerical value

Aqua line: static temperature ratio (T/T^*)

Blue line: static pressure ratio (p/p^*)

Red line: static density ratio (ρ/ρ^*)

Green line: stagnation pressure ratio (p_0/p_0^*)

Fuchsia line: friction factor term ($4fl^*/D$)

Move the mouse cursor over the graph, and slide the black line to the desired Mach number, shock angle, and turning angle is located. As the line is moving, corresponding properties are displayed on right panel interactively. When the desired Mach number is selected, clicking the mouse will cause the result to be recorded in the datasheet and the log file.

Isothermal Flow

This option allows the user to input one of the following four parameters:

1. Mach number (M)
2. Friction factor term ($4fl^*/D$)
3. Stagnation temperature ratio (T_0/T_0^*)
4. Stagnation pressure ratio (p_0/p_0^*)

To enter a value, click on the corresponding EditBox, enter the value, and press ENTER. The corresponding values of the other parameters and of ρ/ρ^* are then displayed. All results will also print on the datasheet at the right, and a log file for later analysis or printout is generated.

Instead of inputting the value in the EditBox, the user can alternatively again pick the input data from a graph. The procedure in this case is as follows:

Click the button, a graph will pop up. The graph properties are as follows:

X-axis: Mach number

Y-axis: numerical value

Aqua line: stagnation temperature ratio (T_0/T_0^*)

Blue line: static pressure ratio (p_0/p_0^*)

Red line: static density ratio (ρ/ρ^*)

Green line: friction factor term ($4fl^*/D$)

Move the mouse cursor over the graph; slide the black line to the desired Mach number. As the line is sliding, corresponding properties for that Mach number is displayed on right panel interactively. When the desired Mach number is selected, the user can do a mouse click and result will be recorded in the datasheet and the log file.

Rayleigh Flow Module

Click the icon on the left panel and the Rayleigh Flow window will pop up.

This module allows the user to input one of the following two parameters:

1. Mach number (M)
2. Stagnation temperature ratio (T_0/T_0^*)

To enter a value, click on the corresponding EditBox, enter the value, and press ENTER. The corresponding values of the parameters M, p/p^*, T/T^*, ρ/ρ^*, p_0/p_0^*, and T_0/T_0^* are then displayed. All results will also print on the datasheet at the right, and a log file for later analysis or printout is generated.

Instead of inputting the value in the EditBox, the user can alternatively again pick the input data from a graph. The procedure in this case is as follows:

Click the button, a graph will pop up. The graph properties are as follows:

X-axis: Mach number

Y-axis: numerical value

Aqua line: static temperature ratio (T/T^*)

Blue line: static pressure ratio (p/p^*)

Red line: static density ratio (ρ/ρ^*)

Green line: stagnation pressure ratio (p_0/p_0^*)

Fuchsia line: stagnation temperature ratio (T_0/T_0^*)

Move the mouse cursor over the graph; slide the black line to the desired Mach number. As the line is sliding, corresponding properties for that Mach number is displayed on right panel interactively. When the desired Mach number is selected, the user can do a mouse click and the result will be recorded in the datasheet and the log file.

Airfoil Module (Supersonic Airfoil Analysis)

Click the icon on the left panel and the Supersonic Airfoil Analysis window will pop up. This module has icons that allow the following:

- Clear all.
- Switch to drawing mode. This allows the user to draw or modify the airfoil layout.
- Switch to a mode in which straight lines connecting points can be drawn.
- Switch to a mode in which spline curves connecting a series of points can be drawn.
- Start the analysis that gives the lift force, the drag force, and the pressure distributions over the airfoil surfaces.
- Save the output bitmap to a file.

There are four steps in supersonic airfoil analysis. These steps are discussed in the next four sections.

Step 1

In this step, the values of three parameters, the airfoil chord length, the Mach number in the flow ahead of the airfoil, and the pressure in the flow ahead of the airfoil head, are inputted. The airfoil chord length must be inputted before proceeding to next step, but it is recommended that the values of all three variables be entered at this stage.

Step 2

Press the line drawing icon and move the mouse pointer over the drawing area. This allows the user to draw the shape of the airfoil. The most basic procedure starts with drawing the so-called chord line, i.e., a straight line connecting the leading edge (the front) of the airfoil to the trailing edge (the back) of the airfoil. When the mouse pointer is moving over the drawing area, the user will notice that two dimension lines follow the line being drawn. These two dimension lines show the x- and y-dimension from the airfoil nose to the point where the mouse pointer is located. Using the displayed x- and y-values, the user can use a left mouse click any time to indicate the point where the line should end. User can draw more points by doing a mouse click again and again. (Points will be automatically connected by lines or curves.) The user can then use the drawing tools to draw the upper and lower surfaces of the airfoil. The user can modify the location of any of the points at a later stage by clicking on the point and then holding on to the mouse button while moving the point to the desired new location.

Sometimes the user will notice that they cannot move the point to the exact location where it is desired to position the point. In this case the user should directly enter the x- and y-values of the required position. This can be done by moving the mouse pointer to the point whose position needs to be changed and doing a right mouse click. This will cause a dialog box that shows the x- and y-values to pop up. The user can then enter the desired values into this dialog box and enter these values by pressing OK.

Step 3

After the shape of the airfoil has been drawn, the user can undertake the analysis of the airfoil by pressing the analysis icon. The drag force and lift force will then be given in the output box and the pressure distribution graph will be displayed on a graph.

The user can then adjust the angle of attack of the airfoil by sliding the track bar. As the angle of attack is changed, the corresponding drag force and lift force values and the pressure distribution graph will be updated interactively.

Step 4

After all parameters are set to the desired values, the user can save the pressure distribution graph by clicking the save icon.

If the user then wants to modify the layout of the airfoil they can do this by clicking the LINE-DRAW icon, which will return the analysis to the beginning of Step 2.

Appendix B: Numerical Examples

The major aim of this book is to help the reader gain a sound understanding of compressible flow phenomena and of the assumptions that are conventionally used in the basic analysis of compressible fluid flows. However, the analyses discussed in this book are, in most cases, based on simplifying assumptions. To illustrate the adequacy of some of these assumptions, the results given by full numerical solutions, obtained using computer fluid dynamics (CFD) methods, of a few typical compressible flow problems will be presented in this appendix and the numerical results will be compared with the results given by the simplified analyses discussed in this book. In addition, an example meant to further illustrate the drag changes that occur in the transonic flow region will also be discussed. The numerical results given in this appendix were obtained using the commercial CFD code FLUENT©. The details of how these numerical results are obtained will not be discussed here.

Numerical Example 1: Stagnation Point Pressure in Supersonic Flow

Here, two-dimensional air flow over a blunt-nosed body will be considered. The flow situation is shown in Figure B.1.

The pressure in the undisturbed flow upstream of the body is 101 kPa. The variation of the pressure at the stagnation point with Mach number for Mach numbers between 0.6 and 1.7 is to be determined. This variation is to be determined using: (1) numerical methods, (2) the approximate method discussed in Chapter 5, and (3) the Bernoulli's equation, i.e., by ignoring compressibility effects.

Solution

In obtaining the numerical solution, the flow is assumed to be steady and two-dimensional, and the effects of viscosity, potential transition to turbulent flow, and fluid property variations are fully accounted for, the air being assumed to behave as a perfect gas. In Chapter 5, it was assumed that when the Mach number was less than one the flow between the undisturbed upstream flow and the stagnation point was isentropic and that when the flow was supersonic the portion of shock wave immediately upstream of the stagnation point (see Figure 5.6) could be assumed to be a normal shock wave and that the flow upstream of and downstream of the shock wave could be assumed to be isentropic. In using Bernoulli's equation, density changes are ignored, and the presence of a shock wave in the flow is therefore ignored.

The variations of the stagnation point pressure with Mach number given by the three approaches discussed above are shown in Figure B.2.

It will be seen that the results given by the approach discussed in Chapter 5 are in excellent agreement with the numerical results indicating that the assumptions on which this analysis is based are valid for the flow situation here being discussed. It will also be seen that, as is to be expected, the results given by the Bernoulli equation approach differ considerably from the numerical results at the higher Mach numbers considered, the difference being almost 45% at the highest Mach number considered. This again illustrates the importance of compressibility effects in high-speed flows. At the lower Mach numbers considered the results given by all three approaches are in relatively close agreement indicating that for the flow situation being considered the effects of compressibility are negligible for Mach numbers less than roughly 0.6.

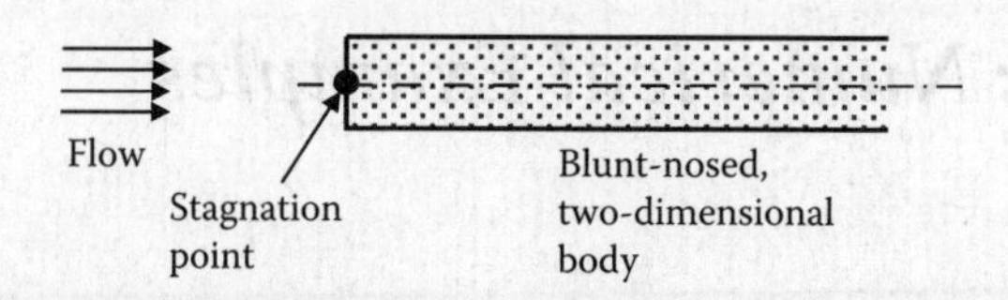

FIGURE B.1
Flow situation being considered.

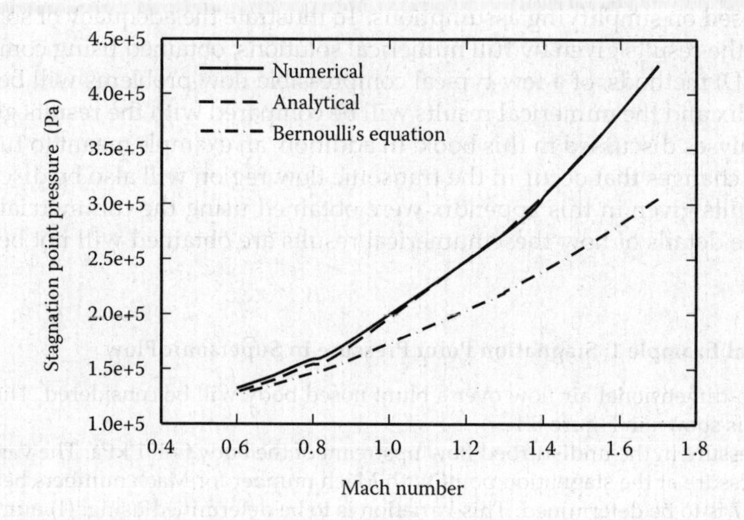

FIGURE B.2
Variations of stagnation point pressure with Mach number given by the numerical solution, the approximate analytical solution, and by ignoring compressibility effects.

Numerical Example 2: Supersonic Flow Over an Inclined Flat Plate

Here, consideration will be given to supersonic air flow over a 1-m-long flat plate set at an angle of 4° to the oncoming flow. The pressure in the undisturbed flow upstream of the plate is 101.3 kPa. The flow situation is therefore as shown in Figure B.3.

The plate can be assumed to be long in the direction normal to the flow, i.e., the flow over the plate can be assumed to be two-dimensional. For Mach numbers between 1.3 and 3.5, the pressures on the upper and lower surfaces of the plate and the lift on the plate are to be determined numerically and using the approach outlined in Chapter 7, and the results given by the two approaches are to be compared.

Solution

In obtaining the numerical solution, the flow has again been assumed to be steady and two-dimensional, and the effects of viscosity, potential transition to turbulent flow, and

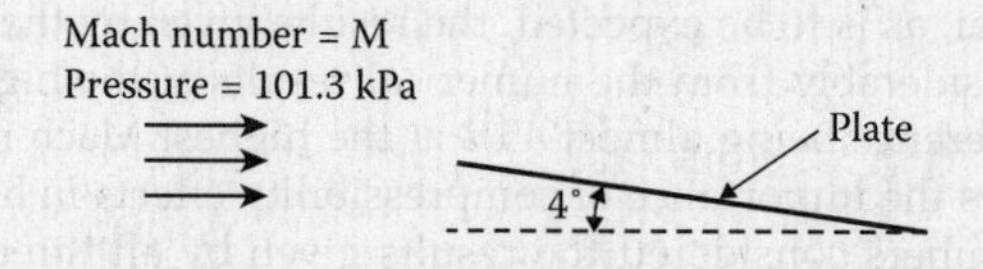

FIGURE B.3
Flow situation considered.

fluid property variations are fully accounted for, the air being assumed to behave as a perfect gas. In Chapter 7, in dealing with this type of problem, it has been assumed that an expansion wave forms at the upstream (leading) edge of the plate on the upper surface and that an oblique shock wave forms at the upstream edge of the plate on the lower surface, i.e., the flow pattern is assumed to be as shown in Figure B.4.

The waves that form at the downstream (trailing) edge of the plate are assumed to have no effect on the pressures on the upper and lower surfaces or therefore on the lift. The pressure on the surface of the plate, and therefore the lift is obtained using the expansion wave and oblique shock relations discussed in Chapters 6 and 7. This approach cannot be used at Mach numbers very close to 1 when the oblique shock wave is detached from the plate.

The variations of the pressures on the upper and lower surfaces of the plate as given by the analytical and numerical approaches are compared in Figure B.5, whereas the variations of the lift with Mach number given by the two approaches are compared in Figure B.6. It will again be seen that the results given by the approach discussed in Chapter 7 are in good agreement with the numerical results, the largest difference between the results being approximately 3%.

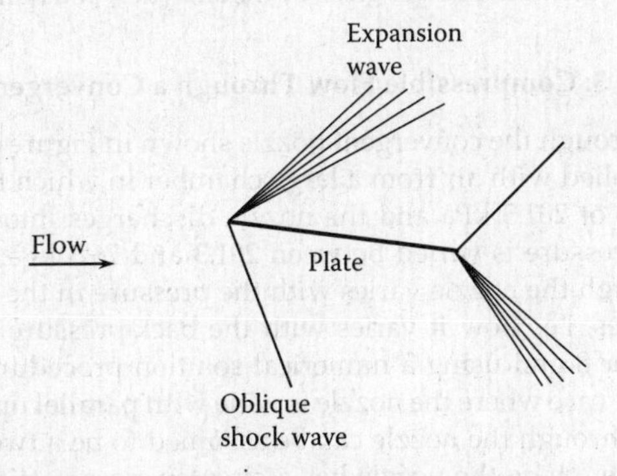

FIGURE B.4
Flow pattern assumed in obtaining the approximate analytical solution.

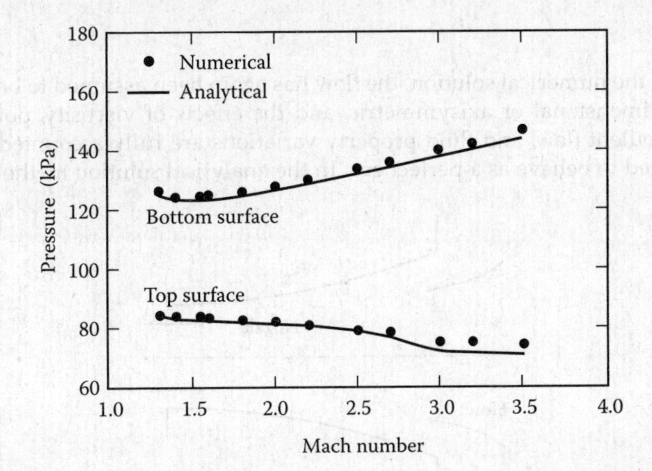

FIGURE B.5
Variations of pressures on the upper and lower surfaces of the plate with Mach number given by the analytical and numerical approaches.

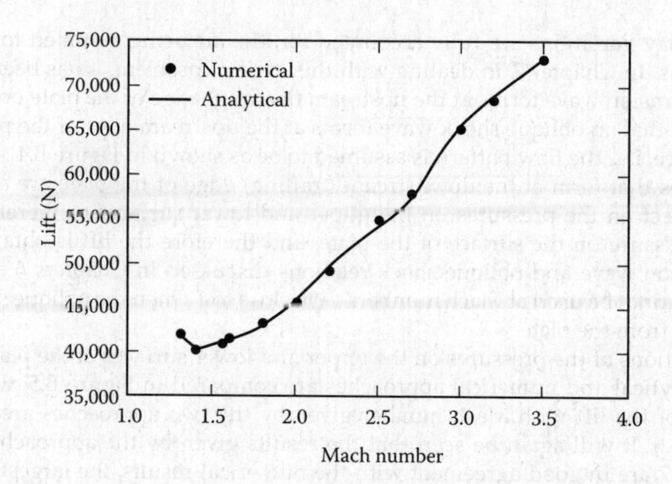

FIGURE B.6
Variations of the lift on plate with Mach number given by the analytical and numerical approaches.

Numerical Example 3: Compressible Flow Through a Convergent Nozzle

Consider air flow through the convergent nozzle shown in Figure B.7.

The nozzle is supplied with air from a large chamber in which the pressure is maintained at a pressure of 201.3 kPa and the nozzle discharges into another large reservoir in which the pressure is varied between 201.3 and 76.3 kPa. Determine how the mass flow rate through the nozzle varies with the pressure in the reservoir into which the nozzle discharges, i.e., how it varies with the back-pressure, using the approach discussed in Chapter 8 and using a numerical solution procedure. Results should be obtained both for the case where the nozzle is wide with parallel upper and lower walls, i.e., where the flow through the nozzle can be assumed to be a two-dimensional plane flow, and for the case where the nozzle has a circular cross-sectional shape at all sections (see Figure B.8). The mass flow rate variation should also be calculated assuming incompressible, one-dimensional flow, i.e., using Bernoulli's equation.

Solution

In obtaining the numerical solution, the flow has again been assumed to be steady and either two-dimensional or axisymmetric, and the effects of viscosity, potential transition to turbulent flow, and fluid property variations are fully accounted for, the air being assumed to behave as a perfect gas. In the analytical solution method discussed

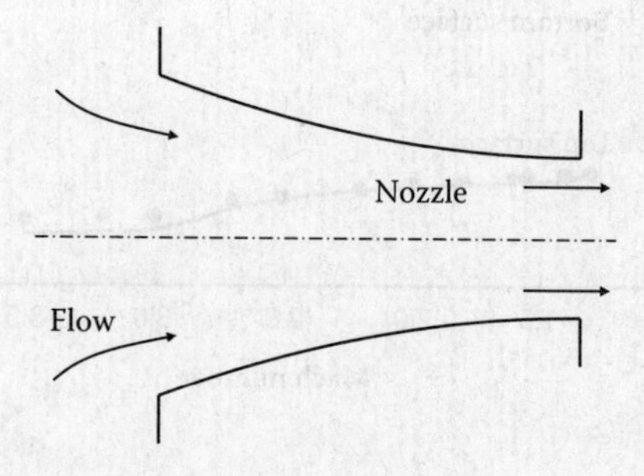

FIGURE B.7
Convergent nozzle shape considered.

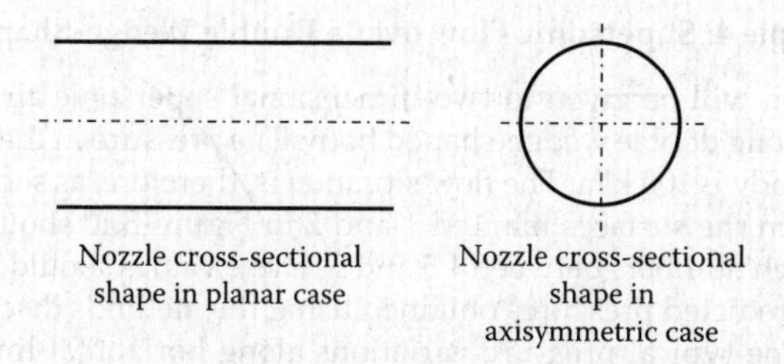

FIGURE B.8
Two-dimensional plane and circular nozzle cross-sectional shapes considered.

in Chapter 8 it is assumed that the flow at all sections is one-dimensional, i.e., that all flow variables have the same values over a given cross section of the nozzle.

The variations of the mass flow rate through the nozzle with the pressure in the reservoir into which the nozzle discharges, i.e., with the back-pressure, for the axisymmetric and two-dimensional cases as given by the numerical solution, by the approach described in Chapter 8, and by ignoring compressibility effects are shown in Figure B.9.

It will be seen that the results given by the one-dimensional approach discussed in Chapter 8 are in good agreement with the numerical results and that both methods show that choking of the flow occurs at essentially the same back-pressure. It will also be seen that, as is to be expected, the results given by ignoring compressibility effects differ considerably from the numerical and analytical results at all but the higher values of the back-pressure considered.

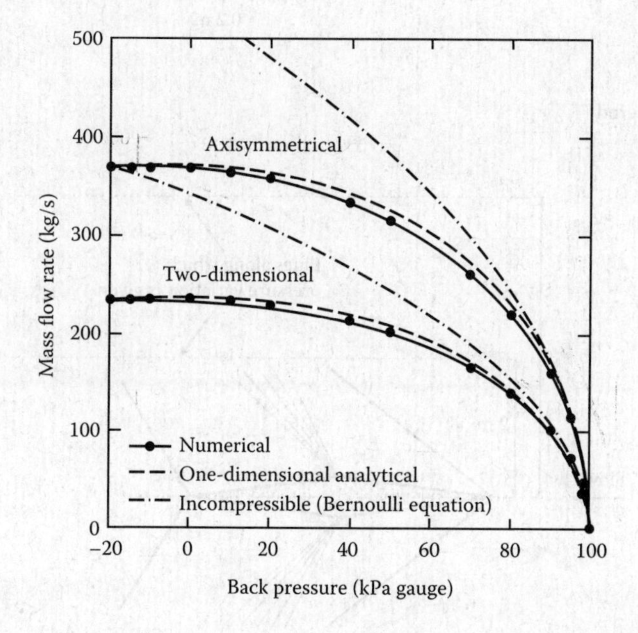

FIGURE B.9
Variations of the mass flow rate through the nozzle with back-pressure for the axisymmetric (circular) and two-dimensional nozzle cases as given by the numerical solution by the approximate analytical solution, and by ignoring compressibility effects.

Numerical Example 4: Supersonic Flow over a Double Wedge-Shaped Body

Here consideration will be given to two-dimensional supersonic air flow over doubly symmetrical 1 m long double wedge-shaped body. The pressure in the undisturbed flow upstream of the body is 100 kPa. The flow situation is, therefore, as shown in Figure B.10.

The pressures on the surfaces marked 1 and 2 in Figure B.10 should be numerically calculated for Mach numbers between 1.5 and 3. These values should be compared with the analytically predicted pressures obtained using the methods discussed in Chapters 6 and 7. Also, some typical pressure variations along horizontal lines at distances of 2 and 7 m from the body, these lines being shown in Figure B.10, should be numerically determined.

Solution

In obtaining the numerical solution, the flow has again been assumed to be steady and two-dimensional, and the effects of viscosity, potential transition to turbulent flow, and fluid property variations are fully accounted for, the air being assumed to behave as a perfect gas. In Chapter 7, in dealing with this type of problem, it has been assumed that oblique shock waves form at the leading edge of the body and that expansion waves form at the turning points that occur at 0.5 m downstream of the leading edge, i.e., the flow pattern is assumed to be as shown in Figure B.11. The waves that form at the trailing edge of the plate are again assumed to have no effect on the pressures on the surfaces of

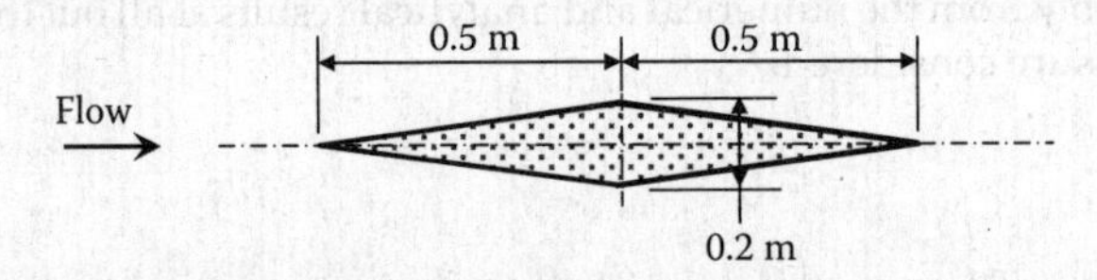

FIGURE B.10
Flow situation considered.

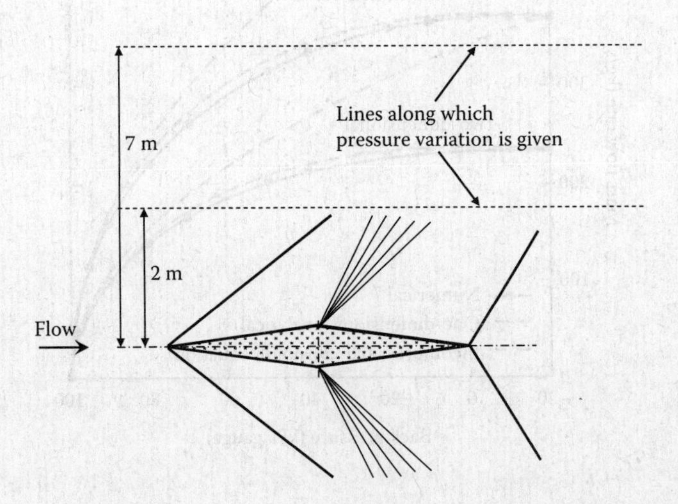

FIGURE B.11
Flow pattern assumed in obtaining the approximate analytical solution and lines used in comparing pressure distributions.

the body. The analytically determined pressures on the surfaces are obtained using the expansion wave and oblique shock relations discussed in Chapters 6 and 7.

The variations of the pressures on the surfaces of the body being considered as given by the analytical and numerical approaches are compared in Figure B.12 and good agreement will be seen to be obtained.

Typical numerically determined variations of pressure along horizontal lines at distances of 2 and 7 m from the body are shown in Figures B.13, B.14, and B.15.

The pressures on these lines will be seen to be much lower than those on the surfaces of the body as a result of the interaction of the oblique shock and the expansion waves. This decrease in the pressure change with distance from a body was used as the basis for assuming in Chapter 3 that the net effect at ground level of an aircraft passing

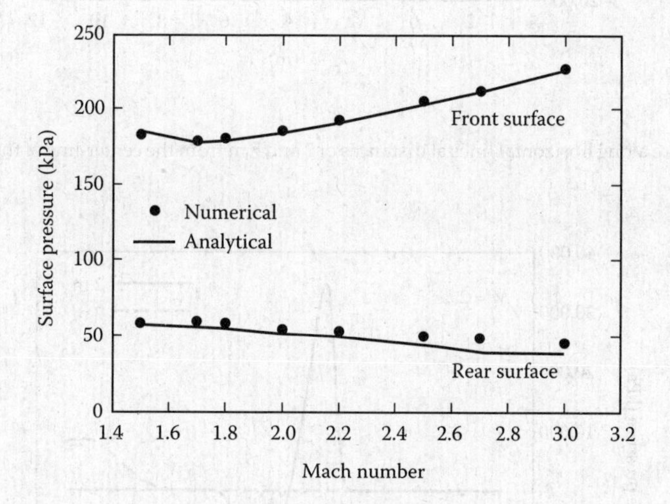

FIGURE B.12
Variations of pressures on the surfaces of the body given by the numerical solution and by the approximate analytical solution.

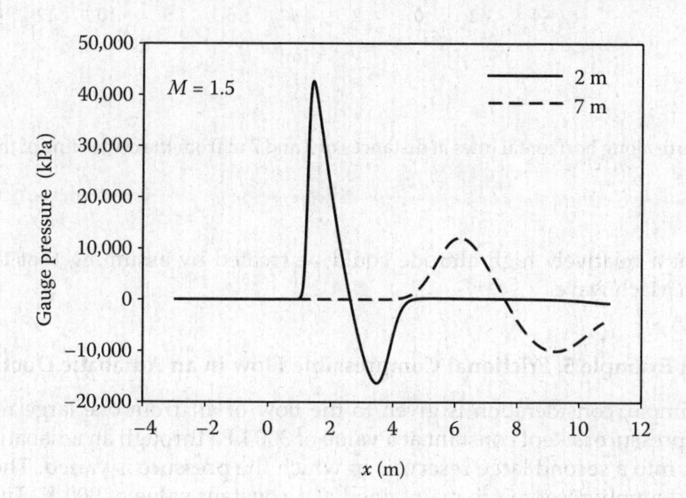

FIGURE B.13
Variations of pressure along horizontal lines at distances of 2 and 7 m from the center-line of the body for Mach 1.5.

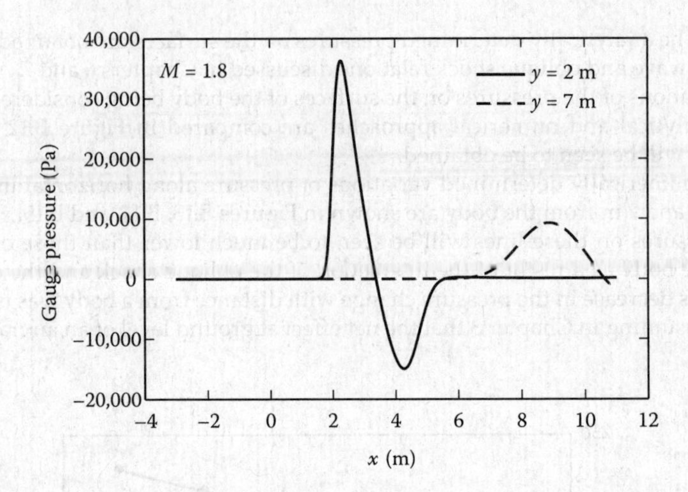

FIGURE B.14
Variations of pressure along horizontal lines at distances of 2 and 7 m from the center-line of the body for Mach 1.8.

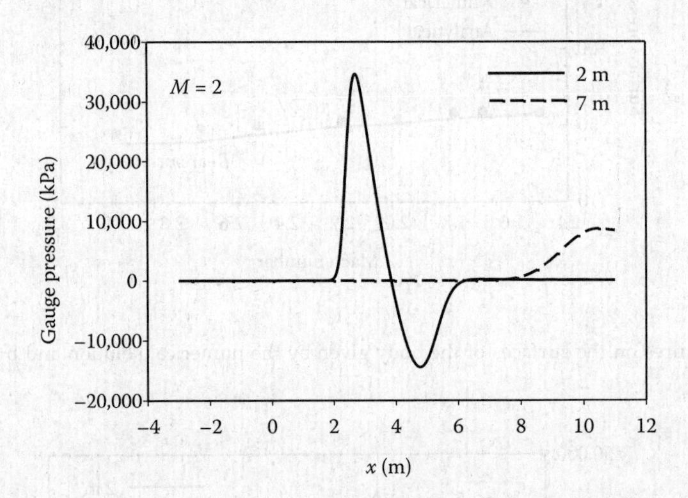

FIGURE B.15
Variations of pressure along horizontal lines at distances of 2 and 7 m from the center-line of the body for Mach 2.

overhead at a relatively high altitude could be treated by assuming that the aircraft generates a Mach wave.

Numerical Example 5: Frictional Compressible Flow in an Adiabatic Duct

In this example, consideration is given to the flow of air from one large reservoir in which the pressure is kept constant at a value of 300 kPa through an adiabatic pipe that discharges into a second large reservoir in which the pressure is varied. The temperature in this supply reservoir is maintained at a constant value of 300 K. The pipe has a diameter of 0.1 m and a length of 10 m. The flow situation is therefore as shown in Figure B.16.

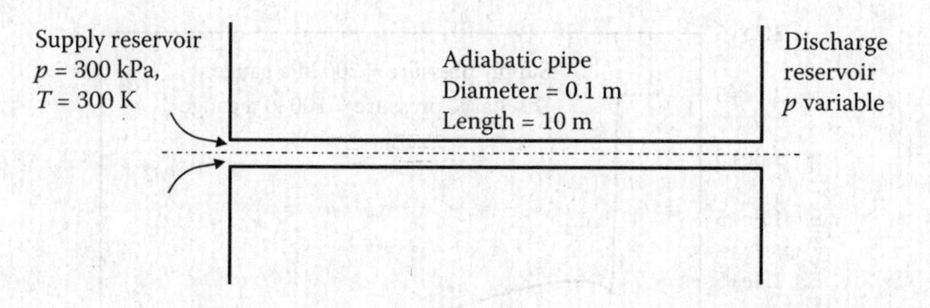

FIGURE B.16
Flow situation considered.

As shown in Figure B.16, there is a sharp inlet to the pipe, i.e., the inlet to the pipe is not rounded. The variation of the mass flow rate through the pipe with decreasing values of the "back-pressure", i.e., with decreasing values of the pressure in the reservoir into which the pipe flow discharges, should be numerically determined. The variation of the mass flow rate through the pipe with decreasing values of the "back-pressure" should then be determined using the approximate analytical approach discussed in Chapter 9. The results given by this approach should be compared with the results obtained numerically and the reasons for any observed differences between the numerical and approximate analytical results should be discussed.

Solution

In obtaining the numerical solution, the flow has been assumed to be steady and axisymmetric and the effects of viscosity, potential transition to turbulent flow, and fluid property variations are fully accounted for, the air being assumed to behave as a perfect gas.

The numerically determined variation of the mass flow rate through the pipe with decreasing back-pressure is shown in Figure B.17.

As expected, the mass flow rate initially increases as the back-pressure is decreased but then reaches a constant value that does not change with further decrease in the back-pressure, the flow then being choked; the Mach number at the exit of the pipe having reached a value of 1. Two typical numerically determined variations of the pressure along the center-line of the pipe are shown in Figures B.18 and B.19. The first of these

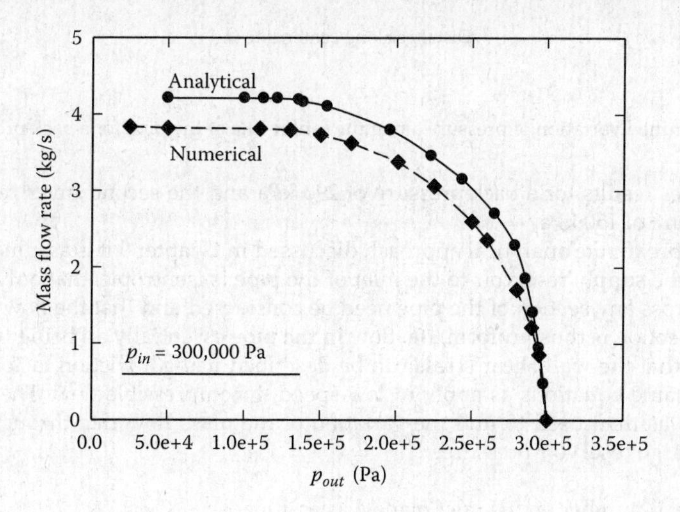

FIGURE B.17
Variation of mass flow rate through the pipe with back pressure.

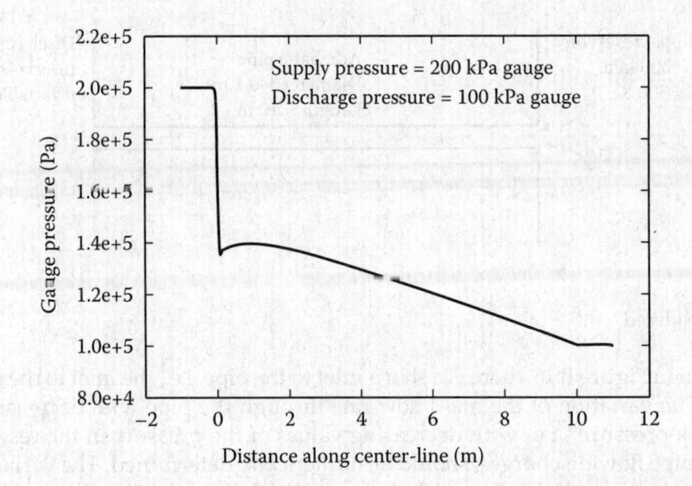

FIGURE B.18
Numerically determined variation of pressure along the center-line of the pipe for a back-pressure of 201 kPa.

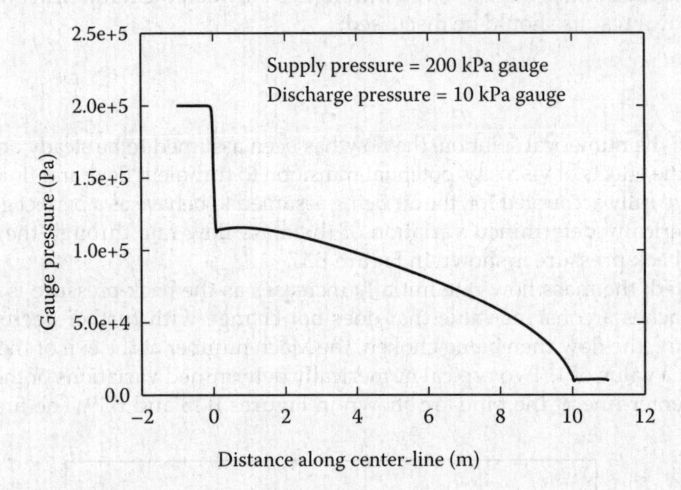

FIGURE B.19
Numerically determined variation of pressure along the center-line of the pipe for a back-pressure of 111 kPa.

figures gives results for a back-pressure of 290 kPa and the second gives results for a back-pressure of 150 kPa.

In the approximate analytical approach discussed in Chapter 9 it is assumed that the flow from the supply reservoir to the inlet of the pipe is isentropic, that only the mean velocity across any section of the pipe need be considered and that the flow across the pipe inlet section is thus uniform, the flow in the pipe essentially all being fully developed, and that the wall shear stress can be described using a friction factor obtained using the same equations as apply in low-speed, incompressible flow. The following approach was then used to find the variation of the mass flow rate through the pipe with discharge reservoir pressure:

1. A Mach number on the inlet plane was assumed.
2. Using the isentropic relations, the pressure, temperature, and velocity on the inlet plane were determined.

3. Using these values of the inlet plane pressure, temperature, and velocity, the Reynolds number and hence the friction factor were determined, a smooth pipe being assumed
4. Using the compressible frictional flow relations or the software COMPROP or tables for frictional flow in a pipe the values of p/p^*, T/T^*, and $4fl^*/D$ corresponding to the assumed inlet Mach number were determined.
5. Using

$$l_o^* = l_i^* - l_{io}, \quad \text{i.e.,} \quad \frac{4fl_o^*}{D} = \frac{4fl_i^*}{D} - \frac{4fl_{io}}{D}$$

where the subscripts i and o refer to conditions on the inlet and outlet of the pipe and l_{io} is the length of the pipe, the value of $4fl_o^*/D$ is calculated, and the values of p/p^* and Mach number on the outlet plane of the pipe are found using compressible frictional flow relations or software or tables for flow in a pipe.
6. If the exit plane Mach number cannot be found using this approach, the process is repeated using a lower assumed value for the inlet Mach number.
7. Using the inlet conditions, the mass flow rate through the pipe is calculated. The exit plane pressure is assumed to be equal to the back-pressure.

The process can only be used for inlet Mach numbers up to that at which choking occurs at the exit of the pipe. For values of the back-pressure lower than that at which choking occurs, the flow in the pipe and therefore the mass flow rate through the pipe will be the same as that at which choking first occurs. Using this procedure the analytically determined variation of the mass flow rate through the pipe with decreasing back-pressure can be found and is also shown in Figure B.17.

It will be seen from Figure B.17 that the value of the back-pressure at which choking occurs, i.e., at which the mass flow rate becomes independent of the back-pressure, given by the numerical results and by the analytical results are in close agreement. However, the mass flow rates predicted by the analytical approach are higher than those obtained numerically. Among the reasons for this are:

1. In the analytical approach, the flow between the supply reservoir and the inlet to the pipe was assumed to be isentropic and the velocity across the pipe inlet section was assumed to be uniform.
2. A constant friction factor was assumed in the analytical approach.
3. A constant velocity was assumed across any section of the pipe in the analytical approach.

The first of these reasons is likely the biggest cause of the difference between the analytical and numerical results. If the pressure distributions shown in Figures 5.3 and 5.4 are considered it will be noted that the pressure passes through a minimum at the inlet of the pipe. This is, of course, associated with the existence of a separated flow region at the inlet of the pipe, which is not accounted for in the analytical approach.

Numerical Example 6: Compressible Flow over an Adiabatic Flat Plate

Here, two-dimensional air flow over a 1-m-long adiabatic flat plate aligned with a steady uniform air flow will be considered, the flow situation being as shown in Figure B.20.

The temperature in the undisturbed flow upstream of the body is 230 K. The variation of the surface temperature of the plate with Mach number for Mach numbers between 0.5 and 1.8 is to be determined. This variation is to be determined using (1) the numerical approach and (2) the approximate analytical method, which was discussed

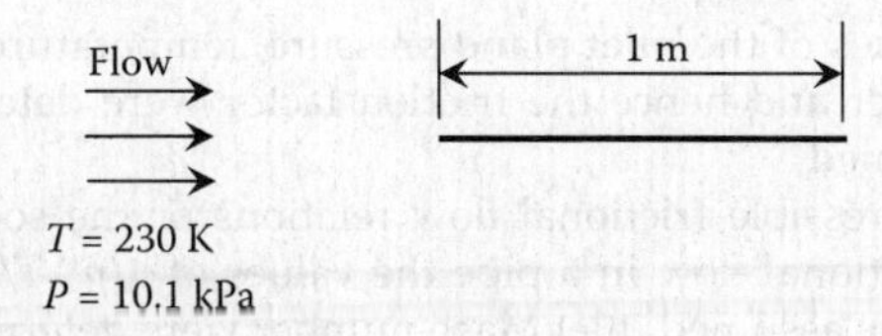

FIGURE B.20
Flow situation considered.

in Chapter 10 and which is based on the use of the assumption that the plate temperature is the same everywhere along the plate (unless there are significant regions of both laminar and turbulent flow) and that the recovery factor, r, is $Pr^{1/2}$ in laminar boundary layer flow and $Pr^{1/3}$ in turbulent boundary layer flow. Pr is the Prandtl number.

Solution

In obtaining the numerical solution, the flow has again been assumed to be steady and two-dimensional and the effects of viscosity, potential transition to turbulent flow, and fluid property variations are fully accounted for, the air being assumed to behave as a perfect gas.

A typical variation of the local plate surface temperature with distance along the plate given by the numerical solution is shown in Figure B.21. Also shown in this figure are the variations of the plate surface temperature for laminar and turbulent boundary flows obtained using the method outlined in Chapter 10.

It will be seen that the numerical method does predict that the surface temperature is essentially independent of the distance along the plate as was assumed in the approximate analytical approach that was discussed in Chapter 10 and that the numerical adiabatic surface temperature results agree with those given by the approximate analytical solution for turbulent flow to better than 1%.

The variation of the mean surface temperature with Mach number as given by the numerical solution and using the method outlined in Chapter 10 for laminar and turbulent boundary flows are shown in Figure B.22. It will again be seen that there is very

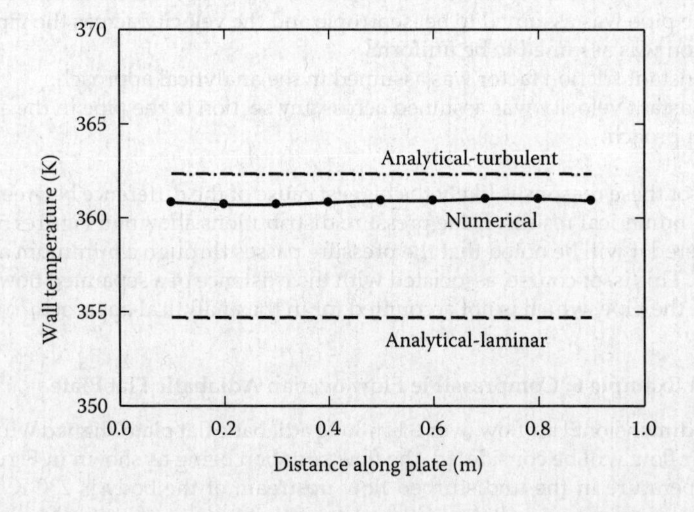

FIGURE B.21
Variation of plate surface temperature with distance along the plate given by the numerical solution and by the approximate analytical solution for a flow with Mach 1.8.

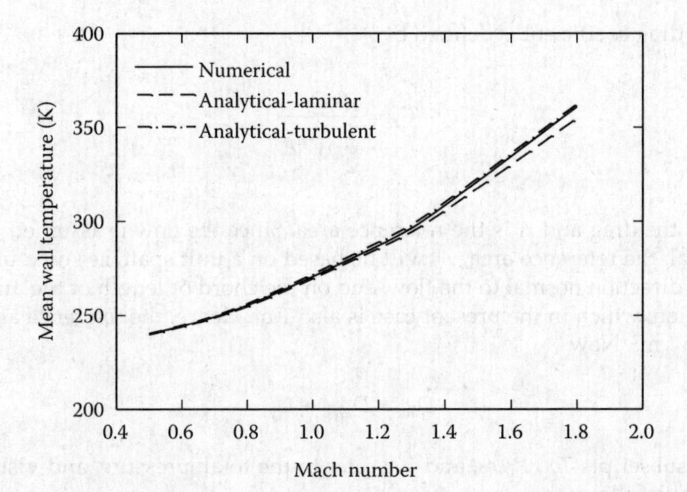

FIGURE B.22
Variation of mean plate surface temperature with Mach number given by the numerical solution and by the approximate analytical solution for a laminar and turbulent boundary layer flow.

close agreement between the numerical results and those given by the approximate analytical solution for turbulent flow.

Numerical Example 7: Drag on a Symmetrical Airfoil in Transonic Flow

The purpose of this example is to demonstrate the drag rise that occurs on a body in transonic flow. Consider two-dimensional flow over the 10% thick symmetrical airfoil whose shape is shown in Figure B.23.

The airfoil has a length in the flow direction, i.e., has a chord length, of 1 m. The pressure and temperature in the undisturbed flow upstream of the airfoil are 101.3 kPa and 15°C, respectively. Numerically calculate the drag on the airfoil for flow Mach numbers of from 0.2 to 1.5. Use these results to determine the variation of the drag coefficient with Mach number.

Solution

In obtaining the numerical solution, the flow has again been assumed to be steady and two-dimensional and the effects of viscosity, potential transition to turbulent flow, and fluid property variations are fully accounted for, the air being assumed to behave as a perfect gas.

Because viscous effects are accounted for in the numerical solution, the drag due to the viscous stresses on the surface of the airfoil, i.e., the viscous drag, and the drag due to the pressure variation over the surface of the airfoil, i.e., the pressure drag, as well as the total drag, which is the sum of the viscous and pressure drags can be obtained.

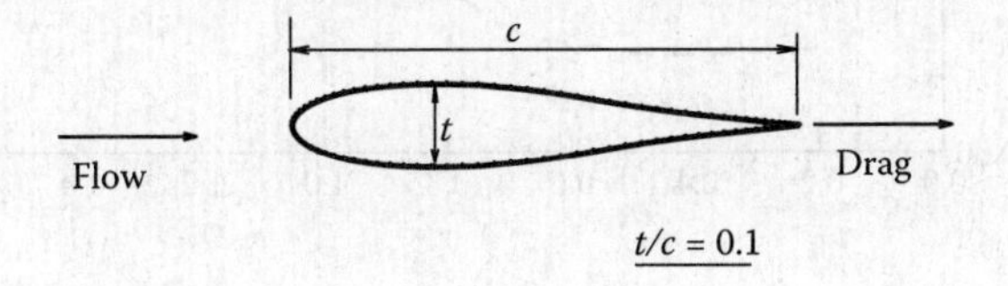

FIGURE B.23
Shape of symmetrical airfoil considered.

Now the drag coefficient is defined by

$$C_D = \frac{D}{\frac{1}{2}\rho V^2 A}$$

where D is the drag and A is the reference area. Since the flow is assumed to be two-dimensional, the reference area, A, will be based on a unit span, i.e., on a unit length, 1 m, in the direction normal to the flow, and on the chord or length of the airfoil in the flow direction, which in the present case is also 1 m. Hence, the reference area, A, will be taken as 1 m^2. Now

$$D_{Tot} = D_{Press} + D_{Vis}$$

where the subscripts *Tot*, *Press*, and *Vis* refer to the total, pressure, and viscous drags respectively, it follows that

$$C_{D_{Tot}} = C_{D_{Press}} + C_{D_{Vis}}$$

The variations of the calculated total drag and the pressure drag coefficients with Mach number are shown in Figure B.24.

It will be seen from these results that at low Mach numbers the pressure drag coefficient is less than half the total drag coefficient, i.e., viscous drag is greater than pressure drag. However, for Mach numbers greater than roughly 0.6, the pressure drag coefficient is more than half the total drag coefficient and the pressure drag becomes increasingly greater than the viscous drag as the Mach number increases. The rapid rise in the pressure and total drag coefficients for Mach numbers greater than 0.7 and the peak in the drag coefficient variations near Mach 1, the "sound barrier", will be noted. Beyond Mach 1 the drag coefficient will be seen to decrease by roughly 20%. This variation of the drag coefficient with Mach number in the transonic flow region was discussed in Chapter 8.

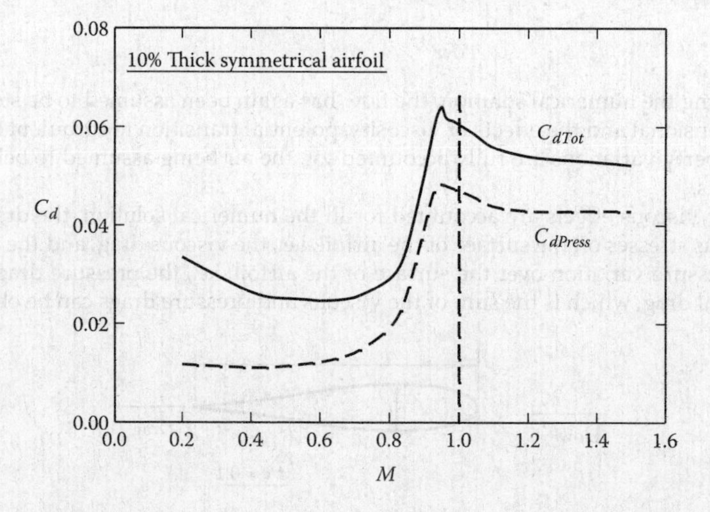

FIGURE B.24
Variations of the numerically predicted total drag and pressure drag coefficients with Mach number.

Appendix C: Mini-Project–Type Assignments

Engineering students seem to gain a much more through understanding of course material when examples of the real-world application of the material are frequently discussed with them. Although this is probably best done by discussing such applications in the classes using both a lecture format and an open classroom discussion format another way of making students aware of the applications of the course material that has proved to be quite successful is to assign them "mini-projects." These reports are usually required to be roughly between two and six pages in length and they may include figures (e.g., downloaded photographs and diagrams) and should reference the books, web pages, etc. from which information was obtained. The students are told that most of the information required to write the mini-reports is available on the web, but that the source must be clearly referenced. The following is a brief list of examples of the type of mini-project topics that have been successfully used in a course on compressible fluid flow.

1. Discuss the X-1 aircraft, the first aircraft to fly supersonically in level flight. Describe how this aircraft was propelled and how it was launched.
2. Give an overview of the life of Ernst Mach and his contributions to our understanding of compressible flow.
3. A system has been proposed by a British research group in which air carrying small amounts of a liquid that is to be injected into a person is accelerated to a supersonic speed that then impinges on the skin and the liquid passes through the skin with little or no sensation of pain occurring. Describe this system.
4. Discuss how a pulse-jet engine works.
5. Discuss how thunder is generated.
6. Discuss how a pipe organ works.
7. Discuss the engine air intake system (the inlet diffuser) that was used in the Concorde supersonic airliner.
8. Discuss the engine air intake system (the inlet diffuser) that was used in the Lockheed SR-71 Blackbird aircraft.
9. Discuss the engine air intake system (the inlet diffuser) used in the F-16 aircraft.
10. Discuss the engine air intake system (the inlet diffuser) used in the Eurofighter Typhoon aircraft.
11. Discuss the engine air intake system (the inlet diffuser) used in the F-35 aircraft.
12. Describe the nozzle that was fitted to the main engine on the Space Shuttle.
13. Vectored thrust jet engine nozzles (not the type used on the Harrier jet) have been developed. Give an example and explain how it works.
14. Discuss what is meant by an afterburner and discuss the type of nozzle usually fitted at the exit from an afterburner.
15. Describe the reentry thermal protection system that was fitted to the Apollo spacecraft.

16. Describe the reentry thermal protection system that was fitted to the Space Shuttle and discuss the causes of the fatal accident involving a Space Shuttle that occurred over Texas during its return to earth following a mission.
17. Discuss what is meant by a de Laval nozzle and discuss who it is named after.
18. Discuss the purpose of and design of blast doors.
19. A person recently reached a supersonic speed in free-fall from a balloon prior to the deployment of a parachute. Discuss the procedure used and any difficulties encountered. Also, discuss the maximum speed achieved.
20. During the construction of the terminal at the airport in Ottawa, Canada, a military aircraft flying at supersonic speed at low altitude over the airport did considerable damage to the terminal. Discuss this incident and the cause and extent of the damage.

Appendix D: Optical Methods in Compressible Flows

Introduction

A number of photographs showing various features of compressible flows, such as shock waves, are given in the main body of this book. An example of such a photograph is shown in Figure D.1. A very brief discussion of the methods used to obtain such photographs will be presented in this appendix.

There are basically three such methods:

1. Shadowgraph
2. Schlieren
3. Interferometer

All of these methods utilize the fact that the speed of light through a gas varies with the density of the gas, i.e., the fact that the refractive index, n, which is the ratio of the speed of light in a vacuum to the speed of light in the gas, is a function of density, i.e.,

$$n = \text{function}\ (\rho) \tag{D.1}$$

where

$$n = \frac{c_0}{c} \tag{D.2}$$

c_0 being the speed of light in a vacuum and c being the speed of light at some point in the gas.

The relation between n and ρ is approximately linear and is usually written as

$$n = 1 + \beta \frac{\rho}{\rho_s} \tag{D.3}$$

where ρ_s is the density of the gas at 0°C and standard atmospheric pressure and β is a constant that depends on the type of gas. Typical values of β are given in Table D.1. These values strictly only apply at a particular wavelength of light.

Equation D.3 is sometimes written in terms of the Gladstone–Dale constant, K, such that

$$n = 1 + K\rho \tag{D.4}$$

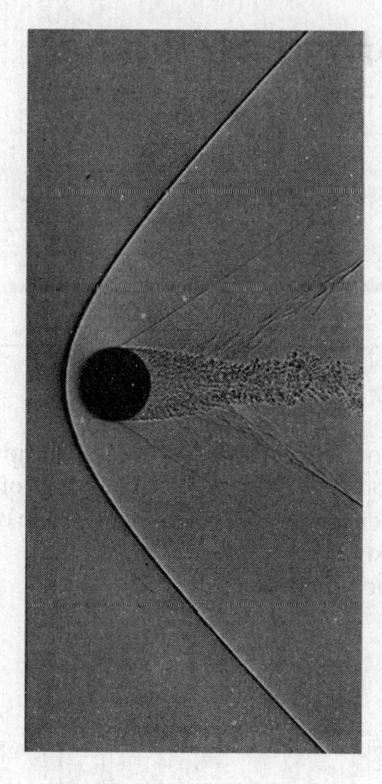

FIGURE D.1
Typical Schlieren photograph of supersonic flow over a body. (Courtesy of NASA.)

TABLE D.1
β Values for Various Gases

Gas	β
Air	0.000292
Nitrogen	0.000297
Oxygen	0.000271
Water vapor	0.000254
Carbon dioxide	0.000451

so

$$K = \beta/\rho_s \tag{D.5}$$

Because the speed of light depends on the density of the gas through which it is passing, it follows that if the density changes in the gas, the speed of light will be different in different parts of the gas. However, there is another related effect produced by the change in refractive index. If a beam of light passes through a gas in which there is a density gradient normal to the direction of the beam, the light will be turned in the direction of increasing density. This is shown schematically in Figure D.2.

The angle through which the light ray is turned is dependent on the gradient of density normal to the direction of the light, i.e., for the situation shown in Figure D.2, on $d\rho/dy$ which by virtue of Equation D.3 will be proportional to the gradient of the refractive index, i.e., on dn/dy.

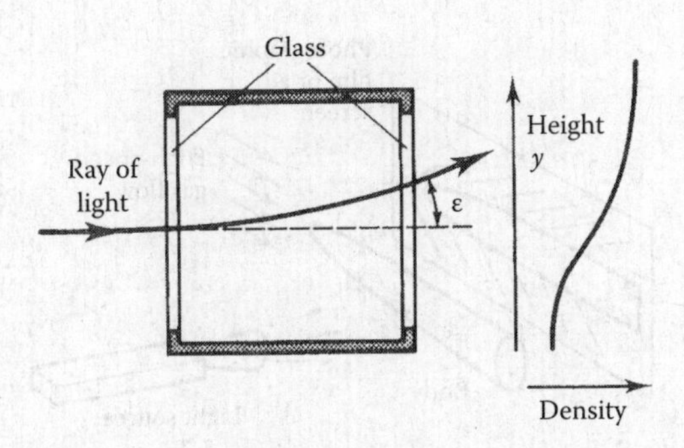

FIGURE D.2
Bending of light beam in the presence of a density gradient.

Shadowgraph System

Consider a series of light rays passing through a gas in which there is a vertical gradient of density as indicated in Figure D.3.

Because of the deflection of the light rays resulting from the density variations, if a screen is placed in such a way that it intercepts the light rays that have passed through the gas, the rays will be crowded together in some places and spread apart in other places as indicated in Figure D.3. When the rays are crowded together, the screen appears lighter than average, whereas when the rays are spread apart, the screen appears darker than average. Hence, because of the density gradients in the gas, regions of light and dark will appear on the screen as indicated in Figure D.4.

If the deflection of the light rays shown in Figure D.3 is considered, it will be seen that if there is a uniform vertical gradient of the density, all of the rays of light will be deflected

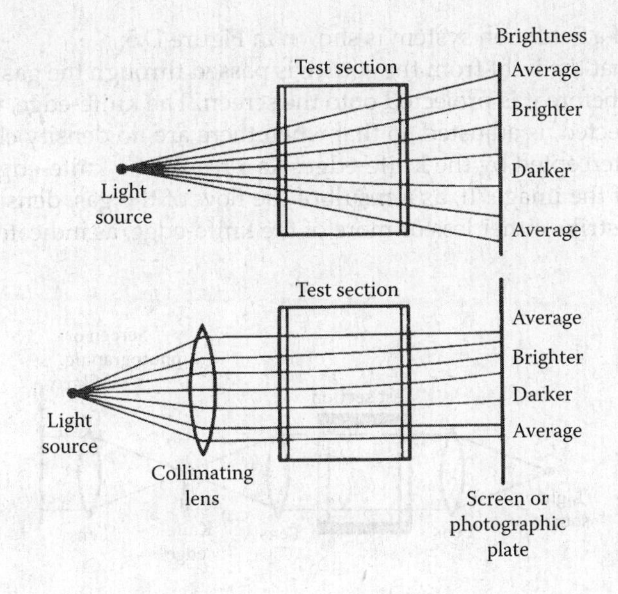

FIGURE D.3
Shadowgraph system.

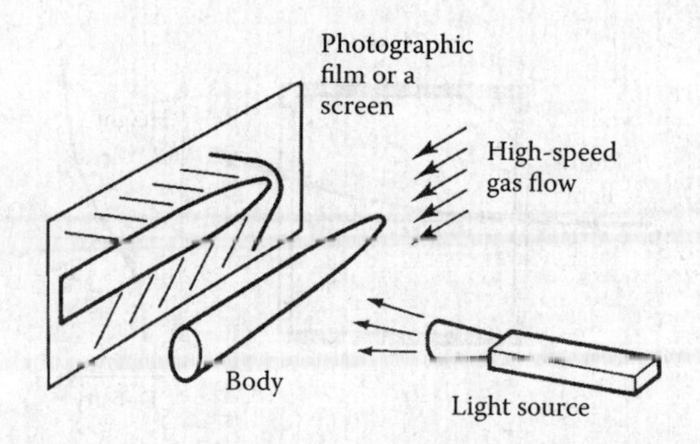

FIGURE D.4
Formation of a shadowgraph.

by the same amount and there will still be a uniform illumination of the screen. The shadowgraph is, therefore, sensitive to the second derivative of density, i.e., to $d^2\rho/dy^2$ and hence to the second derivative of n, i.e., to d^2n/dy^2. Because of this, it is the least sensitive of the three systems considered here. It does, however, give a good indication of regions of high rates of density change, e.g., it gives a clear indication of the presence of shock waves. Of course, density changes in any direction normal to the rays will produce an image. The vertical y-direction was used only for illustrative purposes.

The shimmer about a hot roof in summer and the visible "smokeless" flow of "heat" out of a chimney in winter are basically the result of the shadowgraph effect, the sun in these cases being the light source.

Schlieren System

The basic layout of a Schlieren system is shown in Figure D.5.

It will be seen that the light from the source is passed through the gas and then focused onto a knife-edge before it is projected onto the screen. The knife-edge, whose orientation can usually be selected, is adjusted so that when there are no density changes in the gas, half the light is intercepted by the knife-edge. As a result, the knife-edge produces a uniform darkening of the image. If, as a result of the flow of the gas, density changes occur, the light rays will strike either less or more of the knife-edge, as indicated in Figure D.6.

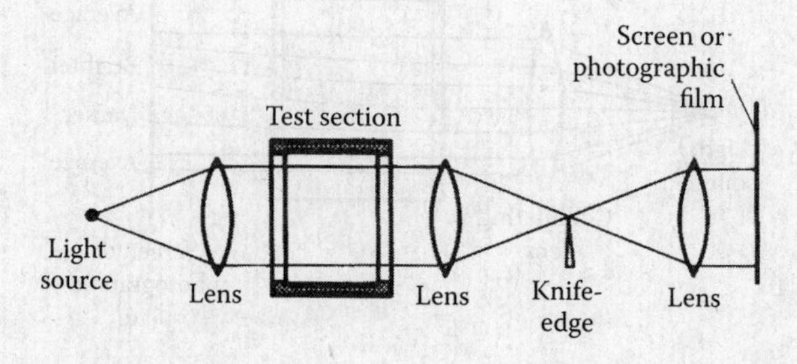

FIGURE D.5
Basic Schlieren system arrangement.

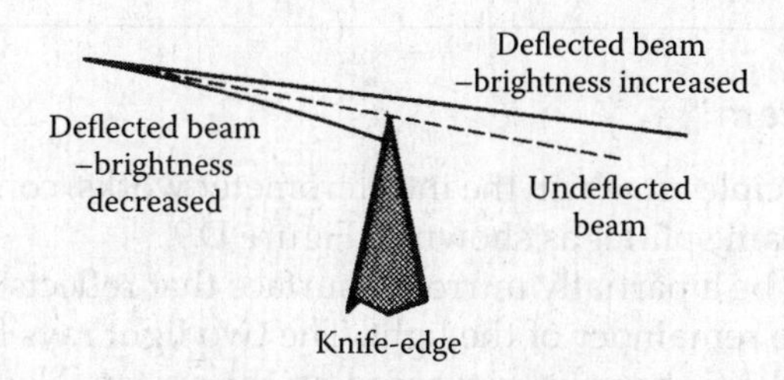

FIGURE D.6
Light rays near knife-edge.

There will, therefore, be either a lightening or a darkening of the image depending on the angle through which the light is turned. The intensity of the image, therefore, will depend on the angle of turning, i.e., on the gradient of density, $d\rho/dy$, i.e., on dn/dy if the knife-edge is horizontal or on $d\rho/dx$, i.e., on dn/dx, if the knife-edge is vertical.

As mentioned above, the direction of the knife-edge can usually be adjusted so that the density gradients in different directions can be examined. This is illustrated in Figure D.7.

In actual Schlieren systems, mirrors rather than lenses are usually used for practical reasons, the system then being as shown in Figure D.8.

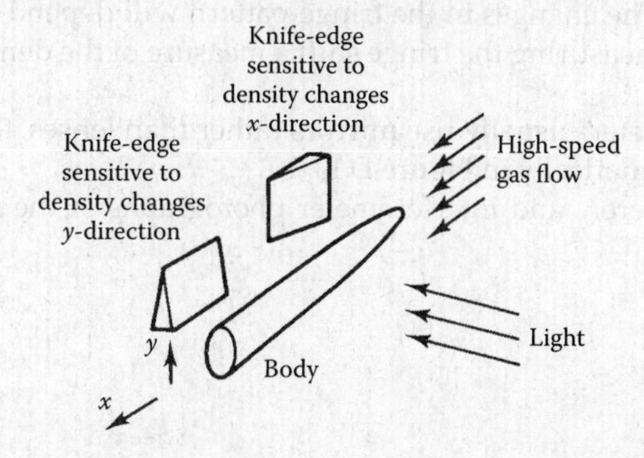

FIGURE D.7
Effect of changing the direction of the knife-edge.

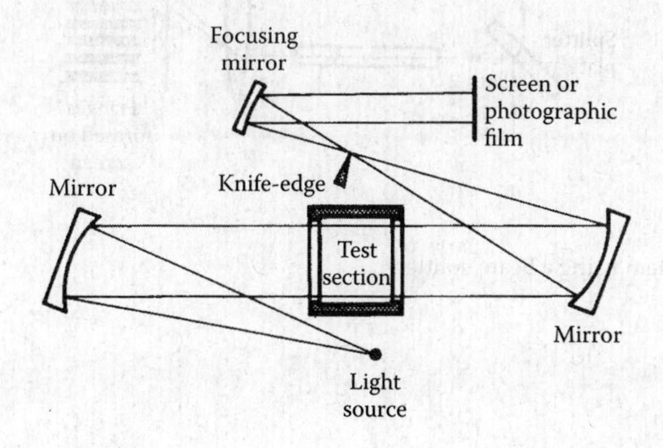

FIGURE D.8
Schlieren system with mirrors.

Interferometer System

To understand the principle on which the interferometer works, consider a ray of light that is split into two by a beam splitter as shown in Figure D.9.

The splitter plate can be a partially mirrored surface that reflects some of the light striking it and transmits the remainder of the light. The two light rays formed in this way follow different paths and are then superimposed on the screen. Now there will be a phase shift between the beams because they follow paths of different length and, as a result, the two rays can reinforce each other giving a bright area on the screen, or they can tend to cancel each other out giving a dark area on the screen. Whether the two rays reinforce or annul each other depends only on the difference between the lengths of the paths taken by the rays if, on these paths, the rays pass through the same media. A series of fringes is, therefore, formed on the screen because the relative lengths of the paths followed by the two rays will depend on the position on the screen being considered as indicated by points q and p shown in Figure D.9.

Now consider what happens if there is a density change in the gas through which one of the rays from the beam splitter passes. The times taken by the two rays to traverse their respective paths now differ from the times taken when the media were the same along the two paths. As a result, there is a change in the fringe pattern due to the density differences along the two paths. The changes in the fringe pattern will depend on the density differences and, in fact, by measuring the fringe shift a measure of the density differences in the flow can be obtained.

Actually, interferometers usually use mirrors rather than lenses, the set-up then resembling that shown schematically in Figure D.10.

Shadowgraph, Schlieren, and interferometer photographs of the same flow are shown in Figure D.11.

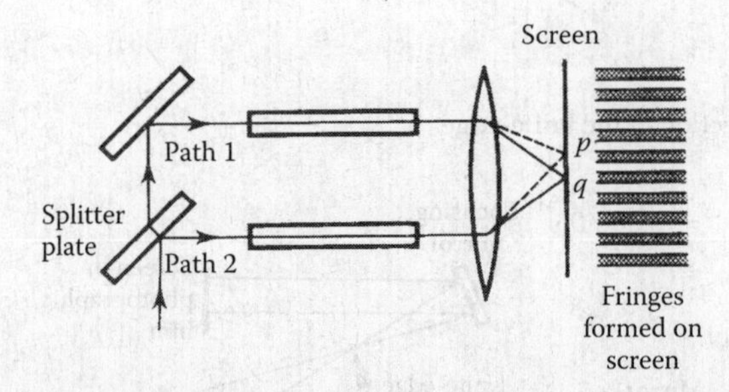

FIGURE D.9
Basic interferometer system using a beam splitter.

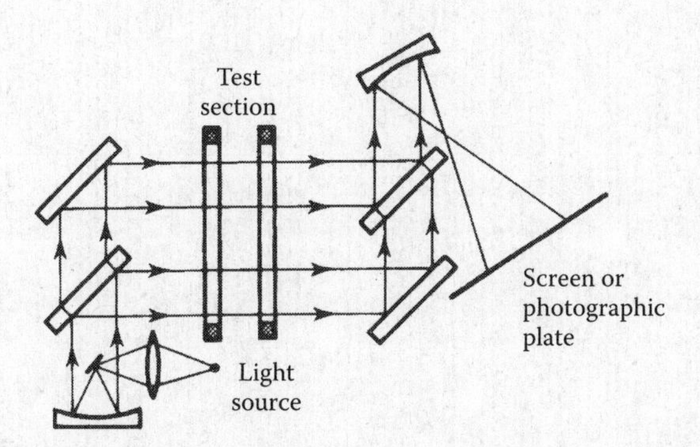

FIGURE D.10
Mach–Zehnder interferometer.

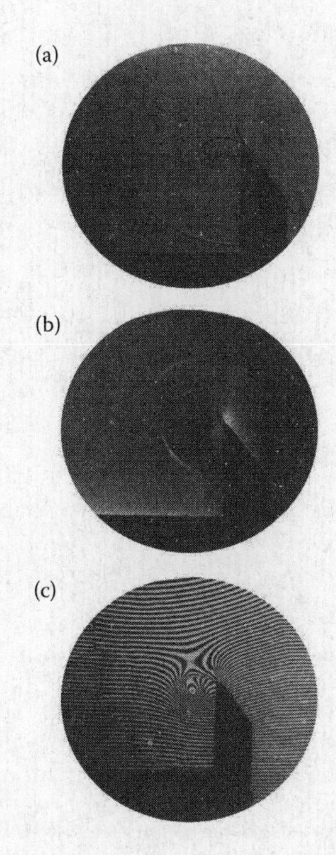

FIGURE D.11
(a) Shadowgraph, (b) Schlieren, and (c) interferometer photographs. (From H.F. Waldron, An Experimental Study of a Spiral Vortex Formed by Shock-Wave Diffraction, Sept. 1954, UTIAS Tech Note 2. University of Toronto Institute for Aerospace Studies. With permission.)

Appendix E: Isentropic Flow Table for $\gamma = 1.4$

M	T_0/T	p_0/p	ρ_0/ρ	a_0/a	A/A^*	θ
0.00	1.00000	1.00000	1.00000	1.00000	—	—
0.02	1.00008	1.00028	1.00020	1.00004	28.94213	—
0.04	1.00032	1.00112	1.00080	1.00016	14.48148	—
0.06	1.00072	1.00252	1.00180	1.00036	9.66591	—
0.08	1.00128	1.00449	1.00320	1.00064	7.26161	—
0.10	1.00200	1.00702	1.00501	1.00100	5.82183	—
0.12	1.00288	1.01012	1.00722	1.00144	4.86432	—
0.14	1.00392	1.01379	1.00983	1.00196	4.18240	—
0.16	1.00512	1.01804	1.01285	1.00256	3.67274	—
0.18	1.00648	1.02286	1.01628	1.00323	3.27793	—
0.20	1.00800	1.02828	1.02012	1.00399	2.96352	—
0.22	1.00968	1.03429	1.02438	1.00483	2.70760	—
0.24	1.01152	1.04090	1.02905	1.00574	2.49556	—
0.26	1.01352	1.04813	1.03414	1.00674	2.31729	—
0.28	1.01568	1.05596	1.03966	1.00781	2.16555	—
0.30	1.01800	1.06443	1.04561	1.00896	2.03507	—
0.32	1.02048	1.07353	1.05199	1.01019	1.92185	—
0.34	1.02312	1.08329	1.05881	1.01149	1.82288	—
0.36	1.02592	1.09370	1.06607	1.01288	1.73578	—
0.38	1.02888	1.10478	1.07377	1.01434	1.65870	—
0.40	1.03200	1.11655	1.08193	1.01587	1.59014	—
0.42	1.03528	1.12902	1.09055	1.01749	1.52891	—
0.44	1.03872	1.14221	1.09963	1.01918	1.47401	—
0.46	1.04232	1.15612	1.10918	1.02094	1.42463	—
0.48	1.04608	1.17078	1.11921	1.02278	1.38010	—
0.50	1.05000	1.18621	1.12973	1.02470	1.33984	—
0.52	1.05408	1.20242	1.14073	1.02668	1.30339	—
0.54	1.05832	1.21944	1.15224	1.02875	1.27032	—
0.56	1.06272	1.23727	1.16425	1.03088	1.24029	—
0.58	1.06728	1.25596	1.17678	1.03309	1.21301	—
0.60	1.07200	1.27550	1.18984	1.03537	1.18820	—
0.62	1.07688	1.29594	1.20342	1.03773	1.16565	—
0.64	1.08192	1.31729	1.21755	1.04015	1.14515	—
0.66	1.08712	1.33959	1.23224	1.04265	1.12654	—
0.68	1.09248	1.36285	1.24748	1.04522	1.10966	—
0.70	1.09800	1.38710	1.26330	1.04785	1.09437	—
0.72	1.10368	1.41238	1.27970	1.05056	1.08057	—
0.74	1.10952	1.43871	1.29669	1.05334	1.06814	—
0.76	1.11552	1.46612	1.31430	1.05618	1.05700	—
0.78	1.12168	1.49466	1.33252	1.05909	1.04705	—
0.80	1.12800	1.52434	1.35136	1.06207	1.03823	—
0.82	1.13448	1.55521	1.37086	1.06512	1.03046	—
0.84	1.14112	1.58730	1.39100	1.06823	1.02370	—

M	T_0/T	p_0/p	ρ_0/ρ	a_0/a	A/A^*	θ
0.86	1.14792	1.62065	1.41182	1.07141	1.01787	—
0.88	1.15488	1.65531	1.43332	1.07465	1.01294	—
0.90	1.16200	1.69130	1.45551	1.07796	1.00886	—
0.92	1.16928	1.72868	1.47841	1.08133	1.00560	—
0.94	1.17672	1.76748	1.50204	1.08477	1.00311	—
0.96	1.18432	1.80776	1.52641	1.08826	1.00136	—
0.98	1.19208	1.84956	1.55154	1.09182	1.00034	—
1.00	1.20000	1.89293	1.57744	1.09544	1.00000	—
1.02	1.20808	1.93791	1.60413	1.09913	1.00033	0.12568
1.04	1.21632	1.98457	1.63162	1.10287	1.00130	0.35097
1.06	1.22472	2.03296	1.65994	1.10667	1.00291	0.63668
1.08	1.23328	2.08313	1.68909	1.11053	1.00512	0.96803
1.10	1.24200	2.13513	1.71911	1.11445	1.00793	1.33619
1.12	1.25088	2.18904	1.75000	1.11843	1.01131	1.73503
1.14	1.25992	2.24492	1.78179	1.12246	1.01527	2.15994
1.16	1.26912	2.30281	1.81450	1.12655	1.01978	2.60733
1.18	1.27848	2.36281	1.84814	1.13070	1.02484	3.07424
1.20	1.28800	2.42496	1.88274	1.13490	1.03044	3.55822
1.22	1.29768	2.48935	1.91831	1.13916	1.03657	4.05718
1.24	1.30752	2.55605	1.95488	1.14347	1.04323	4.56934
1.26	1.31752	2.62512	1.99247	1.14783	1.05041	5.09313
1.28	1.32768	2.69666	2.03111	1.15225	1.05810	5.62717
1.30	1.33800	2.77074	2.07081	1.15672	1.06630	6.17026
1.32	1.34848	2.84744	2.11160	1.16124	1.07502	6.72131
1.34	1.35912	2.92686	2.15350	1.16581	1.08424	7.27934
1.36	1.36992	3.00907	2.19653	1.17044	1.09396	7.84348
1.38	1.38088	3.09418	2.24073	1.17511	1.10419	8.41294
1.40	1.39200	3.18227	2.28611	1.17983	1.11493	8.98700
1.42	1.40328	3.27344	2.33271	1.18460	1.12616	9.56499
1.44	1.41472	3.36780	2.38054	1.18942	1.13790	10.14633
1.46	1.42632	3.46544	2.42964	1.19429	1.15015	10.73047
1.48	1.43808	3.56648	2.48003	1.19920	1.16290	11.31691
1.50	1.45000	3.67103	2.53174	1.20416	1.17617	11.90518
1.52	1.46208	3.77919	2.58480	1.20916	1.18994	12.49486
1.54	1.47432	3.89108	2.63924	1.21422	1.20423	13.08557
1.56	1.48672	4.00683	2.69508	1.21931	1.21904	13.67693
1.58	1.49928	4.12657	2.75237	1.22445	1.23438	14.26862
1.60	1.51200	4.25041	2.81112	1.22963	1.25023	14.86032
1.62	1.52488	4.37849	2.87137	1.23486	1.26662	15.45177
1.64	1.53792	4.51094	2.93315	1.24013	1.28355	16.04268
1.66	1.55112	4.64791	2.99649	1.24544	1.30102	16.63282
1.68	1.56448	4.78954	3.06143	1.25079	1.31904	17.22195
1.70	1.57800	4.93598	3.12800	1.25618	1.33761	17.80988
1.72	1.59168	5.08738	3.19624	1.26162	1.35673	18.39640
1.74	1.60552	5.24390	3.26617	1.26709	1.37643	18.98134
1.76	1.61952	5.40569	3.33784	1.27260	1.39670	19.56453
1.78	1.63368	5.57293	3.41128	1.27815	1.41754	20.14580
1.80	1.64800	5.74578	3.48652	1.28374	1.43898	20.72503

M	T_0/T	p_0/p	ρ_0/ρ	a_0/a	A/A^*	θ
1.82	1.66248	5.92443	3.56361	1.28937	1.46101	21.30208
1.84	1.67712	6.10905	3.64258	1.29504	1.48365	21.87682
1.86	1.69192	6.29982	3.72348	1.30074	1.50689	22.44914
1.88	1.70688	6.49695	3.80633	1.30648	1.53076	23.01893
1.90	1.72200	6.70062	3.89119	1.31225	1.55525	23.58610
1.92	1.73728	6.91104	3.97808	1.31806	1.58039	24.15056
1.94	1.75272	7.12841	4.06706	1.32390	1.60617	24.71223
1.96	1.76832	7.35296	4.15816	1.32978	1.63261	25.27102
1.98	1.78408	7.58489	4.25143	1.33569	1.65971	25.82688
2.00	1.80000	7.82443	4.34691	1.34164	1.68750	26.37973
2.02	1.81608	8.07182	4.44464	1.34762	1.71597	26.92952
2.04	1.83232	8.32729	4.54467	1.35363	1.74514	27.47619
2.06	1.84872	8.59109	4.64705	1.35968	1.77501	28.01970
2.08	1.86528	8.86346	4.75181	1.36575	1.80561	28.56000
2.10	1.88200	9.14466	4.85902	1.37186	1.83694	29.09705
2.12	1.89888	9.43497	4.96870	1.37800	1.86901	29.63082
2.14	1.91592	9.73464	5.08092	1.38417	1.90184	30.16127
2.16	1.93312	10.04396	5.19573	1.39037	1.93543	30.68838
2.18	1.95048	10.36321	5.31316	1.39660	1.96981	31.21212
2.20	1.96800	10.69268	5.43328	1.40285	2.00497	31.73247
2.22	1.98568	11.03269	5.55613	1.40914	2.04094	32.24940
2.24	2.00352	11.38352	5.68177	1.41546	2.07773	32.76291
2.26	2.02152	11.74551	5.81024	1.42180	2.11535	33.27298
2.28	2.03968	12.11898	5.94161	1.42817	2.15381	33.77961
2.30	2.05800	12.50425	6.07593	1.43457	2.19313	34.28276
2.32	2.07648	12.90167	6.21325	1.44100	2.23332	34.78246
2.34	2.09512	13.31159	6.35362	1.44745	2.27440	35.27868
2.36	2.11392	13.73437	6.49711	1.45393	2.31638	35.77143
2.38	2.13288	14.17037	6.64378	1.46044	2.35927	36.26070
2.40	2.15200	14.61998	6.79368	1.46697	2.40310	36.74650
2.42	2.17128	15.08357	6.94686	1.47353	2.44787	37.22883
2.44	2.19072	15.56155	7.10340	1.48011	2.49360	37.70769
2.46	2.21032	16.05432	7.26335	1.48671	2.54031	38.18309
2.48	2.23008	16.56228	7.42677	1.49334	2.58801	38.65504
2.50	2.25000	17.08589	7.59373	1.50000	2.63671	39.12354
2.52	2.27008	17.62555	7.76429	1.50668	2.68645	39.58859
2.54	2.29032	18.18174	7.93852	1.51338	2.73722	40.05023
2.56	2.31072	18.75488	8.11647	1.52010	2.78906	40.50844
2.58	2.33128	19.34557	8.29822	1.52685	2.84197	40.96326
2.60	2.35200	19.95397	8.48384	1.53362	2.89597	41.41468
2.62	2.37288	20.58088	8.67338	1.54041	2.95108	41.86272
2.64	2.39392	21.22670	8.86693	1.54723	3.00733	42.30741
2.66	2.41512	21.89194	9.06454	1.55406	3.06471	42.74874
2.68	2.43648	22.57712	9.26630	1.56092	3.12327	43.18676
2.70	2.45800	23.28280	9.47226	1.56780	3.18300	43.62145
2.72	2.47968	24.00952	9.68251	1.57470	3.24394	44.05285
2.74	2.50152	24.75783	9.89712	1.58162	3.30611	44.48097
2.76	2.52352	25.52832	10.11617	1.58856	3.36951	44.90583

M	T_0/T	p_0/p	ρ_0/ρ	a_0/a	A/A^*	θ
2.78	2.54568	26.32158	10.33971	1.59552	3.43417	45.32746
2.80	2.56800	27.13821	10.56785	1.60250	3.50012	45.74586
2.82	2.59048	27.97882	10.80064	1.60950	3.56736	46.16106
2.84	2.61312	28.84406	11.03818	1.61651	3.63593	46.57309
2.86	2.63592	29.73455	11.28053	1.62355	3.70583	46.98195
2.88	2.65888	30.65097	11.52778	1.63061	3.77711	47.38768
2.90	2.68200	31.59398	11.78002	1.63768	3.84976	47.79020
2.92	2.70528	32.56427	12.03731	1.64477	3.92382	48.18980
2.94	2.72872	33.56255	12.29975	1.65188	3.99931	48.58624
2.96	2.75232	34.58954	12.56742	1.65901	4.07625	48.97962
2.98	2.77608	35.64597	12.84041	1.66616	4.15465	49.36997
3.00	2.80000	36.73260	13.11880	1.67332	4.23456	49.75732
3.05	2.86050	39.58634	13.83897	1.69130	4.44101	50.71267
3.10	2.92200	42.64609	14.59484	1.70938	4.65730	51.64972
3.15	2.98450	45.92497	15.38784	1.72757	4.88382	52.56881
3.20	3.04800	49.43684	16.21945	1.74585	5.12095	53.47031
3.25	3.11250	53.19626	17.09119	1.76423	5.36908	54.35457
3.30	3.17800	57.21857	18.00460	1.78269	5.62863	55.22195
3.35	3.24450	61.51989	18.96131	1.80125	5.90002	56.07281
3.40	3.31200	66.11720	19.96294	1.81989	6.18368	56.90749
3.45	3.38050	71.02834	21.01122	1.83861	6.48006	57.72636
3.50	3.45000	76.27200	22.10785	1.85742	6.78960	58.52974
3.55	3.52050	81.86787	23.25464	1.87630	7.11279	59.31799
3.60	3.59200	87.83657	24.45342	1.89526	7.45009	60.09143
3.65	3.66450	94.19976	25.70606	1.91429	7.80201	60.85041
3.70	3.73800	100.98010	27.01450	1.93339	8.16905	61.59526
3.75	3.81250	108.20136	28.38072	1.95256	8.55172	62.32627
3.80	3.88800	115.88843	29.80673	1.97180	8.95056	63.04378
3.85	3.96449	124.06740	31.29463	1.99110	9.36612	63.74809
3.90	4.04199	132.76549	32.84652	2.01047	9.79895	64.43950
3.95	4.12049	142.01132	34.46462	2.02990	10.24962	65.11831
4.00	4.19999	151.83458	36.15113	2.04939	10.71872	65.78480
4.05	4.28050	162.26674	37.90839	2.06894	11.20686	66.43929
4.10	4.36200	173.34019	39.73871	2.08854	11.71464	67.08201
4.15	4.44450	185.08888	41.64447	2.10820	12.24269	67.71326
4.20	4.52800	197.54826	43.62813	2.12791	12.79166	68.33328
4.25	4.61250	210.75545	45.69219	2.14767	13.36219	68.94234
4.30	4.69801	224.74889	47.83920	2.16749	13.95495	69.54070
4.35	4.78451	239.56898	50.07181	2.18735	14.57063	70.12859
4.40	4.87201	255.25769	52.39268	2.20726	15.20995	70.70626
4.45	4.96051	271.85855	54.80452	2.22722	15.87359	71.27394
4.50	5.05001	289.41721	57.31017	2.24722	16.56231	71.83186
4.60	5.23202	327.59915	62.61428	2.28736	18.01796	72.91930
4.70	5.41802	370.20587	68.32857	2.32766	19.58304	73.97028
4.80	5.60803	417.67252	74.47759	2.36813	21.26398	74.98645
4.90	5.80203	470.46896	81.08690	2.40874	23.06745	75.96935
5.00	6.00004	529.10193	88.18307	2.44950	25.00039	76.92043
5.10	6.20204	594.11725	95.79376	2.49039	27.07004	77.84109

M	T_0/T	p_0/p	ρ_0/ρ	a_0/a	A/A^*	θ
5.20	6.40805	666.10211	103.94769	2.53141	29.28389	78.73267
5.30	6.61806	745.68719	112.67464	2.57256	31.64971	79.59642
5.40	6.83206	833.54962	122.00555	2.61382	34.17557	80.43350
5.50	7.05007	930.41510	131.97247	2.65520	36.86984	81.24507

Appendix F: Normal Shock Table for $\gamma = 1.4$

M_1	M_2	p_2/p_1	T_2/T_1	ρ_2/ρ_1	p_{02}/p_{01}	p_{02}/p_1
1.00	1.00000	1.00000	1.00000	1.00000	1.00000	1.89293
1.02	0.98052	1.04713	1.01325	1.03344	0.99999	1.93790
1.04	0.96203	1.09520	1.02634	1.06709	0.99992	1.98442
1.06	0.94445	1.14420	1.03931	1.10092	0.99975	2.03245
1.08	0.92771	1.19413	1.05217	1.13492	0.99943	2.08194
1.10	0.91177	1.24500	1.06494	1.16908	0.99893	2.13285
1.12	0.89656	1.29680	1.07763	1.20338	0.99821	2.18513
1.14	0.88204	1.34953	1.09027	1.23779	0.99726	2.23877
1.16	0.86816	1.40320	1.10287	1.27231	0.99605	2.29372
1.18	0.85488	1.45780	1.11544	1.30693	0.99457	2.34998
1.20	0.84217	1.51333	1.12799	1.34161	0.99280	2.40750
1.22	0.82999	1.56980	1.14054	1.37636	0.99073	2.46628
1.24	0.81830	1.62720	1.15309	1.41116	0.98836	2.52629
1.26	0.80709	1.68553	1.16566	1.44599	0.98568	2.58753
1.28	0.79631	1.74480	1.17825	1.48084	0.98268	2.64996
1.30	0.78596	1.80500	1.19087	1.51569	0.97937	2.71359
1.32	0.77600	1.86613	1.20353	1.55055	0.97575	2.77840
1.34	0.76641	1.92820	1.21624	1.58538	0.97182	2.84438
1.36	0.75718	1.99120	1.22900	1.62018	0.96758	2.91152
1.38	0.74829	2.05513	1.24181	1.65494	0.96304	2.97980
1.40	0.73971	2.12000	1.25469	1.68965	0.95819	3.04923
1.42	0.73144	2.18580	1.26764	1.72430	0.95306	3.11980
1.44	0.72345	2.25253	1.28066	1.75888	0.94765	3.19149
1.46	0.71574	2.32020	1.29376	1.79337	0.94196	3.26430
1.48	0.70829	2.38880	1.30695	1.82777	0.93600	3.33823
1.50	0.70109	2.45833	1.32022	1.86207	0.92979	3.41327
1.52	0.69413	2.52880	1.33357	1.89626	0.92332	3.48942
1.54	0.68739	2.60020	1.34703	1.93033	0.91662	3.56666
1.56	0.68087	2.67253	1.36057	1.96427	0.90970	3.64501
1.58	0.67455	2.74580	1.37422	1.99808	0.90255	3.72444
1.60	0.66844	2.82000	1.38797	2.03175	0.89520	3.80497
1.62	0.66251	2.89513	1.40182	2.06526	0.88765	3.88658
1.64	0.65677	2.97120	1.41578	2.09863	0.87992	3.96928
1.66	0.65119	3.04820	1.42985	2.13183	0.87201	4.05305
1.68	0.64579	3.12613	1.44403	2.16486	0.86394	4.13790
1.70	0.64054	3.20500	1.45833	2.19772	0.85572	4.22383
1.72	0.63545	3.28480	1.47274	2.23040	0.84736	4.31083
1.74	0.63051	3.36553	1.48727	2.26289	0.83886	4.39890
1.76	0.62570	3.44720	1.50192	2.29520	0.83024	4.48804
1.78	0.62104	3.52980	1.51669	2.32731	0.82151	4.57824
1.80	0.61650	3.61333	1.53158	2.35922	0.81268	4.66951
1.82	0.61209	3.69780	1.54659	2.39093	0.80376	4.76184
1.84	0.60780	3.78320	1.56173	2.42244	0.79476	4.85524

M_1	M_2	p_2/p_1	T_2/T_1	ρ_2/ρ_1	p_{02}/p_{01}	p_{02}/p_1
1.86	0.60363	3.86953	1.57700	2.45373	0.78569	4.94969
1.88	0.59957	3.95680	1.59239	2.48481	0.77655	5.04520
1.90	0.59562	4.04500	1.60791	2.51568	0.76736	5.14177
1.92	0.59177	4.13413	1.62357	2.54632	0.75812	5.23940
1.94	0.58802	4.22420	1.63935	2.57675	0.74884	5.33808
1.96	0.58437	4.31520	1.65527	2.60695	0.73954	5.43781
1.98	0.58082	4.40713	1.67132	2.63692	0.73021	5.53860
2.00	0.57735	4.50000	1.68750	2.66667	0.72087	5.64044
2.02	0.57397	4.59380	1.70382	2.69618	0.71153	5.74332
2.04	0.57068	4.68853	1.72027	2.72546	0.70218	5.84726
2.06	0.56747	4.78419	1.73686	2.75451	0.69284	5.95225
2.08	0.56433	4.88080	1.75359	2.78332	0.68351	6.05829
2.10	0.56128	4.97833	1.77045	2.81190	0.67420	6.16537
2.12	0.55829	5.07679	1.78745	2.84024	0.66492	6.27350
2.14	0.55538	5.17619	1.80459	2.86834	0.65567	6.38268
2.16	0.55254	5.27653	1.82187	2.89621	0.64645	6.49290
2.18	0.54977	5.37779	1.83930	2.92383	0.63727	6.60416
2.20	0.54706	5.47999	1.85686	2.95122	0.62814	6.71647
2.22	0.54441	5.58313	1.87456	2.97836	0.61905	6.82983
2.24	0.54182	5.68719	1.89241	3.00527	0.61002	6.94423
2.26	0.53930	5.79219	1.91040	3.03193	0.60105	7.05967
2.28	0.53683	5.89813	1.92853	3.05836	0.59214	7.17615
2.30	0.53441	6.00499	1.94680	3.08455	0.58330	7.29367
2.32	0.53205	6.11279	1.96522	3.11049	0.57452	7.41224
2.34	0.52974	6.22153	1.98378	3.13620	0.56581	7.53184
2.36	0.52749	6.33119	2.00248	3.16167	0.55718	7.65249
2.38	0.52528	6.44179	2.02133	3.18690	0.54862	7.77418
2.40	0.52312	6.55333	2.04033	3.21189	0.54014	7.89691
2.42	0.52100	6.66579	2.05947	3.23665	0.53175	8.02067
2.44	0.51894	6.77919	2.07876	3.26117	0.52344	8.14548
2.46	0.51691	6.89353	2.09819	3.28546	0.51521	8.27132
2.48	0.51493	7.00879	2.11777	3.30951	0.50707	8.39821
2.50	0.51299	7.12499	2.13750	3.33333	0.49902	8.52613
2.52	0.51109	7.24212	2.15737	3.35692	0.49105	8.65509
2.54	0.50923	7.36019	2.17739	3.38028	0.48318	8.78508
2.56	0.50741	7.47919	2.19756	3.40341	0.47540	8.91612
2.58	0.50562	7.59912	2.21788	3.42631	0.46772	9.04819
2.60	0.50387	7.71999	2.23834	3.44898	0.46012	9.18130
2.62	0.50216	7.84179	2.25895	3.47143	0.45263	9.31544
2.64	0.50048	7.96452	2.27971	3.49365	0.44522	9.45063
2.66	0.49883	8.08819	2.30062	3.51565	0.43792	9.58684
2.68	0.49722	8.21279	2.32168	3.53743	0.43071	9.72410
2.70	0.49563	8.33832	2.34289	3.55899	0.42359	9.86239
2.72	0.49408	8.46479	2.36425	3.58033	0.41657	10.00171
2.74	0.49256	8.59219	2.38575	3.60146	0.40965	10.14208
2.76	0.49107	8.72052	2.40741	3.62237	0.40283	10.28347
2.78	0.48960	8.84979	2.42922	3.64306	0.39610	10.42591
2.80	0.48817	8.97999	2.45117	3.66355	0.38946	10.56937

M_1	M_2	p_2/p_1	T_2/T_1	ρ_2/ρ_1	p_{02}/p_{01}	p_{02}/p_1
2.82	0.48676	9.11112	2.47328	3.68383	0.38293	10.71388
2.84	0.48538	9.24319	2.49553	3.70389	0.37649	10.85941
2.86	0.48402	9.37619	2.51794	3.72375	0.37014	11.00599
2.88	0.48269	9.51012	2.54050	3.74341	0.36389	11.15359
2.90	0.48138	9.64499	2.56321	3.76286	0.35773	11.30223
2.92	0.48010	9.78079	2.58606	3.78211	0.35167	11.45191
2.94	0.47884	9.91752	2.60907	3.80117	0.34570	11.60262
2.96	0.47760	10.05519	2.63223	3.82002	0.33982	11.75436
2.98	0.47638	10.19379	2.65555	3.83868	0.33404	11.90714
3.00	0.47519	10.33332	2.67901	3.85714	0.32834	12.06095
3.05	0.47230	10.68624	2.73833	3.90246	0.31450	12.45000
3.15	0.46689	11.40957	2.85982	3.98961	0.28846	13.24748
3.20	0.46435	11.77998	2.92199	4.03149	0.27623	13.65590
3.25	0.46192	12.15623	2.98511	4.07229	0.26451	14.07078
3.30	0.45959	12.53832	3.04919	4.11202	0.25328	14.49212
3.35	0.45735	12.92623	3.11422	4.15071	0.24252	14.91991
3.40	0.45520	13.31998	3.18020	4.18840	0.23223	15.35415
3.45	0.45314	13.71956	3.24715	4.22511	0.22237	15.79484
3.50	0.45115	14.12498	3.31505	4.26087	0.21295	16.24198
3.55	0.44925	14.53623	3.38391	4.29570	0.20393	16.69557
3.60	0.44741	14.95331	3.45372	4.32962	0.19531	17.15561
3.65	0.44565	15.37623	3.52450	4.36267	0.18707	17.62210
3.70	0.44395	15.80498	3.59624	4.39486	0.17919	18.09504
3.75	0.44231	16.23956	3.66894	4.42623	0.17167	18.57443
3.80	0.44073	16.67998	3.74260	4.45679	0.16447	19.06026
3.85	0.43921	17.12622	3.81722	4.48657	0.15760	19.55254
3.90	0.43774	17.57831	3.89281	4.51558	0.15103	20.05126
3.95	0.43633	18.03622	3.96936	4.54386	0.14475	29.55644
4.00	0.43496	18.49997	4.04687	4.57143	0.13876	21.06805
4.05	0.43364	18.96957	4.12535	4.59829	0.13303	21.58612
4.10	0.43236	19.44499	4.20479	4.62448	0.12756	22.11064
4.15	0.43113	19.92626	4.28520	4.65002	0.12233	22.64161
4.20	0.42994	20.41335	4.36657	4.67491	0.11733	23.17901
4.25	0.42878	20.90628	4.44891	4.69919	0.11256	23.72286
4.30	0.42767	21.40504	4.53222	4.72286	0.10800	24.27316
4.35	0.42659	21.90964	4.61649	4.74595	0.10364	24.82990
4.40	0.42554	22.42006	4.70173	4.76848	0.09948	25.39308
4.45	0.42453	22.93633	4.78793	4.79045	0.09550	25.96270
4.50	0.42355	23.45842	4.87510	4.81188	0.09170	26.53876
4.60	0.42168	24.52012	5.05234	4.85322	0.08459	27.71022
4.70	0.41992	25.60514	5.23346	4.89259	0.07808	28.90745
4.80	0.41826	26.71351	5.41845	4.93011	0.07214	30.13045
4.90	0.41670	27.84520	5.60730	4.96588	0.06670	31.37921
5.00	0.41523	29.00023	5.80004	5.00001	0.06172	32.65373

Appendix G: Table for One-Dimensional Adiabatic Flow with Friction for $\gamma = 1.4$

M	p/p^*	T/T^*	ρ/ρ^*	V/V^*	p_0/p_0^*	$4fl^*/D_H$
0.10	10.94351	1.19760	9.13783	0.10944	5.82183	66.92155
0.12	9.11559	1.19655	7.61820	0.13126	4.86432	45.40796
0.14	7.80932	1.19531	6.53327	0.15306	4.18240	32.51131
0.16	6.82907	1.19389	5.72003	0.17482	3.67274	24.19783
0.18	6.06618	1.19227	5.08791	0.19654	3.27793	18.54265
0.20	5.45545	1.19048	4.58257	0.21822	2.96352	14.53326
0.22	4.95537	1.18850	4.16945	0.23984	2.70760	11.59605
0.24	4.53829	1.18633	3.82547	0.26141	2.49556	9.38648
0.26	4.18505	1.18399	3.53470	0.28291	2.31729	7.68756
0.28	3.88199	1.18147	3.28571	0.30435	2.16555	6.35721
0.30	3.61906	1.17878	3.07017	0.32572	2.03506	5.29925
0.32	3.38874	1.17592	2.88179	0.34701	1.92185	4.44674
0.34	3.18529	1.17288	2.71578	0.36822	1.82288	3.75195
0.36	3.00422	1.16968	2.56841	0.38935	1.73578	3.18012
0.38	2.84200	1.16632	2.43673	0.41039	1.65870	2.70545
0.40	2.69582	1.16279	2.31841	0.43133	1.59014	2.30849
0.42	2.56338	1.15911	2.21151	0.45218	1.52890	1.97437
0.44	2.44281	1.15527	2.11449	0.47293	1.47401	1.69153
0.46	2.33256	1.15128	2.02606	0.49357	1.42463	1.45091
0.48	2.23135	1.14714	1.94514	0.51410	1.38010	1.24534
0.50	2.13809	1.14286	1.87083	0.53452	1.33984	1.06906
0.52	2.05187	1.13843	1.80237	0.55483	1.30339	0.91742
0.54	1.97192	1.13387	1.73910	0.57501	1.27032	0.78663
0.56	1.89755	1.12918	1.68047	0.59507	1.24029	0.67357
0.58	1.82820	1.12435	1.62600	0.61501	1.21301	0.57568
0.60	1.76336	1.11940	1.57527	0.63481	1.18820	0.49082
0.62	1.70261	1.11433	1.52792	0.65448	1.16565	0.41720
0.64	1.64556	1.10914	1.48364	0.67402	1.14515	0.35330
0.66	1.59187	1.10383	1.44213	0.69342	1.12654	0.29785
0.68	1.54126	1.09842	1.40316	0.71268	1.10965	0.24978
0.70	1.49345	1.09290	1.36651	0.73179	1.09437	0.20814
0.72	1.44823	1.08727	1.33198	0.75076	1.08057	0.17215
0.74	1.40537	1.08155	1.29941	0.76958	1.06814	0.14112
0.76	1.36470	1.07573	1.26863	0.78825	1.05700	0.11447
0.78	1.32606	1.06982	1.23951	0.80677	1.04705	0.09167
0.80	1.28928	1.06383	1.21192	0.82514	1.03823	0.07229
0.82	1.25423	1.05775	1.18575	0.84335	1.03046	0.05593
0.84	1.22080	1.05160	1.16090	0.86140	1.02370	0.04226
0.86	1.18888	1.04537	1.13728	0.87929	1.01787	0.03097
0.88	1.15835	1.03907	1.11480	0.89703	1.01294	0.02179
0.90	1.12913	1.03270	1.09338	0.91460	1.00886	0.01451

M	p/p^*	T/T^*	ρ/ρ^*	V/V^*	p_0/p_0^*	$4fl^*/D_H$
0.92	1.10114	1.02627	1.07295	0.93201	1.00560	0.00891
0.94	1.07430	1.01978	1.05346	0.94925	1.00311	0.00482
0.96	1.04854	1.01324	1.03484	0.96633	1.00136	0.00206
0.98	1.02379	1.00664	1.01704	0.98325	1.00034	0.00049
1.00	1.00000	1.00000	1.00000	1.00000	1.00000	0.00000
1.02	0.97711	0.99331	0.98369	1.01658	1.00033	0.00046
1.04	0.95507	0.98658	0.96806	1.03300	1.00130	0.00177
1.06	0.93383	0.97982	0.95306	1.04925	1.00291	0.00384
1.08	0.91335	0.97302	0.93868	1.06533	1.00512	0.00658
1.10	0.89359	0.96618	0.92486	1.08124	1.00793	0.00993
1.12	0.87451	0.95932	0.91159	1.09698	1.01131	0.01382
1.14	0.85608	0.95244	0.89883	1.11256	1.01527	0.01819
1.16	0.83827	0.94554	0.88655	1.12797	1.01978	0.02298
1.18	0.82104	0.93861	0.87473	1.14321	1.02484	0.02814
1.20	0.80436	0.93168	0.86335	1.15828	1.03044	0.03364
1.22	0.78822	0.92473	0.85238	1.17318	1.03657	0.03943
1.24	0.77258	0.91777	0.84181	1.18792	1.04323	0.04547
1.26	0.75743	0.91080	0.83161	1.20249	1.05041	0.05174
1.28	0.74274	0.90383	0.82176	1.21690	1.05810	0.05820
1.30	0.72848	0.89686	0.81226	1.23114	1.06630	0.06483
1.32	0.71465	0.88989	0.80308	1.24521	1.07502	0.07161
1.34	0.70122	0.88292	0.79421	1.25912	1.08424	0.07850
1.36	0.68818	0.87596	0.78563	1.27286	1.09396	0.08550
1.38	0.67551	0.86901	0.77734	1.28645	1.10419	0.09259
1.40	0.66320	0.86207	0.76931	1.29987	1.11493	0.09974
1.42	0.65122	0.85514	0.76154	1.31313	1.12616	0.10694
1.44	0.63958	0.84822	0.75402	1.32623	1.13790	0.11419
1.46	0.62825	0.84133	0.74673	1.33917	1.15015	0.12146
1.48	0.61722	0.83445	0.73967	1.35195	1.16290	0.12875
1.50	0.60648	0.82759	0.73283	1.36458	1.17617	0.13605
1.52	0.59602	0.82075	0.72619	1.37705	1.18994	0.14335
1.54	0.58583	0.81393	0.71976	1.38936	1.20423	0.15063
1.56	0.57591	0.80715	0.71351	1.40152	1.21904	0.15790
1.58	0.56623	0.80038	0.70745	1.41353	1.23437	0.16514
1.60	0.55679	0.79365	0.70156	1.42539	1.25023	0.17236
1.62	0.54759	0.78695	0.69584	1.43710	1.26662	0.17953
1.64	0.53862	0.78028	0.69029	1.44866	1.28355	0.18667
1.66	0.52986	0.77363	0.68490	1.46008	1.30102	0.19377
1.68	0.52131	0.76703	0.67965	1.47135	1.31904	0.20081
1.70	0.51297	0.76046	0.67455	1.48247	1.33761	0.20780
1.72	0.50482	0.75392	0.66959	1.49345	1.35673	0.21474
1.74	0.49686	0.74742	0.66477	1.50429	1.37643	0.22162
1.76	0.48909	0.74096	0.66007	1.51499	1.39670	0.22844
1.78	0.48149	0.73454	0.65550	1.52555	1.41754	0.23519
1.80	0.47407	0.72816	0.65105	1.53598	1.43898	0.24189
1.82	0.46681	0.72181	0.64672	1.54626	1.46101	0.24851
1.84	0.45972	0.71551	0.64250	1.55642	1.48365	0.25507

M	p/p^*	T/T^*	ρ/ρ^*	V/V^*	p_0/p_0^*	$4fl^*/D_H$
1.86	0.45278	0.70925	0.63839	1.56644	1.50689	0.26156
1.88	0.44600	0.70304	0.63439	1.57633	1.53076	0.26798
1.90	0.43936	0.69686	0.63048	1.58609	1.55525	0.27433
1.92	0.43287	0.69074	0.62668	1.59572	1.58039	0.28061
1.94	0.42651	0.68465	0.62297	1.60522	1.60617	0.28681
1.96	0.42030	0.67861	0.61935	1.61460	1.63261	0.29295
1.98	0.41421	0.67262	0.61582	1.62386	1.65971	0.29901
2.00	0.40825	0.66667	0.61237	1.63299	1.68750	0.30500
2.02	0.40241	0.66076	0.60901	1.64200	1.71597	0.31091
2.04	0.39670	0.65491	0.60573	1.65090	1.74514	0.31676
2.06	0.39110	0.64910	0.60253	1.65967	1.77501	0.32253
2.08	0.38562	0.64334	0.59940	1.66833	1.80561	0.32822
2.10	0.38024	0.63762	0.59635	1.67687	1.83694	0.33385
2.12	0.37498	0.63195	0.59337	1.68530	1.86901	0.33940
2.14	0.36982	0.62633	0.59045	1.69362	1.90184	0.34489
2.16	0.36476	0.62076	0.58760	1.70182	1.93543	0.35030
2.18	0.35980	0.61523	0.58482	1.70992	1.96981	0.35564
2.20	0.35494	0.60976	0.58210	1.71791	2.00497	0.36091
2.22	0.35017	0.60433	0.57944	1.72579	2.04094	0.36611
2.24	0.34550	0.59895	0.57684	1.73357	2.07773	0.37124
2.26	0.34091	0.59361	0.57430	1.74125	2.11535	0.37631
2.28	0.33641	0.58833	0.57182	1.74882	2.15381	0.38130
2.30	0.33200	0.58309	0.56938	1.75629	2.19313	0.38623
2.32	0.32767	0.57790	0.56700	1.76366	2.23332	0.39109
2.34	0.32342	0.57276	0.56467	1.77093	2.27440	0.39589
2.36	0.31925	0.56767	0.56240	1.77811	2.31638	0.40062
2.38	0.31516	0.56262	0.56016	1.78519	2.35927	0.40529
2.40	0.31114	0.55762	0.55798	1.79218	2.40310	0.40989
2.42	0.30720	0.55267	0.55584	1.79907	2.44787	0.41443
2.44	0.30332	0.54777	0.55375	1.80587	2.49360	0.41891
2.46	0.29952	0.54291	0.55170	1.81258	2.54030	0.42332
2.48	0.29579	0.53810	0.54969	1.81921	2.58801	0.42768
2.50	0.29212	0.53333	0.54772	1.82574	2.63671	0.43198
2.52	0.28852	0.52862	0.54579	1.83219	2.68645	0.43621
2.54	0.28498	0.52394	0.54391	1.83855	2.73722	0.44039
2.56	0.28150	0.51932	0.54205	1.84483	2.78906	0.44451
2.58	0.27808	0.51474	0.54024	1.85103	2.84197	0.44858
2.60	0.27473	0.51020	0.53846	1.85714	2.89597	0.45259
2.62	0.27143	0.50572	0.53672	1.86318	2.95108	0.45654
2.64	0.26818	0.50127	0.53501	1.86913	3.00733	0.46044
2.66	0.26500	0.49687	0.53333	1.87501	3.06471	0.46429
2.68	0.26186	0.49251	0.53169	1.88081	3.12327	0.46808
2.70	0.25878	0.48820	0.53007	1.88653	3.18300	0.47182
2.72	0.25576	0.48393	0.52849	1.89218	3.24394	0.47551
2.74	0.25278	0.47971	0.52694	1.89775	3.30611	0.47915
2.76	0.24985	0.47553	0.52542	1.90325	3.36951	0.48273
2.78	0.24697	0.47139	0.52392	1.90868	3.43417	0.48627

M	p/p^*	T/T^*	ρ/ρ^*	V/V^*	p_0/p_0^*	$4fl^*/D_H$
2.80	0.24414	0.46729	0.52246	1.91404	3.50012	0.48976
2.82	0.24135	0.46324	0.52102	1.91933	3.56736	0.49321
2.84	0.23861	0.45922	0.51960	1.92455	3.63593	0.49660
2.86	0.23592	0.45525	0.51821	1.92970	3.70583	0.49995
2.88	0.23326	0.45132	0.51685	1.93479	3.77710	0.50326
2.90	0.23066	0.44743	0.51551	1.93981	3.84976	0.50652
2.92	0.22809	0.44050	0.51420	1.94477	3.92382	0.50973
2.94	0.22556	0.43977	0.51291	1.94966	3.99931	0.51290
2.96	0.22307	0.43600	0.51164	1.95449	4.07625	0.51603
2.98	0.22063	0.43226	0.51040	1.95925	4.15465	0.51912
3.00	0.21822	0.42857	0.50918	1.96396	4.23456	0.52216
3.05	0.21236	0.41951	0.50621	1.97547	4.44101	0.52959
3.10	0.20672	0.41068	0.50337	1.98661	4.65730	0.53678
3.15	0.20130	0.40208	0.50065	1.99740	4.88382	0.54372
3.20	0.19608	0.39370	0.49804	2.00786	5.12094	0.55044
3.25	0.19105	0.38554	0.49554	2.01799	5.36908	0.55694
3.30	0.18621	0.37760	0.49314	2.02781	5.62863	0.56323
3.35	0.18154	0.36986	0.49084	2.03733	5.90002	0.56932
3.40	0.17704	0.36232	0.48863	2.04656	6.18368	0.57521
3.45	0.17220	0.35498	0.48650	2.05551	6.48006	0.58091
3.50	0.16851	0.34783	0.48445	2.06419	6.78960	0.58643
3.55	0.16446	0.34086	0.48248	2.07261	7.11279	0.59178
3.60	0.16055	0.33408	0.48059	2.08077	7.45009	0.59695
3.65	0.15678	0.32747	0.47877	2.08870	7.80201	0.60197
3.70	0.15313	0.32103	0.47701	2.09639	8.16904	0.60684
3.75	0.14961	0.31475	0.47532	2.10386	8.55172	0.61155
3.80	0.14620	0.30864	0.47368	2.11111	8.95056	0.61612
3.85	0.14290	0.30269	0.47211	2.11815	9.36612	0.62055
3.90	0.13971	0.29688	0.47059	2.12499	9.79894	0.62485
3.95	0.13662	0.29123	0.46912	2.13163	10.24962	0.62902
4.00	0.13363	0.28571	0.46771	2.13809	10.71872	0.63306
4.05	0.13073	0.28034	0.46634	2.14436	11.20686	0.63699
4.10	0.12793	0.27510	0.46502	2.15046	11.71464	0.64080
4.15	0.12521	0.27000	0.46374	2.15639	12.24269	0.64451
4.20	0.12257	0.26502	0.46250	2.16215	12.79166	0.64810
4.25	0.12001	0.26016	0.46131	2.16776	13.36218	0.65159
4.30	0.11753	0.25543	0.46015	2.17321	13.95495	0.65499
4.35	0.11513	0.25081	0.45903	2.17852	14.57063	0.65828
4.40	0.11279	0.24630	0.45794	2.18368	15.20995	0.66149
4.45	0.11053	0.24191	0.45689	2.18871	15.87359	0.66461
4.50	0.10833	0.23762	0.45587	2.19360	16.56231	0.66764
4.60	0.10411	0.22936	0.45393	2.20300	18.01795	0.67345
4.70	0.10013	0.22148	0.45210	2.21192	19.58303	0.67895
4.80	0.09637	0.21398	0.45037	2.22038	21.26397	0.68417
4.90	0.09281	0.20682	0.44875	2.22843	23.06745	0.68911
5.00	0.08944	0.20000	0.44721	2.23607	25.00040	0.69380
5.10	0.08625	0.19348	0.44576	2.24334	27.07004	0.69826

M	p/p^*	T/T^*	ρ/ρ^*	V/V^*	p_0/p_0^*	$4fl^*/D_H$
5.20	0.08322	0.18726	0.44439	2.25026	29.28388	0.70250
5.30	0.08034	0.18132	0.44309	2.25685	31.64970	0.70652
5.40	0.07761	0.17564	0.44186	2.26314	34.17556	0.71036
5.50	0.07501	0.17021	0.44070	2.26913	36.86983	0.71401

Appendix H: Table for One-Dimensional Flow with Heat Exchange for $\gamma = 1.4$

M	T_0/T_0^*	T/T^*	p/p^*	p_0/p_0^*	V/V^*	ρ/ρ^*
0.06	0.01712	0.02053	2.38796	1.26470	0.00860	116.32407
0.08	0.03022	0.03621	2.37869	1.26226	0.01522	65.68750
0.10	0.04678	0.05602	2.36686	1.25915	0.02367	42.25000
0.12	0.06661	0.07970	2.35257	1.25539	0.03388	29.51852
0.14	0.08947	0.10695	2.33590	1.25103	0.04578	21.84184
0.16	0.11511	0.13743	2.31696	1.24608	0.05931	16.85937
0.18	0.14324	0.17078	2.29586	1.24059	0.07439	13.44341
0.20	0.17355	0.20661	2.27273	1.23460	0.09091	11.00000
0.22	0.20574	0.24452	2.24770	1.22814	0.10879	9.19215
0.24	0.23948	0.28411	2.22091	1.22126	0.12792	7.81713
0.26	0.27446	0.32496	2.19250	1.21400	0.14821	6.74704
0.28	0.31035	0.36667	2.16263	1.20642	0.16955	5.89796
0.30	0.34686	0.40887	2.13144	1.19855	0.19183	5.21296
0.32	0.38369	0.45119	2.09908	1.19045	0.21495	4.65234
0.34	0.42056	0.49327	2.06569	1.18215	0.23879	4.18772
0.36	0.45723	0.53482	2.03142	1.17371	0.26327	3.79835
0.38	0.49346	0.57553	1.99641	1.16517	0.28828	3.46884
0.40	0.52903	0.61515	1.96078	1.15658	0.31373	3.18750
0.42	0.56376	0.65346	1.92468	1.14796	0.33951	2.94539
0.44	0.59748	0.69025	1.88822	1.13936	0.36556	2.73554
0.46	0.63007	0.72538	1.85151	1.13082	0.39178	2.55246
0.48	0.66139	0.75871	1.81466	1.12238	0.41810	2.39178
0.50	0.69136	0.79012	1.77778	1.11405	0.44444	2.25000
0.52	0.71990	0.81955	1.74095	1.10588	0.47075	2.12426
0.54	0.74695	0.84695	1.70426	1.09789	0.49696	2.01223
0.56	0.77249	0.87227	1.66778	1.09011	0.52302	1.91199
0.58	0.79648	0.89552	1.63159	1.08256	0.54887	1.82194
0.60	0.81892	0.91670	1.59575	1.07525	0.57447	1.74074
0.62	0.83982	0.93584	1.56031	1.06822	0.59978	1.66727
0.64	0.85920	0.95298	1.52532	1.06147	0.62477	1.60059
0.66	0.87708	0.96815	1.49083	1.05503	0.64941	1.53987
0.68	0.89350	0.98144	1.45688	1.04890	0.67366	1.48443
0.70	0.90850	0.99290	1.42349	1.04310	0.69751	1.43367
0.72	0.92212	1.00260	1.39069	1.03764	0.72093	1.38709
0.74	0.93442	1.01062	1.35851	1.03253	0.74392	1.34423
0.76	0.94546	1.01706	1.32696	1.02777	0.76645	1.30471
0.78	0.95528	1.02198	1.29606	1.02337	0.78853	1.26819
0.80	0.96395	1.02548	1.26582	1.01934	0.81013	1.23438
0.82	0.97152	1.02763	1.23625	1.01569	0.83125	1.20300
0.84	9.97807	1.02853	1.20734	1.01241	0.85190	1.17385
0.86	0.98363	1.02826	1.17911	1.00951	0.87207	1.14670

M	T_0/T_0^*	T/T^*	p/p^*	p_0/p_0^*	V/V^*	ρ/ρ^*
0.88	0.98828	1.02689	1.15154	1.00699	0.89175	1.12138
0.90	0.99207	1.02452	1.12465	1.00486	0.91096	1.09774
0.92	0.99506	1.02120	1.09842	1.00311	0.92970	1.07561
0.94	0.99729	1.01702	1.07285	1.00175	0.94797	1.05489
0.96	0.99883	1.01205	1.04793	1.00078	0.96577	1.03545
0.98	0.99971	1.00636	1.02365	1.00019	0.98311	1.01718
1.00	1.00000	1.00000	1.00000	1.00000	1.00000	1.00000
1.02	0.99973	0.99304	0.97698	1.00019	1.01645	0.98382
1.04	0.99895	0.98554	0.95456	1.00078	1.03245	0.96857
1.06	0.99769	0.97756	0.93275	1.00175	1.04804	0.95417
1.08	0.99601	0.96913	0.91152	1.00311	1.06320	0.94056
1.10	0.99392	0.96031	0.89087	1.00486	1.07795	0.92769
1.12	0.99148	0.95115	0.87078	1.00699	1.09230	0.91550
1.14	0.98871	0.94169	0.85123	1.00952	1.10626	0.90395
1.16	0.98564	0.93196	0.83222	1.01243	1.11984	0.89298
1.18	0.98230	0.92200	0.81374	1.01573	1.13305	0.88258
1.20	0.97872	0.91185	0.79576	1.01941	1.14589	0.87269
1.22	0.97492	0.90153	0.77827	1.02349	1.15838	0.86328
1.24	0.97092	0.89108	0.76127	1.02795	1.17052	0.85432
1.26	0.96675	0.88052	0.74473	1.03280	1.18233	0.84578
1.28	0.96243	0.86988	0.72865	1.03803	1.19382	0.83765
1.30	0.95798	0.85917	0.71301	1.04366	1.20499	0.82988
1.32	0.95341	0.84843	0.69780	1.04967	1.21585	0.82247
1.34	0.94873	0.83766	0.68301	1.05608	1.22642	0.81538
1.36	0.94398	0.82689	0.66863	1.06288	1.23669	0.80861
1.38	0.93914	0.81613	0.65464	1.07007	1.24669	0.80213
1.40	0.93425	0.80539	0.64103	1.07765	1.25641	0.79592
1.42	0.92931	0.79469	0.62779	1.08563	1.26587	0.78997
1.44	0.92434	0.78405	0.61491	1.09401	1.27507	0.78427
1.46	0.91933	0.77346	0.60237	1.10278	1.28402	0.77880
1.48	0.91431	0.76294	0.59018	1.11196	1.29273	0.77356
1.50	0.90928	0.75250	0.57831	1.12154	1.30120	0.76852
1.52	0.90424	0.74215	0.56677	1.13153	1.30945	0.76368
1.54	0.89921	0.73189	0.55553	1.14193	1.31748	0.75902
1.56	0.89418	0.72174	0.54458	1.15274	1.32530	0.75455
1.58	0.88917	0.71168	0.53393	1.16397	1.33291	0.75024
1.60	0.88419	0.70174	0.52356	1.17561	1.34031	0.74609
1.62	0.87922	0.69190	0.51346	1.18768	1.34753	0.74210
1.64	0.87429	0.68219	0.50363	1.20017	1.35455	0.73825
1.66	0.86939	0.67259	0.49405	1.21309	1.36139	0.73454
1.68	0.86453	0.66312	0.48472	1.22644	1.36806	0.73096
1.70	0.85971	0.65377	0.47562	1.24023	1.37455	0.72751
1.72	0.85493	0.64455	0.46677	1.25447	1.38088	0.72418
1.74	0.85019	0.63545	0.45813	1.26915	1.38705	0.72096
1.76	0.84551	0.62649	0.44972	1.28428	1.39306	0.71785
1.78	0.84087	0.61765	0.44152	1.29987	1.39891	0.71484
1.80	0.83628	0.60894	0.43353	1.31592	1.40462	0.71193

M	T_0/T_0^*	T/T^*	p/p^*	p_0/p_0^*	V/V^*	ρ/ρ^*
1.82	0.83174	0.60036	0.42573	1.33244	1.41019	0.70912
1.84	0.82726	0.59191	0.41813	1.34943	1.41562	0.70640
1.86	0.82283	0.58360	0.41072	1.36690	1.42092	0.70377
1.88	0.81846	0.57540	0.40349	1.38485	1.42608	0.70122
1.90	0.81414	0.56734	0.39643	1.40330	1.43112	0.69875
1.92	0.80987	0.55941	0.38955	1.42224	1.43604	0.69636
1.94	0.80567	0.55160	0.38283	1.44168	1.44083	0.69404
1.96	0.80152	0.54392	0.37628	1.46163	1.44551	0.69180
1.98	0.79742	0.53636	0.36988	1.48210	1.45008	0.68962
2.00	0.79339	0.52893	0.36364	1.50309	1.45455	0.68750
2.02	0.78941	0.52161	0.35754	1.52462	1.45890	0.68545
2.04	0.78549	0.51442	0.35158	1.54668	1.46315	0.68346
2.06	0.78162	0.50735	0.34577	1.56928	1.46731	0.68152
2.08	0.77782	0.50040	0.34009	1.59244	1.47136	0.67964
2.10	0.77406	0.49356	0.33454	1.61616	1.47533	0.67782
2.12	0.77037	0.48684	0.32912	1.64044	1.47920	0.67604
2.14	0.76673	0.48023	0.32382	1.66531	1.48298	0.67432
2.16	0.76314	0.47373	0.31865	1.69076	1.48668	0.67264
2.18	0.75961	0.46734	0.31359	1.71680	1.49029	0.67101
2.20	0.75614	0.46106	0.30864	1.74344	1.49383	0.66942
2.22	0.75271	0.45488	0.30381	1.77070	1.49728	0.66788
2.24	0.74934	0.44882	0.29908	1.79858	1.50066	0.66637
2.26	0.74602	0.44285	0.29446	1.82708	1.50396	0.66491
2.28	0.74276	0.43699	0.28993	1.85622	1.50719	0.66349
2.30	0.73954	0.43122	0.28551	1.88602	1.51035	0.66210
2.32	0.73638	0.42555	0.28118	1.91647	1.51344	0.66075
2.34	0.73326	0.41998	0.27695	1.94759	1.51646	0.65943
2.36	0.73020	0.41451	0.27281	1.97938	1.51942	0.65814
2.38	0.72718	0.40913	0.26875	2.01187	1.52232	0.65689
2.40	0.72421	0.40384	0.26478	2.04505	1.52515	0.65567
2.42	0.72129	0.39864	0.26090	2.07894	1.52793	0.65448
2.44	0.71842	0.39352	0.25710	2.11356	1.53065	0.65332
2.46	0.71559	0.38850	0.25337	2.14890	1.53331	0.65219
2.48	0.71280	0.38356	0.24973	2.18499	1.53591	0.65108
2.50	0.71006	0.37870	0.24615	2.22183	1.53846	0.65000
2.52	0.70736	0.37392	0.24266	2.25943	1.54096	0.64895
2.54	0.70471	0.36923	0.23923	2.29781	1.54341	0.64792
2.56	0.70210	0.36461	0.23587	2.33698	1.54581	0.64691
2.58	0.69953	0.36007	0.23258	2.37695	1.54816	0.64593
2.60	0.69700	0.35561	0.22936	2.41774	1.55046	0.64497
2.62	0.69451	0.35122	0.22620	2.45934	1.55272	0.64403
2.64	0.69206	0.34691	0.22310	2.50179	1.55493	0.64312
2.66	0.68964	0.34266	0.22007	2.54509	1.55710	0.64222
2.68	0.68727	0.33849	0.21709	2.58925	1.55922	0.64135
2.70	0.68494	0.33439	0.21417	2.63428	1.56131	0.64049
2.72	0.68264	0.33035	0.21131	2.68021	1.56335	0.63965
2.74	0.68037	0.32638	0.20850	2.72703	1.56536	0.63883

M	T_0/T_0^*	T/T^*	p/p^*	p_0/p_0^*	V/V^*	ρ/ρ^*
2.76	0.67815	0.32248	0.20575	2.77478	1.56732	0.63803
2.78	0.67595	0.31864	0.20305	2.82345	1.56925	0.63725
2.80	0.67380	0.31486	0.20040	2.87307	1.57114	0.63648
2.82	0.67167	0.31111	0.19780	2.92365	1.57300	0.63573
2.84	0.66958	0.30749	0.19525	2.97520	1.57482	0.63499
2.86	0.66752	0.30389	0.19275	3.02774	1.57661	0.63427
2.88	0.66550	0.30035	0.19029	3.08129	1.57836	0.63357
2.90	0.66350	0.29687	0.18788	3.13585	1.58008	0.63288
2.92	0.66154	0.29344	0.18552	3.19144	1.58177	0.63220
2.94	0.65960	0.29007	0.18319	3.24808	1.58343	0.63154
2.96	0.65770	0.28675	0.18091	3.30578	1.58506	0.63089
2.98	0.65583	0.28349	0.17867	3.36457	1.58666	0.63025
3.00	0.65398	0.28028	0.17647	3.42444	1.58824	0.62963
3.05	0.64949	0.27246	0.17114	3.57904	1.59204	0.62812
3.10	0.64516	0.26495	0.16604	3.74084	1.59568	0.62669
3.15	0.64100	0.25773	0.16117	3.91010	1.59917	0.62533
3.20	0.63699	0.25078	0.15649	4.08711	1.60250	0.62402
3.25	0.63313	0.24410	0.15202	4.27214	1.60570	0.62278
3.30	0.62941	0.23766	0.14773	4.46548	1.60877	0.62159
3.35	0.62582	0.23146	0.14361	4.66743	1.61170	0.62046
3.40	0.62236	0.22549	0.13967	4.87829	1.61453	0.61938
3.45	0.61902	0.21974	0.13587	5.09838	1.61723	0.61834
3.50	0.61581	0.21419	0.13223	5.32802	1.61983	0.61735
3.55	0.61270	0.20885	0.12873	5.56754	1.62233	0.61640
3.60	0.60970	0.20369	0.12537	5.81728	1.62474	0.61548
3.65	0.60681	0.19871	0.12213	6.07759	1.62705	0.61461
3.70	0.60401	0.19390	0.11901	6.34883	1.62928	0.61377
3.75	0.60131	0.18926	0.11601	6.63135	1.63142	0.61296
3.80	0.59870	0.18478	0.11312	6.92555	1.63348	0.61219
3.85	0.59617	0.18045	0.11034	7.23179	1.63547	0.61144
3.90	0.59373	0.17627	0.10765	7.55048	1.63739	0.61073
3.95	0.59137	0.17222	0.10506	7.88202	1.63924	0.61004
4.00	0.58909	0.16831	0.10256	8.22683	1.64103	0.60938
4.05	0.58687	0.16453	0.10015	8.58532	1.64275	0.60874
4.10	0.58473	0.16086	0.09782	8.95794	1.64441	0.60812
4.15	0.58266	0.15732	0.09557	9.34511	1.64602	0.60753
4.20	0.58065	0.15388	0.09340	9.74730	1.64757	0.60695
4.25	0.57870	0.15056	0.09130	10.16496	1.64907	0.60640
4.30	0.57682	0.14734	0.08927	10.59858	1.65052	0.60587
4.35	0.57499	0.14421	0.08730	11.04862	1.65193	0.60535
4.40	0.57321	0.14119	0.08540	11.51560	1.65329	0.60486
4.45	0.57149	0.13825	0.08356	12.00000	1.65460	0.60437
4.50	0.56982	0.13540	0.08177	12.50235	1.65588	0.60391
4.60	0.56663	0.12996	0.07837	13.56300	1.65831	0.60302
4.70	0.56362	0.12483	0.07517	14.70189	1.66059	0.60220
4.80	0.56078	0.11999	0.07217	15.92356	1.66274	0.60142
4.90	0.55809	0.11543	0.06934	17.23269	1.66476	0.60069

M	T_0/T_0^*	T/T^*	p/p^*	p_0/p_0^*	V/V^*	ρ/ρ^*
5.00	0.55556	0.11111	0.06667	18.63419	1.66667	0.60000
5.10	0.55315	0.10703	0.06415	20.13313	1.66847	0.59935
5.20	0.55088	0.10316	0.06177	21.73480	1.67017	0.59874
5.30	0.54872	0.09949	0.05951	23.44467	1.67178	0.59817
5.40	0.54667	0.09602	0.05738	25.26842	1.67330	0.59762
5.50	0.54472	0.09272	0.05536	27.21195	1.67474	0.59711

Appendix I: Table for One-Dimensional Isothermal Flow with Friction for $\gamma = 1.4$

M	p/p^*	T_0/T_0^*	ρ/ρ^*	V/V^*	p_0/p_0^*	$4fl^*/D_H$
0.10	8.45154	0.87675	8.45154	0.11832	5.33336	66.15987
0.12	7.04295	0.87752	7.04295	0.14199	4.45815	44.69912
0.14	6.03682	0.87843	6.03682	0.16565	3.83516	31.84739
0.16	5.28221	0.87948	5.28221	0.18931	3.36982	23.57309
0.18	4.69530	0.88067	4.69530	0.21298	3.00961	17.95273
0.20	4.22577	0.88200	4.22577	0.23664	2.72299	13.97473
0.22	3.84161	0.88347	3.84161	0.26031	2.48992	11.06618
0.24	3.52148	0.88508	3.52148	0.28397	2.29701	8.88303
0.26	3.25059	0.88683	3.25059	0.30764	2.13503	7.20868
0.28	3.01841	0.88872	3.01841	0.33130	1.99735	5.90133
0.30	2.81718	0.89075	2.81718	0.35496	1.87914	4.86503
0.32	2.64111	0.89292	2.64111	0.37863	1.77676	4.03305
0.34	2.48575	0.89523	2.48575	0.40229	1.68744	3.35780
0.36	2.34765	0.89768	2.34765	0.42596	1.60901	2.80463
0.38	2.22409	0.90027	2.22409	0.44962	1.53977	2.34788
0.40	2.11289	0.90300	2.11289	0.47329	1.47837	1.96818
0.42	2.01227	0.90587	2.01227	0.49695	1.42370	1.65071
0.44	1.92081	0.90888	1.92081	0.52061	1.37485	1.38400
0.46	1.83729	0.91203	1.83729	0.54428	1.33110	1.15906
0.48	1.76074	0.91532	1.76074	0.56794	1.29181	0.96873
0.50	1.69031	0.91875	1.69031	0.59161	1.25648	0.80732
0.52	1.62530	0.92232	1.62530	0.61527	1.22467	0.67021
0.54	1.56510	0.92603	1.56510	0.63894	1.19600	0.55364
0.56	1.50920	0.92988	1.50920	0.66260	1.17015	0.45453
0.58	1.45716	0.93387	1.45716	0.68626	1.14686	0.37034
0.60	1.40859	0.93800	1.40859	0.70993	1.12589	0.29895
0.62	1.36315	0.94227	1.36315	0.73359	1.10703	0.23858
0.64	1.32055	0.94668	1.32055	0.75726	1.09010	0.18776
0.66	1.28054	0.95123	1.28054	0.78092	1.07496	0.14522
0.68	1.24287	0.95592	1.24287	0.80459	1.06146	0.10988
0.70	1.20736	0.96075	1.20736	0.82825	1.04948	0.08085
0.72	1.17383	0.96572	1.17383	0.85192	1.03892	0.05733
0.74	1.14210	0.97083	1.14210	0.87558	1.02969	0.03866
0.76	1.11205	0.97608	1.11205	0.89924	1.02170	0.02424
0.78	1.08353	0.98147	1.08353	0.92291	1.01487	0.01359
0.80	1.05644	0.98700	1.05644	0.94657	1.00915	0.00626
0.82	1.03068	0.99267	1.03068	0.97024	1.00448	0.00186
0.84	1.00614	0.99848	1.00614	0.99390	1.00079	0.00008
0.86	0.98274	1.00443	0.98274	1.01757	0.99806	0.00060
0.88	0.96040	1.01052	0.96040	1.04123	0.99623	0.00318
0.90	0.93906	1.01675	0.93906	1.06489	0.99528	0.00759

M	p/p^*	T_0/T_0^*	ρ/ρ^*	V/V^*	p_0/p_0^*	$4fl^*/D_H$
0.92	0.91865	1.02312	0.91865	1.08856	0.99516	0.01362
0.94	0.89910	1.02963	0.89910	1.11222	0.99585	0.02110
0.96	0.88037	1.03628	0.88037	1.13589	0.99732	0.02988
0.98	0.86240	1.04307	0.86240	1.15955	0.99956	0.03980
1.00	0.84515	1.05000	0.84515	1.18322	1.00253	0.05076
1.02	0.82858	1.05707	0.82858	1.20688	1.00623	0.06263
1.04	0.81265	1.06428	0.81265	1.23054	1.01064	0.07531
1.06	0.79732	1.07163	0.79732	1.25421	1.01575	0.08872
1.08	0.78255	1.07912	0.78255	1.27787	1.02154	0.10278
1.10	0.76832	1.08675	0.76832	1.30154	1.02801	0.11741
1.12	0.75460	1.09452	0.75460	1.32520	1.03514	0.13255
1.14	0.74136	1.10243	0.74136	1.34887	1.04294	0.14815
1.16	0.72858	1.11048	0.72858	1.37253	1.05139	0.16414
1.18	0.71623	1.11867	0.71623	1.39619	1.06050	0.18049
1.20	0.70430	1.12700	0.70430	1.41986	1.07026	0.19715
1.22	0.69275	1.13547	0.69275	1.44352	1.08066	0.21407
1.24	0.68158	1.14408	0.68158	1.46719	1.09172	0.23124
1.26	0.67076	1.15283	0.67076	1.49085	1.10343	0.24861
1.28	0.66028	1.16172	0.66028	1.51452	1.11579	0.26616
1.30	0.65012	1.17075	0.65012	1.53818	1.12880	0.28385
1.32	0.64027	1.17992	0.64027	1.56184	1.14247	0.30168
1.34	0.63071	1.18923	0.63071	1.58551	1.15681	0.31961
1.36	0.62144	1.19868	0.62144	1.60917	1.17181	0.33762
1.38	0.61243	1.20827	0.61243	1.63284	1.18749	0.35571
1.40	0.60368	1.21800	0.60368	1.65650	1.20385	0.37385
1.42	0.59518	1.22787	0.59518	1.68017	1.22090	0.39202
1.44	0.58691	1.23788	0.58691	1.70383	1.23865	0.41022
1.46	0.57887	1.24803	0.57887	1.72749	1.25710	0.42844
1.48	0.57105	1.25832	0.57105	1.75116	1.27627	0.44665
1.50	0.56344	1.26875	0.56344	1.77482	1.29617	0.46486
1.52	0.55602	1.27932	0.55602	1.79849	1.31680	0.48305
1.54	0.54880	1.29003	0.54880	1.82215	1.33818	0.50122
1.56	0.54177	1.30088	0.54177	1.84582	1.36032	0.51935
1.58	0.53491	1.31187	0.53491	1.86948	1.38324	0.53745
1.60	0.52822	1.32300	0.52822	1.89314	1.40694	0.55550
1.62	0.52170	1.33427	0.52170	1.91681	1.43144	0.57349
1.64	0.51534	1.34568	0.51534	1.94047	1.45676	0.59144
1.66	0.50913	1.35723	0.50913	1.96414	1.48291	0.60932
1.68	0.50307	1.36892	0.50307	1.98780	1.50990	0.62714
1.70	0.49715	1.38075	0.49715	2.01147	1.53776	0.64489
1.72	0.49137	1.39272	0.49137	2.03513	1.56650	0.66256
1.74	0.48572	1.40483	0.48572	2.05879	1.59613	0.68017
1.76	0.48020	1.41708	0.48020	2.08246	1.62668	0.69769
1.78	0.47481	1.42947	0.47481	2.10612	1.65817	0.71514
1.80	0.46953	1.44200	0.46953	2.12979	1.69060	0.73250
1.82	0.46437	1.45467	0.46437	2.15345	1.72401	0.74978
1.84	0.45932	1.46748	0.45932	2.17712	1.75841	0.76698

M	p/p^*	T_0/T_0^*	ρ/ρ^*	V/V^*	p_0/p_0^*	$4fl^*/D_H$
1.86	0.45438	1.48043	0.45438	2.20078	1.79382	0.78409
1.88	0.44955	1.49352	0.44955	2.22444	1.83027	0.80111
1.90	0.44482	1.50675	0.44482	2.24811	1.86778	0.81804
1.92	0.44018	1.52012	0.44018	2.27177	1.90637	0.83488
1.94	0.43565	1.53363	0.43565	2.29544	1.94606	0.85163
1.96	0.43120	1.54728	0.43120	2.31910	1.98687	0.86829
1.98	0.42685	1.56107	0.42685	2.34277	2.02884	0.88486
2.00	0.42258	1.57500	0.42258	2.36643	2.07199	0.90134
2.02	0.41839	1.58907	0.41839	2.39009	2.11634	0.91772
2.04	0.41429	1.60328	0.41429	2.41376	2.16191	0.93401
2.06	0.41027	1.61763	0.41027	2.43742	2.20874	0.95020
2.08	0.40632	1.63212	0.40632	2.46109	2.25686	0.96631
2.10	0.40245	1.64675	0.40245	2.48475	2.30628	0.98232
2.12	0.39866	1.66152	0.39866	2.50842	2.35705	0.99823
2.14	0.39493	1.67643	0.39493	2.53208	2.40919	1.01405
2.16	0.39128	1.69148	0.39128	2.55574	2.46272	1.02978
2.18	0.38769	1.70667	0.38769	2.57941	2.51769	1.04542
2.20	0.38416	1.72200	0.38416	2.60307	2.57412	1.06097
2.22	0.38070	1.73747	0.38070	2.62674	2.63204	1.07642
2.24	0.37730	1.75308	0.37730	2.65040	2.69149	1.09178
2.26	0.37396	1.76883	0.37396	2.67407	2.75250	1.10705
2.28	0.37068	1.78472	0.37068	2.69773	2.81511	1.12223
2.30	0.36746	1.80075	0.36746	2.72139	2.87935	1.13731
2.32	0.36429	1.81692	0.36429	2.74506	2.94525	1.15231
2.34	0.36118	1.83323	0.36118	2.76872	3.01286	1.16722
2.36	0.35812	1.84968	0.35812	2.79239	3.08220	1.18204
2.38	0.35511	1.86627	0.35511	2.81605	3.15333	1.19677
2.40	0.35215	1.88300	0.35215	2.83972	3.22626	1.21142
2.42	0.34924	1.89987	0.34924	2.86338	3.30106	1.22597
2.44	0.34637	1.91688	0.34637	2.88704	3.77775	1.24044
2.46	0.34356	1.93403	0.34356	2.91071	3.45638	1.25483
2.48	0.34079	1.95132	0.34079	2.93437	3.53698	1.26912
2.50	0.33806	1.96875	0.33806	2.95804	3.61961	1.28334
2.52	0.33538	1.98632	0.33538	2.98170	3.70431	1.29747
2.54	0.33274	2.00403	0.33274	3.00537	3.79111	1.31151
2.56	0.33014	2.02188	0.33014	3.02903	3.88006	1.32548
2.58	0.32758	2.03987	0.32758	3.05270	3.97122	1.33936
2.60	0.32506	2.05800	0.32506	3.07636	4.06462	1.35316
2.62	0.32258	2.07627	0.32258	3.10002	4.16032	1.36688
2.64	0.32013	2.09468	0.32013	3.12369	4.25837	1.38051
2.66	0.31773	2.11323	0.31773	3.14735	4.35880	1.39407
2.68	0.31536	2.13192	0.31536	3.17102	4.46168	1.40755
2.70	0.31302	2.15075	0.31302	3.19468	4.56705	1.42096
2.72	0.31072	2.16972	0.31072	3.21834	4.67497	1.43428
2.74	0.30845	2.18883	0.30845	3.24201	4.78549	1.44753
2.76	0.30622	2.20808	0.30622	3.26567	4.89866	1.46070
2.78	0.30401	2.22747	0.30401	3.28934	5.01455	1.47380

M	p/p^*	T_0/T_0^*	ρ/ρ^*	V/V^*	p_0/p_0^*	$4fl^*/D_H$
2.80	0.30184	2.24700	0.30184	3.31300	5.13319	1.48682
2.82	0.29970	2.26667	0.29970	3.33667	5.25466	1.49976
2.84	0.29759	2.28648	0.29759	3.36033	5.37901	1.51264
2.86	0.29551	2.30643	0.29551	3.38400	5.50630	1.52544
2.88	0.29346	2.32652	0.29346	3.40766	5.63659	1.53817
2.90	0.29143	2.34675	0.29143	3.43132	5.76993	1.55083
2.92	0.28944	2.36712	0.28944	3.45499	5.90640	1.56341
2.94	0.28747	2.38763	0.28747	3.47865	6.04605	1.57593
2.96	0.28553	2.40828	0.28553	3.50232	6.18896	1.58837
2.98	0.28361	2.42907	0.28361	3.52598	6.33518	1.60075
3.00	0.28172	2.45000	0.28172	3.54964	6.48477	1.61306
3.05	0.27710	2.50293	0.27710	3.60881	6.87401	1.64354
3.10	0.27263	2.55675	0.27263	3.66797	7.28588	1.67360
3.15	0.26830	2.61143	0.26830	3.72713	7.72152	1.70326
3.20	0.26411	2.66700	0.26411	3.78629	8.18211	1.73253
3.25	0.26005	2.72343	0.26005	3.84545	8.66887	1.76141
3.30	0.25611	2.78075	0.25611	3.90461	9.18306	1.78991
3.35	0.25229	2.83893	0.25229	3.96377	9.72602	1.81804
3.40	0.24857	2.89800	0.24857	4.02293	10.29912	1.84581
3.45	0.24497	2.95793	0.24497	4.08209	10.90378	1.87323
3.50	0.24147	3.01875	0.24147	4.14125	11.54148	1.90031
3.55	0.23807	3.08043	0.23807	4.20041	12.21377	1.92704
3.60	0.23477	3.14300	0.23477	4.25957	12.92223	1.95345
3.65	0.23155	3.20643	0.23155	4.31873	13.66852	1.97954
3.70	0.22842	3.27075	0.22842	4.37790	14.45435	2.00531
3.75	0.22537	3.33593	0.22537	4.43706	15.28150	2.03078
3.80	0.22241	3.40200	0.22241	4.49622	16.15180	2.05594
3.85	0.21952	3.46893	0.21952	4.55538	17.06717	2.08081
3.90	0.21671	3.53675	0.21671	4.61454	18.02956	2.10539
3.95	0.21396	3.60543	0.21396	4.67370	19.04102	2.12968
4.00	0.21129	3.67500	0.21129	4.73286	20.10366	2.15370
4.05	0.20868	3.74543	0.20868	4.79202	21.21969	2.17745
4.10	0.20614	3.81675	0.20614	4.85118	22.39132	2.20094
4.15	0.20365	3.88894	0.20365	4.91035	23.62089	2.22416
4.20	0.20123	3.96200	0.20123	4.96951	24.91082	2.24713
4.25	0.19886	4.03594	0.19886	5.02867	26.26357	2.26986
4.30	0.19655	4.11075	0.19655	5.08783	27.68171	2.29233
4.35	0.19429	4.18644	0.19429	5.14699	29.16789	2.31457
4.40	0.19208	4.26301	0.19208	5.20616	30.72484	2.33658
4.45	0.18992	4.34045	0.18992	5.26532	32.35537	2.35835
4.50	0.18781	4.41876	0.18781	5.32448	34.06239	2.37990
4.60	0.18373	4.57802	0.18373	5.44281	37.71794	2.42235
4.70	0.17982	4.74077	0.17982	5.56113	41.71655	2.46394
4.80	0.17607	4.90702	0.17607	5.67945	46.08477	2.50471
4.90	0.17248	5.07678	0.17248	5.79778	50.85077	2.54470
5.00	0.16903	5.25003	0.16903	5.91610	56.04435	2.58393
5.10	0.16572	5.42679	0.16572	6.03443	61.69704	2.62242

M	p/p^*	T_0/T_0^*	ρ/ρ^*	V/V^*	p_0/p_0^*	$4fl^*/D_H$
5.20	0.16253	5.60704	0.16253	6.15275	67.84214	2.66021
5.30	0.15946	5.79080	0.15946	6.27108	74.51483	2.69732
5.40	0.15651	5.97805	0.15651	6.38940	81.75220	2.73378
5.50	0.15366	6.16881	0.15366	6.50772	89.59334	2.76959

Appendix J: Oblique Shock Wave Charts for $\gamma = 1.4$

The graphs in this appendix are from NACA Report 1135, *Equations, Tables and Charts for Compressible Flow,* Ames Research Staff, 1953.

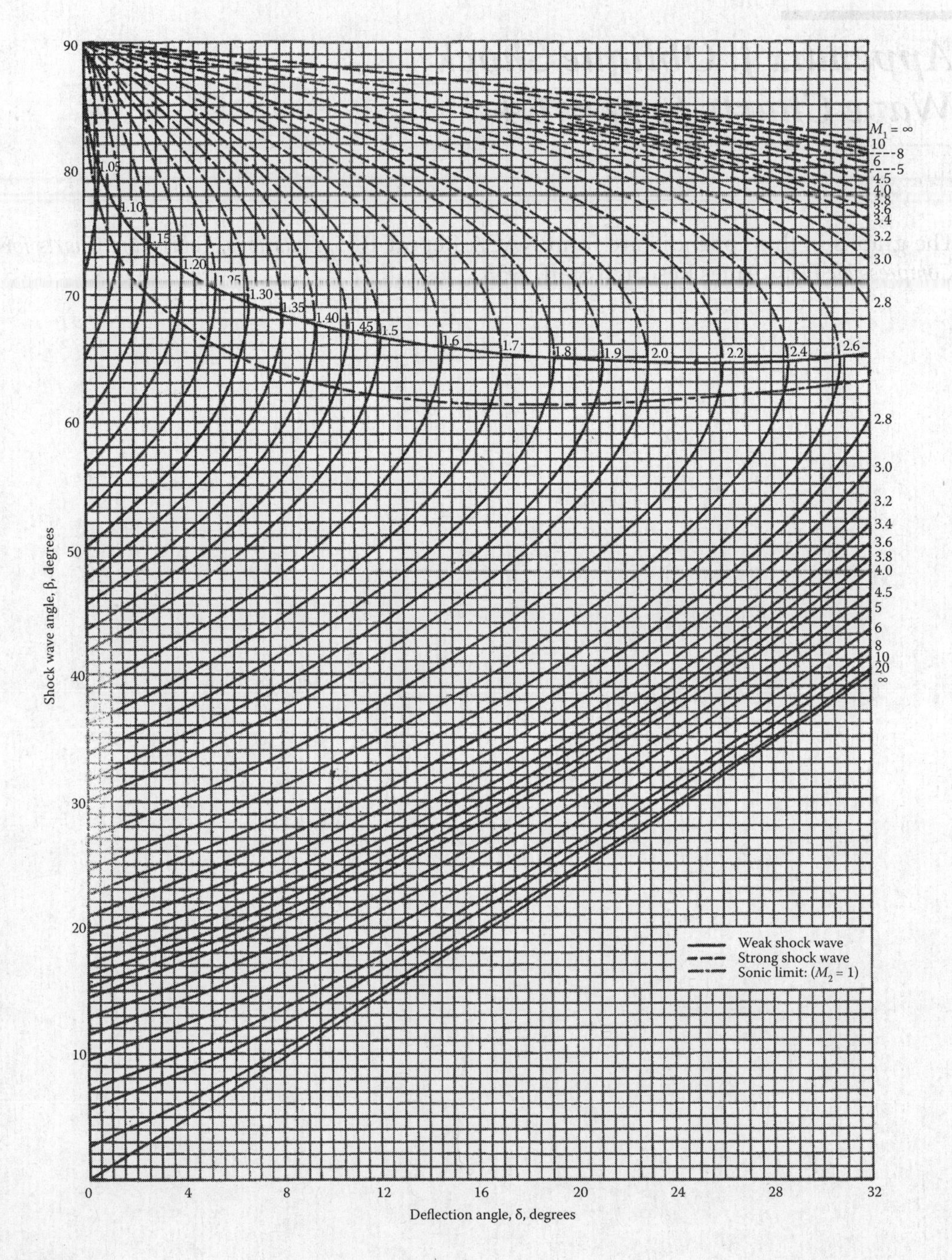

FIGURE J.1
Variation of oblique shock wave angle with flow deflection angle for various upstream Mach numbers.

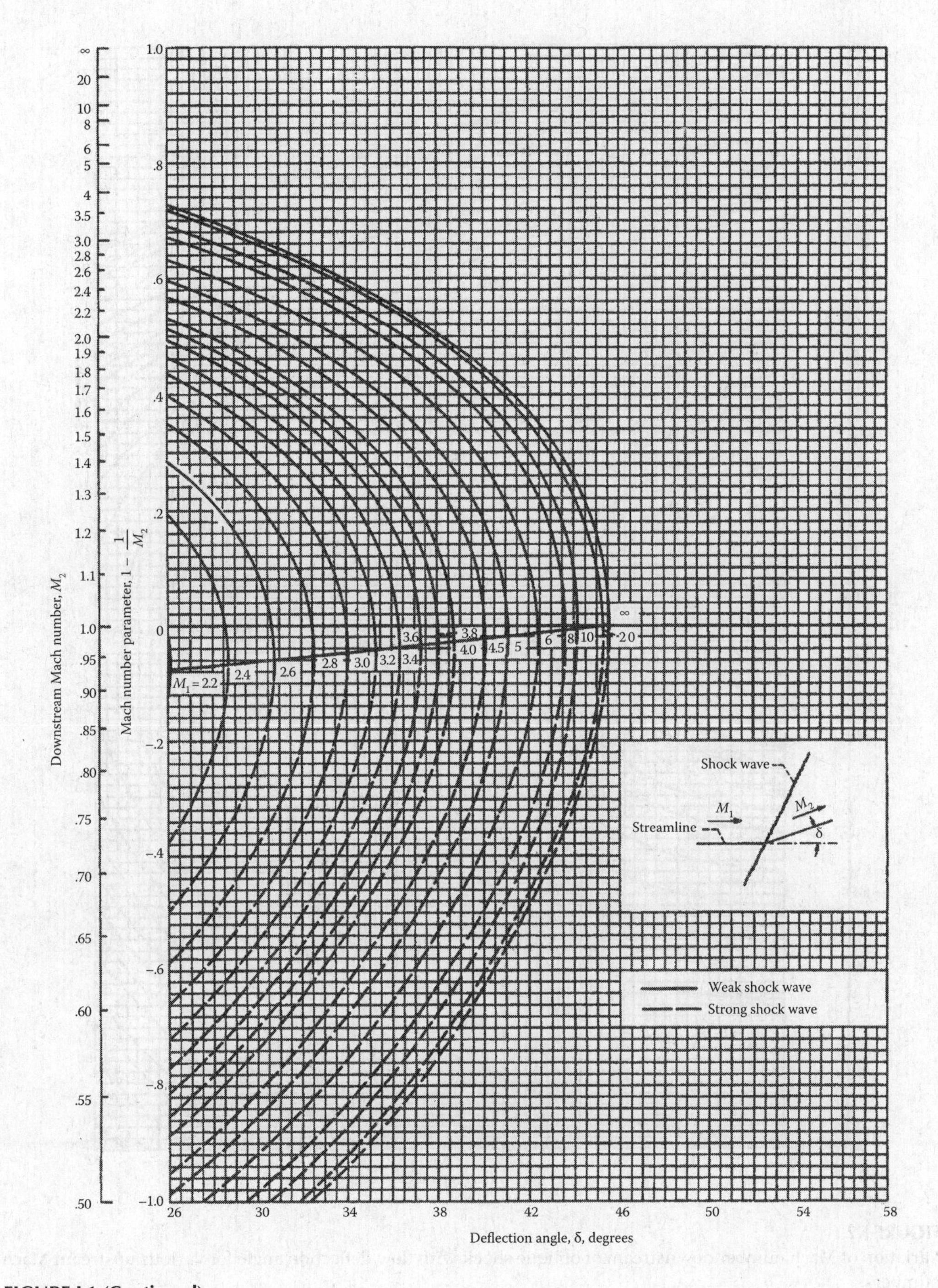

FIGURE J.1 (Continued)
Variation of oblique shock wave angle with flow deflection angle for various upstream Mach numbers.

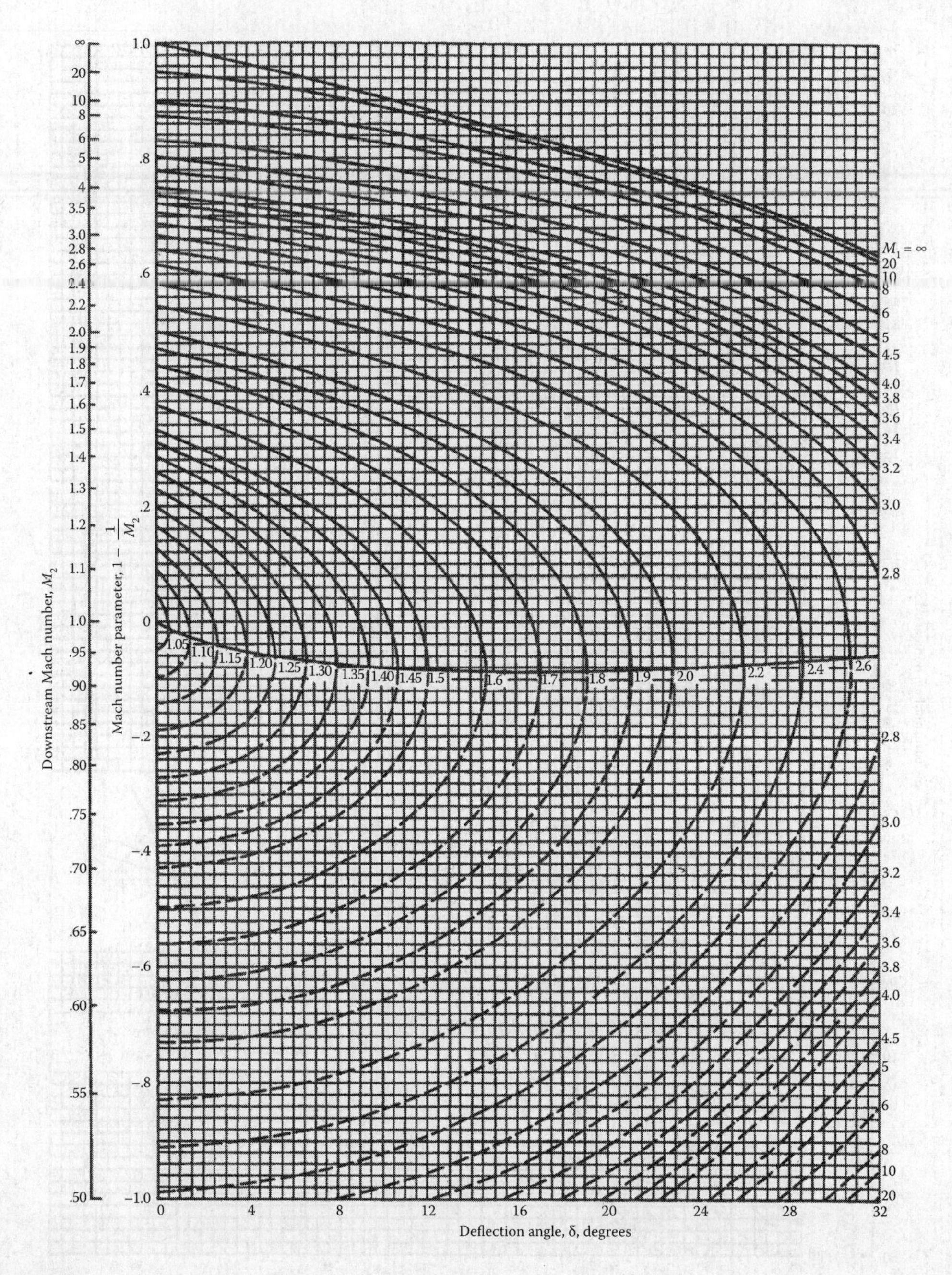

FIGURE J.2
Variation of Mach number downstream of oblique shock with flow deflection angle for various upstream Mach numbers.

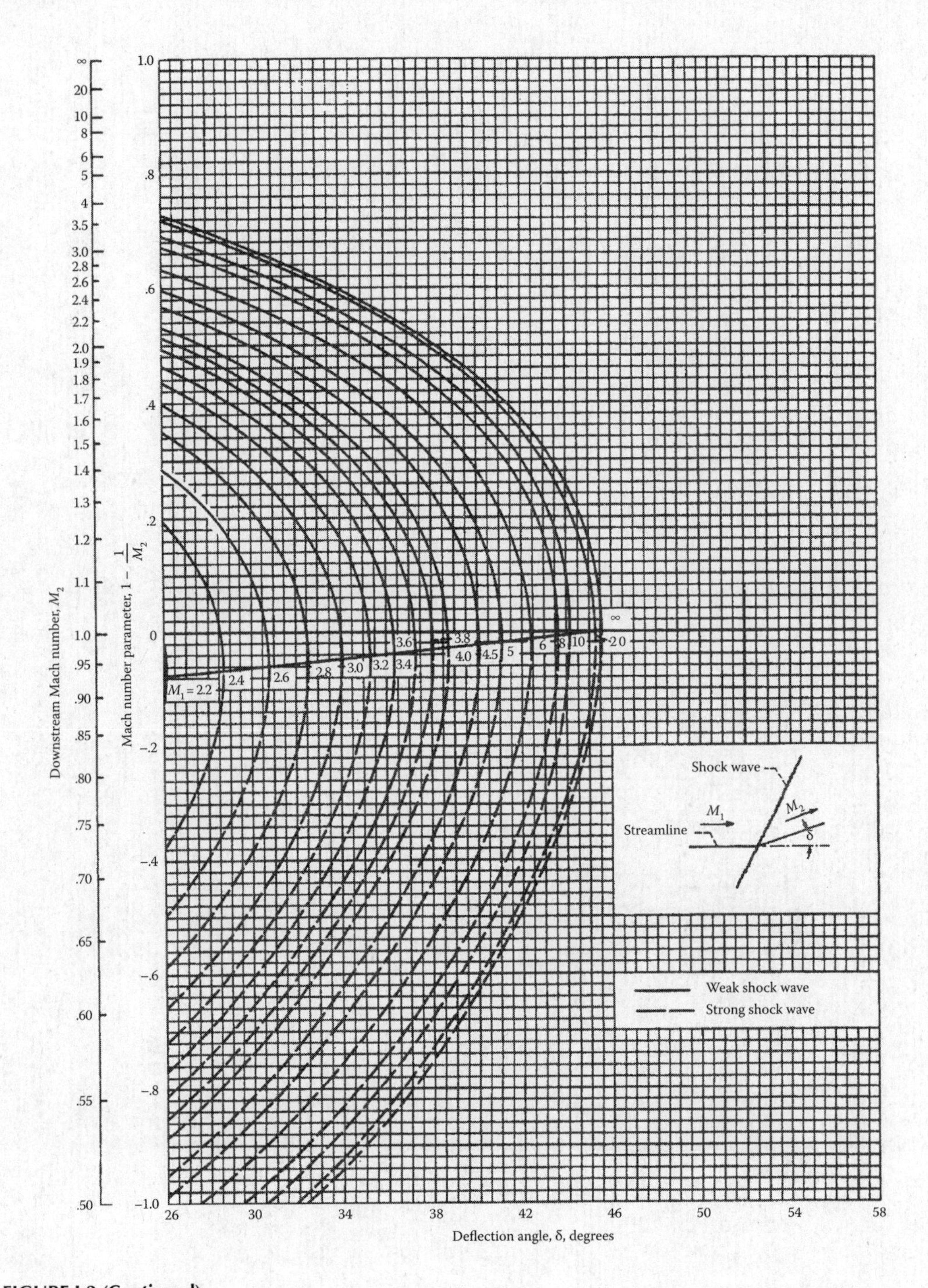

FIGURE J.2 (Continued)
Variation of Mach number downstream of oblique shock with flow deflection angle for various upstream Mach numbers.

Appendix K: Approximate Properties of the Standard Atmosphere

H (m)	T (K)	p (Pa × 10^5)	ρ (kg/m^3)	a (m/s)	ν (m^2/s × 10^{-5})
0	288.16	1.0133	1.2250	340.29	1.4610
200	286.86	0.9895	1.2017	339.52	1.4841
400	285.56	0.9662	1.1787	338.76	1.5077
600	284.26	0.9433	1.1560	337.98	1.5318
800	282.97	0.9209	1.1337	337.21	1.5564
1000	281.67	0.8989	1.1118	336.44	1.5814
1200	280.37	0.8773	1.0901	335.66	1.6070
1400	279.07	0.8561	1.0688	334.88	1.6331
1600	277.77	0.8354	1.0478	334.10	1.6598
1800	276.47	0.8151	1.0271	333.32	1.6870
2000	275.17	0.7952	1.0067	332.54	1.7148
2200	273.87	0.7756	0.9866	331.75	1.7431
2400	272.58	0.7565	0.9669	330.96	1.7721
2600	271.28	0.7377	0.9474	330.17	1.8017
2800	269.98	0.7194	0.9283	329.38	1.8319
3000	268.68	0.7014	0.9094	328.59	1.8627
3200	267.38	0.6837	0.8909	327.79	1.8943
3400	266.08	0.6665	0.8726	327.00	1.9265
3600	264.78	0.6495	0.8546	326.20	1.9594
3800	263.49	0.6330	0.8369	325.40	1.9930
4000	262.19	0.6167	0.8195	324.59	2.0273
4200	260.89	0.6008	0.8023	322.98	2.0983
4600	258.29	0.5701	0.7689	322.17	2.1350
4800	256.99	0.5552	0.7526	321.36	2.1725
5000	255.69	0.5406	0.7365	320.55	2.2109
5200	254.39	0.5263	0.7207	319.73	2.2501
5400	253.10	0.5123	0.7052	318.92	2.2902
5600	251.80	0.4987	0.6899	317.28	2.3732
6000	249.20	0.4722	0.6602	316.45	2.4161
6200	247.90	0.4594	0.6456	315.63	2.4600
6400	246.60	0.4469	0.6314	314.80	2.5049
6600	245.30	0.4347	0.6173	313.97	2.5509
6800	244.01	0.4227	0.6035	313.14	2.5980
7000	242.71	0.4110	0.5900	312.30	2.6462
7200	241.41	0.3996	0.5767	311.47	2.6955
7400	240.11	0.3884	0.5636	310.63	2.7460
7600	238.81	0.3775	0.5507	309.79	2.7977
7800	237.51	0.3668	0.5381	308.94	2.8506
8000	236.21	0.3564	0.5257	308.10	2.9048
8200	234.91	0.3462	0.5135	307.25	2.9604
8400	233.62	0.3363	0.5015	306.40	3.0173

H (m)	*T* (K)	*p* (Pa × 10^5)	ρ (kg/m³)	*a* (m/s)	ν (m²/s × 10^{-5})
8600	232.32	0.3266	0.4898	305.55	3.0756
8800	231.02	0.3171	0.4782	304.69	3.1353
9000	229.72	0.3079	0.4669	303.83	3.1966
9200	228.42	0.2988	0.4448	302.11	3.3207
9600	225.82	0.2814	0.4341	301.25	3.3896
9800	224.53	0.2730	0.4236	300.38	3.4573
10,000	223.23	0.2648	0.4132	299.51	3.5266
10,200	221.93	0.2568	0.4031	298.64	3.5978
10,400	220.63	0.2490	0.3932	297.76	3.6708
10,600	219.33	0.2414	0.3834	296.88	3.7456
10,800	218.03	0.2340	0.3738	296.00	3.8225
11,000	216.73	0.2267	0.3644	295.12	3.9013
11,200	216.66	0.2200	0.3537	295.07	4.0188
11,400	216.66	0.2131	0.3427	295.07	4.1475
11,600	216.66	0.2065	0.3321	295.07	4.2802
11,800	216.66	0.2001	0.3218	295.07	4.4172
12,000	216.66	0.1939	0.3118	295.07	4.5586
12,200	216.66	0.1879	0.3021	295.07	4.7045
12,400	216.66	0.1821	0.2928	295.07	4.8551
12,600	216.66	0.1764	0.2837	295.07	5.0105
12,800	216.66	0.1710	0.2749	295.07	5.1708
13,000	216.66	0.1657	0.2664	295.07	5.3363
13,200	216.66	0.1605	0.2581	295.07	5.5071
13,400	216.66	0.1555	0.2501	295.07	5.6834
13,600	216.66	0.1507	0.2423	295.07	5.8653
13,800	216.66	0.1460	0.2348	295.07	6.0530
14,000	216.66	0.1415	0.2275	295.07	6.2467
14,200	216.66	0.1371	0.2205	295.07	6.4467
14,400	216.66	0.1329	0.2137	295.07	6.6530
14,600	216.66	0.1288	0.2070	295.07	6.8659
14,800	216.66	0.1248	0.2006	295.07	7.0857
15,000	216.66	0.1209	0.1944	295.07	7.3125
15,400	216.66	0.1135	0.1825	295.07	7.7881
15,800	216.66	0.1066	0.1714	295.07	8.2946
16,200	216.66	0.1001	0.1609	295.07	8.8340
16,600	216.66	0.0940	0.1511	295.07	9.4086
17,000	216.66	0.0882	0.1419	295.07	10.0205
17,400	216.66	0.0828	0.1332	295.07	10.6722
17,800	216.66	0.0778	0.1251	295.07	11.3662
18,200	216.66	0.0730	0.1174	295.07	12.1055
18,600	216.66	0.0686	0.1102	295.07	12.8928
19,000	216.66	0.0644	0.1035	295.07	13.7313
19,400	216.66	0.0604	0.0972	295.07	14.6243
19,800	216.66	0.0568	0.0913	295.07	15.5754
20,000	216.66	0.0550	0.0884	295.07	16.0739

Appendix L: Properties of Dry Air at Atmospheric Pressure

T (°C)	ρ (kg/m³)	c_p (kJ/kg · K)	μ (kg/m · s × 10^{-5})	ν (m²/s × 10^{-6})	k (W/m · K)	Pr
−100	2.039	1.010	1.16	5.69	0.0163	0.75
−50	1.582	1.006	1.46	9.25	0.0200	0.73
0	1.292	1.006	1.72	13.31	0.0249	0.72
50	1.092	1.007	1.96	17.92	0.0278	0.71
100	0.946	1.011	2.18	23.02	0.0313	0.70
150	0.834	1.017	2.38	28.58	0.0346	0.70
200	0.746	1.025	2.58	34.57	0.0378	0.70
300	0.616	1.045	2.94	47.72	0.0440	0.70
500	0.457	1.093	3.57	78.22	0.0554	0.70
1000	0.277	1.185	4.82	173	0.0755	0.71

Notes: $Pr = \nu/\alpha$. c_p, μ, and k are approximately independent of pressure. At pressure p, density = density at atmospheric pressure × (p/p_{atm}). At pressure p, $\nu = \nu$ at atmospheric pressure × (p_{atm}/p).

Appendix M: Constants, Conversion Factors, and Units

Constants		
Universal gas constant	$\Re$	= 8.314 kJ/kmol · K
		= 1545.33 ft lbf/lbmol°R
Atmospheric pressure	p_{atm}	= 0.101 325 MPa
		= 101.325 kPa
		= 1.013 25 bar
Speed of light in a vacuum	c_0	= 2.998×10^8 m/s
Gravitational acceleration at sea level	g	= 9.807 m/s^2
		= 32.17 ft/sec^2

Conversion Factors		
Area (A)	1 m^2	= 10.764 ft^2
	1 ft^2	= 0.0929 m^2
Density (ρ)	1 kg/m^3	= 0.062 428 lbm/ft^3
	1 lbm/ft^3	= 16.019 kg/m^3
Energy, work	1 kJ	= 737.56 ft · lbf
	1 ft · lbf	= 1.3558 J
	1 btu	= 778.17 ft · lbf
	1 btu	= 1.0551 kJ
Force	1 N	= 0.224 81 lbf
	1 lbf	= 4.4482 N
Heat flux (q)	1 W/m^2	= 0.317 00 btu/h · ft^2
	1 btu/h · ft^2	= 3.154 69 W/m^2
Heat transfer coefficient (h)	1 W/m^2 · K	= 5.6783 btu/h · ft^2 · °F
	1 btu/h · ft^2 · °F	= 0.1761 W/m^2 · K
Heat transter rate (Q)	1 W	= 3.4118 btu/h
	1 btu/h	= 0.293 07 W
Kinematic viscosity (ν)	1 m^2/s	= 38 750 ft^2/h
	1 ft^2/h	= 2.581×10^{-5} m^2/s
Length	1 m	= 3.2808 ft
	1 ft	= 0.304 80 m
Mass	1 kg	= 2.2046 lbm
	1 lbm	= 0.453 59 kg
Pressure, stress (p, τ)	1 kPa	= 0.145 04 lbf/in^2
	1 lbf/in^2	= 6.894 75 kPa
Power	1 W	= 0.073 756 ft · lbf/s
	1 btu/h	= 0.029 307 W
	1 hp	= 0.074 570 kW
Specific heat (c_p, c_v)	1 kJ/kg · °C	= 0.238 85 btu/lbm · °F
	1 btu/lbm · °F	= 4.1868 kJ/kg · °C
Temperature (T)	T (K)	= T (°C) + 273.15
	T (°C)	= 5/9(T (°F) − 32)
	T (°R)	= T (°F) + 459.67

Conversion Factors		
	T (°F)	= 1.8 T (°C) + 32
Temperature difference (ΔT)	1°C	= 1 K
	1°C	= 1.8°F
	1°F	= 1°R
	1°F	= 0.555 56°C
Thermal conductivity (k)	1 W/m · K	= 0.5782 btu/h · ft · °F
	1 btu/h · ft · °F	= 1.7295 W/m · K
Thermal diffusivity (α)	1 m^2/s	= 38 750 ft^2/h
	1 ft^2/h	= 2.5807×10^{-5} m^2/s
Velocity	1 m/s	= 3.2808 ft/s
	1 ft/s	= 0.304 80 m/s
Viscosity (μ)	1 N · s/m^2	= 2419.1 lbm/ft · h
		= 5.8016×10^{-6} lbf · h/ft^2
Volume (V)	1 m^3	= 35.315 ft^3
	1 ft^3	= 0.028 317 m^3

Bibliography

The items in the following list of books and reports have not been directly referred to in the text. However, their titles in almost all cases make it clear what subject they deal with and the reader who requires more information on any particular topic should have little difficulty in identifying suitable references from this list.

Books and Non-NASA Reports

Anderson, J. D., Jr. (1976) *Gas Dynamic Lasers: An Introduction*. Academic Press, New York.

Anderson, J. D., Jr. (2006) *Hypersonic and High Temperature Gas Dynamics*, 2nd ed. AIAA Education Series. AIAA, Reston, VA.

Anderson, J. D., Jr. (2003) *Modern Compressible Flow with Historical Perspective*, 3rd ed. McGraw-Hill, Boston.

Bai, S. (1959) *Introduction to the Theory of Compressible Flow*. Van Nostrand, Princeton, NJ.

Benedict, R. P. (1983) *Fundamentals of Gas Dynamics*. John Wiley, New York.

Berman, A. I. (1961) *The Physical Principles of Astronautics*. John Wiley, New York.

Bers, L. (1958) *Mathematical Aspects of Subsonic* and *Transonic Gas Dynamics*. John Wiley, New York.

Black, J. (1947) *An Introduction to Aerodynamic Compressibility*. Bunhill Publications, London.

Brower, W. B. (1990) *Theory, Tables, and Data for Compressible Flow*. Hemisphere, New York.

Cambel, A. B. and Jennings, B. H. (1958) *Gas Dynamics*. McGraw-Hill, New York. (Reprinted in 1967 by Dover Publications, New York).

Chapman, A. J. and Walker, W. F. (1971) *Introductory Gas Dynamics*. Holt, Rinehart & Winston, New York.

Cheers, F. (1963) *Elements of Compressible Flow*. John Wiley, New York.

Chernyi, G. G. (1969, 1961) *Introduction to Hypersonic Flow*. Academic Press, New York.

Chushkin, P. I. (1968) *Numerical Method of Characteristics for Three-Dimensional Supersonic Flows. Progress in Aeronautical Sciences*, Vol. 9 (ed. D. Kuchemann), Pergamon, Elmsford, New York. doi:10.1016/0376-0421(68)90004-3.

Courant, R. and Friedrichs, K. O. (1948) *Supersonic Flow* and *Shock Waves*. Interscience, New York (Reprinted in 1999 by Springer-Verlag, New York).

Cox, R. N. and Crabtree, L. F. (1965) *Elements of Hypersonic Aerodynamics*. Academic Press, New York.

Daneshyar, H. (1976) *One-Dimensional Compressible Flow*. Pergamon Press, New York.

Dorrance, W. H. (1962) *Viscous Hypersonic Flow*. McGraw-Hill, New York.

Emanuel, G. (1986) *Gasdynamics: Theory* and *Applications*. AIAA, Washington.

Ferri, A. (1949) *Elements of Aerodynamics of Supersonic Flows*. Macmillan, New York. (Reprinted in 2003 by Dover Publications, Dover Phoenix Editions, Mineola, New York).

Fowler, R. H. and Guggenheim, E. A. (1952) *Statistical Thermodynamics*. Cambridge University Press, New York. Reprinted 1988.

Glass, I. I. (1958) *Theory* and *Performances of Simple Shock Tubes*. Institute of Aerophysics, University of Toronto, UTIAS Review No. 12.

Glass, I. I. (1974) *Shock Waves and Man*. University of Toronto Press, Toronto.

Hall, J. G. and Treanor, C. E. (1968) *Nonequilibrium Effects in Supersonic Nozzle Flows*. NATO AGARDograph No. 124. Advisory Group for Aerospace Research and Development, Neuilly-sur-Seine, France.

Hayes, W. D. and Probstein, R. F. (1966) *Hypersonic Flow Theory*, 2nd ed. Academic Press, New York. Reprinted 1967.

Hill, P. G. and Peterson, C. R. (1992) *Mechanics* and *Thermodynamics of Propulsion*, 2nd ed. Addison-Wesley, Reading, MA. (Reprinted 1998 with corrections).

Hilsenrath, J. et al. (1955) *Tables of Thermodynamic* and *Transport Properties*. NBS Circular 564; reprinted (1960) by Pergamon Press, New York.

Hilton, W. F. (1951) *High-Speed Aerodynamics*. Longman, London.

Hodge, B. K. and Koenig, K. (1995) *Compressible Fluid Dynamics with Personal Computer Applications*. Prentice Hall, Englewood Cliffs, NJ.

Hoerner, S. F. (1992, c1965) *Fluid-Dynamic Drag*, 3rd ed. Hoerner Fluid Dynamics, Bakersfield, CA.

Howarth, L. (editor) (1953) *Modern Developments in Fluid Dynamics High Speed Flows*, Vols. 1 and 2. Oxford University Press, London.

Imrie, B. W. (1974) *Compressible Fluid Flow*. Halstead, New York.

Jeans, J. H. (1904) *The Dynamical Theory of Gases*, 4th ed. Cambridge University Press, New York. (Reprinted in 2012 by RareBooksClub.com) doi:10.5962/bhl.title.24247.

John, J. E. A. and Keith, T. G. (2006) *Gas Dynamics*, 3rd ed. Pearson Prentice Hall, Upper Saddle River, NJ.

Keenan, J. H., Chao, J. and Kaye, J. (1980) *Gas Tables*, 2nd ed. John Wiley, New York.

Kennard, E. H. (1938) *Kinetic Theory of Gases*. McGraw-Hill, New York.

Korobokin, I. and Hastings, S. M. (1957) *Mollier Chart for Air in Dissociated Equilibrium at Temperatures of 2000 K to 15000 K*. NAVORD Report 4446, U. S. Naval Ordnance Lab, White Oak, MD.

Lee, J. F., Sears, F. W. and Turcotte, D. L. (1973) *Statistical Thermodynamics*, 2nd ed. Addison-Wesley, Reading, MA.

Liepmann, H. W. and Roshko, A. (1958) *Elements of Gas Dynamics*. John Wiley, New York. (Reprinted in 2001 by Dover Publications, Mineola).

Marrone, P. V. (1962) *Normal Shock Waves in Air: Equilibrium Composition and Flow Parameters for Velocities From 26,000 to 50,000 ft/s*. Cornell Aeronautical Laboratory Report AG-1729-A-2. Cornell University, New York.

Miles, E. R. (1961, c1950) *Supersonic Aerodynamics*. Dover, New York.

Oswatitsch, K. (1956) *Gas Dynamics*. Academic Press, New York.

Owczarek, K. (1964) *Fundamentals of Gas Dynamics*. International Textbook Company, Scranton, PA.

Pai, S. and Luo, S. (1991) *Theoretical* and *Computational Dynamics of a Compressible Flow*. Van Nostrand Reinhold, New York.

Park, C. (1990) *Nonequilibrium Hypersonic Aerothermodynamics*. John Wiley, New York.

Patterson, G. N. (1956) *Molecular Flow of Gases*. John Wiley, New York.

Pope, A. (1958) *Aerodynamics of Supersonic Flight*, 2nd ed. Pitman, New York.

Pope, A. Y. and Goin, K. L. (1965) *High Speed Wind Tunnel Testing*. John Wiley, New York. (Reprinted in 1978 by R.E. Krieger Pub. Co., Huntington, NY).

Present, R. D. (1958) *Kinetic Theory of Gases*. McGraw-Hill, New York.

Roe, P. L. (1972) Thin shock-layer theory. In *Aerodynamic Problems of Hypersonic Vehicles*. AGARD Lecture Series 42, Vol. 1, 4-1–4-26. Advisory Group for Aerospace Research and Development, Neuilly-sur-Seine, France.

Rosenhead, L. (1954) *A Selection of Graphs for Use in Calculations of Compressible Flow*. Clarendon Press, Oxford.

Saad, M. A. (1992) *Compressible Fluid Flow*, 2nd ed. Prentice-Hall, Englewood Cliffs, NJ.

Sauer, R. (1947) *Introduction to Theoretical Gas Dynamics*. Edwards, Ann Arbor, MI.

Schaaf, S. A. and Chambre, P. L. (1958) Flow of rarefied gases. Section H. In *Fundamentals of Gas Dynamics*, p. 689 (ed. H. W. Emmons), Princeton University Press, Princeton, NJ.

Schreier, S. (1982) *Compressible Flow*. Wiley Interscience, New York.

Sears, F. W. (1956) *An Introduction to Thermodynamics, the Kinetic Theory of Gases, and Statistical Mechanics*, 2nd ed. Addison-Wesley, Reading, MA.

Seddon, J. and Goldsmith, E. L. (1999) *Intake Aerodynamics*, 2nd ed. AE Series, Blackwell Science Ltd., Oxford, UK.

Shames, I. H. (2003) *Mechanics of Fluids*, 4th ed. McGraw-Hill, Boston.

Shapiro, A. H. (1953) *The Dynamics* and *Thermodynamics of Compressible Fluid Flow*, Vols. 1 and 2. Ronald Press, New York.

Sutton, G. P. (1992) *Rocket Propulsion Elements*, 6th ed. John Wiley, New York.

Tannehill, J. C., Anderson, D. A. and Pletcher, R. H. (1997) *Computational Fluid Mechanics and Heat Transfer*. Taylor & Francis, Washington, DC.

Thompson, P. A. (1972) *Compressible-Fluid Dynamics*. McGraw-Hill, New York.

Truitt, R. W. (1959) *Hypersonic Aerodynamics*. Ronald Press, New York.

Vincenti, W. G. and Kruger, C. H. (1986, c1965) *Introduction to Physical Gas Dynamics*, 2nd revised ed. Krieger Publishing Company, Malabar, FL. Reprint with corrections of the 1965 ed. published by Wiley, New York.

Von Mises, R. (1958) *Mathematical Theory of Compressible Flow*. Academic Press, New York. (Reprinted in 2004 by Dover Publications, Dover Books on Engineering, Mineola, New York).

Wittliff, C. E. and Curtiss, J. T. (1961) *Normal Shock Wave Parameters in Equilibrium Air*. Cornell Aeronautical Laboratory Report CAL-111.

Wray, K. L. (1962) Chemical kinetics of high-temperature air. In *Hypersonic Flow Research*, pp. 181–204 (ed. F. Riddell), Academic Press, New York.

Yahya, S. M. (1982) *Fundamentals of Compressible Flow*. Wiley Eastern, New Delhi.

Zel'dovich, Y. B. and Raizer, Y. P. (1968) *Elements of Gasdynamics and the Classical Theory of Shock Waves*. Academic Press, New York.

Zucker, R. D. and Biblarz, O. (2002) *Fundamentals of Gas Dynamics*. Wiley, New York.

Zucrow, M. J. and Hoffman, J. D. (1976) *Gas Dynamics*. John Wiley, New York.

NASA Reports

There are a large number of NASA reports concerned with a variety of aspects of compressible flow. The following list is meant only to give an indication of the types of reports that are available from NASA.

Ames Research Staff (1953) *Equations, Tables* and *Charts for Compressible Flow*. NACA Report 1135.

Baradell, D. L. and Bertram, M. H. (1960) *The Blunt Plate in Hypersonic Flow*. NASA TN-D-408.

Barnhart, P. J. and Greber, I. (1997) *Experimental Investigation of Unsteady Shock Wave Turbulent Boundary Layer Interactions About a Blunt Fin*. NASA CR-202334.

Baty, R. S., Farassat, F. and Hargreaves, J. (2007) *Nonstandard Analysis* and *Shock Wave Jump Conditions in a One-Dimensional Compressible Gas*. NASA LA-14334.

Bryson, A. E., Jr. (1951) *An Experimental Investigation of Transonic Flow Past Two-Dimensional Wedge* and *Circular-Arc Sections Using a Mach–Zehnder Interferometer*. NASA TN 2560.

Davis, D. O., Willis, B. P. and Hingst, W. R. (1998) *Flowfield Measurements in a Slot-Bled Oblique Shock Wave* and *Turbulent Boundary-Layer Interaction*. NASA TM-1998-206974.

Eggers, A. J. and Cyvertson, C. A. (1952) *Inviscid Flow About Airfoils at High Supersonic Speeds*. NACA TN 2646.

Erickson, W. D. (1963) *Vibrational Nonequilibrium Flow of Nitrogen in Hypersonic Nozzles*. NASA TN D-1810.

Gallagher, R. J. (1970) *Investigation of a Digital Simulation of the XB-70 Inlet* and *Its Application to Flight-Experienced Free-Stream Disturbances at Mach Numbers of 2.4 to 2.6*. NASA TN D-5827.

Glass, C. E. (1999) *Numerical Simulation of Low-Density Shock-Wave Interactions*. NASA TM-1999-209358.

Hamed, A. (1995) *An Investigation of Bleed Configurations* and *Their Effect on Shockwave/Boundary Layer Interactions*. NASA CR-199439.

Hansen, C. F. (1959) *Approximation for the Thermodynamic* and *Transport Properties of High Temperature Air*. NASA TR-R-50.

Huang, P. G. and Liou, W. W. (1994) *Numerical Calculations of Shock-Wave/Boundary-Layer Flow Interactions*. NASA TM-106694.

Huber, P. W. (1963) *Hypersonic Shock-Heated Flow Parameters for Velocities to 46,000 Feet per Second* and *Altitudes to 323,000 Feet*. NASA TR R-163.

Inouye, M. and Lomax, H. (1962) *Comparison of Experimental* and *Numerical Results for the Flow of a Perfect Gas About Blunt-Nosed Bodies*. NASA TN D-1426.

Ivey, R. H., Klunker, E. B. and Bowen, E. N. (1948) *A Method for Determining the Aerodynamic Characteristics of Two* and *Three-Dimensional Shapes at Hypersonic Speeds*. NACA TN 1613.

Lillard, R. P. (2011) *Turbulence Modeling for Shock Wave/Turbulent Boundary Layer Interactions*. NASA JSC-CN-25068.

Lomax, H. and Inouye, M. (1964) *Numerical Analysis of Flow Properties About Blunt Bodies Moving at Supersonic Speeds in an Equilibrium Gas*. NASA TR-R-204.

Martin, R. A. (1970) *Dynamic Analysis of XB-70 Inlet Pressure Fluctuations During Takeoff* and *Prior to a Compressor Stall at Mach 2.5*. NASA TN D-5826.

McBride, B. J., Heimel, S., Ehlers, J. G. and Gordon, S. (1963) *Thermodynamic Properties to 6000 K for 210 Substances Involving the First 18 Elements*. NASA SP-3001.

O'Farrell, J. M. and Rieckhoff, T. J. (2011) *Direct Visualization of Shock Waves in Supersonic Space Shuttle Flight*. NASA TM-2011-216455, M-1304.

Rakich, J. V. (1969) *A Method of Characteristics for Steady Three-Dimensional Supersonic Flow with Application to Inclined Bodies of Revolution*. NASA TN D-5341.

Romere, P. O. and Whitnah, A. M. (1983) Space shuttle entry longitudinal aerodynamic comparisons with flights 1–4 with preflight predictions. In *Shuttle Performance: Lessons Learned*, pp. 283–307 (ed. J. P. Arrington and J. J. Jones), NASA CP-2283.

Sarli, V. J., Burwell, W. G. and Zupnik, T. F. (1964) *Investigation of Nonequilibrium Flow Effects in High Expansion Ratio Nozzles*. NASA CR-54221.

Sims, J. (1964) *Tables for Supersonic Flow Around Right Circular Cones at Zero Angle of Attack*. NASA SP-3004.

Tannehill, J. C. and Mugge, P. H. (1974) *Improved Curve Fits for the Thermodynamic Properties of Equilibrium Air Suitable for Numerical Computation Using Time-Dependent or Shock-Capturing Methods*. NASA CR-2470.

Tatum, K. E. (1996) *Computation of Thermally Perfect Properties of Oblique Shock Waves*. NASA CR-4749.

Viegas, J. R. and Howe, J. T. (1962) *Thermodynamic* and *Transport Property Correlation Formulas for Equilibrium Air from 1000 K to 15,000 K*. NASA TN D-1429.

Woods, W. C., Arrington, J. P. and Hamilton, H. H. (1983) A review of real gas effects on space shuttle aerodynamics characteristics. In *Shuttle Performance: Lessons Learned*, pp. 309–46 (ed. J. P. Arrington and J. J. Jones), NASA CP-2283.

Index

Page numbers followed by f and t indicate figures and tables, respectively.

W